The Home Carpenter & Woodworker's

REPAIR MANUAL

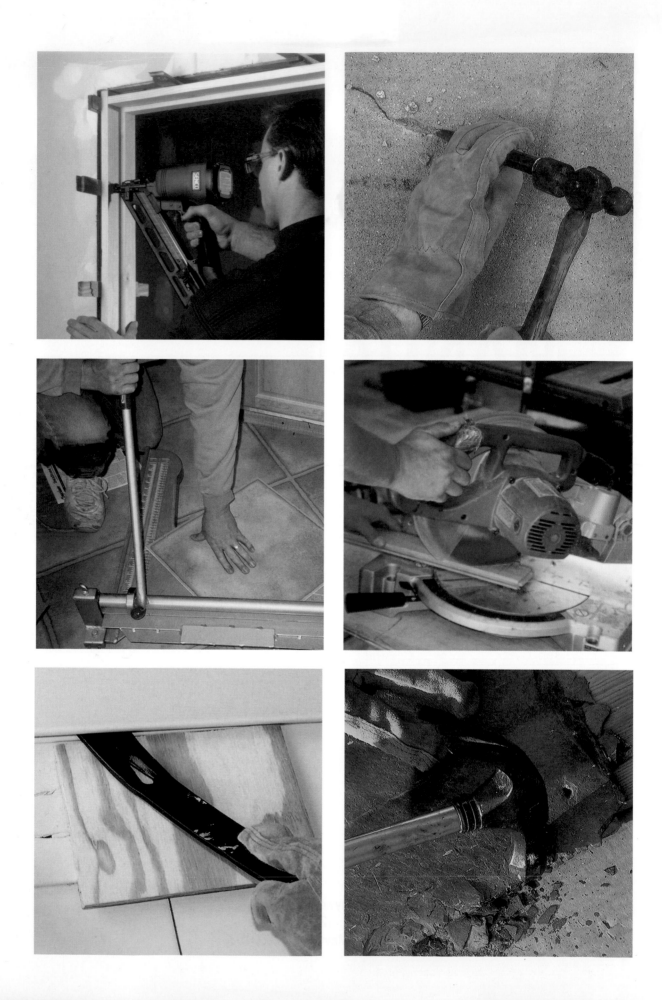

The Home Carpenter & Woodworker's
REPAIR MANUAL

WILLIAM P. SPENCE

Sterling Publishing Co., Inc.
New York

Dedication
To the home craftsmen and women who have the interest, knowledge, and skills to provide much of the maintenance and repairs needed in today's technically complex, multiple material houses. Recognition is also given to those in the skilled construction trades who assist and perform these multifaceted tasks.

Photography and Illustrations by William P. Spence or as credited

Book Design: Jeff Potter, Potter Publishing Studio
Editor: Rodman Pilgrim Neumann

Disclaimer
The author has made every attempt to present safe building practices, but he makes no claim that the information in this book is complete or complies with every local building code. The Publisher does not warrant or guarantee any product described herein, and does not assume, and expressly disclaims, any obligation to obtain information other than that provided by the manufacturer. The reader assumes all risks in connection with instructions herein. The publisher shall not be liable for any damage resulting, in whole or part, from the reader's use of, or reliance upon, this material.

Library of Congress Cataloging-in-Publication Data
Spence, William Perkins, 1925-
  The home carpenters & woodworker's repair manual / William P. Spence.
     p. cm.
  Includes index.
  ISBN 1-4027-1055-0
  1. Carpentry—Amateurs' manuals. 2. Dwellings—Maintenance and repair—Amateurs' manuals. I. Title.

TH5607.S64 2005
694—dc22

                                                    2005002256

10 9 8 7 6 5 4 3 2 1

Published by Sterling Publishing Co., Inc.
387 Park Avenue South, New York, NY 10016
© 2006 by William P. Spence
Distributed in Canada by Sterling Publishing
c/o Canadian Manda Group, 165 Dufferin Street
Toronto, Ontario, Canada M6K 3H6
Distributed in the United Kingdom by GMC Distribution Services,
Castle Place, 166 High Street, Lewes, East Sussex, England BN7 1XU
Distributed in Australia by Capricorn Link (Australia) Pty. Ltd.
P.O. Box 704, Windsor, NSW 2756, Australia

Printed in China
All rights reserved

Sterling ISBN-13: 978-1-4027-1055-1
        ISBN-10: 1-4027-1055-0

For information about custom editions, special sales, premium and corporate purchases, please contact Sterling Special Sales Department at 800-805-5489 or specialsales@sterlingpub.com.

CONTENTS

CHAPTER 1

Safe Working Conditions

There are many home repairs that can be handled by the homeowner; however, they require the use of a wide variety of tools and materials. The chance for an accident is always present. There is no substitute for developing a safety attitude. By this is meant considering what is to be done, what hazards exist, how to avoid an accident, and using personal protective equipment. While personal protective equipment is important, it will not stop an accident from

happening if tools and materials are used in a way that violates the manufacturer's instructions. Both hand and power tools are dangerous, as are some misused materials, especially finishing materials.

General Housekeeping Rules

1. Keep the area in which you are working free of scraps, unused tools, and electric extension wires.
2. Remove nails from lumber that has been pulled loose.
3. Store oily paint and solvent-soaked rags in a non-flammable container. Remove from the house as soon as possible.
4. Keep tools safely stored in proper containers.
5. Store materials so they are protected from moisture and physical damage.

Personal Safety

There are many very fine personal safety products available at the local hardware or building materials supply dealer. Following are some that are frequently used.

Eye protection is something that cannot be overlooked. The cost for eye-protective devices is small. Regular eyeglasses do not provide adequate protection.

When operating any type of power tool, eye protection is essential. Some are available if you do not wear eyeglasses (1-1), while others are made to fit

1-1 This type of safety glasses protects your eyes when you do not wear glasses. *Courtesy Aero Company*

1-2 This is an eyeglass protector that fits over most eyeglasses. Notice the side protection shield.

Courtesy Aero Company

over the eyeglasses (1-2). It is important to keep the eye safety equipment clean. Manufacturers have available lens cleaning systems but at least wash them frequently with water.

If you are tackling a job where there is danger to the face, wear a full face shield (1-3).

Some tasks produce noise at such a level that it could cause temporary **hearing loss,** or with regular occurrence, permanent damage. The simplest protection is a set of reusable earplugs (1-4). Earmuffs will provide even better protection (1-5).

Respiratory inflammation can be caused by many common things, including power sanding and the vapors from various finishing materials. Lightweight disposable masks are good for light sanding of wood and drywall (1-6). After a few hours of use, discard. Do not try to clean them. A better mask is shown in **1-7.** It has a heavy-duty replaceable filter. Some

1-3 A full face shield provides maximum protection and is recommended for use on indoor and outdoor projects. *Courtesy Aero Company*

1-5 Earmuffs provide maximum protection from noise, especially continuous, high-pitched sounds.

Courtesy Aero Company

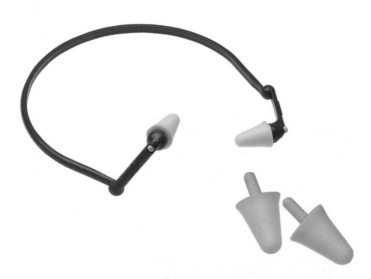

1-4 Earplugs are required when jobs produce high noise levels. *Courtesy Aero Company*

1-6 Lightweight disposable masks are used for light dust-producing activities. Replace them frequently.

extensive finishing operations, especially over a large area such as a floor, usually require a heavy-duty respirator, with a replaceable filter **(1-8).** Replace the filter often.

When handling lumber, metals, and other heavy materials, a good pair of heavy **work gloves (1-9)** is important for the needed protection they provide. Do not wear the work gloves when operating power tools because they could get caught and pull your hand into danger's way. Also, keep available some **rubber** or **plastic gloves** for work with solvents and chemicals **(1-10).**

Consider buying knee pads if you are doing work on floors or installing baseboard. They greatly reduce stress on the knees and are very inexpensive **(1-11).**

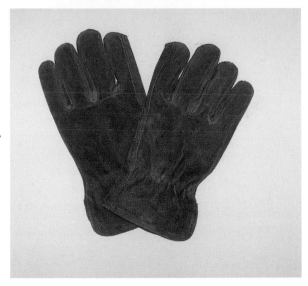

1-9 Heavy-duty work gloves are essential for safe handling of construction materials and heat-producing operations.

1-7 This dust mask has a heavy replaceable filter and is more effective than the low-cost disposable filter.

Courtesy Aero Company

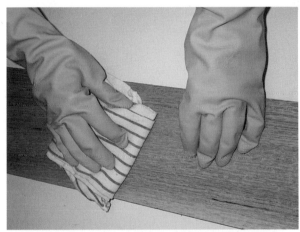

1-10 Rubber and plastic gloves should be readily available for use when handling solvents and chemicals.

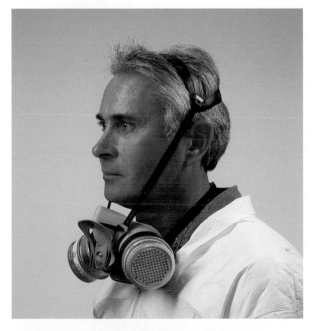

1-8 Severe dust conditions, as well as many finishing operations, such as coating a hardwood floor, require the use of a heavy-duty respirator. It has replaceable filter disks. *Courtesy Aero Company*

1-11 Knee pads cost very little but give great relief and should always be worn when the job requires you to work kneeling.

1-12 Hard hats are required where falling objects or the chance of hitting your head exists. On construction sites they are required for everyone entering the site.
Courtesy Aero Company

Clothing and jewelry can cause accidents. Wear a shirt with sleeves that are back from the wrist. Pants with full-length legs provide considerable protection. Always remove all jewelry—all rings, necklaces, bracelets, watches, as well as ties, and other articles that could dangle into harm's way. Long hair should be confined in a hair net.

For situations where you will be doing a major repair or new construction, a **hard hat** is essential (1-12). The hard hat has a lining that must be adjusted so the hat sets on the band and the band rests squarely around the head. There are two types available. **TYPE 1** has a full brim around the hat, while **TYPE 2** just has a peak on the front. For most purposes a **Class A** hard hat is used for general service where there is no need for voltage protection. Never wear a hard hat over another hat.

Equipment Safety

Some home repairs will require the use of a **ladder** or **scaffold**. Careless installation and use can lead to many unnecessary accidents.

The commonly used ladders are stepladders and extension ladders **(1-13)**. It is important to purchase high-quality ladders. They should be marked indicating they meet the requirements of the Occupational Safety and Health Administration (OSHA).

If a ladder is damaged it should be fully repaired or destroyed. Do not keep a damaged or deteriorating ladder around. Do not overload a ladder. The maximum load is clearly marked on it by the manufacturer.

1-13 The most frequently used ladders are the stepladder and extension ladder. Purchase only a high-quality one.

Using Extension Ladders Safely

Do not use an extension ladder as a work platform for scaffolding. It is not designed to carry horizontal loads. Be certain the feet are placed on a firm and level, nonslippery surface. The extension ladder should have safety feet. If there is a danger of its slipping, block it as shown in **1-14** or use the safety shoe if available. Keep the area at the top and bottom of the ladder clear of debris.

When using an extension ladder to access the roof, adjust it to the correct length to provide a 36-inch extension above the roof edge and place the feet away from the wall of the working distance (**1-15**). If the ladder is adjusted so that it is too short or too long for reaching the work area (**1-16**), climb down and readjust the length so you can reach the work area without leaning or straining.

Be careful when carrying tools or materials up a ladder. Small hand tools can be carried up in a carpenter's belt or by lifting them in a bucket with a rope. Very heavy materials will require a hoist or forklift.

In bad weather avoid using an extension ladder if exposed to high winds. Also remember to clean all mud or snow from your shoes.

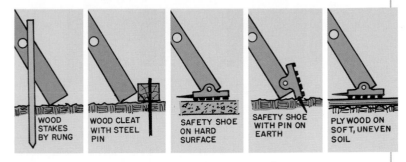

1-14 Make certain the ladder has some means of holding it in place so it does not slip and remains level.

1-15 When climbing onto the roof, set the ladder the proper distance from the edge of the roof and extend it 36 inches above it so you have something to hold to keep your balance.

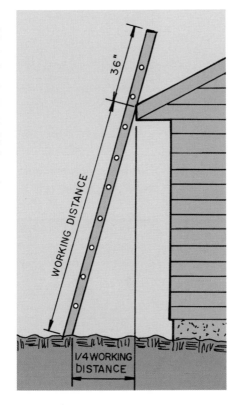

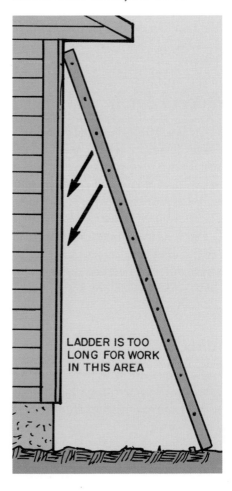

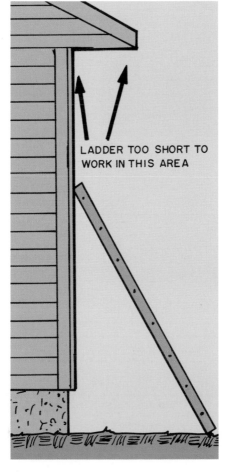

1-16 Adjust the length of the ladder so you can reach the work area without leaning over or stretching. This can throw you off balance.

1-17 Manufactured metal scaffolding is the quickest and safest way to get a working area above the ground. It is available at equipment rental agencies.

1-18 Use level, firmly set, solid blocking below the scaffold legs.

Many ladders that are available are made from aluminum. Since aluminum is a good conductor of electricity, do not use them near an electric wire. Instead use a fiberglass ladder.

Using Stepladders Safely

Possibly the most frequent accident-causing situation is when the stepladder has not been fully opened and the side braces have not been snapped into position. Next is the problem of setting it on a level surface. All four feet should rest on a solid, level surface. If you block up under one or more feet, be certain the blocking is large so the feet will not slip off if the ladder moves a little when you are working on it. Never overload the ladder. While it is designed with a safety margin, it is a good policy not to exceed the design rating.

Using Scaffolds Safely

While scaffolding can be built from construction lumber on the job for the homeowner, the amount of carpentry work and risk involved exceeds the cost of renting metal scaffolding (1-17). The sections are welded into panels which are assembled on the site. It is very important to install the crossbracing or the unit may lean and fall. It is also important that the legs rest on a firm surface. Inside the house the subfloor is a good base. Outside it will be necessary to level the ground and add a solid support upon which the scaffold base plate will rest **(1-18)**. If the scaffold is getting tall, consider tying it to the wall. Once the scaffolding is up, install guardrails above the scaffold planks to help prevent someone from backing off **(1-19)**. Generally scaffolds below six feet high do not require guardrails. Current OSHA regulations are available from the U.S. Department of Labor, OSHA Publications, P.O. Box 37534, Washington, DC 20013-3735.

The scaffold planks are run horizontally over the scaffold rails. Be certain you use approved wood scaffold planks. Typical construction lumber might not span the distances and carry the load of people and materials **(1-20)**.

The ends of the planks overhanging the horizontal end support should be cleated to keep them from sliding off. If several planks are needed to run the length of the scaffold, they should overlap 12 inches and be at least 6 inches on each side of the supporting bar.

1-19 Guardrails are required on scaffolding more than six feet above the ground. Notice the ladder rungs welded into the framing.

1-20 These planks, even if they were originally approved scaffold plank lumber, should be cut up and discarded. The weight of the bricks, mortar, and bricklayers could easily lead to a failure.

1-21 Ladder jacks are a quick way to provide an elevated work platform. The extension ladders must be designed to carry the loads.
Courtesy Werner Ladder Company, 1-724-588-2000

Manufactured scaffolding will have a specified load-bearing capacity, so observe it. It will also have a series of rungs built on one end which you use as a ladder to climb up onto the planks (refer to **1-19**). If such a series of rungs is not on the scaffold, lean an extension ladder against it for mounting. Never climb up the side of the scaffold using the braces. They may bend and destabilize the unit.

Using Ladder Jacks Safely

Ladder jacks are metal frames designed to carry horizontal loads that are fastened to the rungs of extension ladders, as shown in **1-21**. They are usually designed to support one worker who is doing light work. They are not loaded with heavy materials and carry several workers. The ladders used must be high quality and free of defects. The platform is made 12 inches wide so foot space is minimal. The ladders should be spaced not more than eight feet apart to support one person. If they are spaced six feet or less they can usually support two people.

CHAPTER 2

Fasteners

A key to successful home repair and maintenance is to use the best fastener for the job at hand. A visit to the hardware store or building supply dealer will let you examine the many standard fasteners available, as well as look for new products to assist you. As you make choices, consider both mechanical fasteners, such as nails and screws, and bonding fasteners, such as glues and adhesives.

Nails

As you consider using **nails,** hand-driven nails will most likely be used because they are inexpensive, available in a variety of types and sizes, and require only a regular hammer. If you are undertaking a major remodeling or carpentry project it might be worth the expense to rent a power nailer.

There are many different types of nail produced for use in specific applications. Select the nail that is recommended for a particular application. Frequently when you purchase a product requiring nailing, the manufacturer will make specific recommendations.

Nails are made from a variety of metals including steel, stainless steel, aluminum, and copper. If the work is in a moist interior area or any exterior area use aluminum, copper, or stainless steel nails because they will not rust. Galvanized steel nails also resist rust for many years. Cement-coated nails have greater holding power. Nails used in structural framing should have large heads. Those used for finishing jobs, such as interior trim, should have small heads that can be set below the surface of the wood and concealed with a filler. Many types of nail have smooth shanks while others have threaded, barbed, grooved, twisted, or ring shanks which increase their

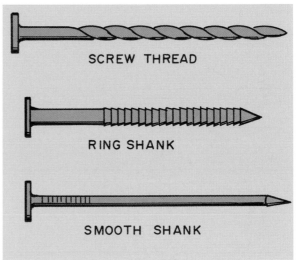

2-1 The most common types of nail shank.

holding power (**2-1**). Examples of exterior and interior nails are shown in **2-2**.

Nail sizes are identified by the term **penny**, which is indicated in writing by a lower case **d**. For example, a 6d nail is 2 inches long. Sizes are shown in **Table 2-1**. The actual diameter of the nail shank will vary for each size, depending upon the type of nail. For example, a 2d common nail has a larger shank than a 2d finish nail. Nails are often sold by the pound. The penny size and type of nail will influence how many

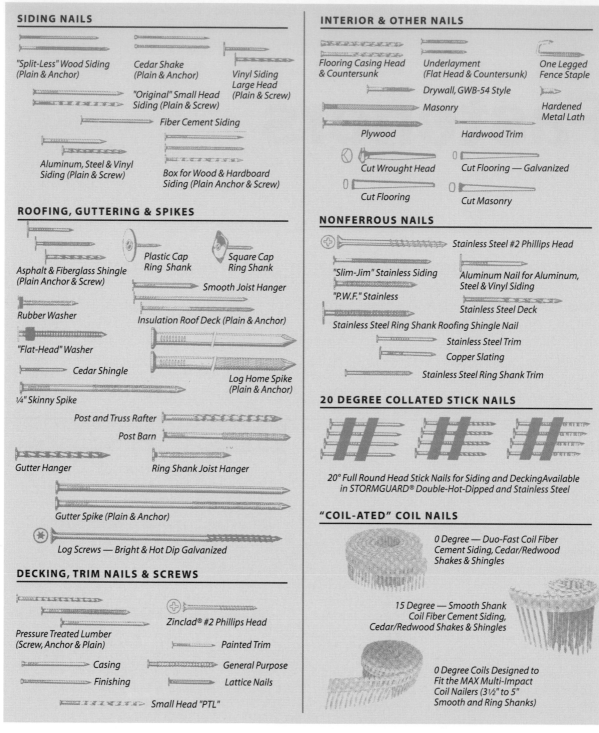

SIDING NAILS

"Split-Less" Wood Siding (Plain & Anchor)

Cedar Shake (Plain & Anchor)

Vinyl Siding Large Head (Plain & Screw)

"Original" Small Head Siding (Plain & Screw)

Fiber Cement Siding

Aluminum, Steel & Vinyl Siding (Plain & Screw)

Box for Wood & Hardboard Siding (Plain Anchor & Screw)

ROOFING, GUTTERING & SPIKES

Asphalt & Fiberglass Shingle (Plain Anchor & Screw)

Plastic Cap Ring Shank

Square Cap Ring Shank

Smooth Joist Hanger

Rubber Washer

Insulation Roof Deck (Plain & Anchor)

"Flat-Head" Washer

Cedar Shingle

Log Home Spike (Plain & Anchor)

¼" Skinny Spike

Post and Truss Rafter

Post Barn

Gutter Hanger

Ring Shank Joist Hanger

Gutter Spike (Plain & Anchor)

Log Screws — Bright & Hot Dip Galvanized

DECKING, TRIM NAILS & SCREWS

Pressure Treated Lumber (Screw, Anchor & Plain)

Zinclad® #2 Phillips Head

Painted Trim

Casing

General Purpose

Finishing

Lattice Nails

Small Head "PTL"

INTERIOR & OTHER NAILS

Flooring Casing Head & Countersunk

Underlayment (Flat Head & Countersunk)

One Legged Fence Staple

Drywall, GWB-54 Style

Masonry

Hardened Metal Lath

Plywood

Hardwood Trim

Cut Wrought Head

Cut Flooring — Galvanized

Cut Flooring

Cut Masonry

NONFERROUS NAILS

Stainless Steel #2 Phillips Head

"Slim-Jim" Stainless Siding

Aluminum Nail for Aluminum, Steel & Vinyl Siding

"P.W.F." Stainless

Stainless Steel Deck

Stainless Steel Ring Shank Roofing Shingle Nail

Stainless Steel Trim

Copper Slating

Stainless Steel Ring Shank Trim

20 DEGREE COLLATED STICK NAILS

20° Full Round Head Stick Nails for Siding and DeckingAvailable in STORMGUARD® Double-Hot-Dipped and Stainless Steel

"COIL-ATED" COIL NAILS

0 Degree — Duo-Fast Coil Fiber Cement Siding, Cedar/Redwood Shakes & Shingles

15 Degree — Smooth Shank Coil Fiber Cement Siding, Cedar/Redwood Shakes & Shingles

0 Degree Coils Designed to Fit the MAX Multi-Impact Coil Nailers (3½" to 5" Smooth and Ring Shanks)

2-2 Nails used in exterior and interior carpentry work. *Courtesy W.H. Maze Company*

Table 2-1 Penny and inch nail sizes.

Penny nail	Inch size	Penny nail	Inch size
2	1	8	2½
3	1¼	9	2¾
4	1½	10	3
5	1¾	12	3¼
6	2	16	3½
7	2¼	20	4

2-4 Power-nailers are driven by compressed air provided by a portable air compressor.

Courtesy Thomas Industries

2-3 Power nailers speed up the installation and are less likely to split the wood. The nails are in clips and are held in the long vertical track shown here next to the wall.

nails you get in a pound. For example, a 2d common nail will have 180 nails per pound, while a 2d finish nail will have 310 nails per pound. Nails are also sold in boxes containing a specified number of nails.

Power Nailers & Staplers

Power nailers (2-3) use compressed air produced by a **portable compressor (2-4)** to drive the fastener. The tools available will drive a variety of fasteners including staples, small brads, finishing nails, and common nails used for wall framing and other large-member nailing jobs. The fasteners are sold in coils or clips which fit into the nailer and are fed automatically **(2-5)**.

Wood Screws & Threaded Fasteners

Wood screws are more difficult to install than nails but generally provide a more secure connection. They also have the advantage of being easily removed.

The most common types have flat, round, or oval heads **(2-6)**. **Flathead** screws are set flush with the surface and are widely used to install hinges and other hardware. The heads of **roundhead** screws remain above the surface and become part of the visible construction. **Oval heads** are partially recessed into the wood and are used where an attractive exposed screw is required. Screw heads are made in a variety of head

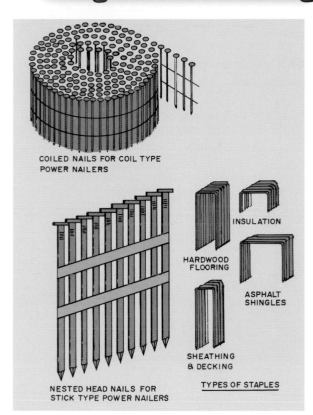

COILED NAILS FOR COIL TYPE POWER NAILERS

INSULATION

HARDWOOD FLOORING

ASPHALT SHINGLES

SHEATHING & DECKING

TYPES OF STAPLES

NESTED HEAD NAILS FOR STICK TYPE POWER NAILERS

2-5 Nails in rolls or bound in a clip are fed into power nailers. Staples are held in series.

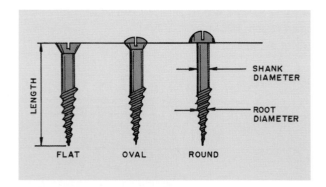

LENGTH

SHANK DIAMETER

ROOT DIAMETER

FLAT   OVAL   ROUND

2-6 Common types of wood screw.

2-7 Types of recess in wood screw heads.

2-8 A special screw designed for use in particleboard.

Table 2-2 Typical sizes of wood screw.

Length (inches)	Shank diameter (wire gauge number)
1/4	2, 4
3/8	2, 3, 4, 5, 6
1/2 and 3/8	2, 3, 4, 5, 6, 7, 8
3/4, 7/8, and 1	4, 5, 6, 7, 8, 9, 10, 11, 12
1 1/4	4, 5, 6, 7, 8, 9, 10, 11, 12, 14, 16
1 1/2	6, 7, 8, 9, 10, 11, 12, 14, 16
1 3/4	6, 8, 9, 10, 12, 14, 16
2	8, 9, 10, 12, 14, 16
2 1/4	10, 12, 14
2 1/2	8, 9, 10, 12, 14, 16
3	10, 12, 14, 16

2-9 Lag screws are used to join thick wood members.

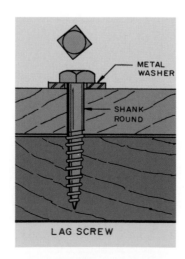

recesses (2-7). The slotted, Phillips, and slotted/Phillips recesses are most commonly used.

The length of wood screws is given in inches. The longer the screw, the larger the wire used to make it. Each length of screw is available in several wire gauge sizes. Commonly available sizes are in **Table 2-2.** There are a number of special screws available. One that could be required frequently is a **particleboard screw (2-8).** Since particleboard is widely used, keep this screw in mind as you do repairs and maintenance.

Lag screws are used to join heavy wood members. They have considerable holding power. A metal washer is placed under the square head (2-9) and the lag screw is installed by drilling a hole for the shank and a smaller hold for the threaded portion. A wrench is used to turn it into the wood.

Dowel screws have wood screw threads on both ends. They are commonly used to join short legs to the furniture frame (2-10). Another similar fastener uses a **hanger bolt** and a **tee nut.** It has machine screws on one end and wood screw threads on the other. The wood screw end is installed in the leg. The

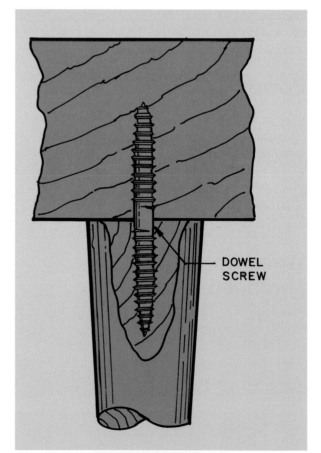

2-10 Dowel screws can be used to join legs to the furniture frame or any other application where a part is rotated to be screwed in place.

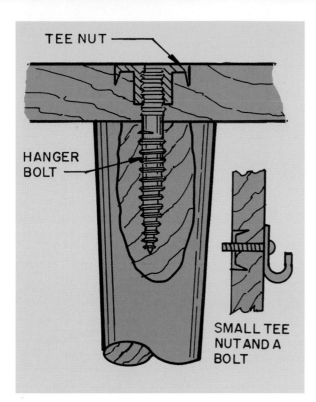

2-11 This hanger bolt uses a tee nut that is set into the wood to enable it to screw the end with the machine screw into the nut. Small tee nuts with bolts are used to secure items to wood panels.

tee nut is tapped into a hole prepared for it. The leg is then screwed into the nut (2-11). Small tee nuts with bolts are used to secure materials to wood panels.

Carriage and **machine bolts** provide a strong connection that can be opened up whenever needed (2-12). The carriage bolt has a rounded head, giving a smooth, exposed surface. Below the head it has a squared shank which keeps it from turning in the hole. The **machine bolt** has a flat, square head (2-12). It should have a washer below it so it sets firmly on the wood. Both should have washers between the nut and the wood.

Screw Eyes & Hook Fasteners

Screw eyes are available in several sizes. The length and size of the eye vary by size. A screw eye is installed by drilling a small hole in the wood and twisting it into the hole (2-13).

Hook fasteners are available with a round cup hook and a square shoulder hook (2-14). They are available in a range of lengths, typically from ½ inch to 1½ inches.

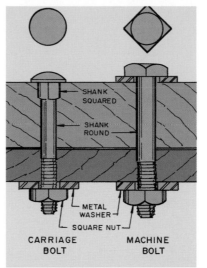

2-12 These bolts are used to secure heavier wood members, such as framing, onto a deck.

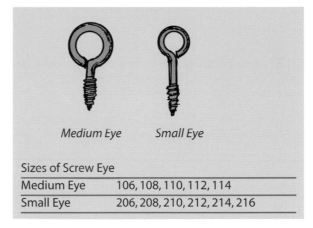

Sizes of Screw Eye	
Medium Eye	106, 108, 110, 112, 114
Small Eye	206, 208, 210, 212, 214, 216

2-13 Common sizes of screw eye.

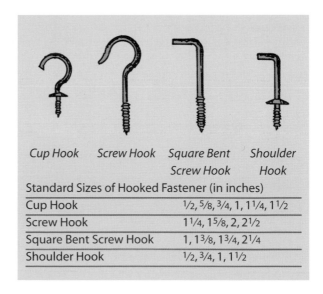

Standard Sizes of Hooked Fastener (in inches)	
Cup Hook	½, ⅝, ¾, 1, 1¼, 1½
Screw Hook	1¼, 1⅝, 2, 2½
Square Bent Screw Hook	1, 1⅜, 1¾, 2¼
Shoulder Hook	½, ¾, 1, 1½

2-14 Typical sizes and types of hook fastener.

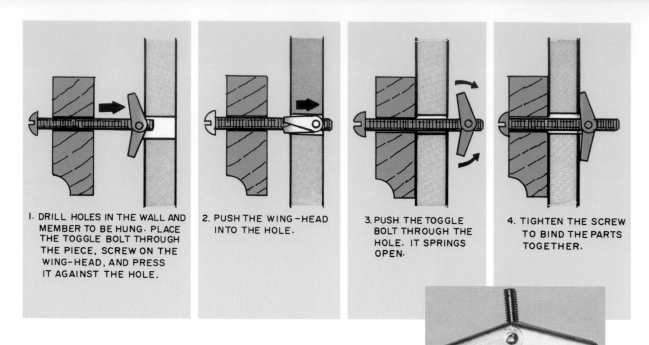

1. DRILL HOLES IN THE WALL AND MEMBER TO BE HUNG. PLACE THE TOGGLE BOLT THROUGH THE PIECE, SCREW ON THE WING-HEAD, AND PRESS IT AGAINST THE HOLE.

2. PUSH THE WING-HEAD INTO THE HOLE.

3. PUSH THE TOGGLE BOLT THROUGH THE HOLE. IT SPRINGS OPEN.

4. TIGHTEN THE SCREW TO BIND THE PARTS TOGETHER.

2-15 Toggle bolts are one device used to mount items on the wall where the finish wall material supports the load.

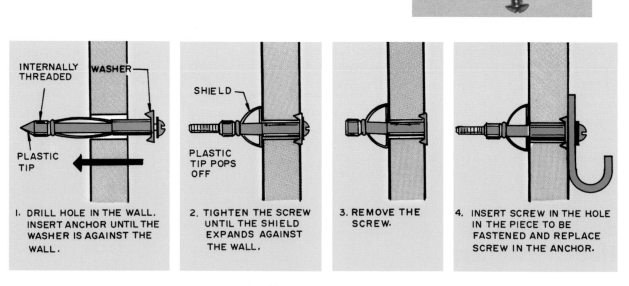

1. DRILL HOLE IN THE WALL. INSERT ANCHOR UNTIL THE WASHER IS AGAINST THE WALL.

2. TIGHTEN THE SCREW UNTIL THE SHIELD EXPANDS AGAINST THE WALL.

3. REMOVE THE SCREW.

4. INSERT SCREW IN THE HOLE IN THE PIECE TO BE FASTENED AND REPLACE SCREW IN THE ANCHOR.

2-16 Hollow wall anchors slide through a small hole in the finish wall material and have a shield that expands, pressing against the wall as the screw is tightened.

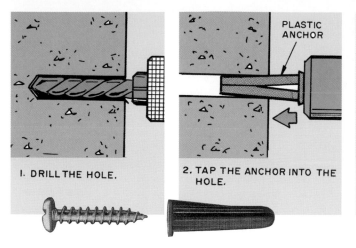
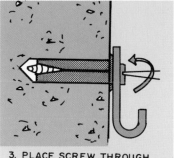

1. DRILL THE HOLE.

2. TAP THE ANCHOR INTO THE HOLE.

3. PLACE SCREW THROUGH THE HOLE IN THE ITEM TO BE ATTACHED, INSERT IT IN THE ANCHOR AND TIGHTEN.

2-17 Plastic anchors are used to support low-weight items hung on a wall.

Wall Fasteners

If the wall is hollow, objects can be anchored to it with several different fasteners.

A **toggle bolt** has a spring-loaded head. A hole large enough for the head to pass through is drilled in the wall covering material. The bolt is passed through the item to be secured to the wall, and the head is screwed on the bolt and pushed through the hole in the wall. Then the screw is tightened (**2-15**).

A **hollow wall anchor** is installed by drilling a hole through the finished wall covering and sliding the anchor into the opening. As the screw is tightened the metal shield flattens out against the inside of the wall. Then the screw is removed, placed through the item to be secured, then screwed back into the anchor (**2-16**).

A **plastic anchor** is a small, tapered plug. A hole sized to give it a very tight fit is drilled in the wall and the plug is tapped in with a hammer. Place the screw through the item to be installed and screw it into the plug. The screw causes the plug to expand and binds it to the sides of the hole (**2-17**). This is used in many materials such as wood, gypsum wallboard, masonry, and concrete.

Concrete & Masonry Anchors

There are a number of products available for anchoring items to concrete or masonry. An **impact anchor** is placed in a hole drilled in the concrete or masonry. The anchor is inserted through the item to be fastened and into the hole. As the screw is driven into the anchor it expands, binding to the sides of the hole (**2-18**).

An **expansion-sleeve masonry anchor** is inserted in the hole and through the item to be fastened. The

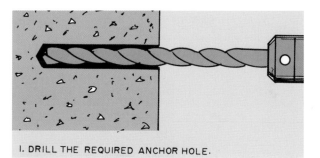

1. DRILL THE REQUIRED ANCHOR HOLE.

2. INSERT THE ANCHOR THROUGH THE ITEM TO BE FASTENED TO THE WALL AND INTO THE HOLE.

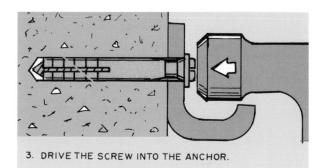

3. DRIVE THE SCREW INTO THE ANCHOR.

2-18 This impact anchor expands as the drive screw is driven into it, thus binding the item held to the wall.

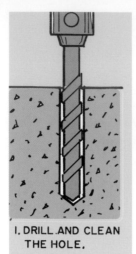

1. DRILL AND CLEAN THE HOLE.

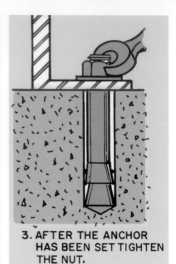

2. INSERT THE ANCHOR IN THE HOLE. PLACE THE ITEM TO FASTENED OVER IT. TAP ON THE NUT TO EXPAND THE ANCHOR.

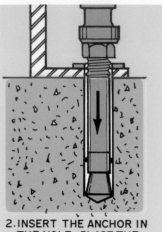

3. AFTER THE ANCHOR HAS BEEN SET TIGHTEN THE NUT.

2-19 This masonry anchor has a sleeve that expands as it is driven into the hole in the masonry.

2-20 This is an aliphatic resin glue used for gluing hard and soft woods.

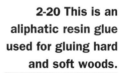

2-21 One important use for contact cement is to bond plastic laminate to countertops.

bolt is inserted into the anchor with the nut on the end. It is driven in place with a hammer. This causes the sleeve to expand. Tighten the nut to anchor the item (2-19).

Glues, Cements & Adhesives

Glues, cements, and adhesives are bonding agents. **Glues** are made from natural materials. **Cements** are rubber-based. **Adhesives** are synthetic materials.

There is a wide range of these products available **(refer also to Table 12-1)**. Some bond many different kinds of materials while others are designed for use on one material, such as wood **(2-20)**. A visit to the hardware dealer or building supply dealer will expose you to this array of bonding agents. Some are used to hold pieces together while mechanical fasteners are being installed or are holding nonload-bearing materials such as insulation or decorative trim. These are referred to as **nonstructural** adhesives. Other adhesives are used to secure materials where no mechanical fasteners are required. These bond load-bearing items and are referred to as **structural adhesives**. These typically include metal, wood, and plastic parts.

Some adhesives cure very rapidly while others require clamping during the set time. New products are on the market from time to time and often have new and improved properties. Manufacturers have available a wide range of **general-purpose** liquid adhesives formulated from materials such as neoprene, nitrile rubber, synthetic resin, and natural rubber. These have a range of materials which they are designed to bond.

In addition manufacturers have a variety of **specialty adhesives** which they design to bond specific materials such as vinyl and rubber, and for applications where waterproof qualities are needed.

Glues

Glues are made from vegetable and animal products. They are no longer widely used since the development of adhesives. They are usually used for cabinet work. **Liquid hide** glue is an animal product in liquid form. It has good joint-filling properties and sets in about two hours. **Casein glue** is made from dry milk curds. It comes in a powder form, to be mixed with water. It has the advantage of being water resistant and can be used for outdoor applications that are not directly exposed to the weather.

Cements

Cements are made from synthetic rubber. Two that are widely used are contact cement and mastic. **Contact cement** is used to bond wood veneers, plastic laminates, and other decorative materials to a wood base. It is applied to both surfaces, allowed to dry, and then the surfaces are brought together (**2-21**). Contact cement can be cleaned off surfaces with lacquer thinner. Since it is flammable, the area in which it is being used should be well ventilated.

Mastic is a thick form of contact cement, usually applied with a caulking gun. A typical job is to bond plywood subfloor to the joists and wall sheathing to the studs (**2-22**).

Adhesives

Adhesives are the most widely used bonding agent (**2-23**). They are classified into two types, **thermoplastics** and **thermosets**.

THERMOPLASTICS

The major types of **thermoplastic** include polyvinyl, aliphatic resin, alpha cyanoacrylate, and hot melts. **Polyvinyl** is commonly called white glue. It sets up in about 30 minutes but takes 24 hours for a complete cure. It is a major bonding agent for wood products.

Aliphatic resin is yellow and is stronger than white polyvinyl adhesives. It is used to bond wood (**2-20**).

Alpha cyanoacrylate, also referred to as Super Glue or Instant Glue, is used to bond nonporous materials such as metal, plastics, rubber, and ceramics

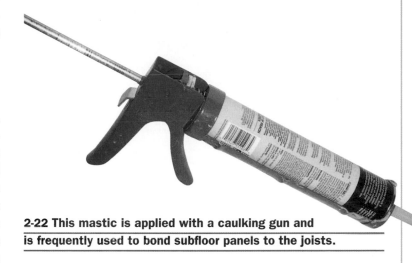

2-22 This mastic is applied with a caulking gun and is frequently used to bond subfloor panels to the joists.

2-23 This adhesive is a waterproof, interior-grade adhesive used for gluing paneling, drywall, foamboard, moldings, plywood, hardboard, and related construction products.

2-24 Super Glue provides an instant bonding that forms a very strong joint. It is used to bond wood, paper, china, rubber, leather, and metal. Keep it off your fingers.

2-25 Hot melt glue guns extrude a melted strip of adhesive from the end of the barrel.

(2-24). They reach handling strength at normal room temperatures in about 10 seconds.

Hot melts glue wood and most other materials. They are heated in a special gun (2-25) and applied from the tip of the barrel. They are quick to use but not very strong. They are used where strength is not a major factor, such as bonding overlay panels to cabinet doors. It sets quickly, so the pieces must be placed together rapidly (2-26).

THERMOSETS

Thermoset adhesives are more resistant to heat and moisture than most bonding agents. These include epoxy, urea-formaldehyde, and resorcinol formaldehyde.

Epoxy can be used to bond almost any kind of material. It produces a very strong joint and is especially helpful when repairing broken furniture, china, and other products. It is packaged in two tubes, one containing the resin and the other the hardener (2-27). These are mixed in equal amounts. Apply a thin coat to both surfaces. Excess material can be removed with nail polish remover or denatured alcohol, while the epoxy is still tacky. Keep it off your fingers.

Urea-formaldehyde is sold in powder form and mixed with water. As soon as it is mixed a chemical reaction begins; therefore do not mix more than you can use in a couple hours. It is water resistant and requires 16 hours of clamping time. In the store it is called **plastic resin.**

Resorcinal-formaldehydes produce a waterproof joint and are used on exposed wood products. They are strong and dependable. They are sold in two containers; one contains the liquid resin and the other the catalyst. These are mixed and stirred. At least 16 hours of clamping time is required.

AEROSOL ADHESIVES

Aerosol adhesives are dispensed in a controlled spray pattern from an aerosol container (2-28). A variety of bonding agents are available for use on different materials such as foam plastic sheets, plastic laminates, and other lightweight materials.

2-26 When using a hot melt glue gun wear gloves because it gets very hot. Eye protection is recommended. The bond is for light-duty connections.

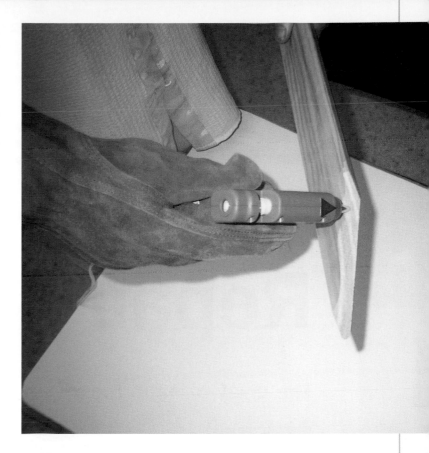

Epoxy Hardener   *Epoxy Resin*

2-27 Epoxy is packaged in two containers, the resin and a hardener. The amount needed is made by mixing equal amounts.

2-28 Aerosol adhesives are used to bond most lightweight materials.

CHAPTER 3

Basic Hand Tools for Home Repair

The array of hand tools available is great and careful consideration of which to buy is important. Over time it is possible to collect quite a few. Consider the cost and quality as you choose. Some, such as cutting tools, should be the best quality because cheap tools will not stay sharp. Others that get less use, such as a framing square or try square, can be among the less costly brands on the market.

3-1 The 6-foot folding rule and a 25-foot tape will handle most measuring jobs.

© 2003 Stanley Tools

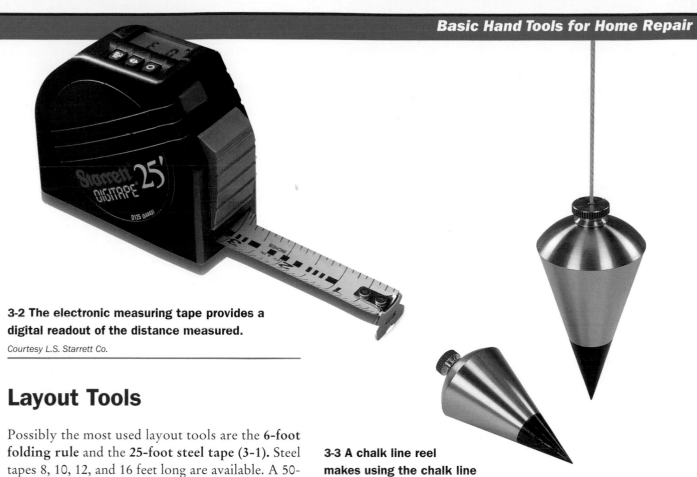

3-2 The electronic measuring tape provides a digital readout of the distance measured.

Courtesy L.S. Starrett Co.

Layout Tools

Possibly the most used layout tools are the **6-foot folding rule** and the **25-foot steel tape (3-1).** Steel tapes 8, 10, 12, and 16 feet long are available. A 50-foot tape is very useful when a job requires long measurements. A very handy tape is an **electronic measuring tape.** It uses a steel tape to span the distance and has a digital readout on the top **(3-2).** You no longer have to strain to read the markings on the tape.

When it is necessary to lay out a long vertical plumb line or a horizontal line, a **chalk line** is an easy and inexpensive tool to use **(3-3).** Get one that is in a metal case so it can wind up easily. Vertical layouts will require you to attach a **plumb bob** to the bottom of the line to pull it tight.

Various types of square are used for checking right angles and layout work **(3-4).** The large **carpenter's square** has inch markings along the edge of the tongue and blade. The markings on the sides are used for rafter layout. The **combination square** has a head that moves along the blade and can be used to lay out 45° and 90° angles. The sliding **T-bevel** can be adjusted to the desired angle and used to lay out that angle or to measure an existing angle. The **try square** is a small tool used to check surfaces for squareness and to lay out 90° angles. **Metal straightedges** are used to measure short distances and to draw short straight lines. A **digital protractor/angle finder** calculates bevel and miter angles for crown molding cuts. It measures the corner angle and computes the miter and bevel angles to set the miter saw **(3-5).**

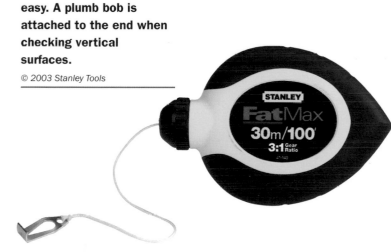

3-3 A chalk line reel makes using the chalk line easy. A plumb bob is attached to the end when checking vertical surfaces.

© 2003 Stanley Tools

Levels can be used to lay out horizontal and vertical lines and to check surfaces to see if they are plumb or horizontal. There are many sizes and designs. Some are as small as 9 inches, while others are 48 inches **(3-6).**

A very useful but more expensive level is a **laser level.** It can be used to locate horizontal and vertical lines on the wall of a room. It can be set on a tripod in the center of the room and rotate as it projects a line onto the walls **(3-7),** or can be mounted on the wall.

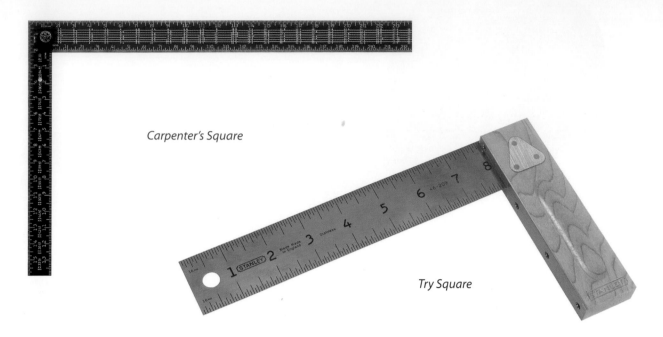

Carpenter's Square

Try Square

3-4 Various types of square and the T-bevel are essential to getting accurate layouts.

© 2003 Stanley Tools

Sliding T-Bevel

Combination Square

3-5 The digital protractor/angle finder is used to calculate bevel and miter angles for crown molding cuts.

3-6 A carpenter' level is used to see if items are plumb and horizontal. © 2003 Stanley Tools

Boring Tools

The most commonly used boring tools include the brace and push drill **(3-8)**. The **brace** holds **auger bits** in the chuck, which bore holes from ¼ inch to ¾ inch. An **expansive bit** is used to bore holes from ⅝ inch to 3 inches. The **push drill** uses special fluted drill points that are supplied with it. They are stored in the handle. After the bit is in the chuck, it is placed on the wood and the handle is pushed down. The up and down motion of the handle rotates the bit.

There are a number of other bits available. The **spade bit** is used with an electric drill. It will produce holes from ⅜ inch to 1½ inches **(3-9).** There are other types of power driven bit available that can be used

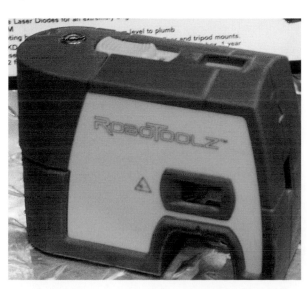

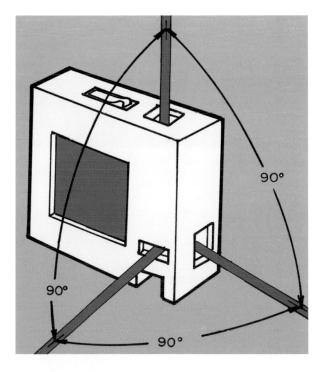

3-7 A small laser level is used to locate horizontal and vertical lines.

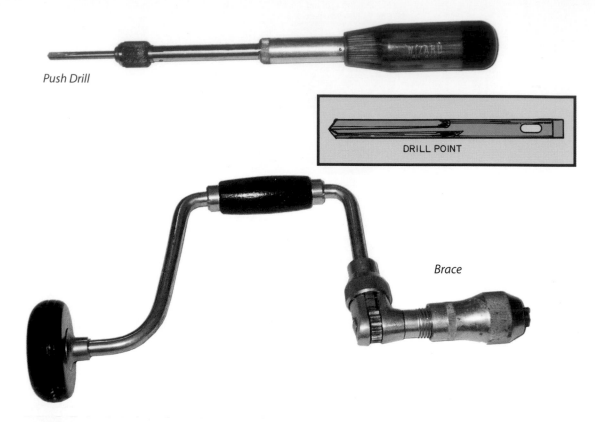

Push Drill

DRILL POINT

Brace

3-8 The push drill and brace are used to bore holes in wood.

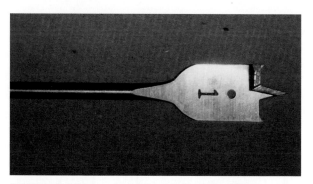

3-9 Spade bits are used in an electric drill.

to bore large-diameter holes in wood and plastic. Examine those on display at your local building supply dealer.

When setting flathead screws in a board you need to countersink the top of the hole. **Countersinks** are available for use in a brace and on an electric drill **(3-10).** They will work on wood, metal, and plastics. A **drill and countersink bit** is a combination of both tools as a single tool. It drills the hole and then countersinks it **(3-11).**

Used in a brace

Used in an electric drill

3-10 Countersinks are used to form a conical depression on the exposed end of a hole so a flathead screw will be flush with the surface of the material. © 2003 Stanley Tools

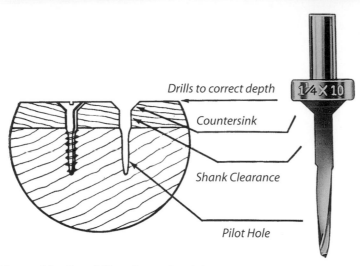

Drills to correct depth

Countersink

Shank Clearance

Pilot Hole

3-11 This combination drill and countersink bores the hole for the body of the screw and countersinks the opening to receive the head of a flathead wood screw. It is used in an electric drill.

© 2003 Stanley Tools

3-12 The average bits in this set range in diameter from 1/4 inch through 3/4 inch. Keep them stored in a protective case so the cutting edges do not get damaged.

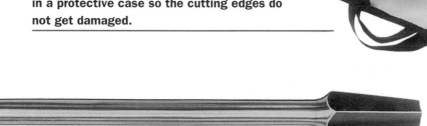

3-13 An expansive bit has cutters that are adjusted to get the diameter of the hole desired. © 2003 Stanley Tools

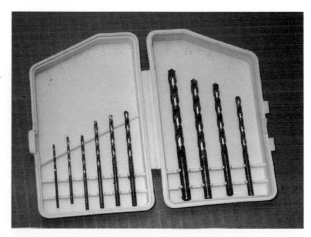

Auger bits are sold in sets as shown in **3-12.** They are used to bore holes in wood that are larger than possible with twist drill sets. They are held in a brace.

Expansive bits are used to bore large-diameter holes in wood. They have adjustable cutters that enable you to bore holes from 5/8 inch to 3 inches in diameter **(3-13).** They are held in a brace.

A set of **straight-shank twist drills** is shown in **3-14.** They drill holes in metal, plastic, and wood. They are used in hand drills and electric drills.

3-14 The straight-shank twist drills in this set range in diameter from 1/16 inch through 1/4 inch. Much larger sets are available.

3-15 A hand saw. © 2003 Stanley Tools

Dovetail Saw

Backsaw

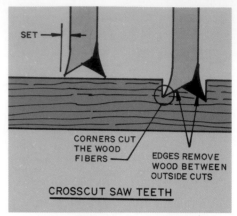

CROSSCUT SAW TEETH

CORNERS CUT THE WOOD FIBERS

EDGES REMOVE WOOD BETWEEN OUTSIDE CUTS

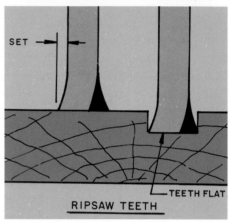

RIPSAW TEETH

TEETH FLAT

3-16 Crosscut saw teeth are like a series of knives. Ripsaw teeth are like a series of chisels.

3-17 Backsaws and dovetail saws have very fine teeth and are used where fine, accurate cuts are needed. © 2003 Stanley Tools

3-18 This small miter box will produce fine, accurate cuts on small stock. © 2003 Stanley Tools

3-19 This large miter box will handle wide moldings. It has a range of preset angles which automatically will accurately set the most frequently used angles. © 2003 Stanley Tools

Cutting Tools

Frequently used cutting tools include hand saws, chisels, planes, and knives. Choose high-quality tools because the cheaper ones will not hold their sharpness as long and will be more difficult to use.

Saws

Hand saws are specified by their length, type of teeth, and the number of teeth per inch. Possibly the 26-inch **crosscut** and **ripsaws** are most widely used. Smaller sizes ranging from 15 to 20 inches are available **(3-15).** Crosscut saws have teeth from 9 to 12 points per inch. The more points, the finer the cut produced. A typical ripsaw will have 5 points per inch. The difference in the crosscut and rip teeth is shown in **3-16.** The crosscut teeth are like a series of knives bent alternately left and right that slice the wood fibers with a cutting action. The ripsaw teeth are chisel-shaped and sever the wood fibers on the bottom edge of the tooth. A **backsaw** is used to make fine, accurate cuts **(3-17).** They are 12 to 18 inches long and have 13 points per inch. They are also used on small **miter boxes (3-18).** Large, heavy-duty miter boxes are sold with a very large backsaw **(3-19).** They are great for professional interior trim carpenters. A **dovetail saw** (refer to **3-17**) is used for very fine, accurate cuts. It is typically 10 inches long and has 13 points per inch. A **compass saw (3-20)** has rip teeth and a narrow pointed blade. It is used to cut curves or pierced openings. A **keyhole saw (3-20)** is a smaller version of the compass saw. To cut very sharp curves and fine work, a **coping saw (3-21)** can be used. It has a very thin blade which can be replaced with blades having different tooth sizes.

Compass Saw

3-20 Compass and keyhole saws are used to cut openings in wood and plastic panels. © 2003 Stanley Tools

Keyhole Saw

33

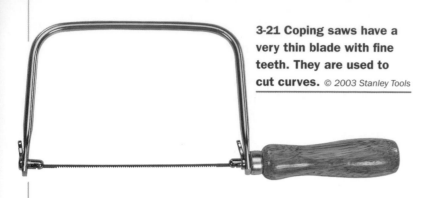

3-21 Coping saws have a very thin blade with fine teeth. They are used to cut curves. © 2003 Stanley Tools

A **hacksaw** is used to cut metal. It has a replaceable blade. Blades with 14 to 32 points per inch are available. The fine-tooth blades are used on thin metal **(3-22)**.

Occasionally you will need to cut some sheet metal, so **sheetmetal shears** will be a valuable addition to your toolbox **(3-23)**.

Chisels, Planes & Knives

Wood chisels are used for trimming wood members and cutting joints. While you can buy individual chisels, it is a good idea to buy a complete set **(3-24)**. This will typically have 4 to 6 chisels with blade widths from ¼ inch to 1½ inches. Be certain to choose a set that has a strong handle firmly mounted on the blade. Since the handle is often tapped with a mallet to make a cut, this is important.

3-22 A hacksaw is used to cut metal and plastic. Blades are available with a range of tooth sizes. The finer teeth are used on thin metals.

© 2003 Stanley Tools

3-23 Sheetmetal shears will cut thin gauges of metal sheets. © 2003 Stanley Tools

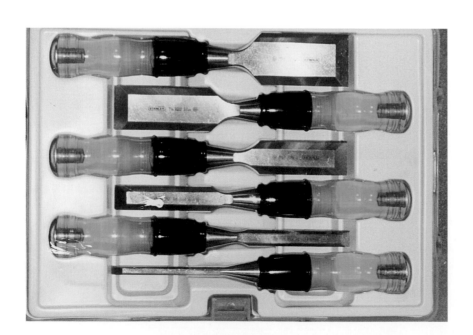

3-24 A typical set of wood chisels will range from $1/4$ inch to $1^1/2$ inches wide.

The most widely used knife is the **utility knife** (3-25). It has extra blades stored in the handle. Packages of replacement blades are available.

A **smooth plane** is typically 9¾ inches long with a plane iron 1¾ to 2 inches wide. It is used to smooth shorter boards. A **jack plane** is generally 14 inches long with a 2-inch plane iron. It is used to smooth longer boards. **Fore** and **jointer** planes are available up to 22 inches long. A **block plane** is 6 to 7 inches long. The plane iron is on a low angle, making it useful for planing end grain and cross grain (3-26).

3-25 The utility knife is possibly the most used tool in your tool chest. It is used to cut, pierce, and trim almost any material except metal.

© 2003 Stanley Tools

Jack Plane

Block Plane

3-26 These are the planes most frequently needed.

© 2003 Stanley Tools

Smooth Plane

35

Flat Wood Rasp

Half-Round Wood Rasp

Flat Mill File

Half-Round Cabinet File

3-27 The mill file has very fine teeth and is designed for smoothing metal. The cabinet file is coarser and used for wood. The rasp is very coarse and rapidly reduces the wood surface.

Three-Square File

Round File

Finishing Tools

Finishing tools are used to bring the wood surface to the final shape and get it ready for the steps to apply the finish coating. **Files** and **rasps** are used to remove small amounts of wood **(3-27)**. Mill files have very fine teeth and produce a fairly smooth surface. Rasps have large teeth and remove large amounts of wood rapidly and leave a rough surface. Both are available in flat, half-round, round, and triangular shapes and a range of lengths. You will need a **file cleaner** to brush over them occasionally to clean wood chips from between the teeth. Be certain to buy some sturdy handles to fit on the pointed tang of the file. The sharp

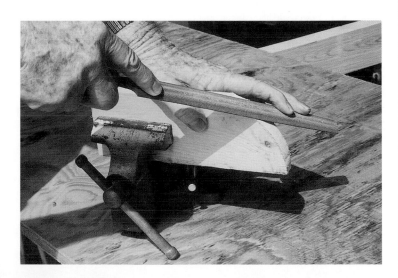

3-28 Convex surfaces are smoothed by filing down the curve with a flat file.

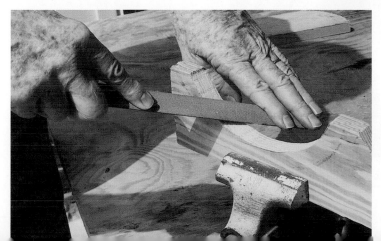

3-29 Concave surfaces are smoothed by filing down the curve with a half-round file.

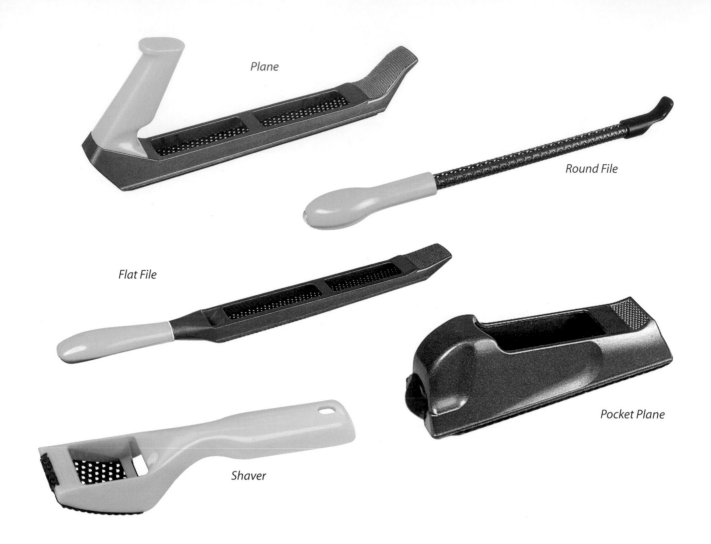

Plane

Round File

Flat File

Pocket Plane

Shaver

3-30 Some of the types of Surform™ filing tools. © 2003 Stanley Tools

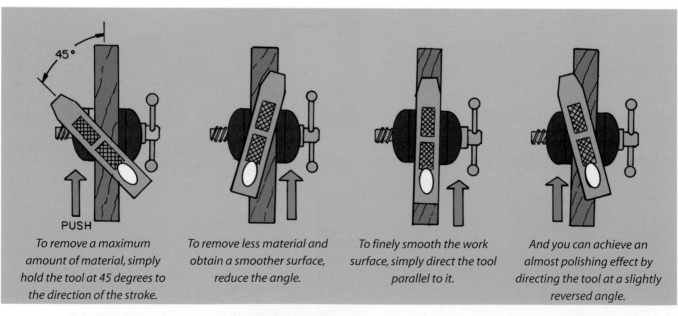

45°

PUSH

To remove a maximum amount of material, simply hold the tool at 45 degrees to the direction of the stroke.

To remove less material and obtain a smoother surface, reduce the angle.

To finely smooth the work surface, simply direct the tool parallel to it.

And you can achieve an almost polishing effect by directing the tool at a slightly reversed angle.

3-31 Recommended ways to use Surform tools. © 2003 Stanley Tools

3-32 Cabinet scrapers are used to put a final smooth finish on a board before it is sanded.

3-33 Paint scrapers remove old paint after it has been softened by a paint remover.

tang can cause injury. **Flat files** are used to smooth convex surfaces (3-28). **Half-round** files are used to smooth concave surfaces (3-29).

Surform™ tools have a series of perforated teeth much like a cheese grater. They remove wood quite rapidly. They can be used to smooth wood, aluminum, vinyl, fiberglass, and rubber (3-30). They are used as shown in 3-31.

Cabinet scrapers produce a very smooth finished surface on wood and get surfaces ready for sanding (3-32). **Paint scrapers** are used to remove old paint

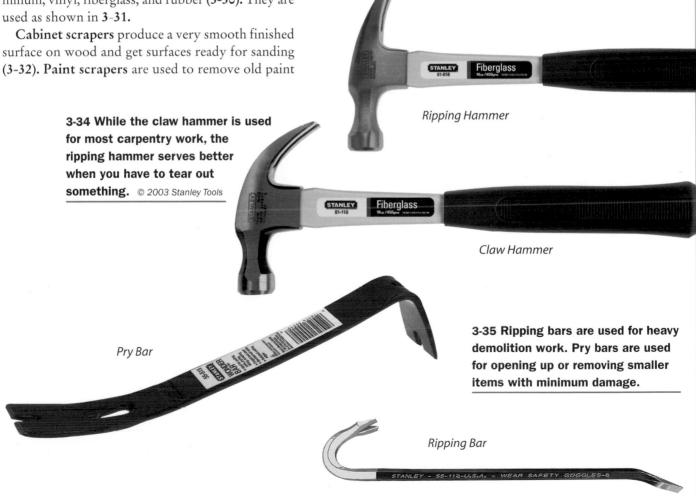

3-34 While the claw hammer is used for most carpentry work, the ripping hammer serves better when you have to tear out something. © 2003 Stanley Tools

Ripping Hammer

Claw Hammer

Pry Bar

3-35 Ripping bars are used for heavy demolition work. Pry bars are used for opening up or removing smaller items with minimum damage.

Ripping Bar

after it has been softened by a paint remover **(3-33)**. Remember, paint used in old houses will most likely contain lead. Lead particles are hazardous to your health. Keep down all paint dust, wear a respirator, and remove all paint remains from the room. There are kits at your local hardware or paint store you can buy to test the paint for lead.

Striking Tools

The **claw** and **ripping hammers** are available in 13-, 16-, 20-, and 22-ounce sizes **(3-34)**. For normal nail driving, a 16-ounce hammer is generally used. If you are nailing house framing, a 22-ounce hammer will make it easier to drive common nails through 2-inch lumber. The **claw hammer** is used for most nail-driving jobs. The claws make it possible to pull nails. The **ripping hammer** has straight claws and can be used to drive behind a board you want to tear off.

There are a variety of **bars** for pulling nails and ripping out old work. Some are shown in **3-35**.

Nail sets are used to set the heads of nails below the surface of the wood **(3-36)**. They are struck with a hammer. The recess created is caulked as the surface is prepared for finishing **(3-37)**. The tips available are $1/32$ through $5/32$ inch in diameter.

3-36 Nail sets are available with a number of different tip sizes. *© 2003 Stanley Tools*

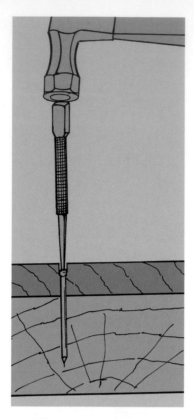

3-37 The nail set is used to drive the head of a finish nail or brad below the surface of the board.

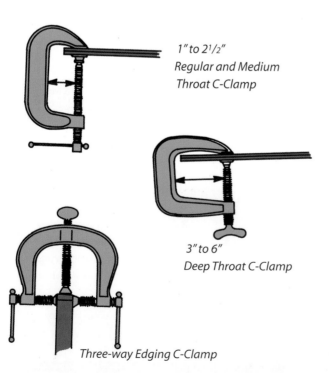

1" to 2$1/2$"
Regular and Medium Throat C-Clamp

3" to 6"
Deep Throat C-Clamp

Three-way Edging C-Clamp

3-38 C-clamps are available in a range of sizes and designs.

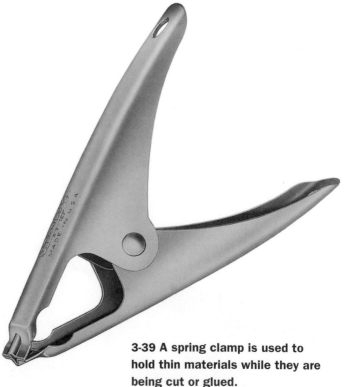

3-39 A spring clamp is used to hold thin materials while they are being cut or glued.

© 2003 Stanley Tools

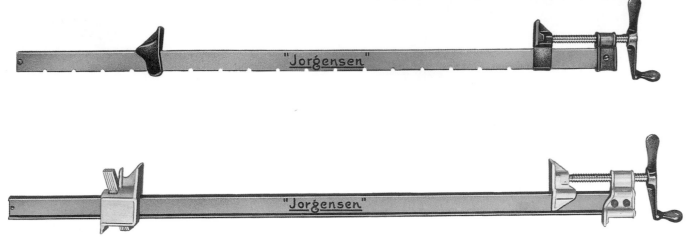

Clamps

The need to hold materials while they are being worked or permanently joined is a constantly occurring event. Manufacturers have developed a fine array of clamping tools from which to choose.

The **C-clamp** is an inexpensive general-purpose tool that should be in every toolbox (**3-38**). A variation of this is the **spring clamp**. The jaws are held closed by a heavy spring in the handle (**3-39**). **Bar clamps** are used to secure long assemblies and are available in a range of sizes and designs (**3-40**).

Fastening Tools

A set of **screwdrivers** with both standard and Phillips tips is important. These will let you drive standard and Phillips head screws. Since the size of the opening in the screwhead gets larger as the screw size increases, the size of the tip has to fit the screw. Phillips head screw points are available from size 0 to size 4. Standard tip sizes run from 0 to 24; however, sizes 0 to 8 will handle a high percentage of the jobs. The length of the blade usually varies with the tip size. Larger tips will have longer blades (**3-41**).

Some of the other screwhead recesses are shown in **3-42**. Screwdrivers are available with tips to fit these recesses.

Many jobs require that bolts and other threaded fasteners be removed or installed. Possibly the universal **vise grip pliers** is a tool that should be in every

3-40 Bar clamps have an adjustable foot on one end and a screw-type clamp on the other end. *Courtesy Adjustable Clamp Company*

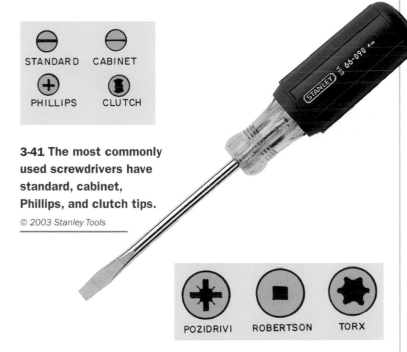

3-41 The most commonly used screwdrivers have standard, cabinet, Phillips, and clutch tips.

© 2003 Stanley Tools

STANDARD CABINET
PHILLIPS CLUTCH

POZIDRIVI ROBERTSON TORX

3-42 Several other types of screw recess requiring special screwdrivers.

3-43 Vise grip pliers are adjustable to fit a wide range of bolts and nuts. They are also useful for gripping other things such as pipe and plumbing fittings.

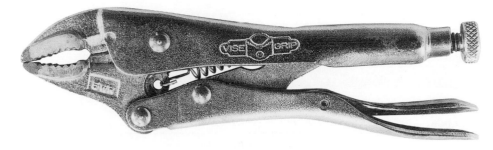

41

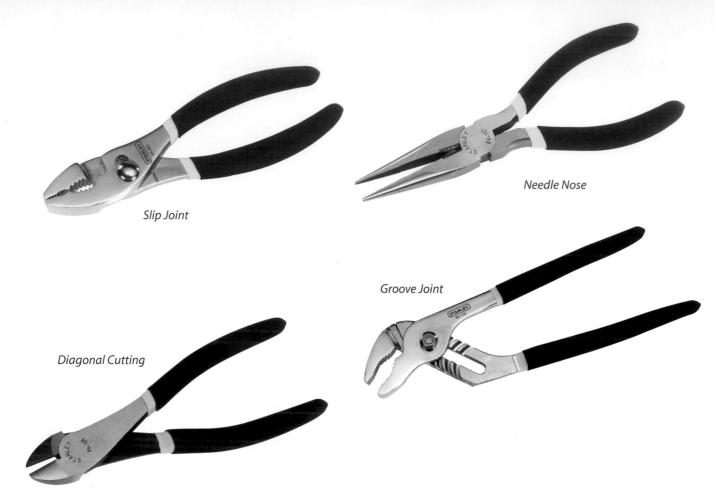

Slip Joint

Needle Nose

Groove Joint

Diagonal Cutting

3-44 These are frequently used pliers.

toolbox. It is adjustable to fit most bolts and nuts commonly used **(3-43).** A set of pliers consisting of **needle nose, diagonal cutting, groove joint,** and **slip joint pliers** are tools often used **(3-44).**

A set of **open-end wrenches** or a set with an open end on one end and a box wrench end on the other would be a good choice **(3-45).** A set with openings from ⅜ to 1 inch serves most needs. Remember, metric wrench sets are also available.

Tools to Rent

There will be times when you need a tool for a few days that is very expensive to buy. Consider renting it. The costs are reasonable and the tool is immediately available. For example, if you need to hammer some holes through a concrete foundation or break up a concrete sidewalk, rent the tool to do the job. If you need a router or power miter saw for a small job, rent it. Scaffolding and long extension ladders are other good items to rent. The list of tools available is long.

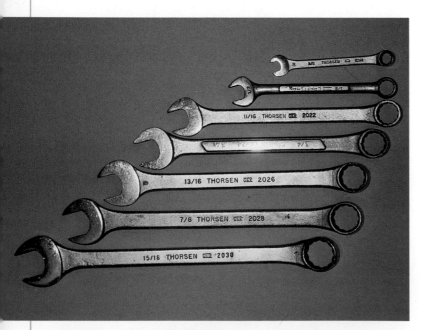

3-45 This is a set of combination open-end and box-end wrenches.

CHAPTER 4

Repairing Interior Walls & Ceilings

The most frequently occurring maintenance tasks to be done on the interior of a house involve repairing and redecorating walls and ceilings. There is a wide range of materials available for each of these tasks. Give careful consideration to the maintenance job about to be done and make a study of the choices of materials available to do it. Then choose the best material that can be afforded and follow the proper procedure for the job.

Working with Gypsum Wallboard

Gypsum wallboard is installed dry, which is why it is also called drywall. It is easier to install than walls covered with plaster, and it is less expensive. It consists of a gypsum core bonded between layers of a specially formulated paper. The side to be exposed is covered with a strong, smooth-faced paper. The back of the panel has a strong, natural-finish paper. The paper is folded along the long edges of the panel, and the ends are square-cut. The most commonly used panel is 4 feet x feet; however, larger panels are available by special order. The kinds of drywall that are normally available include regular, type-X, moisture-resistant, ceiling, and predecorated.

Regular drywall panels are used to finish interior walls. They have either a tapered, rounded edge or a standard taper with a beveled edge. The types of edges available on various panels are shown in **4-1**. Regular panels are available in thicknesses of ⅜, ½, and ⅝ inch. The ½-inch thickness is the most commonly used.

Type-X, fire-rated drywall has a specially formulated gypsum core that contains additives that increase its ability to resist fire.

Moisture-resistant drywall is used on walls and ceilings in areas where there will be higher-than-average moisture. It typically can be identified by the green-color paper. It is not used in areas having direct exposure to water, such as a shower. Use cementitious tile backer board in these areas.

Ceiling drywall is a high-strength ½- and ⅝-inch-thick gypsum panel used on ceilings. The special core is designed to resist sagging that can occur with regular panels on the ceiling over time.

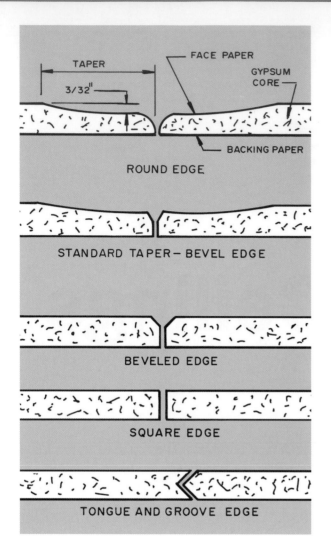

4-1 The common types of edge available on gypsum wallboard.

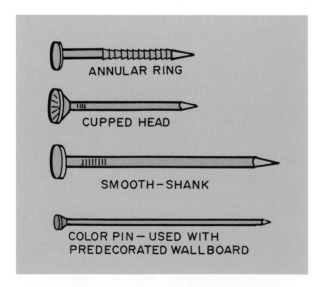

4-2 Nails used to secure gypsum drywall panels to wood framing.

Flexible drywall panels are ¼-inch thick and have heavy paper faces. They are used to bend around concave and convex walls. They bend best in the long direction, so are usually installed with the long edge perpendicular to the wall studs.

Predecorated drywall panels are covered with a decorative material such as fabric or vinyl. They do not require additional finishing. The joints are covered with moldings supplied by the manufacturer, and the nails are colored to blend in with the colors on the panel.

Drywall Fasteners

Annular ring nails provide the best holding power in wood studs. However, **cement-coated nails** are also widely used. **Cooler nails** are used in some parts of the country.

Nails should be long enough to go through the panel and ⅞ inch into the wood stud. Fire-rated wall assemblies require nails that penetrate the stud 1½ inches (4-2).

Drywall screws are used to secure panels to wood and metal studs (4-3). They have greater holding power than nails, thus having fewer pop-outs. They are rapidly installed with a power screwdriver.

Adhesives are used to bond single layers of drywall to framing, masonry, or concrete. They are also used to bond panels together when multilayer assemblies are required. The adhesives are classified into two types. **Stud adhesives** are used to bond drywall panels to the studs. **Laminating adhesives** are used to bond one panel to another, forming a laminated panel.

Installing Drywall Panels

Drywall panels are easy to cut to size. While cutting them, be careful that the face paper is not damaged. Measure carefully. Since many measurements are long, use a tape measure.

4-3 A type-W drywall screw used to secure panels to wood framing.

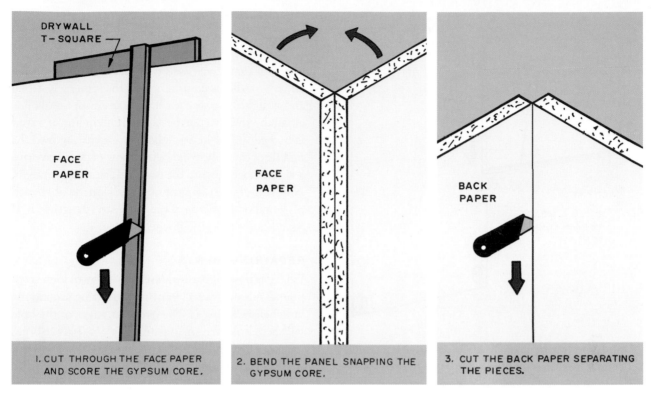

1. CUT THROUGH THE FACE PAPER AND SCORE THE GYPSUM CORE.	2. BEND THE PANEL SNAPPING THE GYPSUM CORE.	3. CUT THE BACK PAPER SEPARATING THE PIECES.

4-4 To cut a gypsum panel, score the core on the face side, snap the core, and cut the backing paper.

CUTTING DRYWALL PANELS

To cut drywall panels, follow these steps (4-4).
1. Measure and locate the line of cut. A tape measure is generally used.
2. Score the drywall panel with a utility knife on the good-paper side. Run the knife along the edge of a drywall T-square. A carpenter's framing square can also be used.
3. Bend the panel away from you and the core will snap along the scored line.
4. Cut the paper covering on the back side to complete the cut.

CUTTING OPENINGS IN PANELS

Openings in the panels for switches, electrical outlets, and other things that penetrate the wall have to be located and cut. One way to locate the openings is to measure their distance from the panel above and from the adjacent panel (4-5). Another way is to cover the edges of the outlet box with chalk and push the panel against it. Then drill holes at each corner, and cut the opening with a compass or electric saber saw (4-6). Another method is to cut the outline with a utility knife, cut diagonals across the opening, tap the pieces toward the back, and then cut the back paper (4-7).

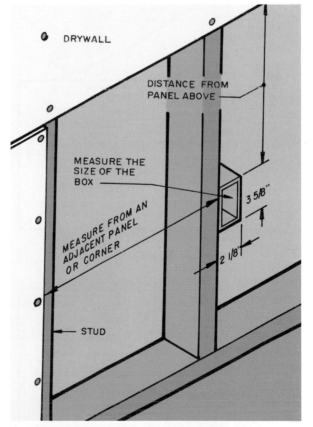

4-5 The openings in the gypsum panel can be located by measuring from adjoining panels or other surfaces.

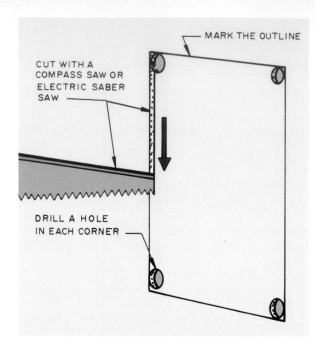

4-6 One way of cutting wall panel openings is to drill a hole in each corner and saw between the corners. A saber saw does a good job.

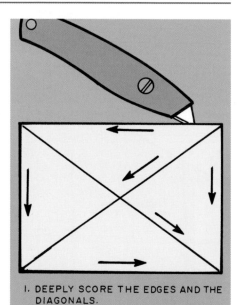

1. DEEPLY SCORE THE EDGES AND THE DIAGONALS.

4-7 Another way to cut an opening in the gypsum panel is to score the edges and diagonals and knock the pieces back with a hammer. Then cut the back paper.

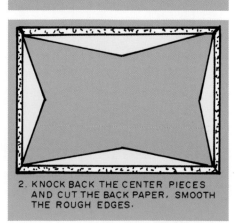

2. KNOCK BACK THE CENTER PIECES AND CUT THE BACK PAPER. SMOOTH THE ROUGH EDGES.

INSTALLING WALL PANELS

A major room repair could require that a wall or ceiling be stripped to the frame and covered with new wallboard or that a new wall be added. Before installing any new wallboard, make certain the framing is strong and straight. To check for straightness, run a chalk line along the studs to see that they are all in line. If a stud is bowed, it can be straightened by cutting into the middle of the hollow side of the bow (4-8). Then drive a wooden wedge into the cut until the stud is straight. Nail scabs (short lengths of boards) made of ½-inch plywood on each side of the wedge to strengthen it. If a stud cannot be straightened, replace it.

REPAIRING DENTS

Fill dents in wallboard with several layers of joint compound. Clean away all loose material from around the hole before filling it. The irregular edges of the hole will help hold the small patch in place. Larger holes are more difficult to repair and require more work.

NAILING THE PANELS

When driving the nails, set them in a small dimple so they can be covered with joint compound (4-9). Do not drive them so hard that they break the paper or

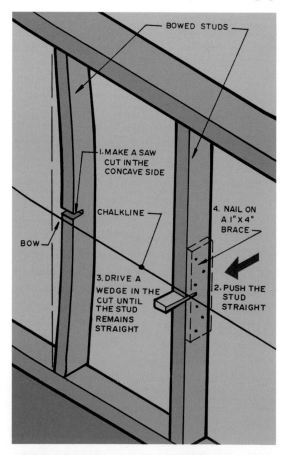

4-8 Bowed studs must be straightened or replaced.

crush the core. If this happens, put another nail about 2 inches away from the first. The panel should be held tightly against the stud and the nail must go straight into it. Use a hammer that has a rounded convex face that leaves a shallow dimple in the panel. A special drywall hammer is available.

PLACING THE PANELS

If the ceiling is the standard 8-foot height, there will be fewer feet available for edge joints if the panels are installed with the long side perpendicular to the studs (**4-10**). This also places the strongest dimension across the framing members. If the ceilings are higher than this (9-foot and 10-foot ceilings are popular), place the narrow strip at the floor (**4-10**). Here it is less noticeable. Panels 10 feet long are available.

The recommended nailing pattern is shown in **4-11**. Double-nailing can be used to reduce nail pop-out. When driving the nail, remember to set it in a dimple.

If an entire wall has to be covered, follow the procedure shown in **4-12**. Start in an upper corner and lay the top row. Size the panels so that if there is a joint it occurs over a door or window. This reduces the length of the joint that will need to be covered. Never have a joint at the corner of an opening. It will tend to open up as the window jamb and framing expand and contract. Some prefer to use the interior corner construction shown in **4-13**. This allows for some wall movement without producing nail pop-out. In addition to nailing the panels, it provides a better job if a bead of stud adhesive is laid on each stud (**4-14**). Press the panel into the adhesive and in the normal manner (**4-15**).

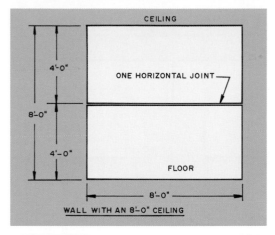

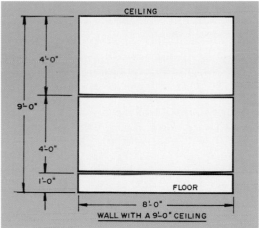

4-10 While gypsum panels can be installed vertically, it is recommended that horizontal installation be used.

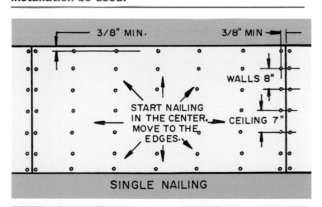

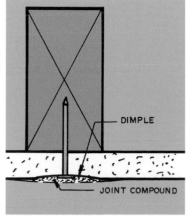

4-9 Set the nail head in a slight dimple in the panel but do not break the paper or crush the gypsum core.

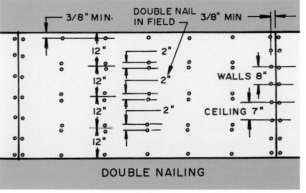

4-11 The recommended placing of fasteners for single- and double-nailed panels.

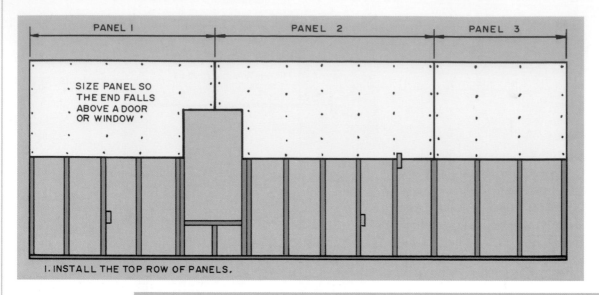

PANEL 1   PANEL 2   PANEL 3

SIZE PANEL SO
THE END FALLS
ABOVE A DOOR
OR WINDOW

1. INSTALL THE TOP ROW OF PANELS.

4-12 When covering a wall, start in a corner and install the top row of panels first. Remember to stagger the joints and butt panels over doors and windows.

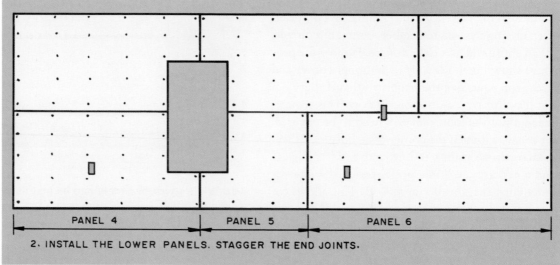

PANEL 4   PANEL 5   PANEL 6

2. INSTALL THE LOWER PANELS. STAGGER THE END JOINTS.

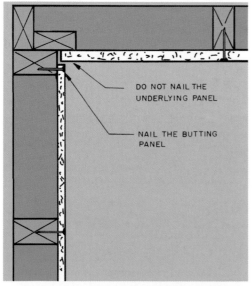

DO NOT NAIL THE
UNDERLYING PANEL

NAIL THE BUTTING
PANEL

4-13 To allow for movement of the walls, some installers omit the fasteners on one of the panels forming an interior corner.

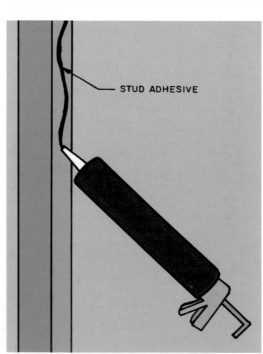

STUD ADHESIVE

4-14 Stud adhesive, plus nails or screws, provides a superior bond between the panel and the stud.

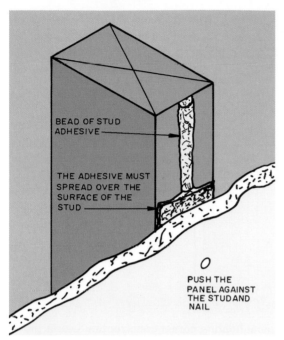

4-15 The panel must be pushed against the adhesive so it spreads over the face of the stud. Nail the panel so it is tight against the stud.

COVERING OLD GYPSUM WALLBOARD

If the old wallboard is sound, although damaged, it need not be removed. Cover it with new ⅜-inch wallboard sheets. To install new wallboard over old, follow these steps:

1. Go over the old wall, looking for loose spots. Renail as needed so the old sheets are firmly in place.
2. Apply an adhesive to the back of a new panel and spread it with a notched trowel (4-16).
3. Press the new sheet against the wall. Use a few nails to hold the sheet in place while the adhesive sets.
4. If you are applying a new sheet to a ceiling, in addition to the adhesive use nails or screws long enough to reach the wood framing.

Installing Ceiling Panels

Ceiling panels should be installed before the wall panels. They are installed as shown in **4-17**. Start in a corner and work across the room. When installing the next row, cut a panel so the joints are staggered. Always place tapered edges together and square-cut ends together. The difference in thickness makes it difficult to tape over the joint. For best results, all edge and end joints should be backed by 2 x 4 blocking. Consider buying panels 12 feet long. While they are heavier, they reduce the number of end joints.

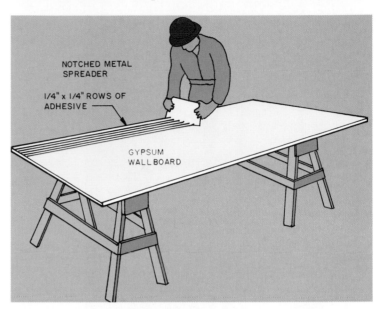

4-16 Double-layer panel application requires the use of a laminating adhesive that is applied to the back of the exposed panel.

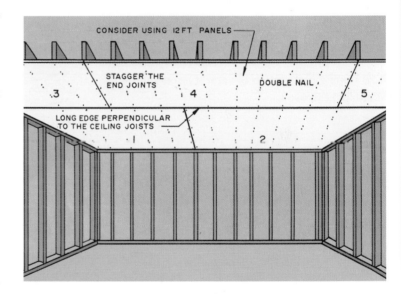

4-17 Begin installing ceiling panels in one corner and work across the room. Then start the next row. Stagger the end joints.

Ceiling panels are difficult to install because they are bulky and ½-inch panels weigh a bit over 50 pounds. It will take two or more people to lift a panel. Strong, stable scaffolding is needed (4-18). Make some T-braces out of 2 x 4 stock, as shown in 4-19. Pad the top of the brace with cloth or plastic so the paper on the panel is not damaged.

While installing the ceiling, consider using floating interior corner construction (4-20). This will allow for movement and reduce nail pop-out.

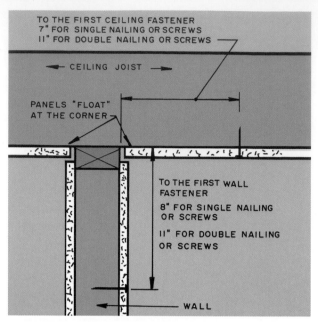

4-20 Typical floating corner construction where the ceiling meets a wall. This allows for movement and reduces cracking at the corner joint.

4-18 Ceiling installation is safer if good-quality scaffolding is used.

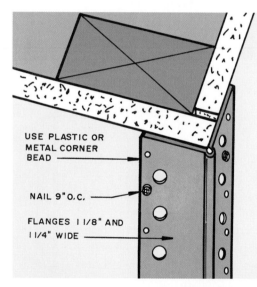

4-21 Metal and plastic corner beads are nailed over external corners. They provide protection from damage.

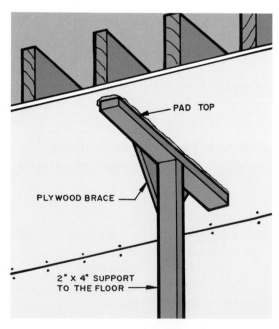

4-19 Make a couple of T-braces to help hold the panels as they are positioned and nailed onto the ceiling.

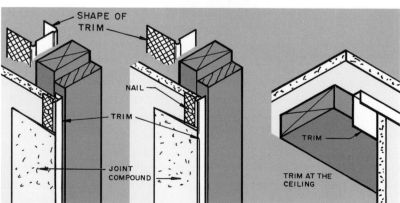

4-22 Metal and plastic trim are available for covering the exposed raw edges of the gypsum panel.

Installing Corner Beads & Trim Casing Bead

External corners should be protected with a metal or plastic **corner bead.** These protect the corner from damage and provide a base for finishing it with joint compound **(4-21).**

Various types of **trim** are also available. They are used to cover any exposed raw edges and provide a base for finishing them with joint compound **(4-22).**

The bead and trim are nailed to the studs every 9 inches with standard annular ring nails or screws.

Taping Tools

The tools needed are 4- and 6-inch **joint taping knives,** a **beveled trowel,** and a 10- or 12-inch **finishing knife (4-23).** In addition, a pan is needed to contain the joint compound as the person taping the joint moves along the joint **(4-24).** Inside and outside-corner finishing tools speed up the taping of corners **(4-25).**

Very fine-grit **drywall sandpaper** is best to use for finishing. It is available in 80, 100, 120, and 150 grit. The larger the number, the finer the abrasive. Start with a coarser abrasive on the first coat and finish with a fine abrasive for the final coat.

Sanding sponges are used to wet-sand the compound. They arc also used to lightly blend the

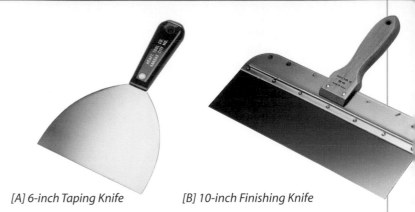

[A] 6-inch Taping Knife   [B] 10-inch Finishing Knife

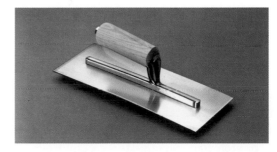

[C] Beveled Trowel

4-23 Taping knives are used to tape the seams and corners and to cover fasteners. The beveled trowel has a slight curve in the blade and is used to feather and finish the seam. *Courtesy Kraft Tool Company.*

4-24 A typical mud pan used to hold the drywall joint compound needed by the person taping the joints.

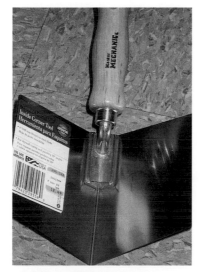

4-25 This inside-corner finishing tool smooths both sides of the corner with one pass.

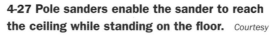

4-26 This is one type of handheld sanding pad. The abrasive paper is held by end clamps and is backed by a rubber pad. *Courtesy Kraft Tool Company*

4-27 Pole sanders enable the sander to reach the ceiling while standing on the floor. *Courtesy Kraft Tool Company*

compounds feathered edge to the surface of the drywall panel.

Handheld sanding pads are used to hold the abrasive. They make it easy to sand and prevent gouging **(4-26)**. **Pole sanders** are available that enable the sander to reach the ceiling from the floor **(4-27)**.

Joint Compounds

Drywall joint compound is available in premixed and dry forms. The dry form has to have the exact amount of water added to get the desired consistency. The premixed comes in a bucket ready to use. When kept properly closed it will keep in a workable condition for a long time. For most people, premixed is the best type to buy **(4-28)**.

There are two types of joint compound available, **drying** and **quick-set.** These provide a range of setting time ranging from 20 minutes to several hours. The use of premixed compounds is recommended.

DRYING JOINT COMPOUNDS

The drying-type joint compound takes up to 24 hours drying time between coats and you should allow even more than this before sanding the last coat. It is of a consistency that makes it easy to apply and sand.

4-28 This is a widely used ready-mixed all-purpose joint compound.

There are several types from which to choose. The **all-purpose** joint compound can be used for all coats required. There is very little cracking or shrinking after they have hardened. It is recommended that the inexperienced person use this premixed, all-purpose joint compound.

The **taping-type** joint compound is used to put the first coat over joints, nails and screws and bond the joint tape to the panel. It can be used for the second coat.

The **topping-type** joint compound is used for the second and third coats over the taping compound. It does not shrink much and provides a firm, crack-free finish. It is easy to apply and sand.

QUICK-SET JOINT COMPOUNDS

The quick-set joint compounds cure rapidly and permit recoating. The curing time varies depending upon the type used. Some cure in 20 minutes, while others take several hours. While they are especially useful for small quick repairs, they are more difficult to sand.

Sanding Safety Techniques

Sanding produces many fine gypsum particles. Wear eye protection, a good respirator, and a hat. See Chapter 1. The white particles will settle on cheeks, arms, and clothing, so stop and clean up frequently. A vacuum is a big help. It can also remove dust that has collected on the floor.

Taping the Joints

Tapered edge joints are covered by laying down a coat of joint compound in the tapered area and pressing the paper tape into it **(4-29).** Keep the knife at a slight angle. Be certain there are no air bubbles under the tape and that the excess compound has been wiped away.

After 24 hours, apply a second coating of compound. Feather it out 4 inches beyond the joint. Keep it smooth and uniform. Use a 6-inch finishing knife for this wider application.

After another 24-hour drying period, apply the third and final finish coat. Use a 10-inch finishing knife. Feather the coat out about six inches or more on each side of the joint **(4-30).**

Let the compound dry at least 24 hours and lightly sand it to get the smooth finish needed. Sponge-sand the feathered edge at this time if this is desired. Try not to dissolve the compound; if you do, another thin

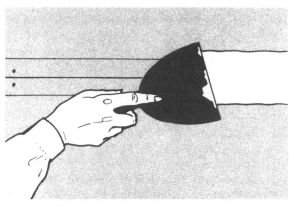

[A] Lay the first coat of joint compound in the recessed area produced by the tapered panel edger.

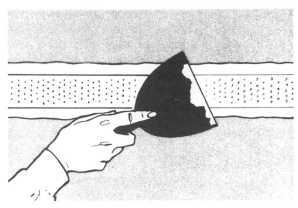

[B] Press the paper tape into the first coat and smooth it out with a trowel, removing excess compound and thinning the compound under the tape.

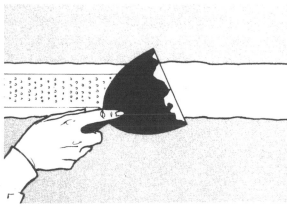

[C] After the first coat has dried lay the second coat over it feathering it, about 4 inches on each side of the joints.

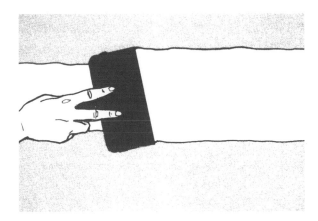

[D] After the second coat has dried apply the third coat, feathering it 7 or 8 inches on each side of the joint.

4-29 Four steps typically used to apply joint compound and tape to gypsum wallboard joints.

Courtesy National Gypsum Company

coating will have to be applied. A finished joint is shown in **4-31.**

Joints formed by **square-cut ends** of panels are more difficult to tape. While they are taped in the same manner as just described, they will produce a hump at the joint. This can be minimized by feathering out each coat of compound wider than that used on tapered edge joints. A typical application will be 18 to 24 inches wide.

4-30 Use an 8- or 10-inch finishing knife to apply the final coats of joint compound.

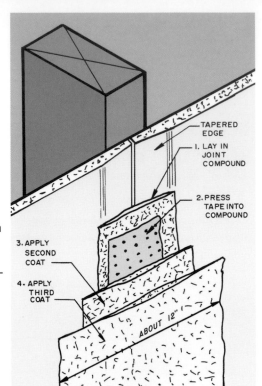

4-31 The parts of a typical finished edge joint.

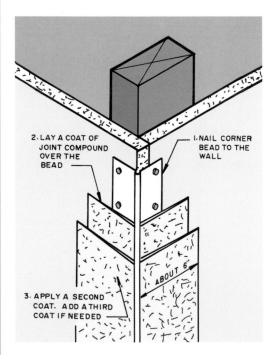

4-32 A finished exterior corner.

4-33 The steps to finish an interior corner.

Taping Corners

Outside corners are taped in the same manner as flat joints. The metal or plastic bead is nailed over the corner and a base coat of joint compound is feathered over it. After it hardens, additional coats are applied; each should extend more over the wall than the original coat—usually about 6 inches on each side (**4-32**).

Inside corners are taped by first laying a thin layer of joint compound down both sides of the corner with a taping knife. Bend a piece of paper joint tape, forming a right angle. Place it on the joint compound and press it into the compound by pulling the knife down the wall (**4-33**). After this hardens, apply additional thin coats and feather them out over the wall about 2 inches. Sand between coats (**4-34**).

Covering Fasteners

The heads of fasteners are covered by troweling a thin layer of joint compound over them (**4-35**). Pick up a quantity of joint compound on the taping knife and wipe it across the head of the fastener. Let this dry 24 hours before applying a second coat. If the surface is bumpy, sand it lightly between coats.

Repairing Drywall Damage

The most common damages requiring attention are nail pops, dents that break the paper surface, holes

1. SPREAD JOINT COMPOUND ON BOTH SIDES OF THE CORNER.

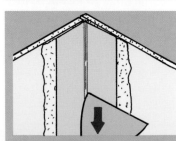

2. FOLD THE TAPE ALONG THE CREASE.

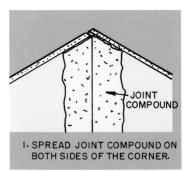

3. PRESS THE FOLDED TAPE INTO THE COMPOUND ON ONE SIDE. BE CAREFUL YOU DO NOT PUNCTURE THE TAPE IN THE CORNER.

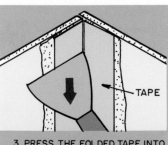

4. THEN PRESS IN THE TAPE ON THE OTHER SIDE. DO NOT DISTURB THE TAPE ON THE FIRST SIDE.

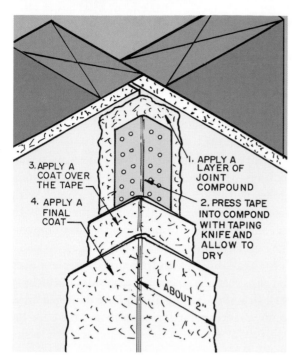

4-34 A finished interior corner.

4-35 Lay the joint compound over the heads of fasteners with a 4-inch taping knife.

broken through the panel, and cracks. They are discussed below.

Repairing Protruding Nails & Screws

Nails or screws protrude because they get squeezed out of the stud as it dries and shrinks. This causes nail pops. To repair this problem, follow these steps (4-36).

1. Hammer the protruding nails back in place, but do not break the paper on the surface of the panel. The hammer blow should form a $\frac{1}{64}$-inch-deep dimple around the nail. Use a hammer with a crowned face to drive the nail.

2. About 2 inches from the first nail, drive a 1¼-inch annular-ring drywall nail or screw.

3. Fill each dimple with joint compound and sand after it hardens. Apply additional coats as needed to get a finished surface. If the nail pop has damaged the drywall around it, remove the loose pieces. Add additional nails or screws as just described. Then lay a piece of adhesive-backed drywall tape or mesh over the damaged area and apply additional coats of compound as needed.

Repairing Fastener Depressions

A fastener depression occurs when the area over a fastener is lower than the wall surface. It may occur when a fastener is driven too deeply or not enough compound is placed over the fastener. Sometimes not

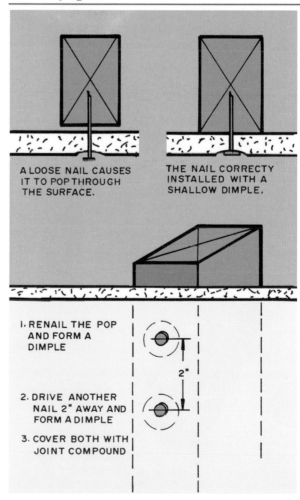

4-36 To correct a protruding nail, reset the loose nail and drive another about 2 inches away. Dimple and coat with joint compound.

enough fasteners were used and the panel is pulling away from the wall. In this case, additional fasteners will have to be added.

To repair a depression, apply additional coats of joint compound over the area and sand as necessary.

Repairing Dents

Fill dents in wallboard with several layers of joint compound. Clean away all loose material from around the hole before filling it. The irregular edges of the hole will help hold the small patch in place. Larger holes are more difficult to repair and require more work.

Repairing Cracks

Often the building will settle a little, causing thin cracks in the drywall. This often occurs at the corners of door and window openings. If you fill the crack with drywall compound, over time the crack will open up again.

Begin by removing all loose material in and on the sides of the crack. You can run the blade of a utility knife down it so it is opened up about ¼- to ¾-inch wide. Lightly sand the painted wall surface along the crack and then fill it with joint compound. Then install a layer of self-adhesive drywall tape (4-37). Finish with two or three coats of drywall compound as discussed earlier for finishing joints. After the last coat has hardened, sand, prime, and repaint the area. Sometimes the paint can be blended in so it is not necessary to repaint the entire wall.

Repairing Small Holes

There are several ways to repair small holes. Building supply dealers have a number of kits that contain all that is needed to make the repair. One kit provides a fiberglass tape patch with an adhesive back (4-38). It is stuck over the hole and covered with tape and joint compound (4-39).

A repair can be made with a piece of scrap drywall by cutting the damaged area into a squared opening with beveled edges and then cutting a patch with beveled edges. Secure the patch in the opening with joint compound and finish with tape and compound in the normal manner (4-40).

A kit available at building supply dealers uses metal clips that are screwed to the sides of the hole (4-41). Begin by cutting around the damaged area, forming a rectangular opening. Install the clips on the sides. Cut a piece of drywall and screw it to the clips. Then tape and finish the edges in the normal manner (4-42).

Repairing Larger Holes

Rather large holes can be repaired by installing wood strips behind the opening with wood screws and cutting and fitting a drywall patch to them as shown in 4-43. Tape and finish the edges in the normal manner.

Holes that are very large are best repaired by cutting the drywall around them from one stud to the other (4-44). Then add wood blocking on all four sides and nail the patch to the blocking. Finish the joints in the normal manner.

4-37 Cracks in the drywall should be cleaned out, covered with adhesive-backed tape, and finished with joint compound.

4-38 One repair kit on the market supplies an adhesive-backed fiberglass patch that is pressed on the drywall over the hole. It is then covered with joint compound.

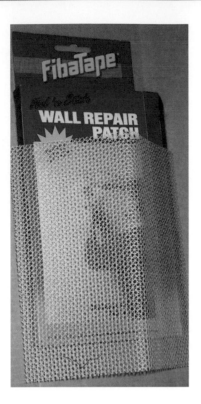

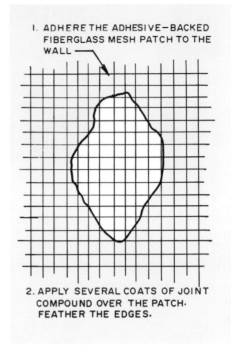

1. ADHERE THE ADHESIVE-BACKED FIBERGLASS MESH PATCH TO THE WALL

2. APPLY SEVERAL COATS OF JOINT COMPOUND OVER THE PATCH. FEATHER THE EDGES.

4-39 The adhesive-backed mesh is applied over the hole and finished with drywall compound.

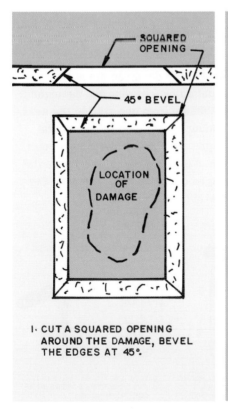

SQUARED OPENING

45° BEVEL

LOCATION OF DAMAGE

1. CUT A SQUARED OPENING AROUND THE DAMAGE, BEVEL THE EDGES AT 45°.

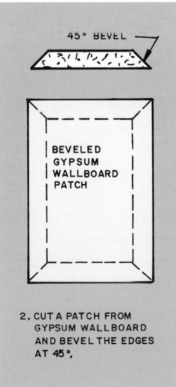

45° BEVEL

BEVELED GYPSUM WALLBOARD PATCH

2. CUT A PATCH FROM GYPSUM WALLBOARD AND BEVEL THE EDGES AT 45°.

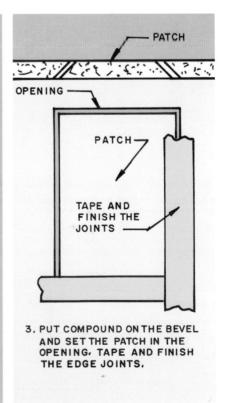

PATCH

OPENING

PATCH

TAPE AND FINISH THE JOINTS

3. PUT COMPOUND ON THE BEVEL AND SET THE PATCH IN THE OPENING. TAPE AND FINISH THE EDGE JOINTS.

4-40 This patch requires that the edges of the squared opening and the edges of the patch be beveled.

4-41 This drywall repair kit has metal clips that attach to the edges of the squared opening and support a drywall patch as shown in 4-42.

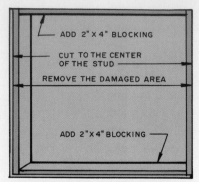

I. CUT AWAY THE DAMAGED AREA. ADD BLOCKING ON THE HORIZONTAL EDGES.

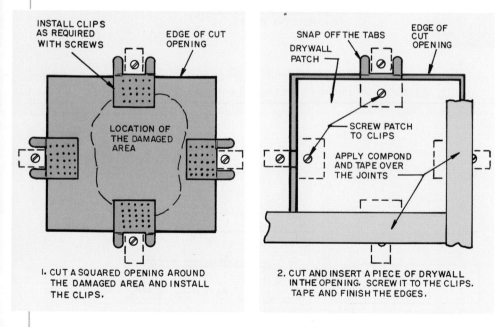

INSTALL CLIPS AS REQUIRED WITH SCREWS

EDGE OF CUT OPENING

LOCATION OF THE DAMAGED AREA

I. CUT A SQUARED OPENING AROUND THE DAMAGED AREA AND INSTALL THE CLIPS.

SNAP OFF THE TABS

DRYWALL PATCH

EDGE OF CUT OPENING

SCREW PATCH TO CLIPS

APPLY COMPOND AND TAPE OVER THE JOINTS

2. CUT AND INSERT A PIECE OF DRYWALL IN THE OPENING. SCREW IT TO THE CLIPS. TAPE AND FINISH THE EDGES.

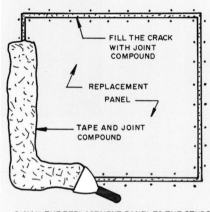

FILL THE CRACK WITH JOINT COMPOUND

REPLACEMENT PANEL

TAPE AND JOINT COMPOUND

2. NAIL THE REPLACEMENT PANEL TO THE STUDS AND BLOCKING.

3. FILL THE CRACK WITH JOINT COMPOUND.

4. TAPE AND FINISH THE JOINTS IN THE NORMAL MANNER.

4-42 These metal repair clips hold the drywall patch in the hole. Then the edges of the joints are taped in the normal manner.

4-44 Large damaged areas are repaired by removing the damaged wallboard from one stud to the next, adding blocking on the horizontal edges, and nailing the replacement panel over the areas. The horizontal edges could be supported with the metal clips shown in 4-41.

4-43 Large holes can be repaired by installing wood strips across the opening and screwing the gypsum drywall patch to them. Then finish the edge joints in the normal manner.

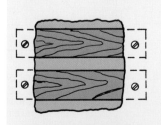

I. CLEAN UP THE EDGES OF THE OPENING. INSTALL WOOD STRIPS TO SUPPORT THE GYPSUM PATCH. SECURE WITH WOOD SCREWS.

2. INSTALL A GYPSUM WALLBOARD PATCH WITH WOOD SCREWS. FILL THE CRACK WITH JOINT COMPOUND.

3. TAPE AND FINISH THE JOINTS AND COVER THE SCREWS WITH JOINT COMPOUND.

Repairing a Water-Damaged Ceiling

If the ceiling drywall is badly damaged and buckling, it will have to be replaced, joints taped, and painted to match the ceiling. Generally it is best to repaint the entire ceiling.

If there is a water spot, first be certain the drywall has completely dried. If the panel is still flat and has no bow, seal the stained area with a primer. Your paint dealer will have several products recommended for this sealing. Once this is dry repaint the spot, or the entire ceiling, if necessary, to get a uniform finish.

If the ceiling has a textured surface that has not been damaged let it dry, prime it, and repaint. If the texture is damaged, and is coming loose, scrape off all loose material with a drywall finish knife (4-45). Then sand the area and feather out the textured material around the edges (4-46). Now cover the cleaned area with a coat of drywall compound. This covers the paper backing on the drywall and brings the base coat up to the level of the surrounding coat (4-47). Let this dry and lightly sand so it is smooth. Then seal the area with a primer to prevent the water stain from bleeding through (4-48). Finally, spray a textured finish using a canned-spray ceiling texture compound. If it has not built up to match that on the ceiling, let it dry and spray a second coat (4-49).

4-46 Carefully sand the exposed area and feather the texture on the edge so it slopes to the center of the area.

4-45 Scrape off all loose texture compound.

4-47 Cover the cleared area with a layer of joint compound. When it is dry, sand smooth.

4-48 After the compound dries and is sanded, spray a primer over it to seal out any possible bleed-through from water damage.

4-49 Spray texture compound over the area. Be careful that it does not build up too thick along the edges of the repaired area. Apply additional coats if needed.

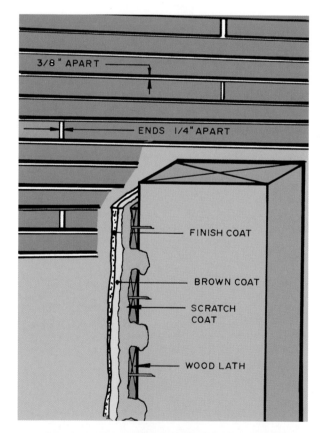

4-50 Wood lath was used for years as the base for a plaster wall finish. It is covered with a three-coat plaster wall system.

Repairing Plaster Walls

Repairs are made using regular gypsum perlite plaster. Perlite is a siliceous rock that is ground and used as a lightweight aggregate in the plaster. It replaces the sand used in standard plaster. For the top finish coat, use a standard plaster finish-coat material.

Houses built many years ago used wood lath. These strips are nailed to the studs with a small space between them. The base coat, called the **scratch coat,** is pressed through the spaces, forming a key that holds the finished plaster to the wall **(4-50).** A second coat, the **brown coat,** is applied over the scratch coat. The **plaster finish coat** is applied over the brown coat.

For many years now, gypsum lath has been used instead of wood lath. It is much like gypsum wallboard. It has a gypsum core covered with multilayer laminated paper that provides for the proper absorption of moisture from the plaster **(4-51).** The panels are ⅜ or ½ inch thick and 16 x 48 inches. The plaster used may be for a one- or two-coat application. The two-coat system uses a special gypsum and wood-

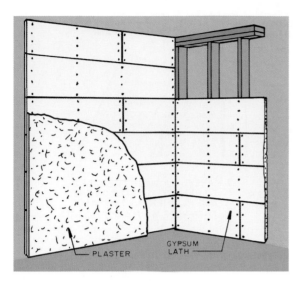

4-51 Gypsum lath is used as the base for plaster walls in residential construction.

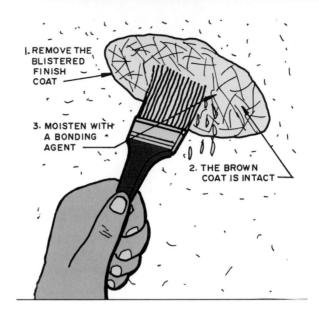

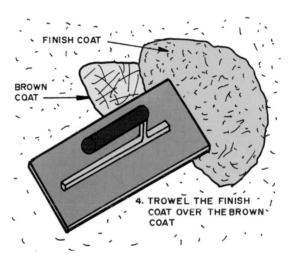

REPAIRING BLISTERS

4-52 A blister on a plaster wall can be repaired by removing the damaged finish coat and applying a new finish coat to the exposed brown coat. A repair on a small damaged area such as that caused by a blow is made in the same way, except that the damaged plaster is cleared down to the lath.

fiber plaster for the first coat and a standard plaster finish coat over it. The one-coat system uses a standard plaster finish material that has extra-high-strength gypsum gauging added.

Metal lath is used on the walls in commercial buildings.

Plaster Failures

Several types of plaster failure require patching. Patches are required when the finish coat blisters and comes loose, when the scratch, brown, or finish coat deteriorates, when the lath and all coats come loose from the studs, and when holes are caused by heavy blows. The first three are usually caused by moisture, so the source of moisture must be detected and corrected before repairs are made.

REPAIRING BLISTERS

If the finish coat blisters and loosens because it is wet, the best solution is to remove the finish coat, dampen the exposed base coat, and apply a new finish coat. Paint the edges of the damaged area with a liquid bonding agent or moisten them with water before applying the new finish (4-52).

REPAIRING CRACKS

Sometimes plaster will develop surface cracks. These are generally caused by vibrations within the building, a weak base coat, or poor structural framing.

Small cracks can be repaired by cleaning them out with a sharp-pointed tool, cutting the crack so the inside is wider than the outside. This produces a wedge to help hold the plaster in place. Fill the crack with a quick-setting patching plaster. Be certain to wet the crack with a small paintbrush before applying

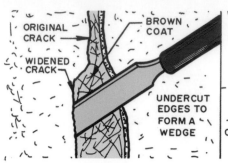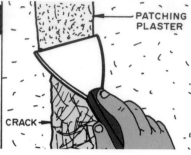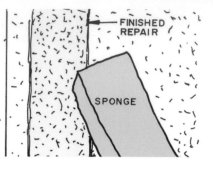

I. WIDEN THE CRACK TO REMOVE ALL LOOSE PLASTER.

2. WET THE CRACK SEVERAL TIMES, APPLY PATCHING PLASTER AFTER SURFACE MOISTURE DISAPPEARS.

3. CAREFULLY WASH THE SIDES TO REMOVE PLASTER ON THE OLD PLASTER. LET DRY AND THEN WIPE WITH A SOFT CLOTH.

4-53 Wide cracks should be widened until all loose finish-coat plaster is removed. Wet and fill the crack with a patching plaster.

the patching plaster. Any dust or particles on the surface can be removed by wiping the area down with a damp cloth. Work carefully, so the surface is not damaged.

Wider cracks should be cleaned out with a chisel, and any loose plaster on each side removed. This may widen a crack to 4 or 5 inches.

If the brown coat is sound, wet the area several times, let the moisture on the surface disappear, and apply a patching plaster over the area (**4-53**). If the lath is damaged, secure a piece of metal lath over it and then apply the scratch and finish coats as required. Plaster will be influenced by the amount of texture material and the coarseness of the original ceiling texture.

Repairing a Damaged Area

Cut away all loose plaster down to the lath. If the lath has not been damaged, replaster over it. First, undercut the edges of the broken area to help hold the patch in place. Then wet the edges with a liquid bonding agent or moisten them with water. If gypsum lath is exposed, do not moisten it. This repair is made like the repair shown in **4-53**.

If the lath has been damaged, it will also have to be replaced. Wood and metal lath are less likely to need replacing than gypsum lath. Water on gypsum lath can cause the paper surfaces to separate from the core.

Damaged lath in large damaged areas should have the repaired area cut back so the lath can be nailed to the studs (**4-54**). Then apply the base and finish coats. It will take two or three coats to build up the plaster layer to be flush with the plaster on the wall. Perlite

plaster will harden in three or four hours, but should not be painted for 24 hours or longer.

Mix the perlite plaster with water until the mixture can be troweled. Do not apply this mixture if the temperature is below 50°F (10°C), and maintain this temperature or a higher one while the plaster is hardening.

If the hole is small and the lath is damaged, cut it away and replace it with a piece of metal lath as shown

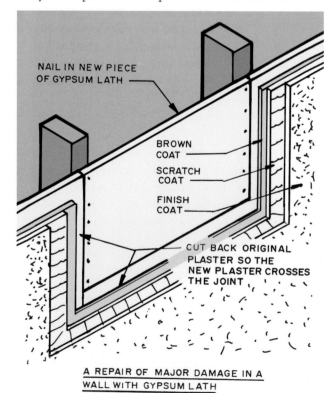

NAIL IN NEW PIECE OF GYPSUM LATH

BROWN COAT

SCRATCH COAT

FINISH COAT

CUT BACK ORIGINAL PLASTER SO THE NEW PLASTER CROSSES THE JOINT

A REPAIR OF MAJOR DAMAGE IN A WALL WITH GYPSUM LATH

4-54 When repairing large, damaged areas in plaster walls, cut out the areas from stud to stud, install new lath, and replaster the areas.

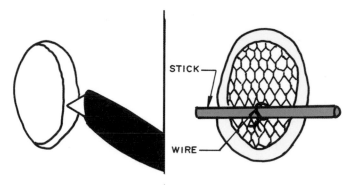

1. CUT OUT THE THE DAMAGED AREA AND SMOOTH THE SIDES SO THEY SLOPE OUT ABOUT 45°.

2. INSERT A PIECE OF METAL LATH AND TIE IT IN PLACE.

3. WET THE EDGE OF THE HOLE. APPLY THE FIRST FILLING THE HOLE ABOUT HALF. SCORE THE SURFACE HORIZONTALLY. WHEN HARD CUT THE WIRE AND APPLY THE FINISH COAT.

STICK

WIRE

4-55 Small damaged areas where the lath has been broken can be repaired using a piece of wire lath.

in **4-55.** Hold it in place with a wire and apply the first coat of plaster. Score this coat horizontally. When it is hard, cut the wire and apply the needed additional coats.

Installing a Suspended Ceiling

There are times when you might prefer to install a new ceiling suspended below an old damaged ceiling. This type of ceiling can also be used where no ceiling exists, as when you are finishing a basement (**4-56**). Before installing over an old ceiling, remove or nail securely any loose pieces of the old ceiling that may fall onto the new ceiling. Sometimes it is best to remove the old, damaged ceiling.

A suspended ceiling is a grid of metal runners that are hung from the ceiling joists on wire hangers. The

4-56 A suspended ceiling can cover all the pipes and wires, producing a beautiful finished enclosure. Notice the recessed lights and heating outlet. *Courtesy Mr. and Mrs. Sherman Creson.*

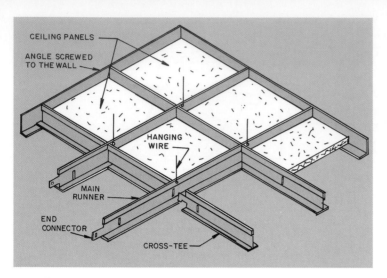

4-57 The frame for a suspended ceiling is hung from the joists by wires. The wires carry the main runner, which is cross-connected by cross-tees. The panels are laid into the grid.

ceiling tile panels are placed in the openings in the grid.

The grid is made of T-shaped metal runners available in 8- and 12-foot lengths. These long pieces run the length of the room. Short pieces called cross-tees run perpendicular to the runners, forming the openings into which the panels are placed (**4-57**). The grid is hung 2 to 4 inches below the old ceiling or joists. The short pieces snap into slots in the long pieces.

Developing the Panel Layout

Make a plan that has the panels at each side wall the same width. Begin by making a drawing of the room. The long side of the panel is usually run parallel with the long side of the room. The main runners are

placed perpendicular to the floor joists. Divide the long side of the room by 48 inches, which is the length of the panel. Add the inches left over to the length of one of the panels and divide by two. This gives the length of the end panels and the location of the first main runner from the outside wall (**4-58**).

Repeat this to find the location of the cross-tees. The only difference is that the panel is 24 inches long instead of 48 inches.

Now it is possible to find the number of panels, the lineal feet of runner, and the number of cross-tees to buy.

An L-shaped runner is needed at the wall. It should be long enough to go around the perimeter of the room.

INSTALLING THE WALL MOLDING

The first step is to nail the wall molding to the wall. It supports the edge of the panel at the wall. Decide the height of the ceiling. Run a chalk line on each wall at this height plus the height of the main runner. Check the chalk line with a line level to see if it is level. Chalk the line and snap a chalk mark on each wall (**4-59**).

Place the top edge of the wall molding on the chalk line and nail it to the studs. Mark each stud on the wall so you can nail into it. Check it with a level while proceeding. Corners are formed as shown in **4-60**.

Another way to rapidly locate the line of the wall molding is with a laser level.

INSTALLING THE MAIN RUNNERS

The main runners are made with a tongue on one end that fits into a slot on the end of the next piece. This splices runners together for long distances. The end of the runner that sits on the wall molding must be

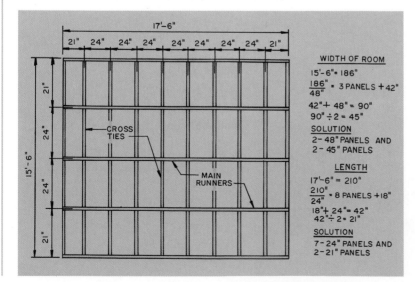

4-58 The layout for a suspended ceiling grid locates the runners and can be used to determine the number of panels and how much runner and cross-tee material is needed.

cut off. Measure from the first cross-tee-connection hole a distance equal to the width of the first panel and cut the main runner (**4-61**). In **4-58** this was 21 inches.

Run a chalk line the length of the room, locating where the first runner will hang, and mark this. Mark where the runner crosses each joist. Install screw eyes at these points on every fourth joist. The screw eyes will be 48 inches apart. Fasten a length of wire to each screw eye. It should be long enough to drop the ceiling the required amount and permit it to be wired to the main runner.

Slip each wire in the opening provided in the main runner (**4-62**). Bend the wire up temporarily until the entire runner is hanging. If a runner is not long

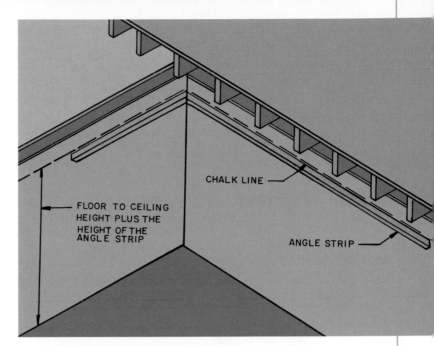

4-59 Locate the wall-mounted channel around the edge of the room with a chalk line.

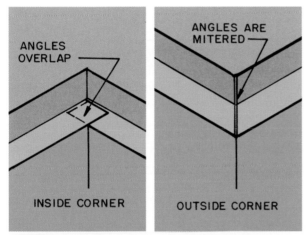

4-60 How to make inside and outside corners on the wall-mounted channel.

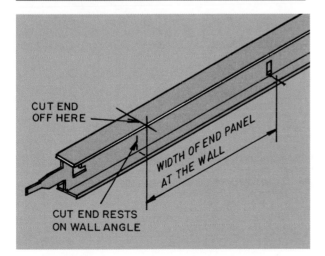

4-61 The end of the main runner is cut off and is supported by the wall-mounted channel.

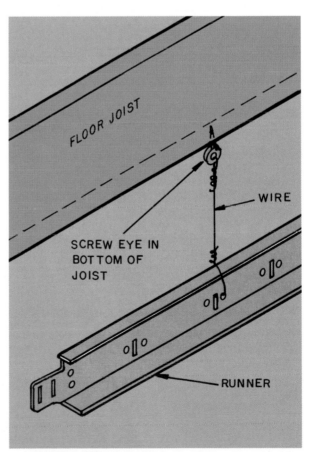

4-62 Hang the main runner from screw eyes in the joist with wire. Be certain to check each for levelness before finally tightening the wire on the runner.

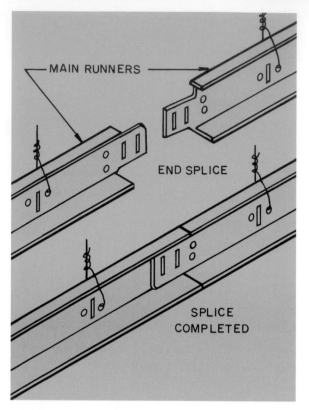

4-63 The main runners have splice connections on the ends so they can be easily joined when installed.

enough, join another piece to it using the splice connection provided **(4-63).** Run a chalk line at the desired ceiling height from one end of the runner to the other. Check to be certain it is level. Then raise or lower the main runner to match the chalk line and wrap the wire around itself to hold the runner in position. Repeat these steps for each main runner.

INSTALLING THE CROSS-TEES

The cross-tees run perpendicular to the main runners. They have a special tab on each end that fits into holes on the main runner. Insert the cross-tees at 4-foot intervals and fasten them as specified by the manufacturer **(4-64).** A finished installation is in **4-65.**

INSTALLING THE CEILING PANELS

The ceiling panels are sold cut to 24 x 48-inch sizes, so they fit into the runners. Slide them through the grid and drop them between the metal runners **(4-66).** If there are pipes, cut the panel in half at the pipe. Then cut the hole for the pipe and slide the two halves around the pipe as they are placed on the runners.

INSTALLING LIGHTS IN SUSPENDED CEILINGS

The manufacturers of suspended systems offer various types of ceiling lights that will fit in place on a

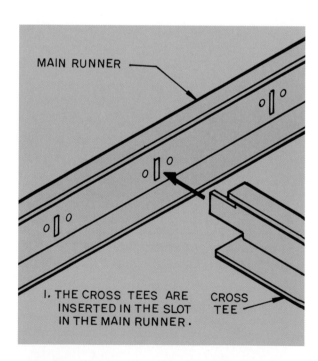

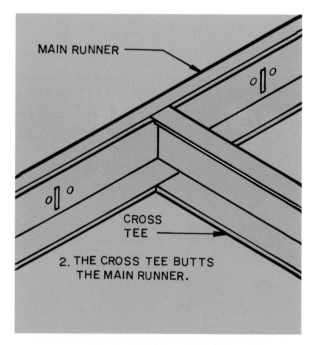

4-64 After the main runners are installed, the cross-tees are placed. They connect to openings provided in the main runner.

ceiling panel. These include fluorescent and incandescent fixtures (4-67). The manufacturer provides installation instructions for installing the lights in the ceiling. It will be necessary to develop an electrical plan to get power to each fixture.

Installing Ceiling Tiles Directly to a Ceiling or the Joists

Ceiling tiles are good to use to cover a damaged flat ceiling or over joists. The tiles are made of fiberboard or foam plastic. One side has a factory-applied finish, so they must be handled carefully. A wide variety of surface textures are available; they also have good acoustical qualities. Typical tiles are ½-inch thick and have tongue-and-groove edges. Typical tile sizes are 12 x 12, 12 x 24, 16 x 16, and 16 x 32 inches.

DEVELOPING A FLAT-CEILING LAYOUT

To plan the layout, make a drawing showing the size of the room (4-68). The tiles next to the walls should be the same size on opposite walls of the room. Avoid using a small 2 x 4-inch strip along the wall.

Proceeding with the layout, divide the width and length of the ceiling by 12 inches or the tile size. This is to indicate that number of tiles needed. In 4-68, tiles were used for the short side, with 6 inches left over. Add 12 inches to the 6 inches and divide by 2. The purpose is to give the size of the outside tiles— which in this example is 9 inches—and to avoid having a narrow strip along the wall.

The room length is figured in the same way. Remember, when using a different tile size, such as 16 x 16 inches, divide the width and length by the actual tile size.

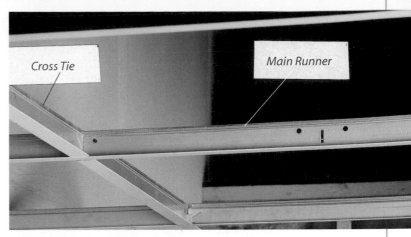

4-65 The metal suspension grid is complete.

4-66 After the grid is finished, the panels can be installed.

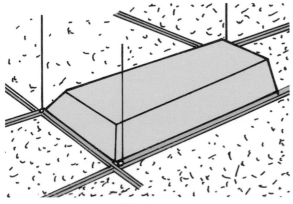

A FLUORESCENT FIXTURE

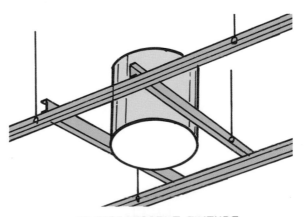

AN INCADESCENT FIXTURE

4-67 Ceiling-systems manufacturers have a variety of light fixtures that fit into the grid, replacing a panel.

4-68 A plan for installing ceiling tile. This locates the wood furring strips on the bottom of the floor joists and indicates how many tiles are needed.

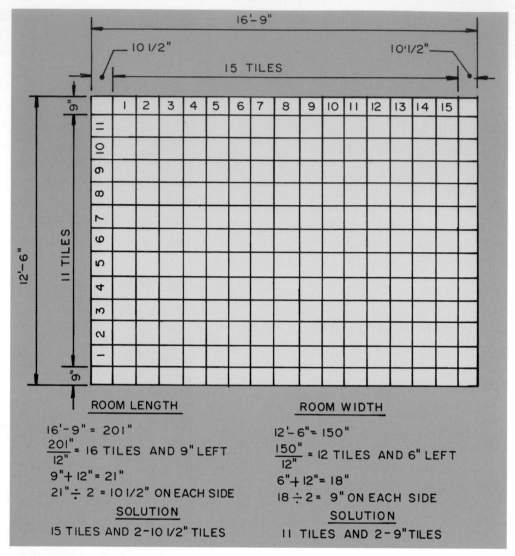

ROOM LENGTH

16'-9" = 201"

$\frac{201"}{12"}$ = 16 TILES AND 9" LEFT

9"+ 12"= 21"

21" ÷ 2 = 10 1/2" ON EACH SIDE

SOLUTION

15 TILES AND 2-10 1/2" TILES

ROOM WIDTH

12'-6"= 150"

$\frac{150"}{12"}$ = 12 TILES AND 6" LEFT

6"+ 12"= 18"

18 ÷ 2 = 9" ON EACH SIDE

SOLUTION

11 TILES AND 2-9"TILES

Installing Tiles over an Old Gypsum Drywall Ceiling

If the old ceiling is damaged but structurally sound, it can be covered by gluing the tiles to it. Go over the old ceiling and nail it to the joist wherever it seems to be weak. Repair any bulges or deteriorated areas. Once the ceiling appears sound, begin the installations. Use the adhesive recommended by the manufacturer.

Refer to the information from the tile layout (4-68). Use chalk lines to locate the edge of each outer row of tiles (4-69). Cut the tiles to the required width. Use a sharp utility knife and cut from the finished face. Use a straightedge to guide the knife (4-70). A sharp, fine-blade saber saw can also be used.

Place adhesive in spots on the back (4-71). Keep adhesive away from the edges of the tiles. Press the tiles against the ceiling and line them up with the mark. Put the cut edges next to the wall. If necessary,

a staple can be driven through the tongue into the ceiling to hold the tiles while the adhesive sets. Continue along the wall. Check constantly to make certain this first row is straight. All other rows depend upon it.

When installing the outside row on the second wall, be certain the chalk lines cross at 90 degrees.

If the ceiling is in bad repair, install wood furring strips over it as described in the next section.

Installing Tiles Using Furring Strips

Furring strips are 1 x 2 or 1 x 3 wood strips that are nailed to the ceiling joists. The ceiling tiles are stapled to the furring strips (4-72). If the joists are not level, place wood shims under the furring strips to bring them level. If the joist extends below the chalk line, plane some of it off (4-73). Check for levelness with a chalk line or laser level.

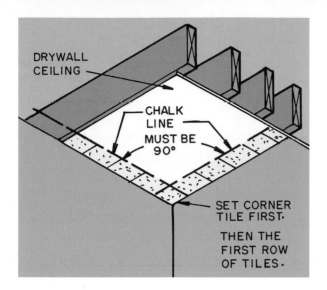

4-69 Use a chalk line to locate the edge of the first row of tiles.

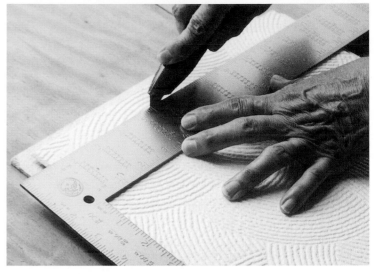

4-70 When cutting ceiling tile, use a straightedge and a very sharp utility knife. Cut with the finished surface on top.

Developing the Furring-Ceiling Layout

This layout is made in the same manner as for the flat ceiling (refer to **4-68**). The chalk lines locate the center of each furring strip. The tiles must meet at the center of each strip (**4-74**).

INSTALLING THE FURRING STRIPS

Nail the first furring strip next to the wall. Measure the width of the first row of tiles in from the wall. Nail the second row of furring strips with this line in the center. Locate each row in this manner. Nail the strips with two 8d box nails to each joist.

FRAMING PROJECTIONS

Occasionally a heat duct or water pipe will extend below the ceiling joists. Projections can be boxed in using 2 x 2 or 2 x 4 framing (**4-75**). The tiles are then stapled to this framing.

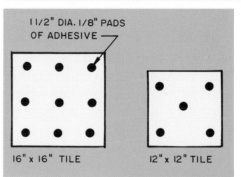

4-71 Place dabs of adhesive on the back of the tile in these patterns.

4-72 Ceiling tiles can be installed by stapling them to wood furring strips nailed to the floor joists. The first row of tiles is located with a chalk line.

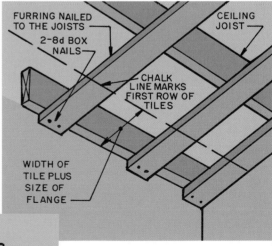

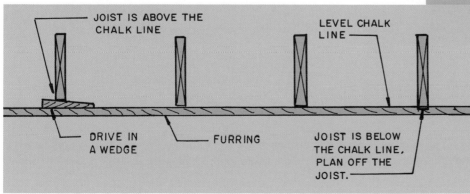

4-73 Adjust the furring so it is level with the length and width of the room. Shim as necessary.

INSTALLING THE TILES

Run chalk lines locating the edges of the tiles on two meeting walls. Be certain they cross at right angles. Cut and install the corner tiles first (4-76). Cut off the side with the groove. The tongue is needed for stapling it to the furring strip. Now cut and install the tiles on the outside row of two walls, meeting in a corner. Staple through the tongue to the furring strip. Usually 9/16 or 5/8 inch is used (4-77). Use three staples on 12-inch tiles and four on 16-inch tiles. Twenty-four-inch panels need five staples. The other two sides of the tile have grooves that fit over the tongue on the next tile (4-78). Push the tiles together and staple to the furring (4-79).

Now begin the second row and continue filling in the ceiling. The last row on the other side will have to be face-nailed or stapled on the edge. Then install some type of molding around the entire ceiling (4-80).

Lights

Recessed and flush-mounted lights are available for installation with the wood-furred ceilings (4-81). Follow the manufacturers installation instructions. Protect the furniture part as you attempt to tap the joint apart.

ADDITIONAL INFORMATION

For more detailed instructions consult the following publication: *Installing & Finishing Drywall*, William P. Spence, Sterling Publishing Co., Inc.

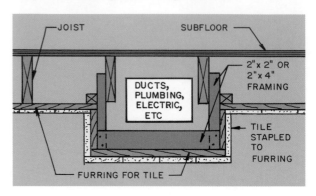

4-75 It is possible to add furring around obstacles in the ceiling and cover the area with the ceiling tile.

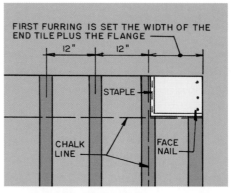

I. SET THE CORNER TILE.

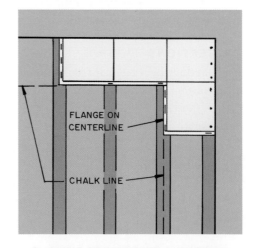

2. STAPLE A ROW OF TILES ALONG EACH WALL. LINE UP WITH THE CHALK LINE.

4-76 Start installing in a corner and complete a row along the two butting outside corners.

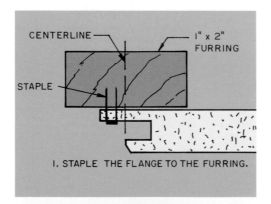

I. STAPLE THE FLANGE TO THE FURRING.

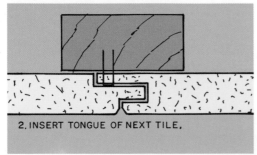

2. INSERT TONGUE OF NEXT TILE.

4-74 The tile is stapled to the furring through the flange. The edge of the finished face falls on the centerline of the furring.

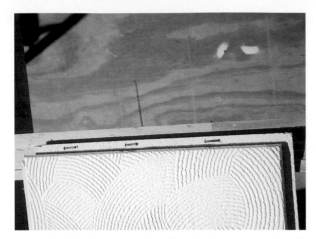

4-77 The tile is secured to the furring by stapling it through the tongue that is on two sides. The other two sides have grooves that fit over the tongue on the next tile. Put three staples in the tongue.

4-79 Push the tiles together and staple the exposed tongue to the furring.

4-78 Each tile has tongues and grooves on the edges that slide together to hold the tiles in place.

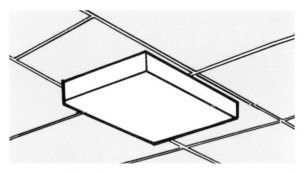

A SURFACE MOUNTED FLUORESCENT FIXTURE

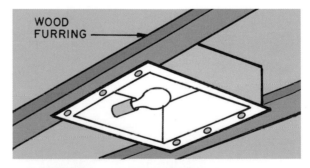

WOOD FURRING

A RECESSED INCADESCENT FIXTURE

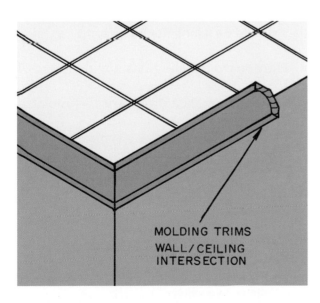

MOLDING TRIMS WALL/CEILING INTERSECTION

4-80 Cover the edge of the tile along the wall with some type of molding.

4-81 Several types of ceiling lights are available.

CHAPTER 5

Installing Paneling & Wainscot

Plywood paneling is available in a variety of wood species on the outer veneer. Typical thicknesses are $5/32$, $1/4$, $5/16$, and $7/16$ inch **(5-1)**. The most commonly used panel size is 4 x 8 feet. Since most panelings are supplied prefinished, they should be handled with care.

Hardboard paneling has a wide range of surface finishes, including various simulated wood grains. The panels are typically $1/4$-inch thick and come in sheets 4 x 8 feet.

5-1 Plywood paneling provides a material that will rapidly cover the wall. It is available in a variety of beautiful wood species.

Courtesy Georgia Pacific Corporation

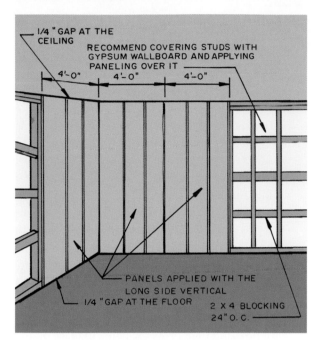

5-2 Plywood and hardboard wall panels are generally applied vertically using the 16-inch spaced studs for support.

5-3 When plywood and hardboard panels are applied horizontally, the edge joints must be nailed to blocking.

Plywood & Hardboard Paneling Installation

The plywood and hardboard panels can be installed with the long side vertical or horizontal. The direction used depends upon the appearance desired on the finished wall. When the panel is installed vertically, the edge of each sheet falls on a stud and is supported by it (5-2). When the panel is applied horizontally, the top edge of the panel must have 2 x 4 blocking installed at the joint (5-3). If the paneling has tongue-and-groove edges, the blocking may not be needed.

If possible, apply paneling over a backing material. The backing material may be ⅜-inch gypsum wallboard, 5⁄16-inch plywood, or wood furring strips. (Furring strips are 1 x 2 or 1 x 3 strips that are nailed to the wall studs.) The backing prevents the paneling from bowing between the studs. If applied directly to the studs, the paneling should be at least ¼-inch thick.

Some building codes require paneling under ¼-inch thick to be installed over a fire-resistant material, such as gypsum wallboard. Check the local building code before starting the job.

Panels can be installed with nails or adhesive. If covering an exterior wall, be sure it has a vapor barrier. (A **vapor barrier** is a plastic sheet secured to the studs to keep interior room moisture from getting into the wall cavity.) Staple plastic sheet to the studs, top plate, and bottom plate.

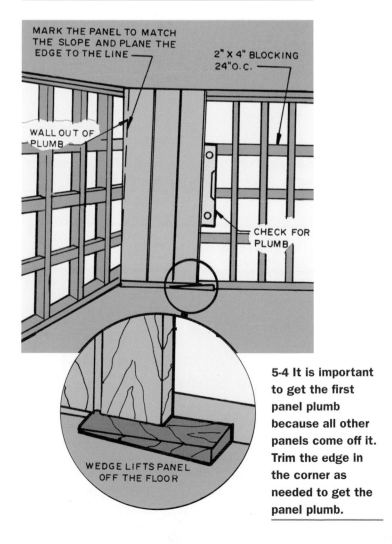

5-4 It is important to get the first panel plumb because all other panels come off it. Trim the edge in the corner as needed to get the panel plumb.

5-5 If the edge of the panel does not rest on the center of the stud, blocking will have to be added to get a proper nailing surface.

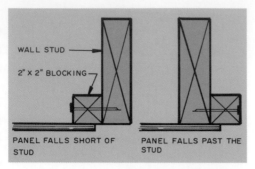

5-6 Specified nail spacing on the interior studs and edges of plywood and hardboard panels.

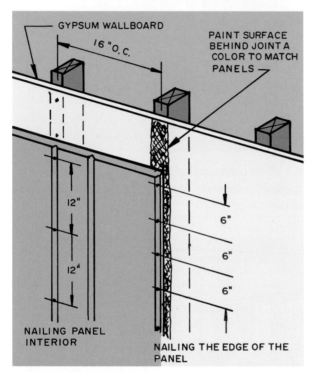

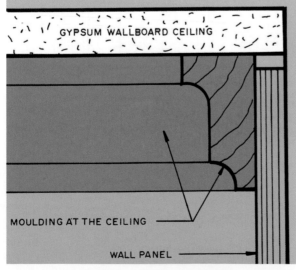

5-7 The space between the panel and the ceiling is covered with some type of molding.

Installing Paneling

1. Begin by fitting a panel in a corner of the room (5-4). Scribe and plane the edge to fit the corner as needed. The panel edge must be plumb (vertical) and centered on a stud. Remember, the proper vertical look of all the other sheets depends upon the first sheets' being plumb. Drive wood wedges between the panel and floor until the panel is plumb and within ¼ inch of the ceiling. If a molding is used at the ceiling, this gap could get larger. While plumbing the panel, it may be necessary to plane parts of the edge in the corner to make it parallel with the intersecting wall. The panel should clear the floor by ¼ to ½ inch to allow for expansion.

2. Mark the openings needed for electrical outlets, doors, and windows. Use chalk around the electrical box, and then press the panel in place on the wall. The outline of the box will show on the back of the panel.

3. Cut the openings marked. Drill a hole large enough to insert the saber-saw blade in the center of the marked area. The drill may cause the surface of the panel to splinter. Drilling in the center of the area will avoid marring the panel.

   If you are cutting the opening from the finished side, use a fine-toothed blade and saw from the finished surface of the panel. If cutting the openings from the back side, putting masking tape on the finished side of the panel along the saw kerf will help avoid splintering. While installing the panel, it may be necessary to trim the edges of the opening to allow the panel to be plumb.

4. Nail or glue the panel to the studs. Some studs may not be accurately spaced to allow the edge of a panel to rest on them. To correct such a condition, nail a 2 x 2 to the inside of the stud (5-5).

5. Nail panels with 1-inch **paneling nails.** Paneling nails have a ring shank and are painted to match the color of the grooves in panels. Use 3d 1¼-inch finishing nails if paneling nails are not available. When nailing over gypsum wallboard, use 1⅝-inch paneling nails or 6d 2-inch finishing nails. Space the nails 6 inches apart on the edges of the panel, and 12 inches apart on the interior of the panel (5-6). Grooves in panels are spaced 16 inches apart, so they should hit a stud. Nail into the grooves to help hide the nails. Leave a hairline crack between plywood panels to allow for expansion.

Hardboard panels expand more than plywood. Leave 1/16 inch between hardboard panels. Paint the wall behind the panel at each joint a color similar to the panel color. This will help disguise the crack.

6. Use wood molding to cover the cracks at the ceiling (**5-7**) and in the corners (**5-8**). At the floor, install the baseboard. Some panel manufacturers provide aluminum corner moldings into which the panels fit (**5-9**). To install the panel, first nail the molding to the wall, properly spaced. Insert one edge of the panel into one molding, bend the panel a bit, and slip the other edge into the other molding. Before doing this, adhesive can be applied to the wall. If preferred, the panel can be installed without adhesive and the panel nailed in place.

7. To fasten a panel with adhesive, lay a continuous 1/8-inch-diameter bead around the perimeter of the panel. On the interior of the panel, lay 3-inch-long beads spaced 6 inches apart on the studs that will hit the interior of the panel as shown in **5-10**. Place the panel against the studs and press it in place. If the wall is covered with gypsum wallboard, place these 3-inch-long adhesive beads on the back of the plywood or hardboard panel where the studs are located behind the drywall. Check it for plumb and add a wedge between the panel and floor to get the proper spacing (refer to **5-4**). Drive a few nails at the top of the panel to hold it to the wall. Some companies recommend pressing the panel to the wall, pulling it out at the bottom for about 10 minutes to help the adhesive set, and then pressing it back against the studs.

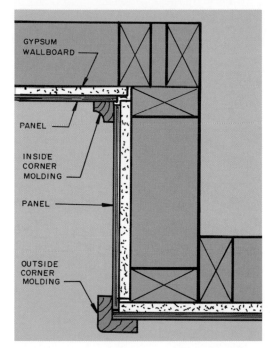

5-8 Inside and outside corners are covered with molding.

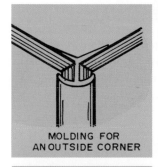

MOLDING FOR AN OUTSIDE CORNER

MOLDING FOR AN INSIDE CORNER

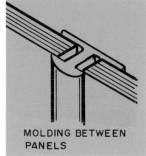

MOLDING BETWEEN PANELS

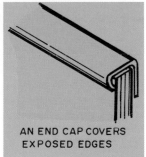

AN END CAP COVERS EXPOSED EDGES

5-9 Vinyl-covered aluminum molding is available to conceal joints between panels.

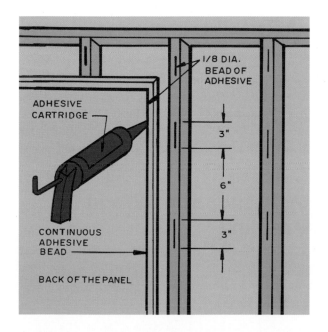

5-10 Adhesive is applied to the back of the panel in continuous beads around the perimeter and in spaced applications on the studs or drywall.

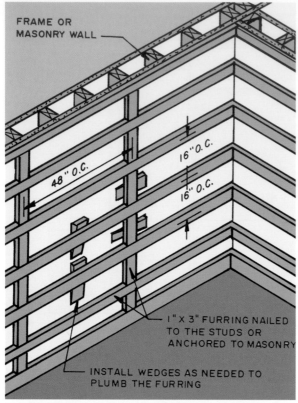

FRAME OR MASONRY WALL

48" O.C.

16" O.C.

16" O.C.

1" X 3" FURRING NAILED TO THE STUDS OR ANCHORED TO MASONRY

INSTALL WEDGES AS NEEDED TO PLUMB THE FURRING

5-11 Damaged and out-of-plumb walls can be covered with paneling by nailing furring strips to them. Masonry walls can be paneled by anchoring furring strips to the masonry.

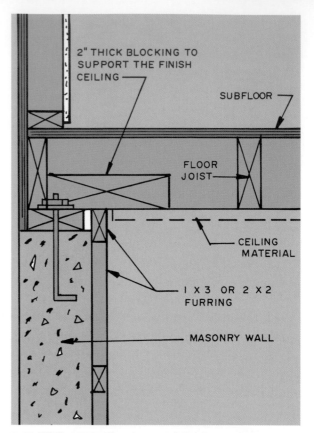

2" THICK BLOCKING TO SUPPORT THE FINISH CEILING

SUBFLOOR

FLOOR JOIST

CEILING MATERIAL

1 X 3 OR 2 X 2 FURRING

MASONRY WALL

5-12 While preparing to panel the basement, remember to add blocking along the masonry wall to support the edge of the ceiling material.

5-13 One way to panel a masonry wall is to construct a stud wall over it. This allows the wall to be insulated.

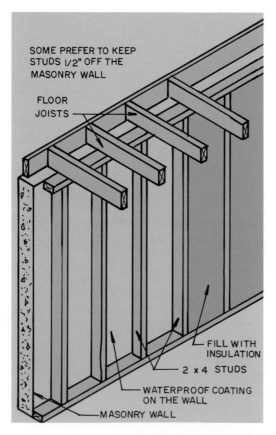

SOME PREFER TO KEEP STUDS 1/2" OFF THE MASONRY WALL

FLOOR JOISTS

FILL WITH INSULATION

2 x 4 STUDS

WATERPROOF COATING ON THE WALL

MASONRY WALL

Paneling over an Existing Wall

When paneling over an existing wall, find each stud and mark its location on the wall. If the wall is somewhat out of plumb, is badly damaged, or is a masonry wall, nail furring strips to it and use wedges to make them plumb (5-11). Then install the paneling.

Special blocking is needed at the ceiling on basement walls where joists run parallel with the wall (5-12). Rigid plastic insulation can be glued to the masonry wall between the furring strips. Another way to panel a masonry wall, such as in a basement, is to frame a stud wall and place it next to the masonry wall (5-13). A stud wall will allow 3½ inches insulation to be installed. Place a waterproof membrane over the wall behind the studs.

Installing Solid-Wood Paneling

Solid-wood paneling is available in a number of wood species and tongue-and-groove profiles. Square-edge boards can be used with a batten or molding at the edge joint (5-14).

5-14 Solid-wood paneling is available in many species of wood and with various decorative machined joints between the panels.

5-15 Wall framing for new construction includes blocking, a vapor barrier, insulation, and gypsum wallboard behind the wood paneling.

The paneling to go over wall studs, which are spaced 16 inches on center, should be at least ⅜-inch thick. It should be kiln-dried to 8 percent moisture. If pieces over 8 inches wide are used, blocking is needed between the wall studs so they can be nailed in the center to reduce cupping.

Check the local building codes to see if a fire-retarding material such as gypsum drywall has to be used behind the paneling. Regardless of the code, this is a smart thing to do.

If the paneling is prefinished, be very careful not to scratch the surface. Store it flat in the room in which it will be installed so it can adjust to the moisture content in the air.

Preparing the Wall

Details for preparing the wall for new construction are shown in **5-15.** Notice that the horizontal blocking is spaced no wider than 24 inches on center. A plastic vapor barrier is placed over the framing; then the gypsum wallboard is applied and, finally, the vertical paneling.

If an existing room is being remodeled, apply horizontal and vertical furring strips on the wall and around doors and windows as shown in **5-11.** This will mean an extension has to be added to the window

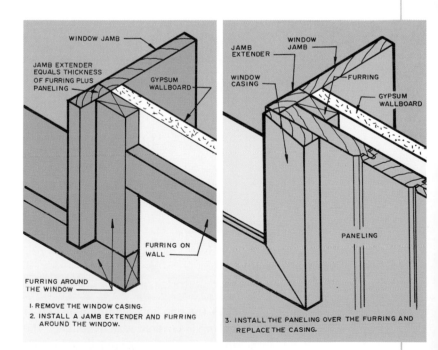

5-16 A jamb extender is needed wherever the paneling butts a window or door.

5-17 Check the wall where the first panel board is installed for plumb. Taper the first board as needed so the exposed edge is plumb. When you get to the other wall, the same check and some correction may be necessary.

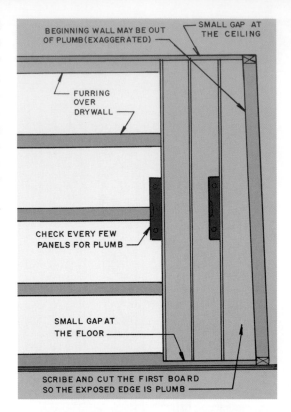

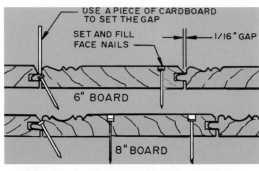

5-18 Recommended spacing and nailing patterns for vertically applied wood paneling.

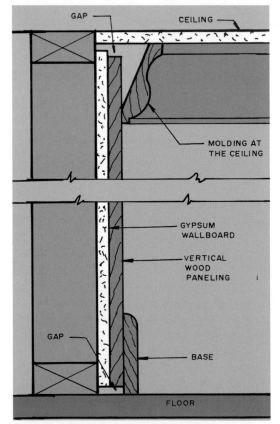

5-19 The gap between the paneling and the ceiling is covered with a molding. The gap at the floor can be covered with a wood base, or carpet can butt the paneling, eliminating the need for a base.

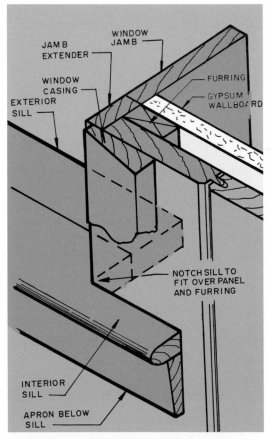

5-20 When the window has an interior sill, it will have to be removed and notched to fit over the jamb extender, which equals the thickness of the panel plus the furring.

jambs so the window casing fits over the paneling (**5-16**).

Now make a plan detailing how to begin and the direction in which to work. Some prefer to start at an outside corner if one exists and work toward inside corners. Then they work from inside corners toward doors and windows or from one inside corner to the next inside corner. Often, the last board on a wall will be narrower than the rest, and this is less noticeable if it is on an inside corner.

Set the first board in a corner and check for plumb. If the wall leans, the slope on the piece will need to be scribed and planed off so the exposed edge is plumb (**5-17**). Continue nailing across the wall, checking frequently for plumb. Leave a small gap at the floor and ceiling to allow for expansion. When arriving at the other wall, measure the space remaining and cut a board to fit this space.

Nailing patterns are shown in **5-18**. Use 6d finish nails. Notice that 6-inch-wide boards use one face nail and one blind nail, while 8-inch-wide boards have two face nails and one blind nail. Set the face nails and fill the holes. Be certain to leave a small gap between panels at each joint to allow for expansion.

The gap at the ceiling is covered with some type of molding, and a baseboard is installed at the floor (**5-19**). Interior and exterior corners do not need to be covered if they have a tight fit. However, if this is a problem, use molding to finish these as shown earlier in **5-8**.

When a door or window is encountered, the window jamb will have to be extended by adding a piece equal to the thickness of the panel plus the furring (refer to **5-16**). If the window has a stool and apron, they will have to be removed. The stool will have to be notched to fit over the paneling and the apron nailed back below it (**5-20**).

Installing Wainscoting

When solid-wood paneling, plywood, or hardboard panels are used to cover the lower part of a wall (**5-21**), this is referred to as **wainscoting**. It is typically installed in the same manner as described for full wall

5-21 This wainscot has a panel of a different species of wood than that used on the frame. Notice the decorative wainscot cap molding.

5-22 This painted wainscot is very durable and especially useful in rooms where there may be items bumping the wall. Kitchens, game rooms, and porches are especially good locations.

Courtesy Georgia-Pacific Corporation

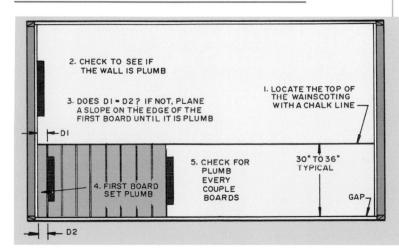

2. CHECK TO SEE IF THE WALL IS PLUMB

3. DOES D1 = D2? IF NOT, PLANE A SLOPE ON THE EDGE OF THE FIRST BOARD UNTIL IT IS PLUMB

1. LOCATE THE TOP OF THE WAINSCOTING WITH A CHALK LINE

D1

4. FIRST BOARD SET PLUMB

5. CHECK FOR PLUMB EVERY COUPLE BOARDS

30" TO 36" TYPICAL

GAP

D2

5-23 The steps for laying and installing a wainscoting.

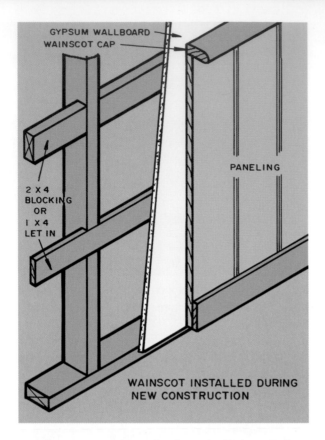

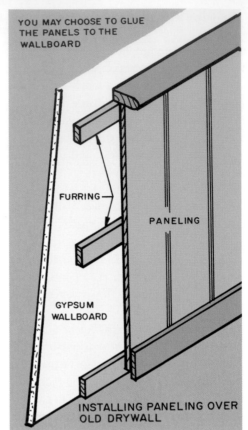

5-24 Two ways wainscoting is typically installed.

paneling. An interesting installation in **5-22** shows the use of plywood paneling in a garage. This is a paint grade paneling and is used in any room where you want to paint the wainscot or paneling.

To begin installation, decide how high the wainscoting will be and run a chalk line along the wall (**5-23**). This locates the top of the paneling. Start installation in one corner and proceed across the room. If this is over old gypsum wallboard, furring can be added to the wall and the panels nailed. If the wall is sound, paneling can be glued to it with paneling adhesive (**5-24**). Finally, install the wainscot cap. This can take several forms, and one can be made using stock molding (**5-25**).

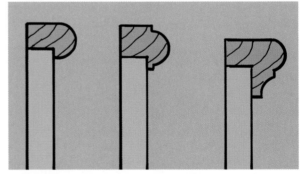

5-25 Various wainscoting caps are available. If desired, one can be custom designed and machined.

CHAPTER 6

Repairing & Replacing Interior Doors

Interior doors often develop problems over time. They may stick and rub on the door frame. The latch may not fit into the strike so it no longer holds the door closed. These and other problems are not difficult to fix.

Doors That Stick

If the door sticks, begin by finding where it is rubbing the door frame. An examination will usually show scrape marks on the frame and edge of the door. If this is not obvious, slide a piece of heavy paper between the edge of the door and the jamb. Mark wherever it will not pass with a piece of tape **(6-1)**. This is an area needing attention. There are a number of things that might cause this and each must be considered. Possibly the door has swelled due to absorption of moisture. First remove the door to an area where it can dry. Once it has returned to normal, replace it in the door frame. It may swing clear and no longer stick. If this is the case, seal all the edges with the paint or clear finish used on the door. Be certain to seal all the edges—top, bottom, and sides.

If the door still sticks after drying, check the hinges to see if they are loose. Pull up on the door. If the hinges are loose they will move a little. A quick repair try is to simply tighten the screws. If they will

6-1 If a door is sticking on the frame, the point of contact can be found by sliding a piece of stiff paper along the edge in the crack. Mark the rubbing area with tape.

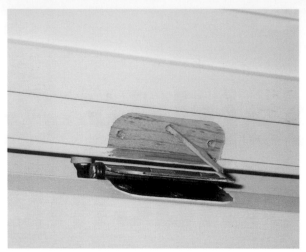

6-2 Glue small wood pins in the enlarged screw holes. After the glue has set reinstall the screws.

6-3 Cardboard shims can be placed under the hinge to move the top or bottom edge out, possibly stopping it from scraping on the door frame.

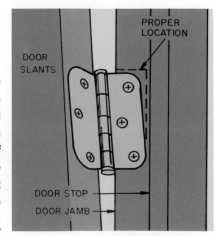

6-4 If the hinge edge is not installed parallel with the edge of the jamb, the door will not operate properly.

not screw up tight, make some small wood plugs and glue them in each screw hole (6-2). After the glue has set, reinstall the screws. Sometimes installing longer screws will solve the problem.

If the tightened screws do not even up a door so it swings clear, consider placing cardboard shims under the top, bottom, or possibly all three hinges (6-3). Following are typical examples.

• If the top outside edge of the door strikes the jamb, put one or more shims under the bottom hinge and as needed under the center hinge.
• If the bottom outside edge of the door strikes the jamb, put one or more shims under the top hinge and as needed under the center hinge.
• If the door binds the jamb along the hinge side, shim all three hinges.
• If the door binds the jamb along the hinge side at the top, shim the top hinge, or if at the bottom only, shim the bottom hinge.

Sometimes the hinge leaves have been recessed too deep. This puts great pressure on the edge of the door as it is closed. Shim the leaves with cardboard until they are flush with the surfaces of the jamb and edge of the door. Likewise, if the hinges were installed without being recessed, they protrude, which may interfere with a door operating easily.

Doors That Won't Close Properly

If the hinges are placed a little out of alignment as shown in 6-4, the door will not close properly and the latch will most likely not fit into the strike plate. The hinges must be installed parallel with the edge of the jamb and they must line up vertically. Reset any hinges that are out of line.

Sometimes old doors have been painted so many times the paint can keep the door from swinging freely. If it is a small amount, sand the paint, leaving a thin film to protect from moisture (6-5). Some will then rub a wax on the edge. If necessary cut back to the bare wood and recoat.

In older houses there may be settling of the floor, which causes the door frame to be out of square. When you look at the opening around the door you can see the variance in the width of the crack around it. Check the frame with a carpenter's square to verify this (6-6). In this case it will be necessary to plane a little off the edges of the door. Mark the area that is binding as shown in 6-1. Most of the time it is possible to plane the edges without removing the door. A block or smoothing plane works well because you only need

6-5 If the edge of the door has multiple coats of paint, try sanding the top coat off and see if the door will stop rubbing the jamb.

6-6 Check the door jambs to see if they are square.

to remove a small amount (6-7). When planing the top or bottom, remember to plane from the outside edge toward the center of the door (6-8). If you plane toward the edge you might split off pieces of the edge. To plane the bottom edge you will have to remove the door by pulling the hinge pins (6-9).

Removing a Door from its Hinges

To remove the pin, put the tip of an old screwdriver under the head of the pin and tap it up with a hammer. Since the door loosens up when a pin is removed, pull the center pin first. Then tap the top and bottom up but not quite out. Hold the door firmly and pull the pins by hand.

If you plane near the latch or hinges you will have to remove them. It may be necessary to cut their

6-7 The edges of the door can be trimmed a little with a block or smoothing plane.

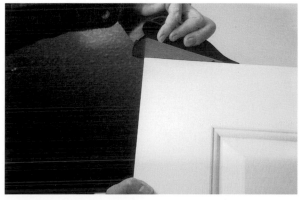

6-8 When planing the top and bottom edges of the door, plane from the outside edge toward the center.

6-9 To remove an interior door remove the hinge pins.

recesses a bit deeper so they are flush with the surface when reinstalled. As you plane, close the door frequently until it swings clear. Remember to seal the planed area to keep out moisture.

Fixing the Strike
So a Door Will Latch

Sometimes the latch will not engage and hold the door closed. Examine it carefully as you close the door to see if it fits into the opening in the strike (6-10). If it is rubbing a little against the sides of the opening, remove the strike and file the opening a bit larger. If the contact is large, remove the strike; move the strike up or down or sideways as needed, cut the recess to the new location, and reinstall it. You may have to enlarge the mortise in the door jamb into which the strike fits (6-11). If the strike has been recessed too deep, place one or more cardboard shims behind it to bring it up flush with the surface of the jamb (6-12).

One way to get the strike in the desired position is to place it over the latch and close the door. You

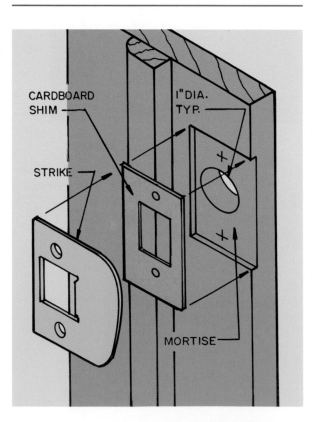

6-10 The latch must slide into the opening in the strike.

DOOR STOP

STRIKE

T. REMOVE THE STRIKE.

ENLARGE THE RECESS

ENLARGE THE LATCH HOLE

2. REPOSITION THE STRIKE AND ADJUST THE RECESS AND THE LATCH HOLE.

6-11 It may be necessary to move the strike. Cut the recess larger to allow it to be moved.

3. REINSTALL THE STRIKE.

CARDBOARD SHIM

1" DIA. TYP.

STRIKE

MORTISE

6-12 If the strike mortise is cut too deep, insert cardboard shims behind it to bring it flush with the face of the jamb.

can move it a bit and mark the apparent location. Then temporarily screw it to the jamb with one screw. Try the door to see if the latch falls inside the opening. Move it if necessary. Once it is in place mark around it, remove the screw, and enlarge the mortise with a wood chisel and install the strike.

Another factor that may make it hard to close the door so the latch enters the strike is that the door-stop may be pressing against it. If it is a nailed stop, it requires removing and renailing it, leaving a ⅛-inch gap between it and the door. An easy way to set the gap is to put a piece of cardboard between the stop and door, and then to nail it. It is usually nailed with 3d (1½-inch) finishing nails (**6-13**). Set the nails and caulk and you are ready to paint.

Fixing Squeaky Hinges

Squeaky hinges are annoying and can easily be silenced. First try placing a few drops of machine oil along the head of the pin. If this fails, raise the pin a little and run oil along it (**6-14**). Swing the door so the oil is worked down the pin. Keep a cloth under the pin to catch the oil that may run through.

Settling of the House

Finally, if the house has settled so the door frame is quite a bit out of square, the solution involves removing the casing and frame and installing a new frame and possibly a new door. This is explained later in this chapter.

Adjusting Bypass Doors

Bypass doors are used on closets. They may have two or three doors that slide past each other (**6-15**). They have wheels attached at the top of the doors. The wheels ride in a metal track that is screwed to the top door frame (**6-16**). Each door runs on a separate track (**6-17**). There is a guide attached to the floor to keep the doors in line. Over time they may not run smoothly and require some adjustment. Following are several possible problems.

The track or floor guide has gotten bent. As you slide the doors back and forth you may notice that the wheel does not roll easily in a particular spot, causing it to stick. Using a pliers, try to bend the metal track enough to allow the wheel to pass smoothly along the track.

The wheel attachment may be bent. If the metal attachment that holds the wheel to the door is bent,

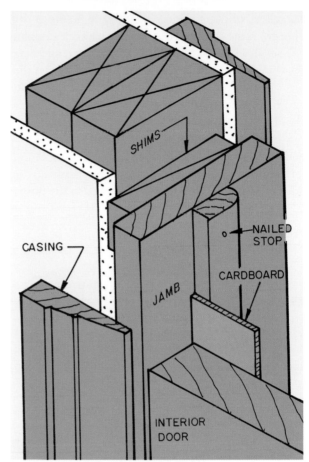

6-13 Space the nailed doorstop about ⅛ inch from the door. A thick cardboard shim can serve to establish the space.

6-14 Squeaky hinges can be silenced quickly by running a little machine oil down the pin.

the wheel will be running on an angle. Remove the door from the track and bend the attachment straight.

The wheel adjustment screw may be loose, allowing the door to drop and drag on the floor. Loosen the screw, raise the door until it is parallel with the track, and tighten the screw as shown in **6-18.** The doors should not rub on the floor guide.

Something has gotten in the track. This simply means the track must be kept clean.

The floor guide is loose or out of line, causing the doors to rub as they pass. Set the track so the doors pass without rubbing and tighten the screw. If the screw will not tighten, remove it and glue in a wood pin. Reset the screw after the glue has dried.

Adjusting Bifold Doors

Bifold doors are used on closets. They may be two or four panel sets of doors. They are hinged along the adjoining edges **(6-19).** There are guide pins in the top edge of the door that run in a metal track. At the floor is a metal pivot plate that is secured to the

6-15 Bypass sliding doors open over each other, exposing half the closet opening.

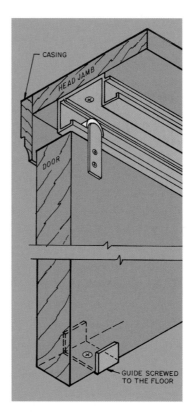

6-16 Bypass sliding doors run on a metal track mounted to the head jamb. A guide is attached to the floor to keep the doors running straight.

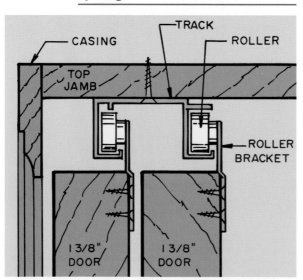

6-17 Typical hardware for the installation of two bypass doors. When you select the hardware be certain it will handle doors the thickness you plan to use.

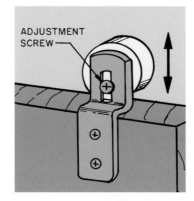

6-18 Some bypass door hardware has an adjustment screw that permits the roller to be raised and lowered so the door hangs level.

side door jamb (6-20). The door swings on this plate and a pin in the top edge of the door next to the side jamb. A top pin is located on the edge of the second door. It slides in the track, guiding the door in a parallel motion (6-21).

Following are some adjustments that may be needed.

The top of the door may rub on the track or the pins on top may occasionally slip out of the track. To correct this you can adjust the height of the door by turning the threaded pivot at the floor. The pivot may have a nut you turn, or a fixed flange, requiring that you turn the entire threaded pivot.

The bottoms of the doors may drag on the carpet. It may only require you to raise the door a little by adjusting the pivot as just described. If there is no room at the top to raise the door, you may have to raise the pivot plate so it is above the carpet as shown in 6-20, and trim a small amount off the bottom of the door.

If the top pin on the outside door comes out of the track as the door is closed, it means the door is not

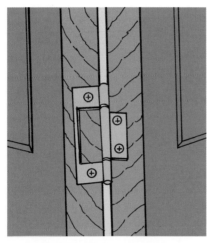

6-19 Bifold doors are hinged with nonmortise hinges screwed to the edges of the adjoining doors.

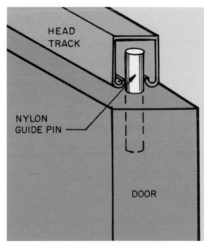

6-21 The guide pin is placed in the top of the second door near the end. It slides in the channel guiding the movement of the door.

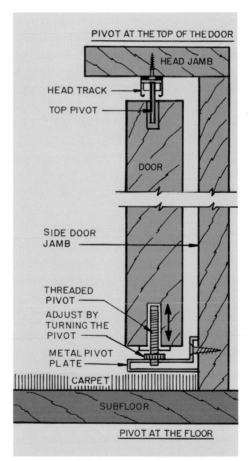

6-20 The bifold doors swing on pivots located at the side jamb.

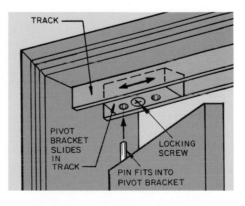

6-22 The top pivot pin fits into the pivot bracket, which can be moved in the track to get the door parallel with the track.

parallel with the side jamb. The top pin at the side jamb is held in place with a pivot bracket. Loosen the screw and slide the plate which pushes the top pin until the edge of the door is parallel with the side jamb. Tighten the screw (6-22).

If the doors seem loose where they join, check the hinges. Reset any loose screws.

Lockset Problems

Locksets are one thing that cause very little trouble for many years. If yours is in really bad shape, consider replacing it. Try to get one that will fit in the openings already cut in the door for the old lockset.

Most common problems can be easily solved. One that occurs after a few years is that the handle and lock seem loose. This is corrected by tightening the bolts used to join the two halves of the lock. Exterior locks will sometimes get a bit difficult to unlock. Squirt a little graphite in the cylinder and work the key in and out and then rotate it. Do not insert oil or grease.

Problems with the latch not engaging the strike plate are covered earlier in this chapter.

If you consider replacing the lockset, examine the types commonly available. The dealer selling the lockset often has someone who can set the pins in the cylinder to match the key used for the old lock. Then you will not have to carry an extra key. Some prefer to have all exterior locks set the same so only one key is needed. If the dealer cannot do this, a locksmith can do the job.

Types of Lockset

The two types of lockset are tubular and cylindrical. Locksets used on interior doors are the **tubular type** (6-23). They are available with a lock button which provides privacy in bathrooms and bedrooms. They have a tiny hole in the outside knob into which a strong wire can be inserted, providing an emergency way of unlocking the door from the outside. The tubular lock in **6-24** is used on exterior doors. It is opened from the outside with a key.

The **cylindrical lockset** provides greater security than the tubular type. It has the mechanism in a large cylinder from which the latch bolt extends (6-25). The one with the lever type handle is easier for most people to open, especially children and those who have some difficulty with turning a round knob.

A **dead bolt** is much like the cylindrical lock except it can be opened by a key from the outside

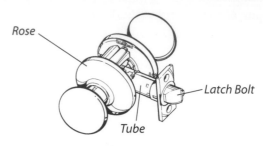

NONLOCKING

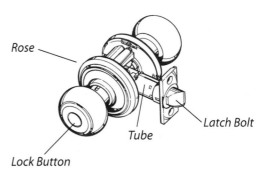

LOCKING

6-23 Tubular-type interior locksets are available as locking and nonlocking types. *Courtesy Arrow Lock Manufacturing Company*

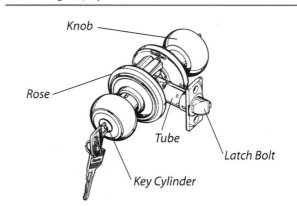

6-24 This is a tubular-type exterior lockset. It has a key cylinder in the outside knob and a lever to control the lock on the inside knob. *Courtesy Arrow Lock Manufacturing Company*

and a lever-type handle on the inside. It has no knobs. Its purpose is to provide extra security. The large latch protrudes into the door frame and cannot be moved without a key (6-26). It should be noted that some dead bolts need to be opened from the inside with a key. This is very dangerous and banned in many places. Should there be an emergency where you need to get out of the house quickly, you have to find the key and this can delay your exit.

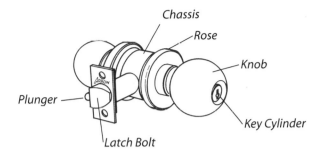

Chassis
Rose
Knob
Plunger
Key Cylinder
Latch Bolt

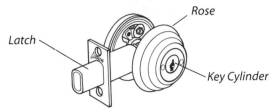

Rose
Latch
Key Cylinder

6-26 A dead bolt provides additional security. *Courtesy Arrow Lock Manufacturing Company*

6-27 A small dead bolt lock next to the latch blocks attempts to slide the latch back with a plastic card. *Courtesy Arrow Lock Manufacturing Company*

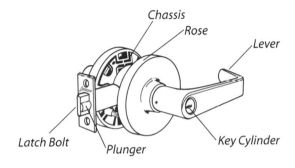

Chassis
Rose
Lever
Latch Bolt
Plunger
Key Cylinder

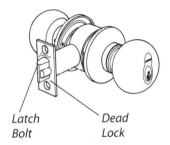

Latch Bolt
Dead Lock

6-25 Cylindrical locksets provide excellent security for exterior doors. *Courtesy Arrow Lock Manufacturing Company*

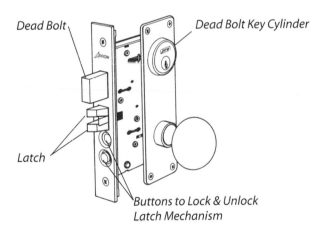

Dead Bolt
Dead Bolt Key Cylinder
Latch
Buttons to Lock & Unlock Latch Mechanism

6-28 A mortise lock has a latch bolt and a dead bolt. *Courtesy Arrow Lock Manufacturing Company*

6-29 These exterior door entry locksets include a separate deadbolt and a latching handle set assembly.

Some **entry locksets** have a small dead bolt that runs alongside the latch (**6-27**). This prevents someone from sliding the latch back with a plastic card.

A **mortise lockset** is a high-quality entry lockset (**6-28**). Its size and design differ a great deal from the others which are set in a hole bored in the door. It requires a deep mortise to be cut into the edge of the door to receive the mechanism. In addition to great security, it adds considerably to the appearance of the door (**6-29**).

Installing a New Door Frame

If a door frame is badly warped or twisted it is a good idea to remove the casing, pull it out, and replace it. After it has been removed, check the framing of the rough opening to see if the studs are plumb and the header is level (6-30). Now check the size of the rough opening (6-31). You will need the width and height so that when you buy a new door frame it will fit in the opening and leave space around it to plumb and level it. If you are going to reuse the old door, the frame must accommodate it and allow for a ⅛-inch space on the sides and top. At the floor it must allow for the existing flooring. If the door is a little large it can be trimmed.

Should you decide to replace the door and frame, buy a prehung door-and-frame unit. The unit is shipped with the hinges in place and the holes in the door bored for a tubular-type lock (6-32). Since the opening is already there, select a unit that will fit the opening as shown in 6-31. Often on the cardboard

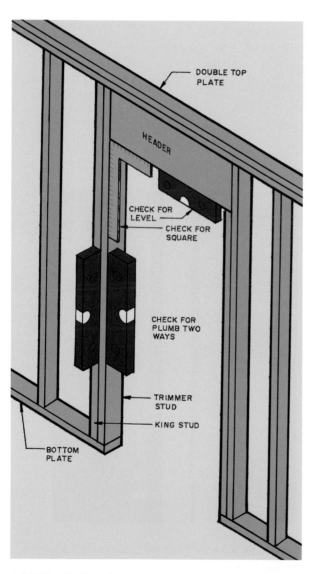

6-30 Check that the rough opening is level, plumb, and square.

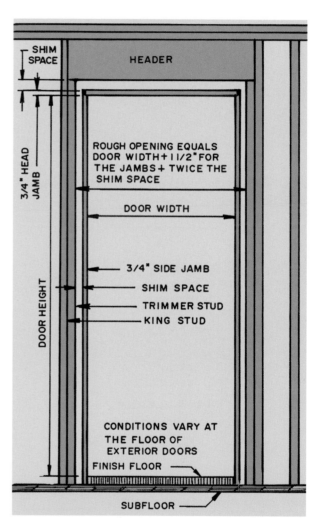

6-31 Check the size of the rough opening to be sure the door frame will have room to be installed, leveled, and plumbed. This is a typical layout for an interior door.

6-32 This prehung door in a frame is protected with heavy cardboard edges. Notice the hole has been bored for the lockset.

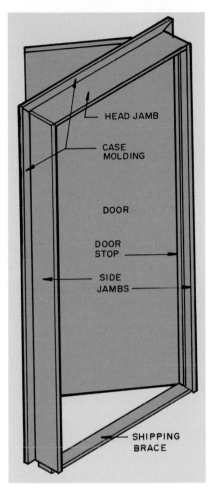

HEAD JAMB

CASE MOLDING

DOOR

DOOR STOP

SIDE JAMBS

SHIPPING BRACE

6-33 A typical prehung interior door unit has an assembled frame, casing, and door. The casing is nailed to one side. The casing for the other side is shipped unattached and is installed after the unit is set in the door opening.

Table 6-1 What to Note When Ordering Prehung Interior Doors.

Interior Door Specifications

TYPE OF DOOR

   design of door

   surface material of door

   core construction

DIMENSIONS OF DOOR

   door size

   door thickness

Door jamb

   type of jamb

   width of jamb

Door casing

   type of casing

   size of casing

Hand of door

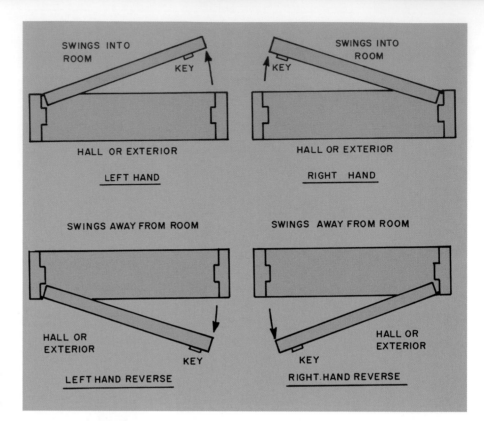

6-34 The hand of doors used in residential construction is determined from the key or outside face.

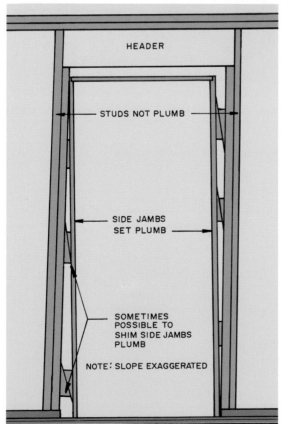

6-35 This rough opening has the studs leaning to one side forming a rhombus. The side jambs can be set plumb by shimming if the studs are not too out of plumb.

covering protecting the door and frame, the manufacturer will have printed the installation instructions.

Should this be new construction, carpenters often will frame the new rough opening to be ½ to 1 inch wider than the outside width of the frame. If the frame is not on the job, add 1½ inches to the width of the door to be used; this allows for the ¾-inch side jambs. Then add the ½-to 1-inch space desired, to allow space for plumbing the door **(6-31)**.

The height of the header will include the door height, ¾-inch top frame, 1 to 1½-inch shim space at the header, plus an allowance for the finish flooring. Refer to **6-31**.

A typical fully assembled prehung door and frame are shown in **6-33**. It will usually have the casing installed on one side. When ordering the prehung door you will have to specify the things listed in **Table 6-1.** Designating the hand of the door is important.

Specifying the Hand of the Door

The **hand of a door** refers which side will have the hinges; the door may be further designated according to which way it is to swing. A frequently used technique for specifying the hand is shown in **6-34.** This system requires the door be viewed from outside the room. For example, a bedroom door is viewed from

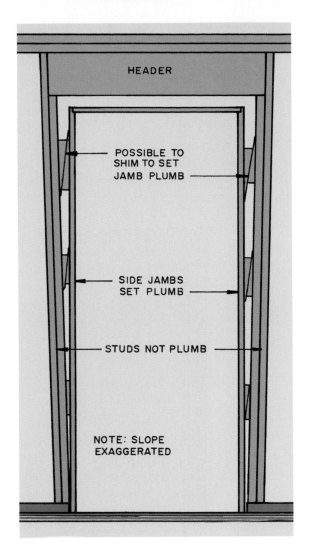

HEADER

POSSIBLE TO
SHIM TO SET
JAMB PLUMB

SIDE JAMBS
SET PLUMB

STUDS NOT PLUMB

NOTE: SLOPE
EXAGGERATED

6-36 The studs on this rough opening are out of plumb and sloping in opposite directions. If the slope is not too large, the door frame can be set plumb and secured with shims.

6-37 Set the bottom of the unit into the rough opening.

6-38 Place the casing against the drywall.

the hall to determine the hand. An exterior door is viewed from outside the house.

When you view the door from the outside, the right- and left-hand doors will **swing away from you.** If the hinges are on the right side, it is a right-hand door. If the door swings toward you, such as into a hall or porch, it is designated a **reverse door.** Therefore if the hinges are on the right side, it is a **right-hand reverse door.**

Installing Prehung Interior Doors

Some prehung doors will come with the frame knocked down, to be assembled on the site. Usually the hinges are installed on the jamb and door. Others are fully assembled as shown in **6-33.**

Do not move new doors into a house until the drywall has been hung, taped, dried, and sanded. Install

as soon as possible thereafter. Prime all surfaces and edges, including the top and bottom.

If the rough opening is the right size and the studs are fairly plumb, you can begin the installation. If the jamb is a little out of plumb this can be handled using double shims as shown in **6-35** and **6-36.** Check to see if the wall leans into the room. If it does, the door will always hang open. If it leans away from the room it will always swing closed and never stand open. Should this condition exist, the plumb of the wall should be corrected.

After you unpack the prehung door, move it to the opening. Center it by first setting the bottom in place (**6-37**) and then lean the casing against the drywall (**6-38**). Use a pry bar or blocking below the door to raise it as needed for the flooring. Place a level against the casing to be certain it is plumb (**6-39**).

Now examine the gap at the head jamb. This gap should be uniform. If it is not, move the hinge side

6-39 Adjust the height with a pry bar or blocking below the door. Check the casing with a level and adjust until it is plumb. Nail at the upper left corner to hold it in position.

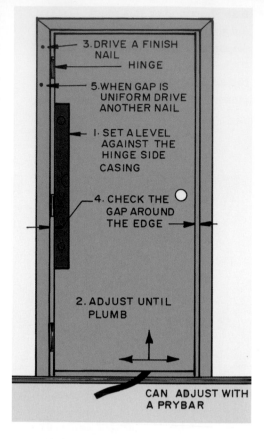

3. DRIVE A FINISH NAIL

HINGE

5. WHEN GAP IS UNIFORM DRIVE ANOTHER NAIL

1. SET A LEVEL AGAINST THE HINGE SIDE CASING

4. CHECK THE GAP AROUND THE EDGE

2. ADJUST UNTIL PLUMB

CAN ADJUST WITH A PRYBAR

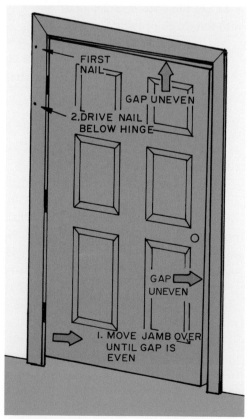

FIRST NAIL

GAP UNEVEN

2. DRIVE NAIL BELOW HINGE

GAP UNEVEN

1. MOVE JAMB OVER UNTIL GAP IS EVEN

6-40 Adjust the gap at the head and side jambs by moving them with the shims. Then nail through the casing.

jamb out with the shims until the gap is uniform. This will also adjust the gap between the latch side edge of the door and side jamb (**6-40**). If the gap at the head jamb is even and the one at the side jamb is not, move the side jamb over by adjusting the shims. When it is in place drive a finishing nail through the casing a few inches below the corner miter. Use 7d (2¼-inch) finish nails. It helps if you have someone help hold as you nail it in place.

Now drive another nail a few inches below the top hinge. Then swing the door to see if it is clear. If all

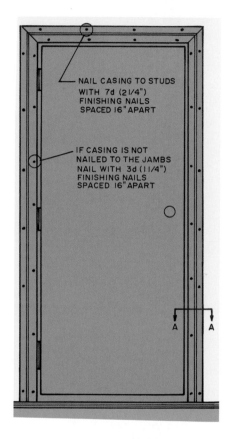

NAIL CASING TO STUDS WITH 7d (2 1/4") FINISHING NAILS SPACED 16" APART

IF CASING IS NOT NAILED TO THE JAMBS NAIL WITH 3d (1 1/4") FINISHING NAILS SPACED 16" APART

A   A

6-41 Recommendations for nailing the casing to the jambs and to the studs that form the rough opening.

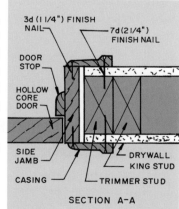

3d (1 1/4") FINISH NAIL

7d (2 1/4") FINISH NAIL

DOOR STOP

HOLLOW CORE DOOR

SIDE JAMB

CASING

DRYWALL

KING STUD

TRIMMER STUD

SECTION A-A

is right, nail the hinge side casing to the wall, spacing the nails every 12 to 16 inches. If all is still in order, nail the top casing and then the latch side casing as shown in **6-41.**

At this point some consider the door adequately installed. Others prefer to shim between the stud and side jambs and nail through this into the stud **(6-42).** Locate the nails so they are covered by the door stop. Blocking is installed at the location of the lock and behind each hinge. If the frame is slightly bowed and the door rubs, you can put some blocking behind the bow and pull it out by nailing into the stud.

Finally, install the casing on the other side of the frame and nail the miters **(6-43).** Then set the nails so the painter can cover them with caulking **(6-44).** If the doorstop has not been installed, install as described earlier in this chapter.

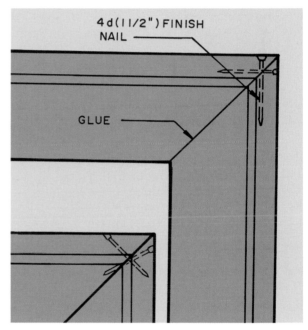

6-42 Prehung hollow-core doors can be installed only by nailing the casing to the studs. Adding wood shims and nailing through the jamb and shims strengthens the installation and reduces the chances of the jambs' bowing.

6-43 When installing casing on the jambs, it is good practice to glue and nail the miter.

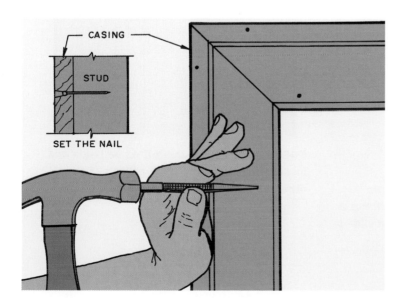

6-44 After the door has been installed, set the heads of the nails so the painter can cover them before painting the casing.

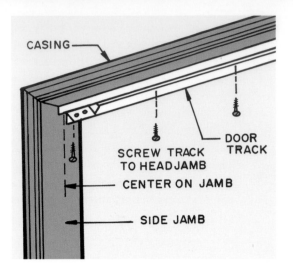

6-45 Screw the track to the head jamb. Locate its center on the centerline of the jamb.

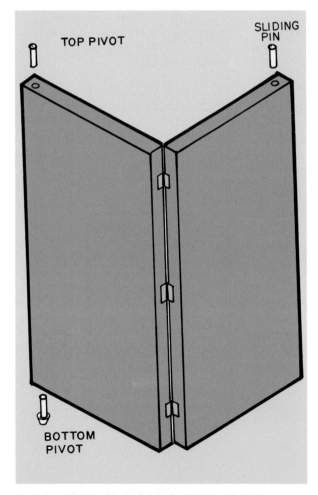

6-46 Insert provided pivots in top and bottom corners of the door's side, next to the jamb. Put a pin in the corner of the other door so it can slide in the track.

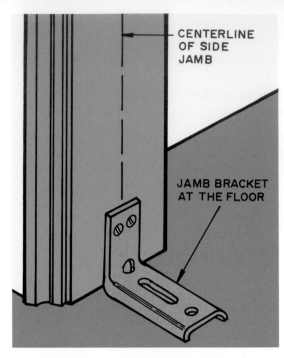

6-47 Install the jamb bracket on the centerline or the side jamb. It must line up vertically with the centerline of the track and be above the finished floor.

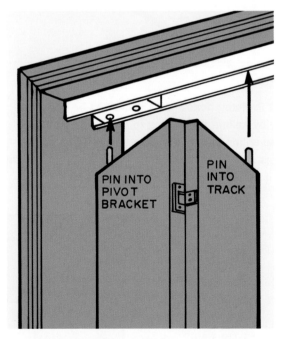

6-48 To install the door, lift it on an angle and insert one pin in the pivot bracket and the other pin in the track.

Installing Bifold Doors

The hardware supplied by the manufacturers of bifold doors will vary; however, the following examples are typical. Detailed installation instructions are supplied with each set of doors.

Check the side jambs for plumb and the head jamb to be certain it is level. The bifold doors will not function properly if the door frame is not true. Check the size of the framed opening with the size specified for the doors to be installed. It must be exactly as specified.

Begin by cutting the track to length, with a metal-cutting hacksaw. Install the track to the head jamb, which will be predrilled for screws supplied with the hardware (6-45). Install it on the centerline of the head jamb. Install the top pivot bracket in the end of the track next to the side jamb. It is secured with a set screw which, when loosened, permits the bracket to slide in the track, allowing the top of the door to be spaced from the jamb as necessary.

Bore holes in the top corners of the doors and tap the plastic pivots in them. Install the bottom threaded pivot in a hole as directed by the manufacturer (6-46).

Then install the jamb bracket to the side jamb. Position it so it is just above the finished floor. Remember, if the floor is to have carpet it must be above the carpet (6-47). It is centered on the centerline of the jamb so as to be perfectly aligned with the track. Secure it to the jamb with screws.

To install the door insert the one top pin in the top pivot bracket and the pin on the end of the other door in the track (6-48). Lift the doors and move them over the bottom pivot bracket. Set the bottom threaded pivot into the bottom pivot bracket (6-49). Then turn the threaded pivot until the door is the desired height and moves freely. If the door is not plumb (parallel with the side jamb), lift it out of the floor bracket and move it toward or away from the side jamb as needed. When the edge is plumb, lower the threaded pivot back into the slot in the floor bracket (6-50).

Now add the desired door pull. It can match those used on the doors in the room but it is just the handle that is screwed to the face of the door (6-51).

If it is a four-door installation, as shown in 6-51, add door liners on the inside of the abutting doors about 12 inches above the floor. Adjust them back and forth on the slotted holes until the doors remain tightly in line when closed (6-52).

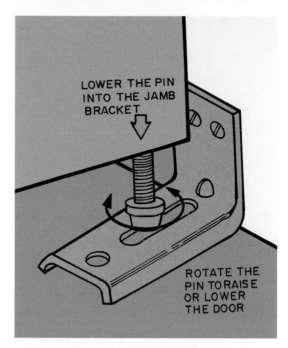

6-49 After the pins on the top of the door are in the track, place the bottom pivot over the jamb bracket at the floor. Lower it into the bracket. Turn the pin to raise or lower the door as needed.

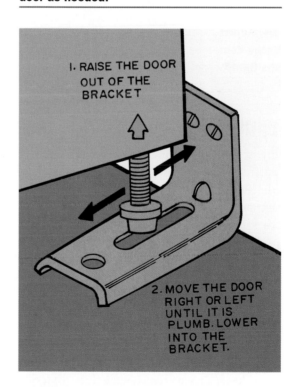

6-50 To plumb the door, lift it out of the jamb bracket and move it right or left as needed and set it back in the bracket.

6-51 This finished four-door installation has door pulls matching those used on the locksets of the swinging doors.

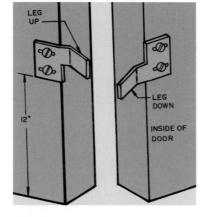

6-52 Install door aligners on the inside of the doors. Adjust until the door closes tightly.

Installing Hinges

Should you buy a replacement door or replace parts of or an entire door frame, it will be necessary to install the hinges. Since hinges do wear over the years, it is recommended you buy new ones. Be certain you get door hinges. A typical butt hinge used to hang interior doors is in **6-53.** Some have square corners but most used today have round corners. Hinges with a square leaf and a round leaf are often used on fiberglass and metal doors **(6-54).**

Butt hinges used to hang doors are **swagged,** meaning they have the leaves bent so that there is only a small space between them when the door is closed **(6-55).** Hinges used on interior doors have a loose pin that makes it easy to remove the door. Exterior doors have a fixed pin which cannot be removed. This is a security feature.

It is important to choose the proper size hinges for the door. Notice in **Table 6-2** that the hinge size changes for various door thicknesses. The hinge must be wide enough to permit the door to clear the casing, so the

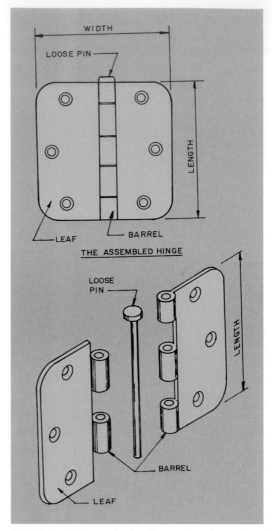

6-53 A butt hinge typically used to hang interior doors. Some have square corners. This shows a loose-pin hinge typically used on interior doors. Exterior door hinges have a fixed pin that cannot be removed.

type and size of casing must be considered as well. Mortise hinges installed on 1⅜ inch and 1¾ inch doors are shown in **6-56.** The clearance depends upon the door thickness and hinge width shown in **Table 6-2.** This is based on setting the hinge ¼ inch from the back edge of the door as shown in **6-57.**

It is recommended that three hinges be placed on both hollow-core and solid core doors. The top hinge is located 7 inches from the top and the bottom hinge 10 inches from the bottom of the door. The third hinge is centered between the top and bottom hinges **(6-58).**

6-54 Steel and fiberglass doors are often hung with a hinge that has a round corner leaf that mounts on the door frame and a square leaf hinge that mounts on the door.

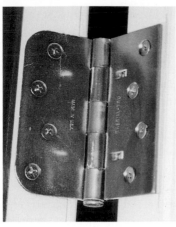

Hinge on Metal Door

Hinge on Fiberglass Door

Table 6-2 Selecting the Proper Width Hinge.

Door thickness (inches)	Hinge width when open (inches)	Clearance* of door from wall (inches)
1 3/8	3 1/2	1 1/4
	4	1 3/4
1 3/4	4	1
	4 1/2	1 1/2
	5	2
2 1/4	5	1
	6	2

* Based on providing the hinge with a mortise stopping 1/4 inch from the edge of the door. The specific clearance needed must be determined by taking into consideration the type and size of the door casing, which the door must clear.

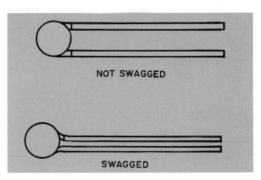

NOT SWAGGED

SWAGGED

6-55 Door hinges are swagged, meaning that the leaves are bent in such a way as to reduce the gap between the leaves.

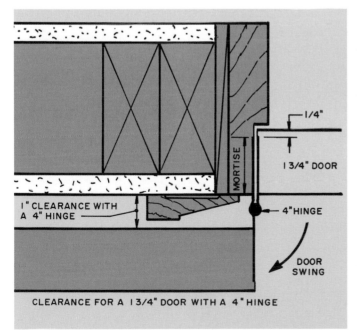

MORTISE

1/4"

1 3/4" DOOR

1" CLEARANCE WITH A 4" HINGE

4" HINGE

DOOR SWING

CLEARANCE FOR A 1 3/4" DOOR WITH A 4" HINGE

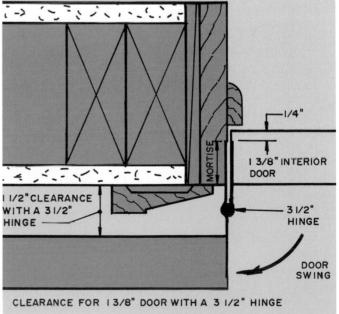

MORTISE

1/4"

1 3/8" INTERIOR DOOR

1 1/2" CLEARANCE WITH A 3 1/2" HINGE

3 1/2" HINGE

DOOR SWING

CLEARANCE FOR 1 3/8" DOOR WITH A 3 1/2" HINGE

6-56 When selecting the hinge to use, the door width and the required clearance must be considered.

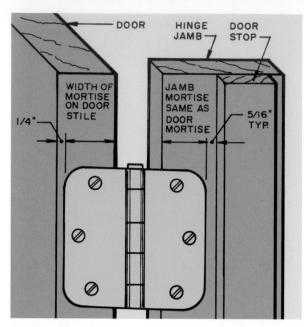

6-57 The hinge is set in ¼ inch from the back side of the door. The mortise on the jamb is the same width as the mortise on the door.

6-58 Typical locations for hinges, locksets, and dead bolts.

6-60 The door is held on its side with door holders. This enables the mortise templates to be installed on it. *Courtesy Carey Template Company*

Installing Mortise Hinges in the Door

With mortise hinges, one leaf is set in a mortise cut in the edge of the door (6-59). The other is set in a mortise in the jamb.

Finish carpenters will cut the mortise with a mortise template and an electric router. The door is held on edge by several metal door holders with the edge to be mortised facing up (6-60). A mortise template is clamped to the door (6-61). The opening is adjusted to the size of the hinge leaf. The electric router is lowered over the opening and the router bit cuts the mortise following the template (6-62). This gives the rounded corner needed by the most commonly used hinges. If a square corner is needed, the round is cut away with a wood chisel (6-63).

If you plan to cut the mortise manually, follow the steps in 6-64. Begin by marking the edges of the hinge on the surface and mark the thickness of the hinge on the face of the door and jamb. Bore a shallow hole on each corner. Typically a one-inch hole will be used. Check the hinge to be certain. Use a wood chisel to remove the wood in the leaf area, being careful not to cut too deep. If you do cut too deep put cardboard shims behind the leaf to bring it flush with the surface. If the hinge has square corner leaves, cut as shown in 6-63.

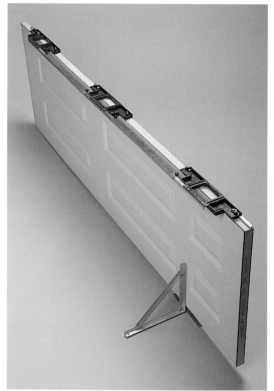

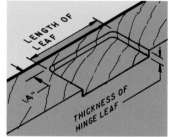

6-59 A mortise is a recessed area cut into the door and jamb into which the hinge fits, setting it flush with the surface of the wood.

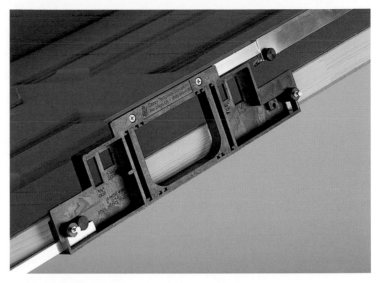

6-62 Install the required diameter bit in the router. Lower it onto the template and follow the sides of the opening. Set the depth of cut to the thickness of the hinge leaf. *Courtesy Carey Template Company*

6-61 Clamp the hinge mortise template to the door. Adjust it to produce the appropriate diameter of rounded corner. For square corners use a small-diameter router bit and clean up the corner with a chisel. *Courtesy Carey Template Company*

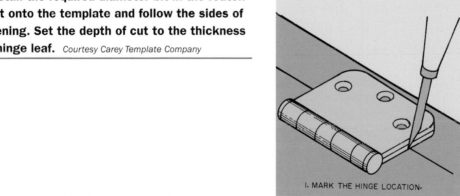

1. MARK THE HINGE LOCATION.

2. MARK THE DEPTH OF THE MORTISE. BORE HOLES AT EACH CORNER FOR THE ROUND HINGE CORNERS. CUT THE OUTLINE TO DEPTH.

CHISEL BEVEL FACING IN

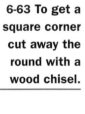 **6-63 To get a square corner cut away the round with a wood chisel.**

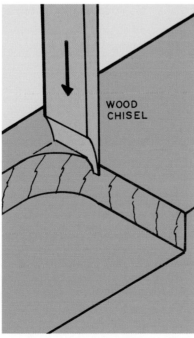

WOOD CHISEL

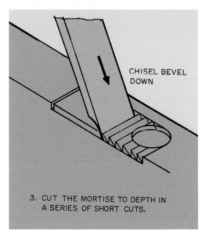

3. CUT THE MORTISE TO DEPTH IN A SERIES OF SHORT CUTS.

CHISEL BEVEL DOWN

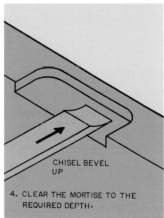

4. CLEAR THE MORTISE TO THE REQUIRED DEPTH.

CHISEL BEVEL UP

6-64 The steps to hand-cut a mortise for a round corner hinge.

Installing Mortise Hinges in the Side Jamb

After the mortises on the door are cut, install the hinges on the door. Be certain to install the hinges with the loose pin up or it will eventually fall out as the door is used. Temporarily tack the doorstop in place to support the door. Locate the stop by measuring in from the front of the jamb the thickness of the door plus ¹⁄₁₆ inch (6-65).

Put the door in the opening against the stops (6-66). Adjust it to the desired position. Wedges under it at the floor may help hold it at the desired height. Now mark the top and bottom edges of the hinge where it touches the side jamb (6-67). This locates the position of the leaf on the side jamb. The back edge of the leaf is kept ⁵⁄₁₆ inch from the stop (refer to 6-57). Mark the outline of the leaf on the jamb. You can cut the mortise as described for cutting the door mortise.

Now install the leaf on the jamb. Slide the door in place and insert the pins. Swing it carefully to see if it swings easily. Make adjustments if necessary.

6-65 To locate the doorstops measure in from the edge of the side frame. Temporarily tack in place.

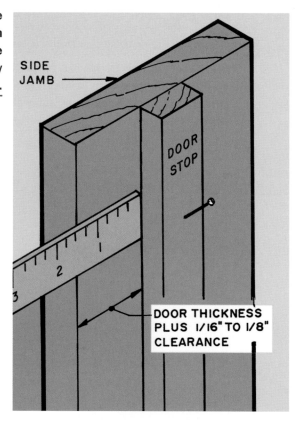

SIDE JAMB

DOOR STOP

DOOR THICKNESS PLUS 1/16" TO 1/8" CLEARANCE

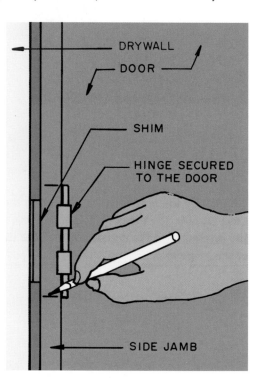

DRYWALL

DOOR

SHIM

HINGE SECURED TO THE DOOR

SIDE JAMB

6-67 Once the door is in the correct position, mark the top and bottom edges of the leaves of each hinge on the side jamb.

6-66 Place the door against the doorstops. Raise it to the desired height and mark the hinge locations.

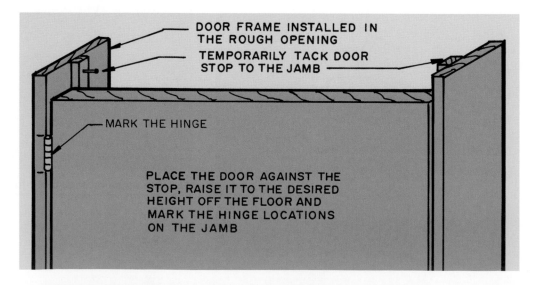

DOOR FRAME INSTALLED IN THE ROUGH OPENING

TEMPORARILY TACK DOOR STOP TO THE JAMB

MARK THE HINGE

PLACE THE DOOR AGAINST THE STOP, RAISE IT TO THE DESIRED HEIGHT OFF THE FLOOR AND MARK THE HINGE LOCATIONS ON THE JAMB

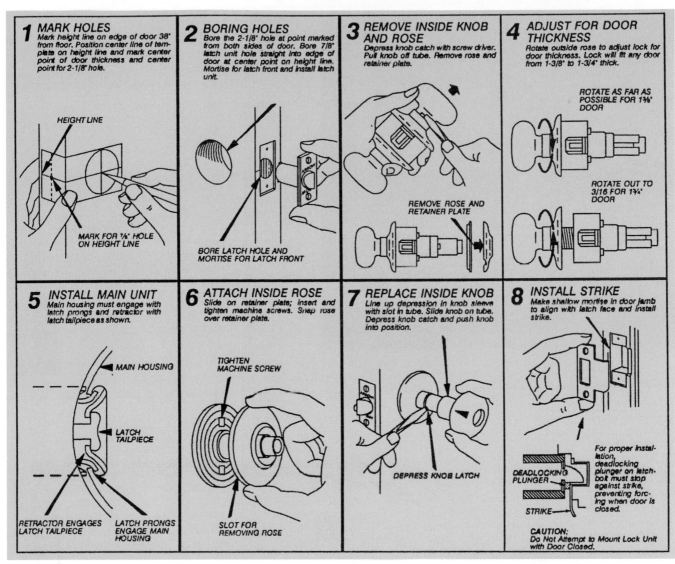

1 MARK HOLES
Mark height line on edge of door 38" from floor. Position center line of template on height line and mark center point of door thickness and center point for 2-1/8" hole.

HEIGHT LINE

MARK FOR 7/8" HOLE ON HEIGHT LINE

2 BORING HOLES
Bore the 2-1/8" hole at point marked from both sides of door. Bore 7/8" latch unit hole straight into edge of door at center point on height line. Mortise for latch front and install latch unit.

BORE LATCH HOLE AND MORTISE FOR LATCH FRONT

3 REMOVE INSIDE KNOB AND ROSE
Depress knob catch with screw driver. Pull knob off tube. Remove rose and retainer plate.

REMOVE ROSE AND RETAINER PLATE

4 ADJUST FOR DOOR THICKNESS
Rotate outside rose to adjust lock for door thickness. Lock will fit any door from 1-3/8" to 1-3/4" thick.

ROTATE AS FAR AS POSSIBLE FOR 1⅜" DOOR

ROTATE OUT TO 3/16 FOR 1¾" DOOR

5 INSTALL MAIN UNIT
Main housing must engage with latch prongs and retractor with latch tailpiece as shown.

MAIN HOUSING

LATCH TAILPIECE

RETRACTOR ENGAGES LATCH TAILPIECE

LATCH PRONGS ENGAGE MAIN HOUSING

6 ATTACH INSIDE ROSE
Slide on retainer plate; insert and tighten machine screws. Snap rose over retainer plate.

TIGHTEN MACHINE SCREW

SLOT FOR REMOVING ROSE

7 REPLACE INSIDE KNOB
Line up depression in knob sleeve with slot in tube. Slide knob on tube. Depress knob catch and push knob into position.

DEPRESS KNOB LATCH

8 INSTALL STRIKE
Make shallow mortise in door jamb to align with latch face and install strike.

For proper installation, deadlocking plunger on latch bolt must stop against strike, preventing forcing when door is closed.

DEADLOCKING PLUNGER

STRIKE

CAUTION: Do Not Attempt to Mount Lock Unit with Door Closed.

6-68 These are typical instructions provided with the lockset showing how to install it. Be sure to read through all of the steps that will be required before you actually mark or bore the holes and install the lockset.

Courtesy Arrow Lock Manufacturing Company

Installing Interior Locksets

If you are replacing a lockset, make an effort to buy one that fits the existing openings in the door. It is very difficult to try to fill in or alter existing openings.

If you are installing a new door, the new lockset will come with detailed installation instructions as shown in **6-68.** It is possible that the door you bought may already have holes pre-bored for the lockset. If this is the case you simply have to install the new lockset. If the holes were not prebored you will first need to locate them. Use the template supplied by the lockset manufacturer to locate the centers of the holes on the edge and face of the door. This will also show the required diameters.

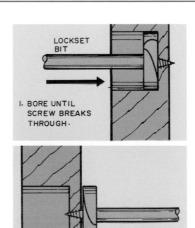

LOCKSET BIT

1. BORE UNTIL SCREW BREAKS THROUGH.

2. COMPLETE THE HOLE BY BORING FROM THE OTHER SIDE.

6-69 Bore the lockset hole from one side until the screw breaks through. Then finish boring the hole from the other side.

6-70 After the lockset hole has been bored, locate and bore the latch hole in the edge of the door.

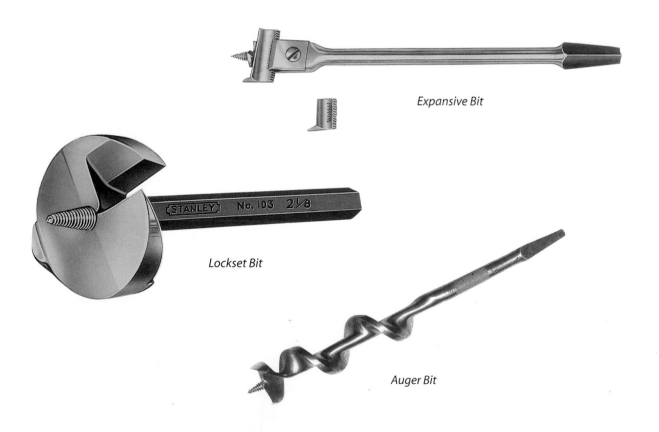

Expansive Bit

Lockset Bit

STANLEY No. 103 2⅛

Auger Bit

6-71 Bits used to bore the holes for the lockset and latch.

Generally the center of the lockset is located 36 inches above the bottom edge of the door.

Boring the Holes

When you bore the face hole that goes all the way through, do not bore it through from one side because it will split out the wood on the other side. First bore in until the screw breaks through the other side. Then complete the hole by boring from that side (**6-69**). Next bore the hole in the edge of the door. A finished job is shown in **6-70.** Notice the hole in the edge of the door has a mortise cut around it. This sets the end plate of the lock latch flush with the surface. The large-diameter holes are usually bored with a lockset bit or an expansive bit. The end hole is made with an auger bit (**6-71**). Now install the lockset as directed by the manufacturer's instructions (**6-72**).

1. Place the latch into the hole in the edge of the door.

3. Place one half of the lockset into the largest hole.

2. Secure the latch with screws.

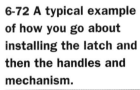

6-72 A typical example of how you go about installing the latch and then the handles and mechanism.

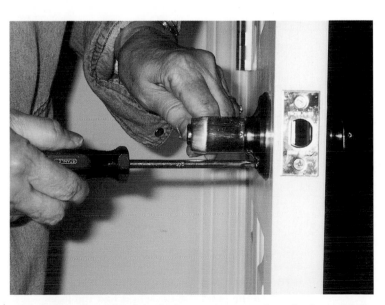

4. Insert the other side of the lockset and connect it to the first with the bolts provided.

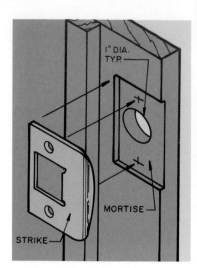

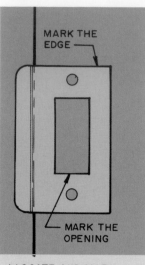

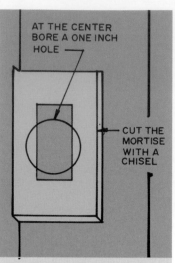

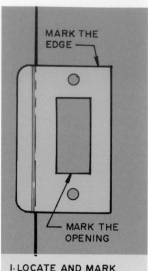

6-73 The strike is set in a shallow mortise so it is flush with the face of the jamb.

6-74 Locate the strike on the jamb, bore the hole for the latch, and cut the mortise for the strike.

6-75 This nonmortise hinge is used on bifold doors. It does not require a mortise and leaves only a narrow space between the closed doors.

6-76 This non-mortise hinge can be used to hang doors to the jamb. Be certain it has the load-carrying capacity to carry the door you plan to use.

Placing the Strike

After the lockset has been installed in the door it is time to locate and install the strike. The strike is installed on the door jamb. It is located opposite the faceplate on the latch and set so the latch enters the center of strike opening. The strike is set in a mortise so it is flush with the face of the jamb (6-73).

To locate the strike, place it over the latch and carefully close the door. Be careful you do not damage the edge of the jamb. Hold the strike in place so the latch is centered in it and mark the top and bottom edges. Open the door and place the strike on the marks and draw around it and the latch opening (6-74). Cut a mortise so the strike is flush with the surface of the jamb. Then bore a large-diameter hole in the center of the opening. It should be deep enough so the latch does not hit the bottom. Install the strike. Close the door and see if the latch enters the strike and the door is held closed with a minimum of movement. It may require some adjustment to get the desired fit.

Nonmortise Hinges

Lightweight doors such as those used on bifold doors often use nonmortise hinges (6-75 and 6-76). One leaf fits into an opening in the other leaf. This reduces the thickness greatly and leaves a space between the doors equal to the thickness of only one leaf.

CHAPTER 7

Installing Interior Trim

Whether you are remodeling or planning new construction, the choice of interior trim will directly influence the style and character of a room. If the house is an old classical style, the trim you replace or install should reflect the moldings used during that period. This can include the use of pilasters, arches, pediments, and large multipiece crown moldings found in 1700s and 1800s classical English and Georgian houses (7-1). Other periods, such as the Federal style, relied on architecturefrom ancient Greece,while later, Victorian-era houses had elaborate baseboard and crown moldings (7-2).

Modern America has in general used smaller moldings of a simpler design, yet these were influenced by various architectural movements (7-3). The demand for less expensive houses that are built faster has led to even simpler moldings and less interior trim. Basically it involves covering the cracks around windows and doors and finishing the wall at the floor. This does not give the house a finished appearance of beauty, character, and substance.

7-1 This pilaster and wood paneling reflects interior trim used in early English and Georgian houses.

7-2 Elaborate multipiece crown molding cornices were popular in the Victorian era.

7-3 These cherrywood French doors are the focal point of the room. As is typical of today's houses, narrower and simpler interior trim moldings have been used. *Courtesy Weather Shield Manufacturing Company*

7-4 The chair rail provides the opportunity to use related but distinctly different wall finishes above and below it. *Courtesy Mr. and Mrs. James Sazama*

7-5 This wide baseboard wrapped around the base of the fluted pilaster adds stability to the wall and is very attractive.

Character-Building Applications

Chair rails not only protect the wall from damage, they also provide a way to divide the wall finish, providing a base area below the molding and a wall area above that can be decorated to complement the character of the room **(7-4)**.

A wide **baseboard** not only blends the wall to the floor but provides a decorative base for the wall **(7-5)**. It adds a lot to use **crown molding** to provide a transition between the wall and ceiling **(7-6)**. **Wall frames (7-7)** and **wainscoting (7-8)** form a major decorative feature of a room. A fireplace surround or mantel can enhance an otherwise rather bland fireplace **(7-9)**.

Trim can also be used to change the character of a room. For example, a chair rail or wainscot can make a small room seem larger by placing emphasis on the horizontal. A room with a low ceiling can be made to appear a little higher by using vertically applied moldings such as pilasters and paneling with a vertical emphasis (refer to **7-1**). Ceiling beams give emphasis to the length of the room and add warmth and coziness **(7-10)**. If you have a high ceiling, consider the use of columns, which will add a touch of class. A **niche** can be used to enhance a bland stretch of wall **(7-11)**.

7-6 A simple crown molding is an effective way to blend the wall and ceiling intersection.

7-7 Wall frames are made by mounting molding on the wall. Typically the area inside the frame is decorated in a manner different from the wall. These are polymer moldings. *Courtesy Architectural Products by Outwater, LLC, 1-800-835-4400*

7-8 Wainscot is installed on the lower wall area. It provides the major architectural feature of the room. This is polymer wainscot. *Courtesy Architectural Products by Outwater, LLC, 1-800-835-4400*

7-9 A wood fireplace surround using a number of moldings of different species of wood and an overlay form the focal point of the room.

7-10 Wood ceiling beams add warmth and a rustic appearance to a room.
Courtesy Mr. and Mrs. James Sazama

7-11 A niche recessed in the wall becomes a focal point for displaying a sculpture, vase. or other art object. This is a polymer niche. *Courtesy Architectural Products by Outwater, LLC, 1-800-835-4400*

Selecting a Molding

The first decision is whether the molding will be painted, finished natural, or stained. When remodeling a classical old house, it is necessary to match the existing trim—which may turn out to be an expensive proposition. It could require custom-milled profiles and an expensive wood, such as cherry or walnut.

When building a new house, you have a lot more freedom, with regard to the appearance desired and the cost. If using stock moldings you have a choice of profiles and sizes and the possibility of combining several to produce a more detailed molding. You will be limited on choice of materials, though.

The best overall appearance is produced if all the trim in a room represents one style. Mixing a colonial trim and something with a Greek or Roman origin does not produce the best results.

If all the trim in the room is new, you can check the thicknesses before you buy to be certain that where pieces butt together, they will be flush on the surface. If it is a remodeling job and only some of the trim is replaced, take a piece of the old trim with you as you select the new material. A typical problem occurs where a base butts the casing on a door. This will also let you be certain the new trim has the same profile as that which will remain in the room.

Before you actually buy the trim be certain how it will be finished. If it is to be painted the choice is easy. If a clear or stained finish is desired, the species of wood must be chosen—and don't forget to get the cost. If using a polymer molding get the manufacturer's finishing instructions.

Wood Trim

While wood trim is available in a number of species, frequently the choice at the local building supply dealer will be between a soft wood and oak. Manufacturers will produce molding in various hardwoods by special order.

Availability is another factor to consider. The local building supply dealer will most likely have a limited supply of stock wood moldings. Ascertain the possibility of getting more detailed profiles of hardwoods. These will have to be ordered. What is the chance that the dealer can actually secure these by the time you need to install them? What will they cost?

Softwood trim is easier to cut and nail. Some hardwoods require a pilot hole to be drilled for each nail. Power nailers make nailing both types of wood easier (**7-12**).

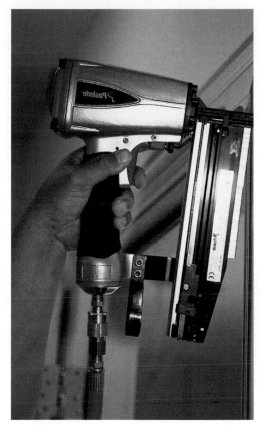

7-12 Trim is rapidly installed using a power nailer. It is also less likely to split the molding.

Courtesy Paslode, An Illinois Tool Works Company

Open-grain hardwoods, such as oak, will require that a paste wood filler be applied to fill the pores before it is finished. Also consider the location. If the trim is to be in an area of high humidity, use wood species that are less affected by moisture. Birch, cherry, and poplar have low checking and warping properties, whereas beech, chestnut, and oak have high checking and warping properties. Moldings made from polymers are not affected by moisture and serve well in moist locations.

STOCK & CUSTOM WOOD TRIM

One choice to be made as the trim is chosen is whether the stock moldings available will produce the image you desire or if they must be custom made to your design. In some cases a finish carpenter will have the equipment and knowledge to custom-make the trim for you. Local woodworking shops can also produce moldings. Profiles for a few of the wood stock moldings commonly available are in **7-13**.

If stock moldings available appear satisfactory, profiles can be combined to produce a larger, more detailed molding (**7-14**). You can also combine a custom molding that has a special feature with one or more stock moldings.

Manufacturers produce a wide range of special and classical wood moldings (**7-15**). These provide that

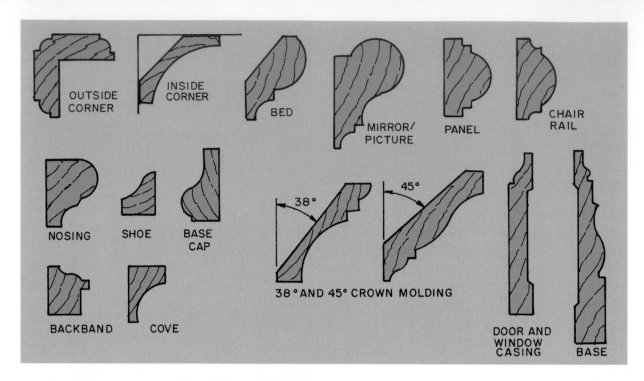

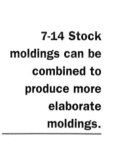

7-13 Example of the commonly available stock moldings.

7-14 Stock moldings can be combined to produce more elaborate moldings.

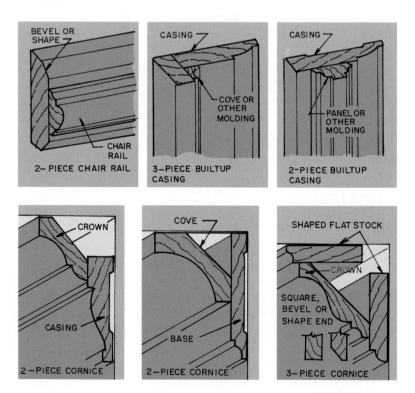

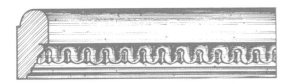

[A] Picture & Mirror Hanging Molding

[B] Chair Rail

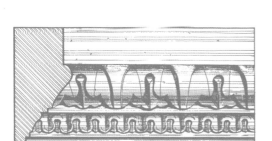

[C] Door Trim

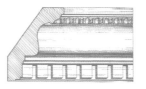

[D] Crown Moldings

7-15 These are just a few of the many high-quality, hand-carved moldings available for classic homes. They are available in hardwoods such as walnut, mahogany, cherry, oak, and poplar. *Courtesy Architectural Paneling Inc.*

special look for the room and are used to trim the room in the classic style of a past era. They are made from select hardwoods such as walnut, mahogany, cherry, poplar, and oak.

Wood trim that is to be painted must be made from a wood that machines and sands to a smooth finish. Pine is probably the most commonly used wood. Both white and yellow pine are available in various sections of the country. Poplar, a hardwood, is also available because it is inexpensive and wears well. In western states fir is often used for paint-grade, natural finish, and stain-grade molding. Paint-grade molding may have minor blemishes requiring the painter to caulk them, set nails, and even caulk poorly fitting joints.

Hardwood molding requires a clear surface and perfectly fitting joints.

A paint-grade pine molding made of finger-jointed short pieces is less costly than clear single pieces. When sanded carefully and painted, the joints are seldom seen. Finger-jointed wood molding is also available covered with a solid-wood veneer or a wood-grained vinyl sheet. These are cut and installed the same as clear wood. However, care must be taken so the veneer or vinyl covering is not damaged (**7-16**).

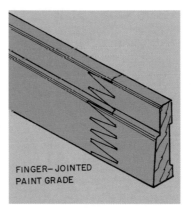

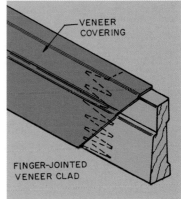

7-16 Finger-jointed wood molding is good for trim that will be painted. Veneer-and vinyl-covered molding present the look of solid hardwood products.

Hardwood trim is often available in oak, poplar, cherry, walnut, and mahogany. However most molding manufacturers will machine out molding in any type of hardwood you choose. Obviously, these are beautiful but more expensive than paint-grade molding. In addition, the manufacturer will produce custom profiles in any of these woods.

If you are going to machine the trim yourself, a good grade of lumber is necessary. It must be straight, kiln dried for interior use, and free of knots, pin holes, splits, and checks.

WOOD TRIM GRADES

Trim grades for wood products indicate the quality of the material. While the terms used to identify each grade may vary across the country, basically they fall into two broad categories: **paint-grade** and **stain-grade.** Each grade is then divided into several subcategories. The stain-grades are high-quality products generally reserved for interior trim. They are more expensive than paint-grades. If you can find a top quality paint-grade it might be stained for use in some rooms and reduce the cost. Basically for top-quality work areas where clear or stained trim is desired, it is best to use stain-grade trim.

BUYING WOOD MOLDING

Often the moldings ordered from the local building supply dealer will have pieces that are in a condition not as good as you expected. They are often stored in outbuildings where there is no moisture control. They get banged around as people look through the supply, put them on the truck, and deliver them to the site, so expect some damage. If you go to the dealer and are permitted to select the pieces you want, check to see that they are not warped or twisted. Place the molding in the house after the drywall has been taped and dried. Place it flat on the floor and put sticks between the pieces of molding so air can circulate. Leave the molding there a week or so to let it reach the moisture equilibrium content of the air in the house. The house should be closed and if possible the heating/cooling system operational to control moisture in the air.

Since the actual size of the molding may vary somewhat from the given size, it is wise to try to buy all the material at once so the molding received was all run at the same time and most likely will be the same size.

Plastic Moldings

Interior trim is available molded from polymers. Typically, fiberglass-reinforced polyester and extruded polystyrene are used. They are available with a factory-applied prime coating ready for painting or prefinished with an artificial wood-grain pattern. If painted, use a good-quality oil-based or latex paint. They are never coated with a lacquer-based product. Plastic moldings are available in complex, detailed profiles representing many of the classical moldings built of wood in early buildings. Since it is a single piece it is easier to install than a multipiece wood molding of the same design. Rigid polymer moldings are lightweight and cut with regular woodworking tools. They are bonded to the wall with an adhesive supplied by the manufacturer. Sometimes they are held with a few nails until the adhesive cures. While sometimes more expensive than wood, polymer moldings save considerable time for installation.

BUYING PLASTIC MOLDING

Plastic molding is stocked by the local building supply dealer. As is true with wood, the sizes and profiles available are extensive and not every available molding will be in stock. Give the dealer advance notice so the desired profiles can be ordered from the plastic molding company. Plastic molding is available in lengths from 6 to 16 feet, depending on the profile and manufacturer.

FLEXIBLE MOLDING

Some types of plastic molding are very flexible and may be bent to the desired contours. They are easily formed around arched openings and large-diameter openings and large-diameter columns. They are easily installed as baseboards, crowns, and casings on walls having rounded surfaces.

Some types are more flexible than others, so consider this factor when making a selection. They are nailed with pneumatic nailers and are hard enough so that the surface is not damaged. They can be cut with a saw or a knife. Some are installed with a manufacturer-supplied adhesive. They are available in stain-grade and paint-grade and with an embossed wood grain.

Medium-Density Fiberboard Molding

Medium-density fiberboard (MDF) is made from a mixture of wood fibers, wax, and a resin binder and is available in a range of molding types and profiles. MDF is compressed into the molding profile under high pressure. Then the finished molding is factory primed, making it easy to paint. It produces a nice, finished appearance.

MDF is very hard and should be cut with carbide-tipped saws. When being cut, it produces a fine sawdust, so be sure to wear a respirator to filter what you breathe in and some goggles to protect your eyes.

Since it is so hard it cannot be hand-nailed. A pneumatic nailer is required, or it can be secured with screws. When nailed it tends to push up around the nail and this has to be chiseled or sanded smooth before the hole is caulked.

Estimating Trim Quantities

Base, shoe, ceiling molding, and chair rail quantities are determined by finding the lineal feet in the perimeter of the room plus any related areas, such as a closet or foyer, and adding 10 percent for waste. Do not deduct anything for door openings. Moldings are available in lengths typically from 6 to 16 feet. Buy the longest pieces possible.

Quantities of door and window casing can be figured by calculating the lineal feet around each opening and adding 20 percent for waste. It is possible to buy the casing in precut kits ready to install with no cutting.

When to Paint Walls & Trim

When to paint the walls and trim is another decision. While it is best to prime the wood trim on the front and back before installation, this is often not done. It can be painted or sealed in the garage or a room that has only the subflooring exposed.

Some prefer to paint the interior walls before the trim is installed. Then the painters return and paint the trim or finish the natural or stained trim. This is a bit slower than installing the trim to be painted before painting the walls, but generally produces a neater job. If the walls and trim are painted after the trim has been installed, the latex paint on the walls is often sprayed or rolled on, which requires the trim be masked so it is protected. Then the trim is painted with some form of enamel. Sometimes the trim has the final finish coat applied before it is installed. After it has been installed the nails are set and the holes are filled and touched up with the paint. There may also be chips at the miter joints and areas needing caulking, which have to be painted by hand.

Making Your Own Molding

Begin by choosing the best quality wood in the species to be used. It should be straight, kiln dried for interior use, which is generally 6 to 8 percent moisture, and as free from pin holes, splits, and checks as possible. Remember, paint-grade trim can have minor defects filled before painting. This can be done with hardwoods but it is difficult to get the filler to match the final color of the stain.

7-17 A heavy-duty router can be used to shape moldings. *Courtesy Porter-Cable Corporation*

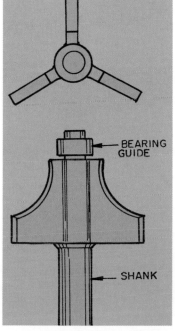

7-18 A typical router cutter.

BEARING GUIDE

SHANK

Moldings can be made using your router or shaper. If you use a **router,** a heavy-duty unit having 1½ to 1¾ horsepower is recommended (**7-17**). Routers also come with bits having a ½-inch shank, which is less likely to vibrate, bend, or break under heavy, continuous loads (**7-18**). A router will produce its best work if you mount it on a router table. In this position it operates like a wood shaper (**7-19**).

A **wood shaper** is a heavy-duty tool and is best to use if you plan to run a lot of molding in your home workshop (**7-20**). Since it has a more powerful motor, it will make deeper cuts than the router, reducing the number of passes necessary to shape the molding.

In both cases be certain you understand how to use these machines and observe all safety recommendations. Wear eye protection and respiratory protection. Some also prefer to wear earplugs.

7-19 When the router is mounted on the router table it operates like a wood shaper.

Courtesy Porter-Cable Corporation

7-20 A wood shaper is used to shape the profile of wood members. *Courtesy Delta International Machinery Company*

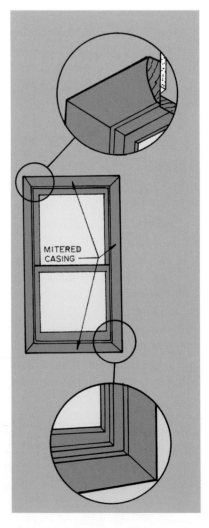

MITERED CASING

7-21 This mitered window casing is installed on all four sides of the window.

Window Casings

The most commonly used style of window casing is some form of mitered frame. The windows can be framed with casing on all four sides (**7-21**) or on three sides with a stool and apron at the bottom (**7-22**).

Another style uses rectangular casing and joins the side and casing in a butt joint. In **7-23** are several variations. One type has a flat head casing extending beyond the side casing. Another type uses some form of molding to enclose the trim. Rosettes are typically used when a house has a traditional style.

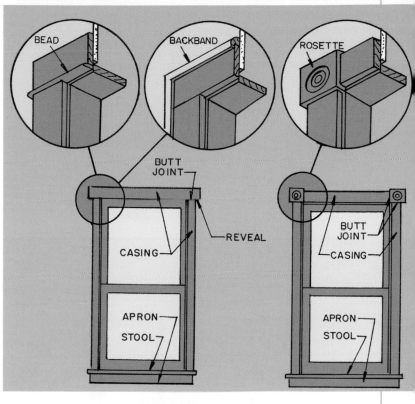

7-23 Butted window casing can have the head casing decorated with a molding. Rosettes also add to the appearance.

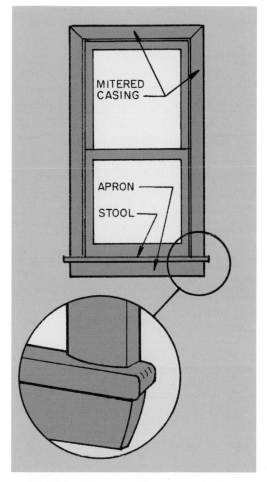

7-22 This mitered window casing is on three sides of the window and has a stool and apron on the bottom.

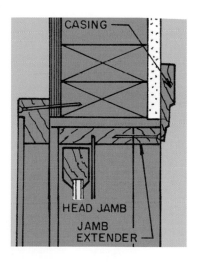

7-24 If the window jamb is below the drywall, a jamb extender will be added to bring it flush.

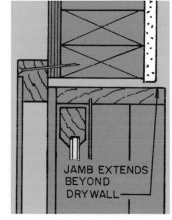

7-25 If the window jamb extends beyond the drywall it will have to be planed smooth.

117

7-26 If the drywall extends a little beyond the window frame remove it with a Surform tool.

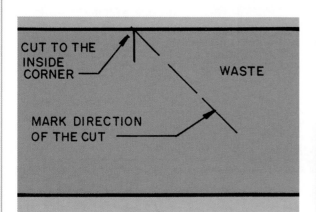

7-27 This high-quality miter box will produce accurate, smooth cuts. © 2003 Stanley Tools

Preparing to Install the Window Casing

Begin by checking the window frame to see if it is level, plumb, and square. If it is not, it will take some corrections or force you to cut miters that are not at a 45-degree angle. Check to see that the window opens easily. Now is the time to fix that before the casing is installed. Be certain the window jambs are flush with the surface of the drywall. If they are below the surface, a wood strip will have to be added (**7-24**). This strip is called a **jamb extender**. If the jamb extends out past the drywall, it will have to be planed flush (**7-25**). If the drywall is just a small amount past the frame, it can be smoothed back with a Surform tool (**7-26**).

If the space between the window frame and the side studs is open, fill it with insulation.

Cutting the Miter

Miters are 45-degree cuts on the ends of butting casing, which, when brought together, form a 90-degree corner. They can be cut with a hand-powered miter box or a power miter saw.

USING A HAND-POWERED MITER BOX

A good quality hand-miter box will produce accurate cuts (**7-27**). The saw has a scale in degrees. Line up the blade with the angle you want and lock the saw in this position. Place the casing with the mark indicating the inside corner of the miter so the saw

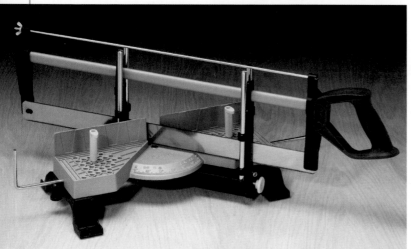

7-28 Mark the inside point of the miter and lightly sketch a line showing the direction of the cut.

7-29 This inexpensive miter box has a plastic case with slots permitting the cutting of miters on 45 degrees and 90 degrees and face angles at 45 degrees and at 22½ degrees.

trims the edge of the mark. Note if the cut is to the right or left of the mark. This is usually sketched on the casing with a light pencil mark (**7-28**). Move the saw across the stock but do not press down or try to force it to cut faster. The smoothest cut occurs when it goes at its own rate.

There are small, plastic miter boxes that cut on 90- and 45-degree angles and make a 45-degree bevel cut. They use a small, inexpensive back saw. They are fine for small moldings (**7-29**).

USING A POWER MITER SAW

A power miter saw will cut the casing accurately and rapidly from 45 degrees left to right. The point marked 0 degrees is actually 90 degrees with the fence. A typical scale is shown in **7-30**. If you want to cut a 45-degree angle, set the saw on the 45-degree mark.

If you want some other angle, such as 60 degrees, move the pointer to the 30-degree mark. Since the 0 mark is 90 degrees, you have to subtract the angle wanted, 60 degrees from the 90-degree setting on the scale.

There are a number of other settings and adjustments necessary when using a power miter saw. Carefully read the manual that comes with the machine. The manual will also have an extensive list of safety procedures which you need to observe.

Once the saw is set on the required angle, place the casing on the table and hold it against the fence. Place the side with the mark indicating the inside corner of the cut against the fence. Check to see that the saw is set to cut the angle in the desired direction. Turn on the power, let the blade reach full speed, and lower the saw toward the casing. Move the casing until the saw will cut at the corner mark (**7-31**). Keep the hand holding the casing against the fence well away from the cutting area. Wear eye protection. When the cut is finished raise the saw, turn it off, and let the blade stop rotating. Some saws have a braking mechanism. Always keep the guard over the blade. Do not reach in and try to remove scrap pieces of wood while the

7-30   The miter scale on a typical power miter saw. It is set on zero, which is 90 degrees to the fence.

7-31   A completed miter cut on a power miter.

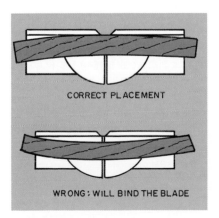

CORRECT PLACEMENT

WRONG: WILL BIND THE BLADE

7-32   If the stock is bowed, place the crown of the bow against the fence of the miter saw.

blade is rotating. Wait until the saw is in the up position and the blade has stopped rotating.

Another danger to watch out for is when cutting casing that is bowed. Badly bowed stock should be discarded. If it is necessary to cut a bowed piece, place the bow against the fence as shown in **7-32**. If the bow is away from the fence the stock may pinch the blade, causing it to kick back the material.

Detailed information on the safe use of power woodworking tools is available from The Power Tool Institute, 1300 Summer Ave., Cleveland, Ohio 44115-2851.

ADJUSTING THE MITERS

The miters will not always fit tightly on the first try, and the casing will have to be removed and the miter adjusted. In **7-33** are the types of corrections that will

need to be made. The top drawing shows a miter that has no gaps and is square. This is how all miters should appear, but because of variances in the wall surface or the window frame, some adjustments to the miter often need to be made. The casing must always keep the reveal uniform on all sides of the window.

The second drawing (**7-33**) shows a miter with the toe touching, but open at the heel. The third shows the miter touching at the heel and open at the toe. In these drawings the amount of gap is greatly exaggerated; in actual practice the gap will be much smaller. Usually the gap is in the range of $\frac{1}{64}$ to $\frac{1}{32}$ of an inch.

CLOSING A GAP IN THE MITER

The gap in a miter may be closed by removing wood from one side of the miter. This may be done several ways. One way is to place the casing on the miter saw,

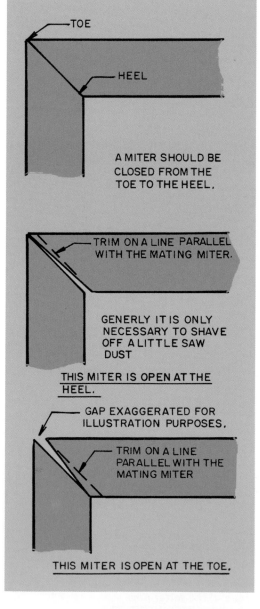

7-33 Typical problems that occur as you try to install mitered casing. The amount of the opening shown is exaggerated. Typical openings will be much smaller but must be corrected.

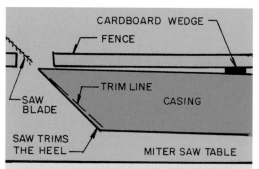

PLACE THE WEDGE AWAY FROM THE BLADE TO CORRECT A GAP AT THE TOE.

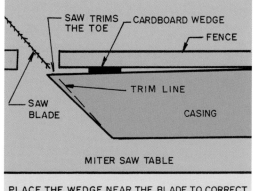

PLACE THE WEDGE NEAR THE BLADE TO CORRECT A GAP AT THE HEEL.

7-34 A miter can be adjusted by trimming a bit of sawdust off the toe or heel as needed to make the miter close.

place a small wedge between the casing and the fence, and make a very fine cut across the miter. All you will get will be some sawdust off of the end to be lowered (**7-34**). The amount to remove is judged by examining the joint and observing the size of the gap. Usually a thin piece of cardboard gives enough change in the angle of the miter to close the joint. Check the joint after cutting, and remove more if needed.

Another way to lightly trim a miter is to use a block plane, as shown in **7-35.** Plane down the slope. Slant the plane a little across the surface, and move it with a slicing motion. Take very light cuts, and be certain the plane is very sharp. Caution must be exercised to never slant the surface to the front; however, it can be slanted a little to the back surface. This will actually help close the miter. Often a miter will have par-

allel edges, but will not close. This is an indication that the surfaces of the miter slant toward the front. Remove some wood off the back edge until the miter closes.

Installing a Picture Frame Mitered Casing

A picture frame casing has the casing installed on all four sides with the corner usually mitered (refer to **7-21**).

Begin by marking the **reveal** on the corners of the casing. Typically a reveal is ³⁄₁₆ or ¼ inch. A reveal is the amount the casing is set back from the edge of the frame. An easy way to do this is to set a combination square with the blade extending out the amount of the reveal (**7-36**). Make a pencil mark on each corner and several between the corners (**7-37**).

7-35 When adjusting a miter with a block plane, plane it on a steady surface, on a slight angle across and down the slope of the mitered surface.

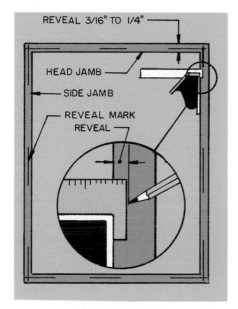

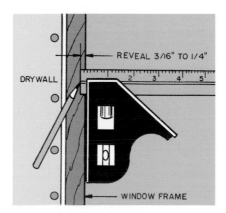

7-36 The reveals can be marked using a combination square. Reveals are typically ³/16 inch.

7-37 Mark the reveal at each corner and along the jambs between the corners.

A picture frame miter casing has two lengths of casing. The length of each can be measured on the short or long side of the casing. In **7-38** the lengths are taken to the **short side** of the casing. This includes the distance inside the jambs plus twice the reveal. When laying out the head casing, mark the point of intersection of the reveal lines. Then lightly mark a pencil line indicating the direction of cut. Do not press so hard that you score the surface of the casing. This helps you cut in the proper direction when you get to the miter saw (**7-39**).

Picture frame mitered casings can be installed in different ways. Following is one suggested procedure. Start the installation by nailing the head casing to the jamb edge with 3d or 4d finishing nails. Do not set the nails firmly, because it may be necessary to adjust the casing to get a tightly closed miter. Be cer-

tain it fits flat to the surface of the drywall. If it does not, it may be necessary to file off some of the drywall or plane a little off the jamb. Be certain the casing lines up with the reveal marks; it should just cover the marks (**7-40**).

Now install the side casings, making sure that they line up with the reveal marks and that the miters close. If this is the case, nail them to the frame. Check again to be certain the miters are closed after nailing. If the miters do not close, adjust them as described in the next paragraph. Finally, cut and install the sill casing. You can measure the actual length required, or cut a miter on one end, place the stock across the bottom, and mark the location of the other end. Cut and install the bottom casing. If all miters are closed, finish nailing by placing nails about 8 to 10 inches apart along the edge of the frame. Then using 6d or 8d fin-

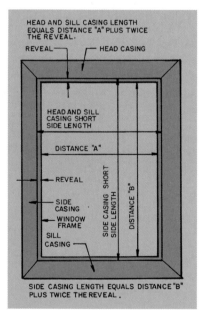

7-38 The short side of the casing is equal to the distance between the jambs plus twice the reveal.

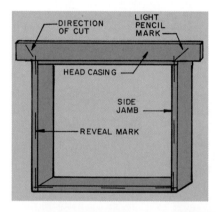

7-39 To lay out the head casing on the short side, mark the point of intersection at each corner. Lightly sketch a line showing the direction of cut.

7-40 A typical procedure for installing a picture frame casing.

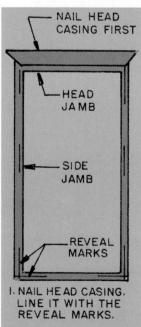

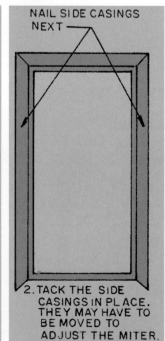

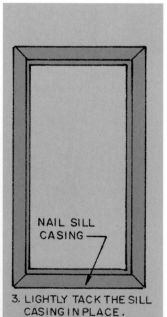

ishing nails, nail the thick side of the casing to the header or stud. Suggested nailing patterns are shown in **7-41.** When installing hardwood casing it may be necessary to drill small holes for each nail. They should be a little smaller than the diameter of the nail.

The use of a power nailer greatly speeds up the work. It also leaves one hand free to hold the casing in place (refer to **7-12),** and is less likely to split the casing.

Installing the Stool & Apron

Windows can also be trimmed using a stool and apron at the sill. Typical stool profiles are shown in **7-42.** They are available as stock material from building materials suppliers. In some cases the window manufacturers have stool material for their specific window units available. A stool with an angled bottom surface is used on windows made with a sloping sill.

FITTING THE STOOL

First cut the stool to its finish length. This includes the width between the window side jambs, plus two times the reveal, plus two times the width of the casing, plus two times the overhang. The **horn** is equal to the width of the casing plus the reveal and any extension allowed beyond the casing. This is typically ½ to 1 inch. It is simply decorative so can be any distance desired **(7-43).** Then measure the window unit to see how deep a notch is needed to form each horn, and lay this out on the stool. Cut the horn notches, and fit the stool to the window frame. Trim and adjust, as necessary, to get the required fit as shown in

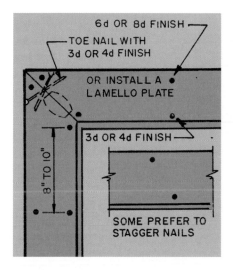

7-41 A suggested nailing pattern for installing casing. Some prefer to stagger the nails. Notice how the miters can be fastened.

6d OR 8d FINISH
TOE NAIL WITH 3d OR 4d FINISH
OR INSTALL A LAMELLO PLATE
3d OR 4d FINISH
8" TO 10"
SOME PREFER TO STAGGER NAILS

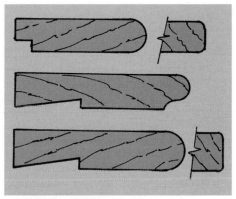

7-42   Typical stock stools.

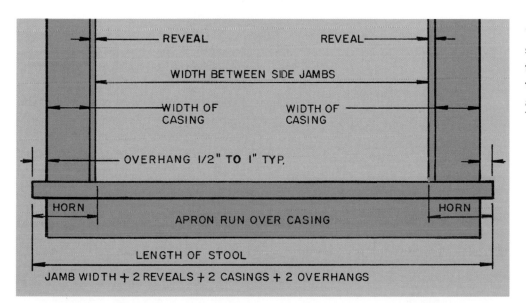

REVEAL — REVEAL
WIDTH BETWEEN SIDE JAMBS
WIDTH OF CASING   WIDTH OF CASING
OVERHANG 1/2" TO 1" TYP.
HORN   HORN
APRON RUN OVER CASING
LENGTH OF STOOL
JAMB WIDTH + 2 REVEALS + 2 CASINGS + 2 OVERHANGS

7-43   A layout showing how to find the length of the stool, apron, and horn.

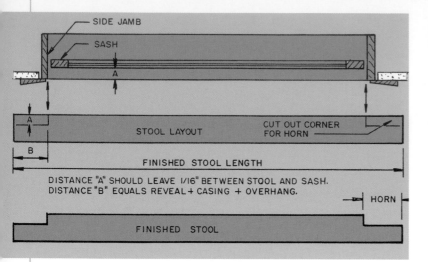

STOOL LAYOUT

CUT OUT CORNER FOR HORN

FINISHED STOOL LENGTH

DISTANCE "A" SHOULD LEAVE 1/16" BETWEEN STOOL AND SASH.
DISTANCE "B" EQUALS REVEAL + CASING + OVERHANG.

HORN

FINISHED STOOL

7-44 A typical layout for a window stool. This may vary a bit, depending upon the window being used.

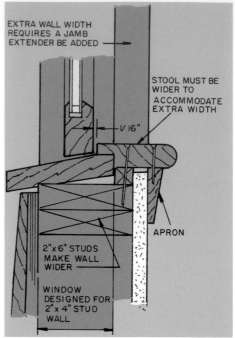

EXTRA WALL WIDTH REQUIRES A JAMB EXTENDER BE ADDED

STOOL MUST BE WIDER TO ACCOMMODATE EXTRA WIDTH

1/16"

APRON

2"x 6" STUDS MAKE WALL WIDER

WINDOW DESIGNED FOR 2"x 4" STUD WALL

7-45 The stool must be wider when the wall width requires added. Nail the stool to the sill and apron. When possible, nail into the rough 2-inch sill framing below.

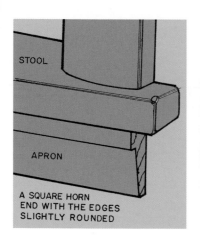

STOOL

APRON

A SQUARE HORN END WITH THE EDGES SLIGHTLY ROUNDED

STOOL

APRON

A HORN WITH A ROUNDED END

7-46 The horn can be sloped to provide the appearance desired.

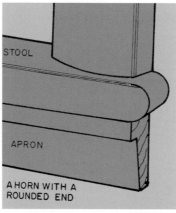

GLUE 8 NAIL

STOOL

SIDE GRAIN

APRON

A MITERED RETURN

7-44. Be certain the stool is level. If it is not, it may be necessary to shim it. The stool must be $\frac{1}{16}$ to $\frac{1}{8}$ inch clear of the sash.

If the window needs a jamb extension, some prefer to install the stool first and then add a jamb extension. Others prefer to add the jamb extension and cut the stool to it (**7-45**).

After you are satisfied with the fit of the stool, consider shaping the ends of the horn (**7-46**). They can be left square, as cut, if you sand out the saw marks. Some prefer to round the horn or shape a profile. The stool with the mitered return provides face grain on the exposed end. The other two expose end grain. Face grain is especially nice when the stool is a hardwood and a smooth return surface is desired.

Now nail the stool in place. Nail through the stool into the rough sill (if possible) using 8d finishing nails. After the apron is installed, nail the stool to it with 6d finishing nails (refer to **7-45**). Since the actual design of window frames varies, the stool may have to be nailed into the sill below it. On some windows, it is not possible to install a stool.

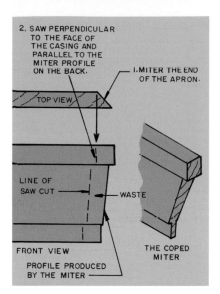

7-47 The end of the apron can be coped to reflect the profile of the casing.

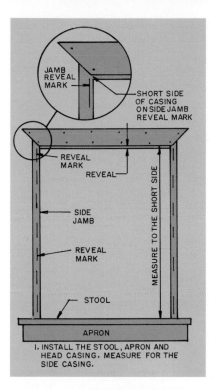

7-49 Install the stool and apron before the head and side casing.

7-48 When hardwood casing is used many prefer to miter the end of the apron and attach a return piece to cover the end grain.

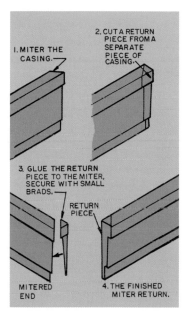

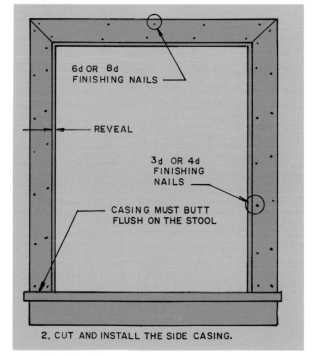

INSTALLING THE APRON

The **apron** is usually made from the same material that is used to trim the window. It is placed below the stool, with the thick edge up against it (refer to **7-45** and **7-46**). The apron covers any opening between the sill and the drywall. The length of the apron is usually made the same as the overall width from one side of the window casing to the other. This permits the stool horn to extend beyond the apron (refer to **7-43**). Some prefer to extend the apron a little beyond the casing.

The ends of the apron may be finished in several ways. The easiest is to cut them square. Another is to cut the profile of the face design on the end of coping (**7-47**). A third way is to miter the ends and glue a return (**7-48**).

Installing Mitered Casing with a Stool

A mitered window casing with a stool and apron is shown earlier in **7-22.** Begin by installing the stool as described earlier. The top and side casing are installed as described for picture frame casing; however, the side casing is cut square on the bottom where it butts the stool.

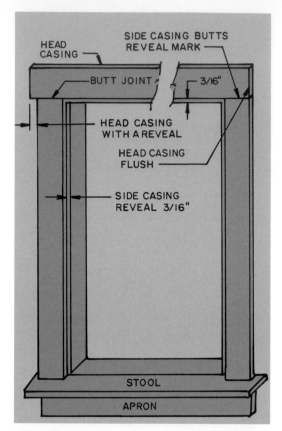

7-50 Flat casing can be installed with a reveal, or the side casing can be flush with the end of the head casing.

After the stool is installed cut and install the head casing, lining it up with the reveal marks. Then measure the length of each side casing. Measure both sides because there is often a slight difference in the length. Some measure to the short side while others measure the long side. Place the side casing on the reveal marks (7-49). Check the miter and adjust it if not tight. Check the bottom to be certain it is square with the stool. Nail the casing and the corners as shown earlier in 7-41.

Installing Butted Casing with a Stool

The butted window casing is installed by first cutting the side casings to length. Both ends are cut square. The top end should touch the reveal marks. Be certain that the end butting the stool has a tight, closed joint when the casing is in line with the reveal marks.

The head casing may be cut so it is flush with the outside edges of the side casing or cut the same length as the stool, providing a small reveal (7-50). However, the head casing may be even wider and have various moldings applied to it. Often the door and window head casings are of the same design.

7-51 Rosettes are often used to form the corners of butted casing.

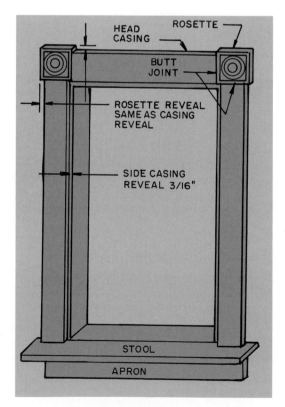

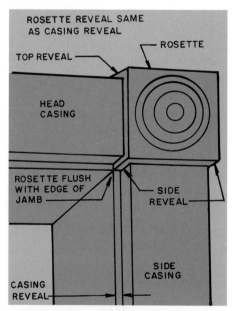

7-52 The edge of the rosette is in line with the inside face of the head and side jambs, and reveals are the same as the casing reveal.

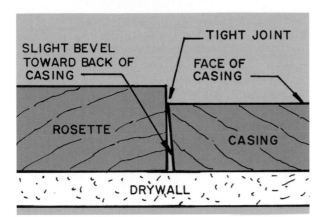

7-53 If a very slight bevel is cut on one of two butting pieces it helps produce a tightly closed joint.

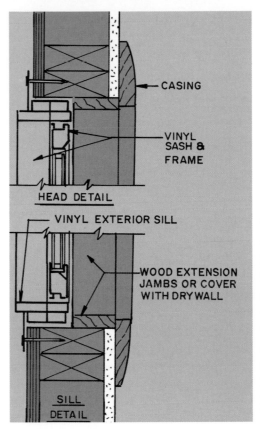

7-54 When vinyl windows are installed the interior of the rough window opening is trimmed with wood extension jambs and the casing is installed in the normal manner.

Using Rosettes

Butted casings on some traditional houses use rosettes. The rosettes, also called corner blocks, are available from building materials suppliers. They are typically wider than the casing and overhang it the same amount as the casing reveal (7-51).

Begin by installing the sill and apron as described earlier. Then install the side casings as described for butted casings. Next nail the rosettes on top of the side casing. Some prefer to align one side of the rosette with the edge of the casing, while others align it with the window jamb (7-52). Then measure the distance between the rosette and cut the head casing. It must butt tightly against each rosette. This will usually take some minor adjustments. It helps if the butting ends can have a slight bevel to the back of the casing (7-53). This makes it easier to get a tight fit.

Casing Vinyl Windows

The exact design of vinyl windows varies but the following is typical. Vinyl windows are installed with a nailing fin. This leaves some of the rough opening exposed. This can be covered with wood trim as shown in 7-54 or covered with drywall. After the jamb extension trim has been installed, the casing is placed in the manner described earlier.

Door Casing

Door casing is typically the same molding as that used to case the windows; however, on a major door, such as the front entrance, a wider, more detailed casing would enhance the entrance and foyer (7-55). Door

7-55 A major door, such as the door into the foyer, often has more detailed casing and glazing. The beautiful glazing on this entrance is laminated glass, which is more break-resistant than tempered glass. *Courtesy ODL*

casing is joined at the top corners in the same manner as the window casing shown earlier in **7-23**.

Preparing to Install the Door Casing

Basically the preliminary preparation is the same as for installing window casing. Check the door frame for squareness and plumb and to make sure that the door jamb is flush with the drywall as described for window preparation.

INSTALLING DOOR CASING

In new construction or a remodeling job where the old door and frame are to be replaced, the new unit chosen typically will have the frame assembled, door installed, and casing applied on one side. The casing for the other side will be precut and nailed in place after the frame has been installed. This procedure is discussed in Chapter 6.

If you are replacing the casing on an existing door you can sometimes find precut casing kits at the local building supply dealer. If they have a casing pattern suitable for you, this saves the problem of cutting the miters.

Often to get the casing pattern desired you have to buy the molding in long strips, cut the miters, and then cut the pieces to length.

INSTALLING A MITERED CASING

Begin by marking the reveal on the head and side jambs. The easiest way to do this is with a combination square **(7-56)**. The reveal most commonly used is 3/16 to 1/4 inch.

While different procedures are used to install casing, the following is typical.

First cut and install the head casing. Measure from the intersection of the reveal marks on each side **(7-57)**. Cut the miters on each end. Place the casing on the reveal marks and tack it to the jamb. Use 3d or 4d (1¼ or 1½-inch) finish nails. Leave the nails sticking out a bit so they can be pulled if it is necessary to move the casing. Place about every 12 inches. When using hardwood casing, pilot holes will most likely be drilled for each nail.

Next place a piece of casing along the side jambs and mark the corner of the reveal on it and the angle of cut **(7-58)**. Cut the miters and place against the jamb along the reveal marks. Check the miter. If it is tight, nail the side casing to the jamb and stud **(7-59)**.

Some place glue in the miter joint before the casing is nailed in place. Then nail the miter as shown earlier

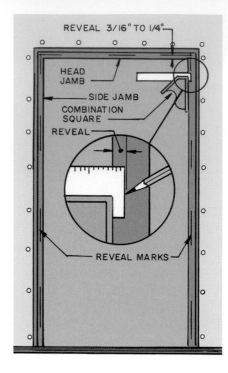

7-56 Mark the reveal along the side and head jambs.

in **7-41**. This pulls the miter together and helps keep it closed over the years.

Since the miter will often not close tightly, some adjustments must be made as mentioned earlier in this chapter.

Once the miters are closed and the casings are tacked in place, drive the nails close to the surface and set them below the surface of the casing. The painter will fill these and sand the surface in preparation for the final finish.

Often it is necessary to trim the bottom of the side casing so the flooring, made of wood, laminate, or vinyl, can slide under it. If you know the thickness of the floor this amount can be trimmed off before the casing is installed. Often it is necessary to trim it later. To do this place a piece of wood that is the thickness of the finished floor next to the casing. Use it as a guide to trim off the bottom of the casing. Use a fine-tooth saw **(7-60)**.

The previous discussion relates to hand-nailing the casing to the jamb and trimmer stud. A power nailer will greatly speed up the work and leave one's hands free to hold the casing as it is nailed. Use 1½-slight-headed finish nails (refer to **7-12**).

Information on laying out and cutting miters is presented earlier in this chapter.

BUTTED CASINGS

Mark the reveal on the head and side jambs as shown for mitered casing **(7-56)**. Cut the side casing to length. Mark it where the head and side jamb reveal marks

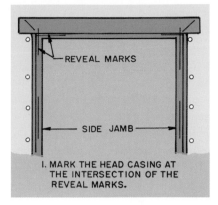

1. MARK THE HEAD CASING AT THE INTERSECTION OF THE REVEAL MARKS.

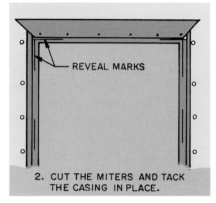

2. CUT THE MITERS AND TACK THE CASING IN PLACE.

7-57 Mark the inside edge of the casing on the intersection of the reveal marks. Miter and install with the inside point at this intersection.

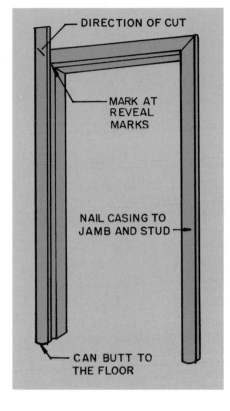

7-58 Place the casing along the side jambs and mark and cut the miter. Then install the casing along the reveal marks.

7-59 The side casing is nailed to the jamb and the trimmer stud.

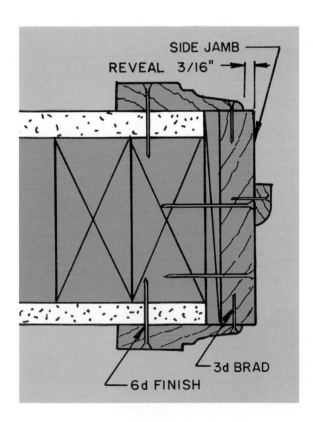

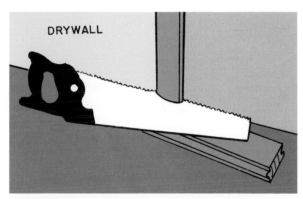

7-60 Place a piece of the flooring next to the door casing. Mark the line of cut. Place a wood block that thick next to the casing and use it as a guide for the saw.

129

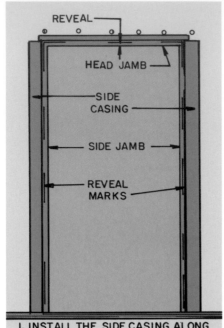

I. INSTALL THE SIDE CASING ALONG THE REVEAL MARKS. CUT FLUSH WITH THE HEAD JAMB REVEAL MARKS.

7-61 When installing door casing with butted joints, cut the side casings to length and install them. Then cut and install the head casing.

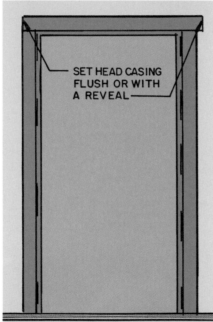

2. MARK AND CUT THE HEAD CASING TO LENGTH. INSTALL ON TOP OF THE SIDE CASING.

cross. Be certain the casing is square with the floor (7-61). If the finished floor has been installed before the door casing, you can get a tighter fit with the floor if you cut the bottom end of the casing on a slight slant toward the back.

After the side casings are nailed to the jamb the head casing is marked and cut to length. It may be cut flush with the casing or have a slight reveal (7-61). If the head will have moldings attached creating a decorative head, it should be assembled before it is installed.

Butted casings are nailed as shown in 7-62.

Sometimes the head casing is made from stock thicker than the side casing. This produces a decorative reveal at the corner (7-63).

Butted casings with plinths and rosettes are installed much the same way as described for plain butted casing. The exact sequence of events will vary but the following is typical. Review the use of rosettes on windows earlier in this chapter.

First install the plinth blocks. The plinth blocks are installed flush with the edge of the door frame (7-64). They are usually wider than the casing and have a reveal on all sides. Next cut the side casings to

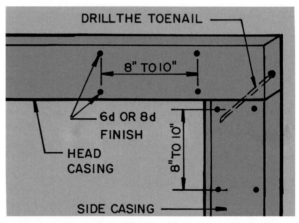

7-62 A typical nailing pattern for installing butted casing.

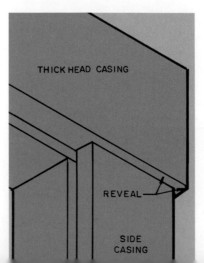

7-63 The appearance of a butted casing can be improved by using a thicker head casing, producing a reveal.

7-64 The plinth is set flush with the edge of the door jamb.

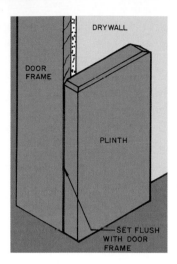

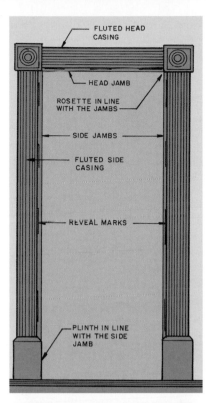

7-65 First install the plinth blocks, then the side casing, followed by the rosettes. Finally, install the head casing.

length and install on the reveal marks on the jamb. Then install the rosettes. Set them flush with the edge of the head and side jamb. Finally, cut and install the head casing between the rosettes. Measure and cut carefully. You can slightly bevel the square end to get a tighter fit (**7-65**). It should be noted that butted casings may use only plinth blocks and have mitered or butted head casing or use only corner blocks and let the side casing run to the floor.

Installing Baseboard & Shoe Molding

Do you install the baseboard before or after the finish flooring has been installed? In general it depends upon the type of flooring to be laid.

Many prefer to have **unfinished hardwood flooring** laid before the baseboard. The flooring can be sanded up to the wall with no danger of damaging the baseboard. If **prefinished wood flooring** is to be used, generally it is laid after the baseboard is installed. This reduces the chance of its being damaged (**7-66**). If prefinished flooring is installed first cover, the work areas with red rosin paper or some other protective material. **Vinyl floor** covering is usually installed after the baseboard is in place. **Ceramic tile** flooring can be installed either before or after the baseboard is in place. **Laminate flooring**, also referred to as a floating

7-66 This prefinished hardwood flooring is being laid after the trim has been installed and the walls and trim painted.

7-67 Laminate floating flooring is laid to within ⅜ inch of the wall before the baseboard is installed.

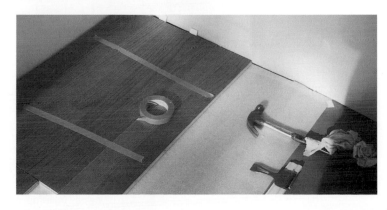

131

floor, is laid to within ⅜ inch of the wall before the baseboard is installed (7-67). **Carpet** is usually laid after the baseboard has been installed.

Installing Baseboard

The door casing should be in place, as should cabinets that may be butted by the baseboard.

Begin by planning the sequence to be used for installing the base. While carpenters use different procedures, the plan in **7-68** is typical. Plan so each piece has a square cut on one end and a cope or miter on the other. Generally the first piece will be the one on the longest wall. This piece may have square cuts on each end and will usually require a splice with a scarf joint. This is discussed below (**7-70**). Outside corners will be mitered and inside corners coped as discussed in **7-72**.

Some work to their left around the room, while others prefer to work to the right. Some prefer to measure and mark the length of each piece, cut them all to length and then begin installation, while others mark, cut and install each piece separately. When measuring to the end to be mitered or coped, mark the direction of the cut on the stock to remind you which way to set the saw. If the cut is to be a square cut for a butt joint, mark the location with an S (square). If a corner is found to be not square (90 degrees), mark the size of the angle required on the back of the baseboard. Note whether it is an inside or outside corner.

In **7-68** the decision to work to the left was made. The longest piece, No. 1, was spliced and butted to the wall. The adjoining pieces are coped on one end and square cut for a butt on the other. Where the base

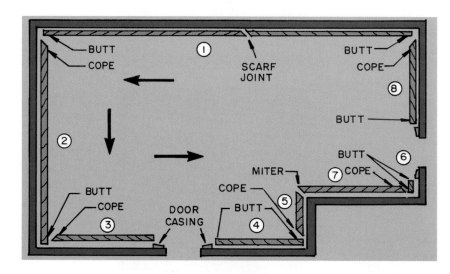

7-68 A typical baseboard layout plan working from the right to the left. Some prefer to work in the other direction.

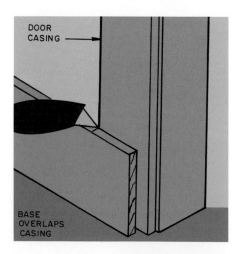

7-69 The length can be accurately marked by placing the baseboard against the wall and marking it with a utility knife.

7-70 Baseboard is usually spliced by mitering the butting ends, forming a scarf joint.

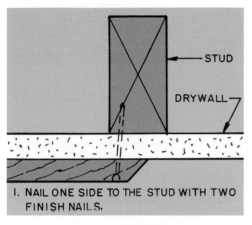

I. NAIL ONE SIDE TO THE STUD WITH TWO FINISH NAILS.

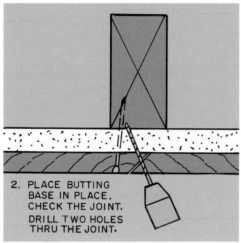

2. PLACE BUTTING BASE IN PLACE, CHECK THE JOINT. DRILL TWO HOLES THRU THE JOINT.

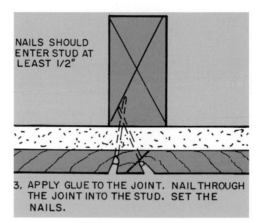

NAILS SHOULD ENTER STUD AT LEAST 1/2"

3. APPLY GLUE TO THE JOINT. NAIL THROUGH THE JOINT INTO THE STUD. SET THE NAILS.

7-71 Locate the scarf joint at a stud so the joint can be nailed into a solid backing.

butts the door casing is another square cut. Notice the outside corner is mitered.

MEASURING THE LENGTH

Some prefer to cut long baseboard to the exact length, while others will cut it ⅟₁₆ to ⅛ inch longer to get a tight fit. Shorter pieces can be cut the exact length or just a bit shorter. It is important to measure the length to within ⅟₃₂ to ⅟₁₆ inch. Long lengths will require two people to handle the tape. The length can also be measured by placing a piece of baseboard along the wall and marking the length. A utility knife (**7-69**) can be useful for making accurate marks.

Since the baseboard butts against the door casing, cabinets, bookcases, fireplace surrounds, window seats, and other built-in items must be installed before the baseboard. Stair skirtboards often are tied into the baseboard. If there are heat registers mounted on the wall they hopefully are in place. If not, the heating contractor should locate them exactly where they will be installed. Observe if there are any changes in floor level. For example, a solid-wood floor may butt a vinyl covered floor. Provision must be made for the difference in thickness if it is desired to keep the top of the baseboard level. If the difference is

small, the baseboard can be trimmed narrower where needed.

SPLICING MOLDING

When molding has to be spliced it makes the best finished appearance if it is cut on a 45-degree angle, forming a scarf joint (**7-70**). It can be joined with a butt joint, filled with caulking, and will be fairly hidden when painted; however, over time the molding will change in length and the joint may open. If a scarf joint opens a little it is not as noticeable.

The joint should be located at a stud so there is adequate nailing surface to close and hold the joint (**7-71**). One side is nailed to the stud. The other is nailed through the joint into the stud. Use two nails in each piece. It helps to lay a bead of carpenter's glue in the joint.

MAKING INSIDE CORNERS

Inside corners may be formed by mitering the joining moldings or by coping one of them. Coping does a better job because as the moldings move, any crack at the corner is not apparent, while an open miter is obvious. Coped corners are used on all types of molding such as baseboards, shoes, crown, and chair rails.

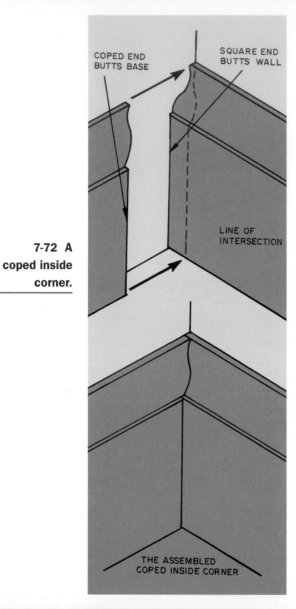

COPED END
BUTTS BASE

SQUARE END
BUTTS WALL

LINE OF
INTERSECTION

**7-72 A
coped inside
corner.**

THE ASSEMBLED
COPED INSIDE CORNER

One of the moldings is butted to the wall. The other has the end coped, which fits over the profile of the first piece (7-72).

Begin by cutting a miter on the end to be coped (7-73). Cut the molding an inch or so longer than required. This provides some wood so the coping saw can get started to cut on the waste side of the edge, forming the profile. It also provides an allowance if it becomes necessary to trim the cope. The miter should slope in the same direction as it would if an inside miter were to be cut.

Next cut the profile using a coping saw (7-74). The edge formed by the miter is the line of the profile to be cut. The coping saw is kept approximately **perpendicular** to the front of the molding and as it saws it follows the profile formed when the molding was mitered.

It helps as you cut the profile to make cuts from the back of the molding up to the profile (7-75). This allows the first portion cut to fall away and makes it easier to cut the next part of the profile. Be careful as you cut to keep the edge of the profile sharp and free of any chips or saw marks.

Continue cutting to the profile and making back relief cuts until the cope is finished. Then place it against the other molding to see how it fits. It will often require a little filing to make a close fit. Use half-round and round files of different sizes so they fit the curves being adjusted. Some prefer to angle the cope a little more so the back edge, which is not seen,

**7-73 Begin by mitering
the end to be coped.**

is slightly behind the front profile edge. This allows the front profile edge to rest firmly against the other molding. The finished coped joint will have a sharp profile (refer to **7-72**).

Place the coped end tightly against the adjoining baseboard. Since the coped piece is cut a little longer than needed it can be pressed against the wall, forcing the joint closed. Nail it to the studs and bottom plate. It can be glued if you want.

MITERING OUTSIDE CORNERS

The common way to finish outside corners is with a miter joint (**7-76**). This is used with all kinds of molding.

7-74 Start coping at the narrow shaped edge of the baseboard.

7-75 It helps to make relief cuts, removing parts of the profile that have been cut.

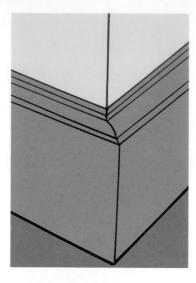

7-76 Miter joints are used on outside corners and are cut on 45 degrees.

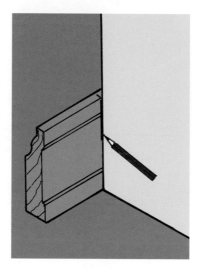

7-77 Lay the baseboard along the wall and mark the length. This mark is the inside corner of the miter and is placed against the miter saw fence.

When marking the length of a piece that ends in an outside corner, prepare the end that butts the wall. Depending upon the layout, this could be a butt joint or a coped end. Then with this piece in place and extending beyond the outside corner, mark the location of the inside edge of the miter with a knife or sharp lead pencil (**7-77**). The distance could also be measured with a tape. This mark is on the side of the baseboard next to the wall (**7-78**). Mark both pieces that are forming the corner.

When cutting the miter, place the inside face of the baseboard against the fence and set the saw on the desired angle.

Since many corners are not perfectly square, usually some adjustments are necessary. Some prefer to cut some scrap stock to check the corner or check it with a square. Slight adjustments in the angles may be necessary when cutting the finished baseboard. After cutting the baseboard make a trial fit (**7-79**). Trim if necessary to get a closed point. Then nail one side in place. Nail into the wall studs. You can find them by tapping on the wall or looking for some slight dimples in the drywall where the nails were placed. Lay a bead of carpenter's glue along the miter joint and install the adjoining piece. Some prefer to lay a bead of carpenter's glue on the mitered end and pin it

7-78 Mark the baseboard on the back side next to the wall. Mark the direction of the miter.

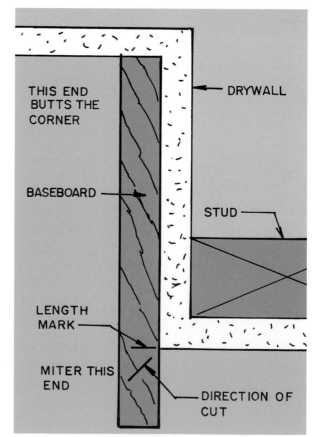

THIS END BUTTS THE CORNER

DRYWALL

BASEBOARD

STUD

LENGTH MARK

MITER THIS END

DIRECTION OF CUT

together with a brad power nailer before nailing the second piece to the wall.

Installing Miterless Baseboard

A wood product that eliminates the need for miters or coping uses internal and external corner pieces against which the baseboard is butted (**7-80**). The corner pieces are glued and nailed to the wall. The baseboard is cut to length with 90-degree ends and fitted between the corner pieces. The nails are countersunk and filled. This product can be stained, painted, or varnished.

Installing Square Edge Base

Square edge base is installed by mitering exterior corners and either butting or mitering inside corners. If base cap and shoe moldings are used, they are mitered on exterior corners and coped on inside corners.

FASTENING THE BASEBOARD

Softwood base can be installed by nailing through the finish wall material into the stud with two finish nails long enough to give good penetration into the stud and plate. Usually 6d or 8d finishing nails are ade-

quate. Place one nail about ½ inch from the bottom into the bottom wall plate (**7-81**). This will be covered with the shoe molding or carpet. The other nails will have to be set and the holes filled. Nail into each stud and near the end of each piece. If splitting is a problem, drill small holes for the nails. The procedure for nailing wide base is shown in **7-82**. Hardwoods often require that the nails have predrilled holes. This helps prevent splitting and bent nails. The use of a power nailer reduces splitting and speeds up the installation (**7-83**).

Before starting to nail you must locate each stud. There are a number of ways this can be done. A stud finder tool can be used. Also you can tap on the drywall until you get a solid sound. Then drive a nail at this point to see if it hits the solid stud or breaks through the drywall. If it misses move the nail over and try again until the stud is hit. Try to get near the center. Mark it on the wall or floor. This will leave a series of nail holes in the wall but they will be covered by the baseboard. After you locate one stud you can measure every 16 inches and you will be very near the other studs. If there is an electrical box it is usually mounted on a stud.

If the wall has a very slight curvature the base can usually be pulled tight to it as it is nailed. An extra nail or two may help.

7-79 After cutting the miter, place it against the corner to check the fit. It may require some minor adjustment before it is nailed.

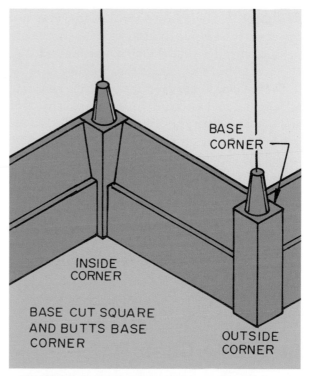

7-80 Miterless baseboard uses corner pieces on the inside and outside corners. The baseboard is cut to length to fit between them.

BASE CORNER

INSIDE CORNER

BASE CUT SQUARE AND BUTTS BASE CORNER

OUTSIDE CORNER

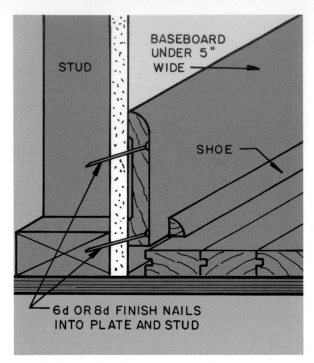

7-81 Baseboards 5 inches and under are nailed to the stud and plate.

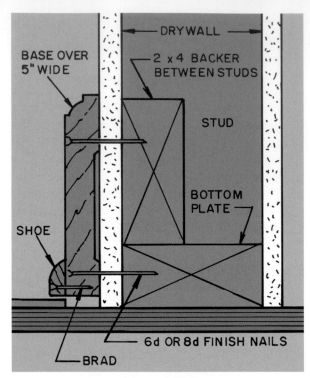

7-82 Baseboards over 5 inches wide should be nailed to backer boards placed between the studs.

Installing Shoe Molding

Shoe molding is installed after the finish flooring is in place. If the floor is wood and has not been sanded and finished, cut and fit the molding and lightly tack it in place. The floor finishers will remove it so they can sand closer to the baseboard. It is permanently installed after the wood floor has been finished. Not all flooring materials require shoe molding.

The molding is measured and cut in the same way as described for baseboard. Inside corners are coped and outside corners are mitered. It is nailed to the baseboard with brads as shown in **7-82.** Most prefer to nail the molding to the baseboard rather than to the floor. If it is nailed to the floor and the baseboard moves a crack, unfinished wood above the shoe will be exposed. If it is nailed to the baseboard a crack can appear at the floor; however, this is not very noticeable. When nailed with power-driven brads the shoe molding is less likely to split and the work goes must faster.

When the shoe molding meets the door casing the end can be mitered (**7-84**).

Installing Chair Rails

Chair rails are available in various stock profiles; however, if something larger or different is wanted it can be made by assembling two or more moldings.

7-83 Baseboard can be hand- or power-nailed.

If a stock chair rail molding is used it will generally be rather narrow. It is recommended that blocking be let in between the studs before the drywall is installed to provide a solid nailing base (**7-85**). Narrow chair rails can bow slightly, leaving small gaps along the wall. Larger moldings made from an assembly of moldings or a wide custom-made molding may be strong enough to be nailed to the studs (**7-86**).

Chair rails are generally installed 32 to 36 inches above the floor as shown in **7-85**.

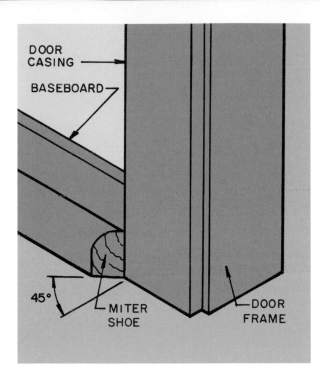

7-84 The shoe molding should be mitered when it butts a door casing.

7-85 Narrow chair rail requires blocking be let into the studs to provide an adequate nailing surface.

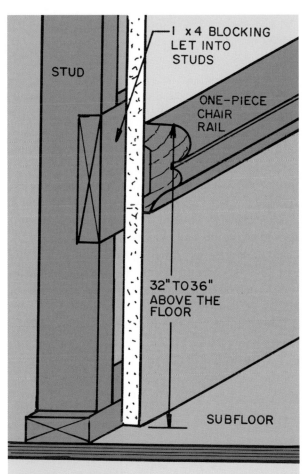

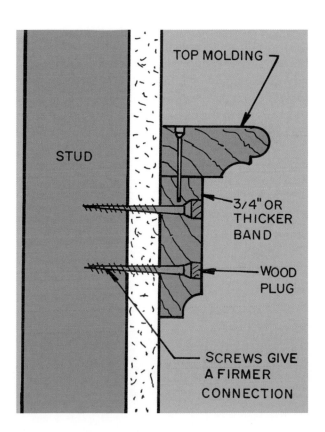

7-86 Wide and multiple-piece chair rails usually are strong enough so that blocking between the studs is not required.

7-87 Chair rail and picture molding locations can be marked by snapping a chalk line.

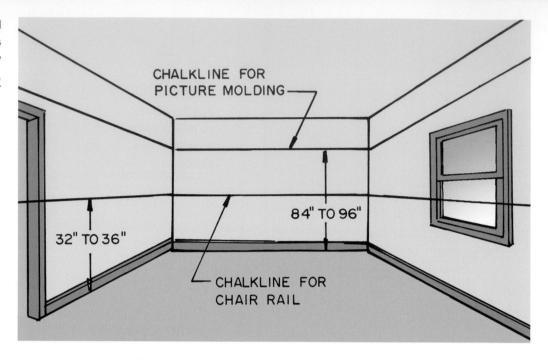

CHALKLINE FOR PICTURE MOLDING

84" TO 96"

32" TO 36"

CHALKLINE FOR CHAIR RAIL

Begin by locating a level line around the room at the desired height. This can be done with a laser level or a chalk line (7-87). Place the single molding chair rail on the line and tack it to the blocking every 8 to 10 inches with small finish nails. Two nails at each location are usually required. If a two-piece rail, as shown in 7-86, is used, first place the band on the line and nail it. Then install the cap molding. Often there is some small waviness in the drywall. The painters should caulk this gap before finishing. Usually the chair rail parts are primed before they are cut and installed. Be certain to prime the back. This blocks moisture from entering the molding, reducing the chance of its warping over time.

The chair rail will usually meet a door or window casing or a cabinet. At a cabinet it can simply butt it. If the chair rail is narrower than the casing it can also butt it. If it is wider it can be notched around it (7-88).

Installing Picture Molding

Picture molding is installed in the same manner as chair rail molding. It must have blocking let in between the studs, which must be installed before the drywall installers appear (7-89). The picture molding is usually installed 84 to 96 inches above the floor (7-87). If the room has 8-foot ceilings the picture molding also serves as a small crown molding. Since they support heavy pictures they need to be securely nailed or, better still, screwed to the blocking. Screws with oval heads can be left exposed. Flathead screws should be set below the surface but make sure that the molding chosen is thick enough to allow this to be done.

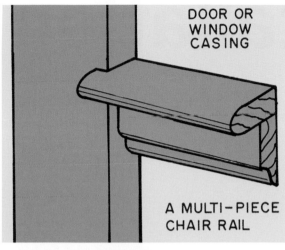

DOOR OR
WINDOW
CASING

A MULTI-PIECE
CHAIR RAIL

7-88 When a chair rail meets a casing, the thickness of the molding influences how you will butt it.

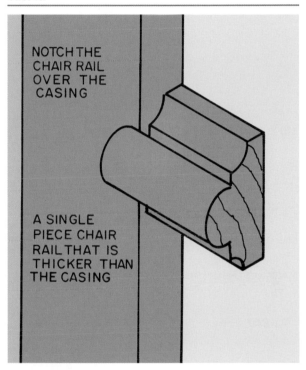

NOTCH THE
CHAIR RAIL
OVER THE
CASING

A SINGLE
PIECE CHAIR
RAIL THAT IS
THICKER THAN
THE CASING

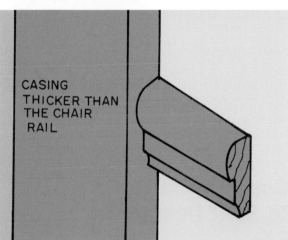

CASING
THICKER THAN
THE CHAIR
RAIL

7-89 Picture molding should have blocking let into the studs so it can be securely fastened, enabling it to carry heavy pictures.

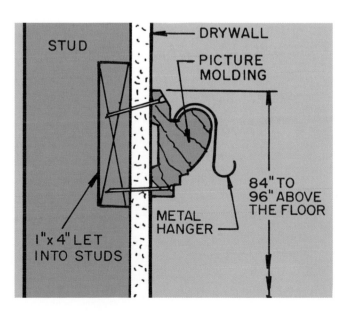

STUD

DRYWALL

PICTURE
MOLDING

84" TO
96" ABOVE
THE FLOOR

METAL
HANGER

1"x 4" LET
INTO STUDS

CHAPTER 8

Stair Installation & Repair

When you are planning to build new stairs or to do some extensive remodeling and repair, you must observe the requirements of the local building code. The code is designed to make certain the stairs are safe. The wrong tread or riser size can be the cause of serious falls. The following discussion is typical of codes currently in use. Consult your local building inspection department for specific local requirements.

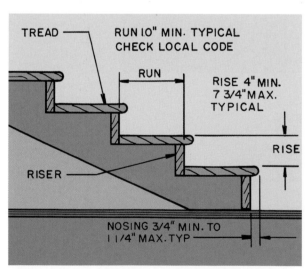

8-1 These are typical riser and tread requirements for stairs in residences. Check your local building code for requirements for your house.

Codes Related to Stair Design

You must be certain your stairs are designed to meet local building codes. The codes are prepared to set standards that are determined necessary for safe stairs. If the rise is too high or the tread too narrow, it could cause a dangerous fall.

While codes may vary a little in general they have the same basic set of requirements. Following are specifications typically found in many codes. Consult your local codes so your stairs will pass inspection.

Rise & Run

The rise and run of each stair must be the same for the entire length of the stairs. Codes permit a small variance which can be caused by cutting, but try to keep it to ¹⁄₁₆ inch—about the width of a saw kerf. A ³⁄₁₆-inch difference may pass inspection, but try to work closer. The commonly used **maximum riser** is

7¾ inches. The commonly used **minimum riser** is **4 inches (8-1).**

The commonly used **unit run** is 10 inches. Check your local code. While these sizes can vary slightly depending on the code, a rule of thumb to find a comfortable combination is that the sum of 2 risers plus 1 tread without the 1-inch nosing should not be less than 24 inches or more than 25 inches. Some examples are shown in **Table 8-1.**

Stair Width

Residential stairs must be at least 36 inches wide if the building has an occupancy load of 49 people or less. This would apply to single-family and multi-family residences. If a building has an occupancy load of 50 or more, the stair must be 44 inches wide **(8-2).**

Handrails

Handrails are carefully regulated by local building codes. They are very important because they prevent falls. The handrails must be securely fastened to the wall framing. Standard handrails and hanging devices are widely available.

Residential stairs must be at least 36 inches wide and have one handrail **(8-3).**

The handrail must be located 30 to 38 inches measured vertically above the tread and have at least 1½ inches clear space between it and the wall. It should not extend more than 3½ inches into the required stair width. In addition, the handrail must run continuously the entire length of the stairs and extend 12 inches beyond the top riser and one tread width beyond the bottom riser. The handrail should be 1¼ to 2⅝ inches in cross-sectional areas and have all smooth edges **(8-4).**

Slope of a Stair

The stair should be designed so it is on an angle of 30 to 35 degrees. Angles to 40 degrees can be used if the rise and run limits are met.

Landings

Landings should be as wide or wider than the stair. The width in the direction of travel need not be more than 4 feet.

Table 8-1   Examples of Acceptable Rise and Run Proportions.

Unit Rise (inches)	Unit Run (inches)	Slope (degrees– minutes)	Total (2 risers plus 1 unit run = 24 to 25 inches)
6 5/8	11 3/8	30°–30'	24 5/8
6 3/4	11 1/4	31°–30'	24 3/4
7	10 1/2	33°–35'	24 1/2
7 1/2	10	38°–45'	25
7 3/4	10	37°–30'	25 1/2

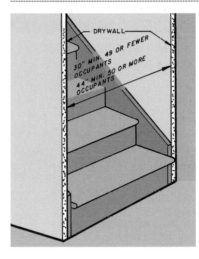

8-2 The minimum stair width is regulated by the number of occupants in the building who use the stairs. Check your local building code.

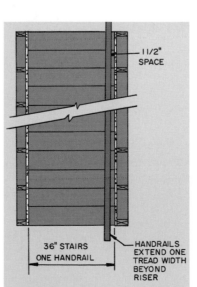

8-3 Typical requirements for the use of a handrail in a residence.

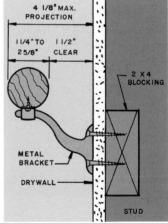

8-4 Handrails mounted on a wall should allow at least 1½ inches of clearance between the wall and the handrail.

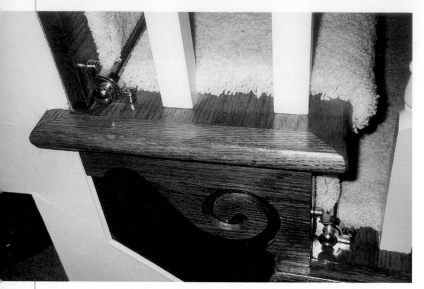

8-5 The space between the balusters is regulated by code. Codes typically allow a maximum of 4 inches.

Balusters

The spacing between **balusters** is generally limited by code to 4 inches. The actual spacing should permit two balusters to rest on each tread (**8-5**).

Stair Terminology

The major parts of a straight stair are shown in **8-6**. The unit rise is the vertical height of one step and the unit run is the horizontal size of one step. The total rise is the height of one rise multiplied by the number of risers. You must consider the situation from one floor to the next as it relates to the subfloor and any finished flooring to be used. A typical example is shown in **8-7**. Here both floors are to be finished with solid-wood flooring. An example with vinyl floor covering and carpet is shown in **8-8**. If you are going to replace damaged **carriages** or **stringers**, these fac-

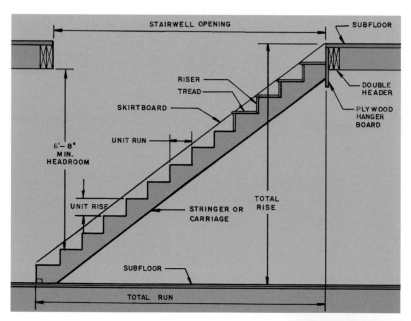

8-6 These are the main parts of a stairway.

8-7 When both floors have the same thickness finish flooring, the subfloor to subfloor and finish floor to finish floor measurements the same.

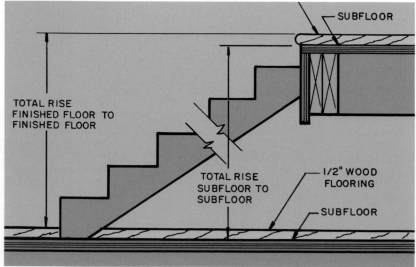

tors must be considered. If the floors will not be changed, check the old carriage to see if it meets the code. If so, it can serve as the pattern for the new stringer. If it does not, you will have to design one using an approved unit rise and run.

A finished stair is shown in **8-9**. The handrail is supported by balusters. The **balustrade** is the assembly of the handrail and balusters.

Renovating an Old Stair

Old stairs are built in a number of different ways. As you begin repairs, examine them to see what type of construction was used. This will influence how you will handle the repair. Typical problems include squeaky treads, sagging carriages, shaky balusters, and loose handrails.

The main structural supports are the stringers or carriages. It should be made clear that both stringers and carriages support the treads and risers. The difference is that stringers have mortises cut into them into which the tread and riser are inserted. The treads and risers are housed into the stringer (**8-10**). Carriages are notched members that in effect carry the treads. They are cut in a sawtooth pattern (**8-11**).

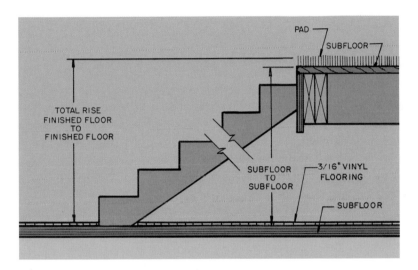

8-8 If the floors have different finished flooring, subtract the thickness of the lower finished floor and add the thickness of the upper finished floor.

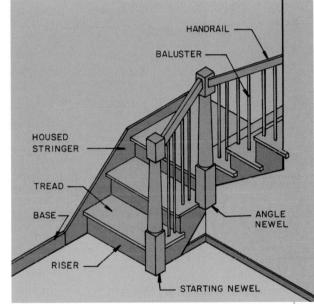

8-9 Typical finished stairs showing the balustrade assembly consisting of the handrail, balusters, and newel posts.

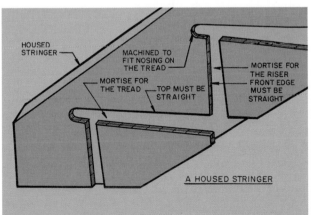

8-10 A typical housed stringer.

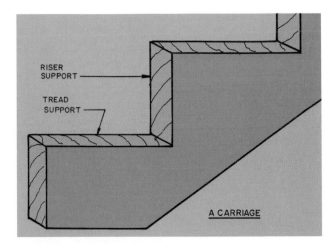

8-11 A typical carriage.

Squeaky Treads

The squeaks are caused by the loose tread sliding up and down on the nails securing it to the carriage or by it rubbing against the riser. If the underside of the stair is covered you will have to secure the tread to the stringer or carriage and riser with nails or screws. Nails provide a quick fix but will fail in a few years. Wood screws provide a permanent fix. If the tread is to be covered with carpet, the fasteners can be secured with the heads flush or just a little below the surface of the wood. They will be covered by the pad and carpet. If the wood tread is to be exposed to view,

the nails can be set and a wood filler of a color matching the tread can cover the nails. If a screw is used, drill a counterbore for the head (**8-12**) and glue a plug over it (**8-13**). The plug can be of the same wood or of a different wood, giving a bit of decoration (**8-14**). Plugs can be cut from the wood available with a plug cutter (**8-15**). Plugs can also be purchased at the local building supply dealer.

If the stair is to be covered with carpet, first staple the carpet pad to the treads (**8-16**). Then install the carpet over it (**8-17**). Carpet can also be installed without a pad but if this stair is heavily used the carpet will be crushed sooner than if a pad was used (**8-18**).

Another thing to do if you have access to the underside of the stair is to nail, or better still, to glue and screw the riser to the tread (**8-19**). Then nail glue blocks to the tread and riser as shown in **8-20**.

If you can get behind the stair you will find that some treads are set into mortises cut into a stringer (refer to **8-10**). The stringer is secured to the wall

8-12 Drill counterbored holes in the tread. The counterbore bit produces the hole for the screw and the counterbored section for the head and plug.

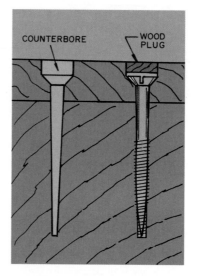

8-13 The heads of wood screws can be hidden by setting them in a counterbore and gluing a wood plug in it.

8-14 The wood plug over a counterbore can be a contrasting type of wood, giving a decorative feature.

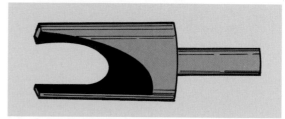

8-15 A plug cutter.

8-16 For the best results install a carpet pad on the treads before installing the carpet.

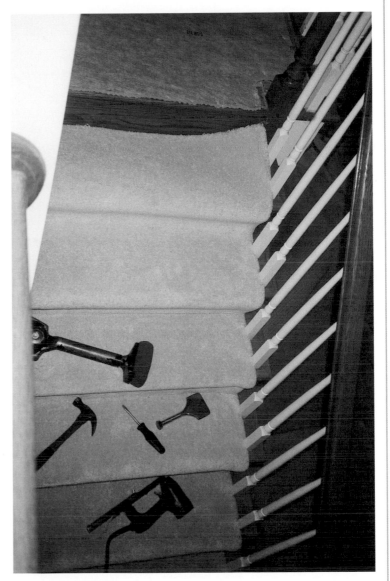

8-17 Lay the carpet over the tread and up the riser. Roll under the exposed edge next to the balustrade.

8-18 This stair carpet was laid without a carpet pad.

147

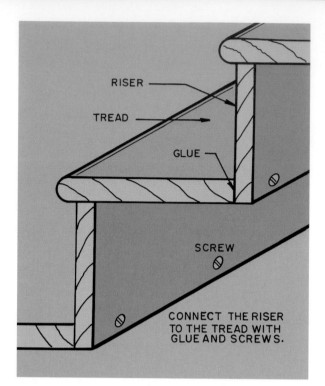

RISER

TREAD

GLUE

SCREW

CONNECT THE RISER TO THE TREAD WITH GLUE AND SCREWS.

8-19 Gluing and screwing the riser to the tread will reduce squeaks.

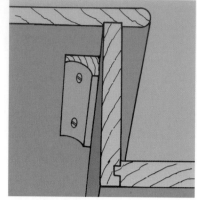

8-20 Connecting the tread and riser with glue blocks reduces movement and squeaks.

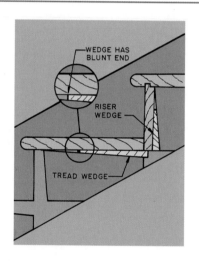

8-21 The blunt end wedges are glued and driven in the mortise behind the treads and risers.

WEDGE HAS BLUNT END

RISER WEDGE

TREAD WEDGE

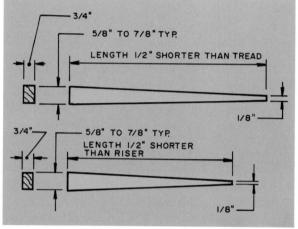

3/4"

5/8" TO 7/8" TYP.

LENGTH 1/2" SHORTER THAN TREAD

1/8"

3/4"

5/8" TO 7/8" TYP.

LENGTH 1/2" SHORTER THAN RISER

1/8"

8-22 Typical wood wedges used to secure the treads and risers in housed stringers.

framing. It has mortises into which the treads and risers fit. They are secured to the stringer by wood wedges that are glued and driven in place as shown in **8-21.** Typical wedges used are shown in **8-22.** Sometimes the wedges are sound and can be reglued and tapped into place. Otherwise cut out the old wedges and insert new ones. Sometimes a very thin wedge can be driven on top of or below an old wedge.

If these wedges still permit a squeak or movement in the center of the stairs, it will be necessary to cut and install a carriage down the center or reinforce the existing center carriage if one exists. This will correct the sag in the center of the stairs.

Sagging Stairs

Old stairs that still squeak after you have tried to screw the treads to the carriages or risers or have movement as you walk up them will need some structural reinforcing so they are stiff and do not sag. To make such repairs it is necessary to remove the finished surface material behind the stairs. As you cut and pry loose this material, be careful you do not cut into the stringer, carriage, treads, or risers. This is a messy operation so be certain to remove all waste material before you begin repairs. If there are any decorative moldings that you may want to reuse, pry them off carefully.

Examine the structure for rot or termites. If these exist the damaged material must be removed and termite treatment is essential. If the rot is serious replace all the members that no longer have structural value. If some sag occurs due to age or warp but the carriages and stringers are basically sound, leave them but add reinforcing material. How this is done depends upon the situation but the following ideas are typical.

If there is no center carriage, add one. Jack the stair straight and work in the new carriage (8-23). If the sag is small, sometimes jacking the carriage straight and reinforcing it by nailing a 2 x 4 or wider notched stick on each side of the center carriage will hold the stair straight (8-24). Another possibility is to cut a new full carriage and install it next to the original carriage.

If the stair is no longer level it could be caused by the carriage or stringer pulling away from the wall. If it is sound you can jack it in place and re-nail it to the studs (8-25). You can also install some 2 x 4 supports to the studs between the carriage or stringer and the floor (8-26). If the sag is due to rotten or cracked carriages, stringers, or wall studs, a big tear-out and rebuild job will be necessary. Sometimes the wall is covered with drywall before the stairs are

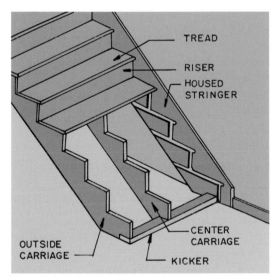

8-23 A center carriage adds considerable stiffness to a stairway and should always be installed.

8-24 The view from behind the stairs shows 2 x 4 stiffeners nailed on each side of the center carriage.

8-25 If the carriage or stringer along a wall is loose but sound, it can be jacked into place and renailed to the wall.

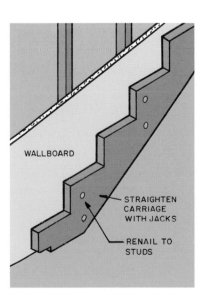

8-26 A loose or sagging carriage or stringer can be supported by nailing blocking to the studs below it.

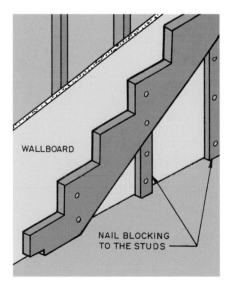

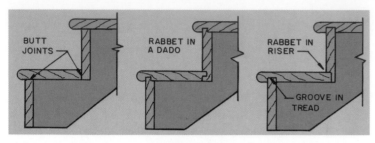

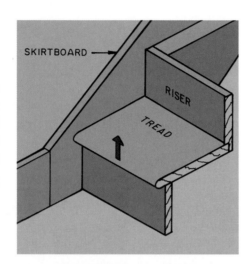

8-28 **If you pry loose a tread on a carriage, be careful not to scratch the skirtboard or damage the riser.**

8-27 **Common ways used to join treads and risers.**

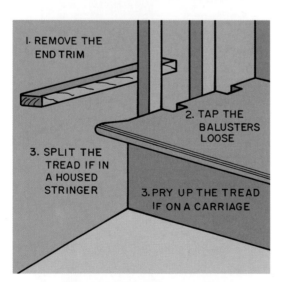

8-29 **Loosen the balustrades before removing the tread. Sometimes the tread must be split into narrow pieces to get it out of the mortise on a housed stringer.**

installed and it will take some removal if you want to check the condition of the studs.

If the sag is considerable and replacing structural members is necessary, you might have to remove the treads and risers and possibly disassemble the balusters. Since many of these were cut to fit a specific location, mark them so they can be repositioned. Put a piece of masking tape on each piece and write on it with a black-ink felt-tip pen.

Replacing Treads

Treads that are cracked or warped should be replaced. The immediate problem is to know how they were installed. In **8-27** are the most commonly used constructions.

Removing a tread can be a difficult task. If the stair uses carriages, the tread will be nailed to them and usually butt the skirtboard. As you pry up the tread try to avoid scratching the skirtboard (**8-28**). If a stringer is used, remember the tread is inserted in a mortise and cannot be lifted out.

If the tread has balusters in an end they may be set in with dowels. Drill holes through the tread in line with the dowels and split the tread with a wood chisel. Then a split to the dowel will usually free it. If the balusters are set in notches in the end of the treads, first remove the end trim and carefully tap the balusters free from the tread. Then drill some holes in the tread and split it apart with a wood chisel (**8-29**). Sometimes it is possible to pry up the tread but be careful not to damage the riser.

Usually the riser is secured to the tread on the back edges. You may have to open that joint and cut the nails or screws with a metal-cutting saw.

Cut the new tread to fit the area. Bore holes for the balusters with dowels or notches for those with a tenon. Trim and fit as necessary. Then install by screwing to the carriage or wedging in the mortise if a stringer exists. Finally reinstall the balusters.

Stabilizing a Landing

A landing is a platform installed in a flight of stairs that breaks it into two or more flights. It is used to give a resting spot in a long flight or to change the direction of the stairs. Landings are used on straight, L-shaped, and U-shaped stairs.

Typical landing construction is shown in 8-30. The platform is framed with wood joists just like the house floor. It is supported by stud walls on open sides and is secured to interior partitions, if it butts against any (8-31).

If, as the platform ages, it sags or allows movement, it is necessary to reinforce the structural framing. This may include adding additional joists, resecuring to wall framing, and adding additional studs on the open sides. If there is a bit of racking (sideways movement) get it level and plumb and nail diagonal 2 x 4 members across the studs running from the header to the bottom plate at the floor.

Replacing Damaged Balusters

Balusters may be secured to the tread by a dowel on the bottom or a tenon. The first problem is trying to find new balusters of the same design. If they are old you will most likely have to take one to a woodworking shop and have them turn a new one. Sometimes it is possible to put the damaged baluster together with dowels or metal pins, caulk any cracks left, and reinstall and paint it. If a lot of them have been damaged consider replacing all of them.

To remove a doweled baluster saw it in half and carefully rock each piece loose. You could cut near the tread and bore out the part left in the dowel hole.

Tenoned balusters can be removed by removing the trim piece as shown in 8-29. They can then be carefully tapped loose without damaging the tread.

To reinstall a doweled baluster put the small end in the hole in the handrail and slide the bottom across

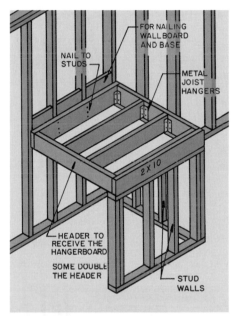

8-30 Typical framing for a stairway landing.

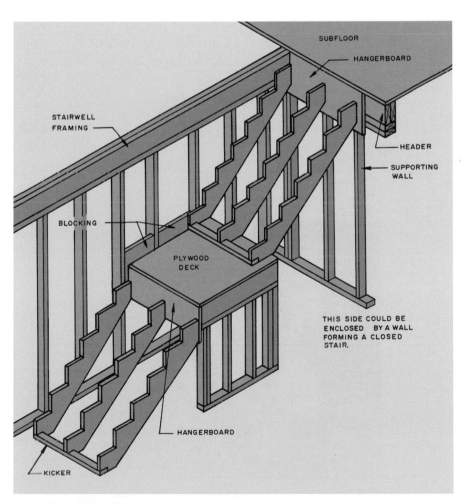

8-31 This landing is for an a straight stairway. Notice the use of a hangerboard.

the tread. Put glue in the bottom hole and push the bottom dowel into it. Position so the square base is parallel with the end of the tread.

The baluster with the tenon is glued into the hole in the handrail and the notch. The end molding is then nailed over the end of the tread.

Installing a Disappearing Set of Stairs

A disappearing set of stairs is used to provide access to areas, such as as attics, which can be used for light storage and will not be accessed frequently. They do not take up floor space and are less expensive to build than permanent ones. They are typically located in the garage ceiling or in a hall. Disappearing stairs do not meet building codes for access to living areas **(8-32)**.

There are two types of disappearing stairs, folding and sliding. The **folding type** is in three hinged sections. The **sliding type** slides up into the attic as a one-piece unit. When the stairs move into the attic the opening is covered with a wood panel.

Folding Disappearing Stairs

The most commonly used disappearing stairs used in residential construction have three ladder-like sections. One section is secured to a plywood ceiling panel and the other two unfold **(8-33)**.

8-32 Disappearing stairs provide an economical way to access storage space in the attic and do not permanently occupy floor space. *Courtesy Werner Ladder Company, 1-724-588-2000*

The ceiling panel is held in the closed position by strong springs on each side. You open the panel by pulling on a cord attached to it. The ceiling panel swings on a piano-type hinge. As you lower the panel, grasp the folded sections and pull them out and unfold them until they are in a straight position **(8-34)**.

Folding disappearing stairs are available in wood and aluminum as shown in **8-35** and **8-36**.

As you select a set of disappearing stairs, examine several types. The size of the opening in the ceiling will vary and may influence the choice. Typically it requires an opening in the ceiling 22 to 30 inches wide

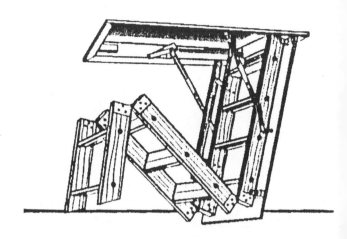

8-33 The folding disappearing stairs have three sections that fold up on top of the ceiling panel.
Courtesy Werner Ladder Company, 1-724-588-2000

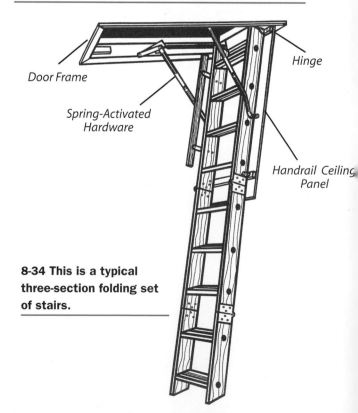

Hinge

Door Frame

Spring-Activated Hardware

Handrail Ceiling Panel

8-34 This is a typical three-section folding set of stairs.

and 54 to 72 inches long. The required opening is larger than the size of the panel. The opening must accommodate the jambs, springs, and hardware needed to install it. The manufacturer's instructions will specify the size of the rough opening.

Observe the labels and warnings that come with the stairs. For example, there are weight limits on what you can carry on them. A typical limit is 300 pounds.

Since disappearing stairs are steeper than regular stairs, manufacturers recommend you do not climb these stairs with an object in your hands. You need

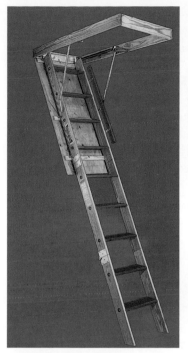

8-35 The wood folding disappearing stairs are available in several widths and have a weight-carrying capacity of 300 lbs. They come with a handrail and slip-resistant treads. *Courtesy Werner Ladder Company, 1-724-588-2000*

8-36 The aluminum folding disappearing stairs are lightweight, easy to operate, and unaffected by moisture. They have a weight-carrying capacity of 300 lbs., and come with a handrail and slip-resistant treads. *Courtesy Werner Ladder Company, 1-724-588-2000*

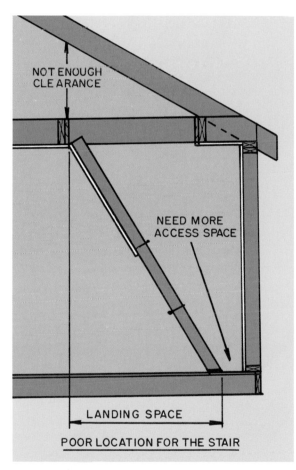

8-37 Locate the stair so there is plenty of room to mount it and so you have adequate headroom in the attic.

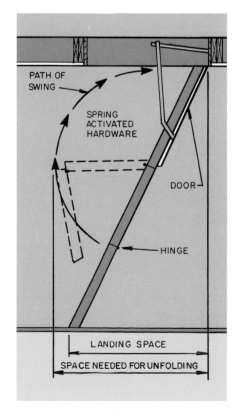

8-38 When locating the folding disappearing stairs, leave enough room so they can be folded and swing clear at the ceiling.

both hands on the side rails as shown in **8-32.** Instead have someone help you to move items up and down from the attic.

LOCATING THE STAIRS

Before you order a disappearing stair, measure the height from the floor to the ceiling. Stairs are available for ceiling heights from 7 feet to 10 feet 10 inches. Also check in the attic to see if the roof will give you the headroom needed to use the stairway. You need as much headroom in the attic as possible. Placing the stairs near the center of the house will obviously give you as much headroom as possible **(8-37).** You

also need room on the floor for the required landing space **(8-37).** In addition, space is needed between the end of the stairs and a wall or other obstacle so you can mount the stairs safely. Finally, the stairs require space so they can swing out and unfold **(8-38)**; this is specified by the manufacturer.

THE ROUGH OPENING

The instructions that come with the stairs will specify the size of the rough opening. The rough opening usually specified is about ½ inch wider than the actual size of the framing for the top of the stairs.

1. Chisel away some of the ceiling in the center of the stair location.

2. Cut away a part of the ceiling until you find a joist.

3. Draw a rectangle the size of the required rough opening. Cut away the ceiling wallboard exposing the joists.

8-39 After locating the opening for the stair, cut away the wallboard.

Courtesy Werner Ladder Company, 1-724-588-2000

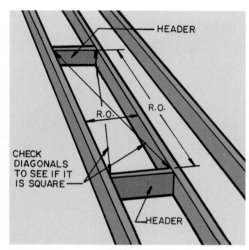

8-40 Stairs requiring a 22¹/₂-inch rough opening can fit between joists spaced 24 inches O.C. without cutting a joist. *Courtesy Werner Ladder Company, 1-724-588-2000*

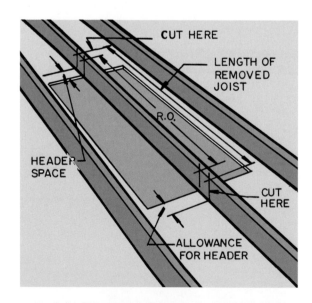

8-41 A layout when joists are spaced 16 inches O.C. After cutting the opening, mark the joist to be cut. Remember to allow for a double header on each end. *Courtesy Werner Ladder Company, 1-724-588-2000*

If the stairway is to be installed in existing construction, locate it on the ceiling and cut away the wallboard (8-39). Check the location of the ceiling joists so one side of the opening will be an existing joist. As you cut away the wallboard wear eye protection and a dust mask. Also check in the attic to be certain there are no electric wires or pipes that could be cut by mistake. Once the wallboard has been removed you will need to reinforce the ceiling joists and install headers.

It greatly simplifies installation if the stairs can be installed parallel with the joists. This requires cutting at most one ceiling joist.

If the stairs are located parallel with the ceiling joists that are installed 24 inches O.C. and you use a stair requiring a 22-inch wide rough opening, you will not have to cut a joist. Install headers as shown in 8-40.

Wider stairs and ceilings with joists spaced 16 inches O.C. will require you to cut one joist as shown in 8-41. After you cut the opening in the ceiling, nail 2 x 6 braces to the joists to be cut and at least one joist on each side. This prevents the loose wallboard from cracking during installation (8-42).

Now cut the joist and install double headers on each end (8-43). Then install a stringer on one side to

8-42 Before cutting the joist, add 2 x 6 reinforcements on each end to hold the cut joist in place. Otherwise you will likely crack the ceiling wallboard. *Courtesy Werner Ladder Company, 1-724-588-2000*

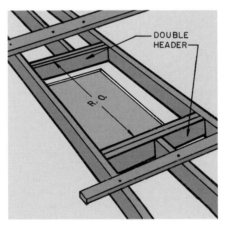

8-43 After you cut the joist, install the double headers on each end. *Courtesy Werner Ladder Company, 1-724-588-2000*

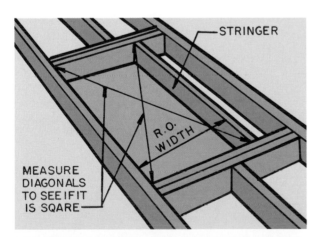

8-44 After the headers are in place nail the stringer to establish the width of the rough opening. Nail the wallboard ceiling to the headers and stringer. *Courtesy Werner Ladder Company, 1-724-588-2000*

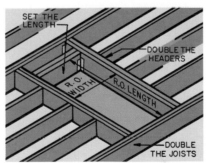

8-45 If you run the stair perpendicular to the ceiling joists, you will have to cut several joists. The ceiling will need to be supported by beams and posts below. *Courtesy Werner Ladder Company, 1-724-588-2000*

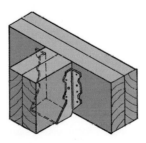

8-46 Headers and joists can be hung with metal joist hangers.

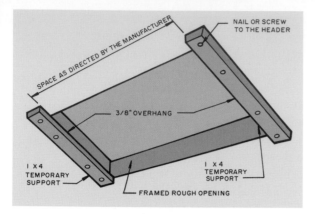

8-47 This is the rough framed opening from below. The temporary supports hold the stair frame in place as it is leveled and secured to the rough ceiling framing.

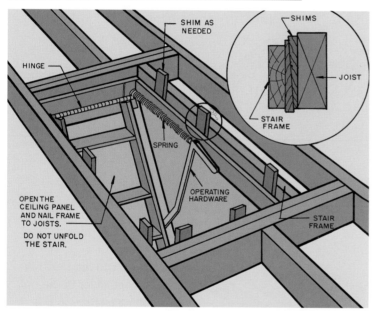

8-48 After setting the folded stair in the opening, secure it to the ceiling joists and headers with nails or screws. You can lower the ceiling panel to give room to install the fasteners. Use wedges to keep the sides of the stair frame straight.

Courtesy Werner Ladder Company, 1-724-588-2000

establish the width of the rough opening **(8-44)**. Nail the wallboard to the headers and stringer and remove the temporary supports.

If the stairs are installed perpendicular to the joists the layout is the same. You will have to cut several joists and secure to double headers on each side of the opening **(8-45)**. Connect the headers to the joists with metal joist hangers **(8-46)**. You will have to support the ceiling from below with posts until the cut joists are secured to the header.

INSTALLING THE STAIRS

Follow the manufacturer's installation instructions. The following suggestions are typical.

Begin by installing temporary bracing on each end of the rough opening **(8-47)**. It projects about one inch into the rough opening. If secured with wood screws it can be easily removed when the installation is finished.

Now with several helpers raise the stairway into the opening. You will need several strong, steady stepladders. Keep the stairs in the folded position on the plywood panel. The side with the hinge must be on the side of the opening from which you want the stairs to pivot. Rest the frame on the temporary supports. It helps if there is someone in the attic to maneuver the stairs in the opening. Be certain to have subfloor installed in the attic around the opening before starting the installation.

Next open the panel with the stairs folded on top. Level the frame and tack it to the sides of the rough opening. Do not drive the nails completely in because it may be necessary to remove some to readjust and level the frame.

Check to be certain the frame is still square. You can do this by measuring the diagonals or checking each corner with a square. Now tap wood shims between the stair frame and the joists as shown in **8-48**. Drive the nails home but do not overdrive or

8-49 The ladder on the wood disappearing stair must have the length adjusted so that each hinged joint closes and the feet rest solidly on the floor. *Courtesy Werner Ladder Company, 1-724-588-2000*

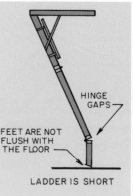

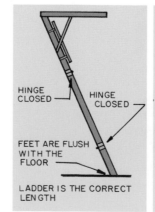

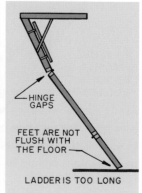

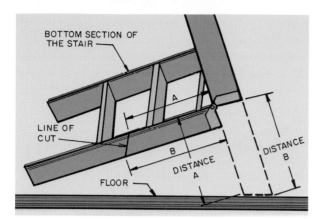

8-50 To adjust the ladder to the correct length, measure the distances A and B, lay them out on the bottom section, and cut on the line located. *Courtesy Werner Ladder Company, 1-724-588-2000*

8-51 Aluminum disappearing stair ladders have adjustable feet that slide until the ladder is the proper length; they are then bolted to the leg. *Courtesy Werner Ladder Company, 1-724-588-2000*

the frame will become bowed and the ladder will rub on the frame. Secure the frame to the rough opening with 16d box nails or 3-inch wood screws. If using screws drill pilot holes through the frame and anchor holes in the joists. Now screw the main hinge at the top of the stairs through the frame and into the rough opening framing. Observe any other instructions given by the manufacturer. Lower the ladder to see if it will move freely.

ADJUSTING THE LADDER

After installation the ladder will most likely need some adjustments. Extend it to the floor and check the length (**8-49**).

If the ladder is too long, measure the distance from the floor to the second section of the ladder as shown in **8-50**. It is suggested that you measure each leg separately because frequently the distance from the ceiling to the floor is not the same on both sides of the ladder. Mark the trim line and cut to it. Lower the third section to the floor and test it for fit.

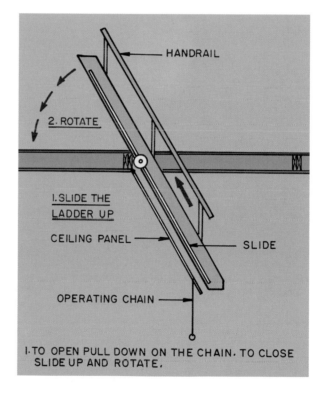

8-52 This aluminum sliding disappearing stair slides up into the attic and folds down against the top of the ceiling joists. *Courtesy Werner Ladder Company, 1-724-588-2000*

157

If the ladder is short you can add on pieces to each side of the third piece, measure, and cut to length.

Aluminum stairs have adjustable feet (8-51). Loosen the screws and slide the feet until the ladder is the correct length. Be certain the screws are securely tightened.

Sliding Disappearing Stairs

Sliding disappearing stairways have solid stringers. The stairs slide on guide bars carried by spring-loaded cables. In the closed position the stairs slide over the top of the ceiling joists and the ceiling panel (8-52). The stairway is pulled down by pulling on a hanging cord. It then rears up into the attic and slides down the ceiling panel to the floor.

Sliding disappearing stairways are available in the same sizes as folding stairs and have load-carrying capacities of 400 to 800 pounds. They tend to be sturdier than folding stairs (8-53).

Sliding disappearing stairs are installed in much the same way as folding stairs. Observe the manufacturer's detailed instructions.

8-53 A wood sliding disappearing stair has the ladder as a single unit. It uses special hardware to slide into the attic and fold down against the joists. *Courtesy Werner Ladder Company, 1-724-588-2000*

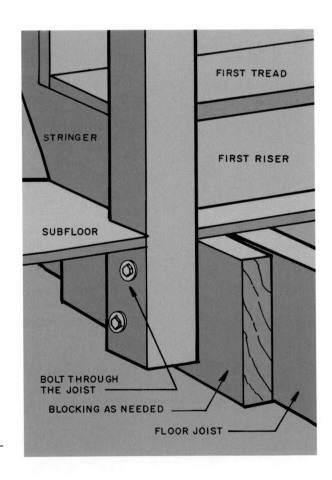

FIRST TREAD

STRINGER

FIRST RISER

SUBFLOOR

BOLT THROUGH THE JOIST

BLOCKING AS NEEDED

FLOOR JOIST

8-54 Run the newel post through the subfloor and bolt it to a floor joist if one is close.

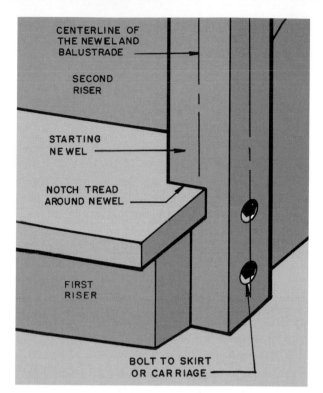

CENTERLINE OF
THE NEWEL AND
BALUSTRADE

SECOND
RISER

STARTING
NEWEL

NOTCH TREAD
AROUND NEWEL

FIRST
RISER

BOLT TO SKIRT
OR CARRIAGE

8-55 This newel post is bolted to the carriage and is notched to line it up with the centerline of the balustrade. Notice the tread is notched around the newel.

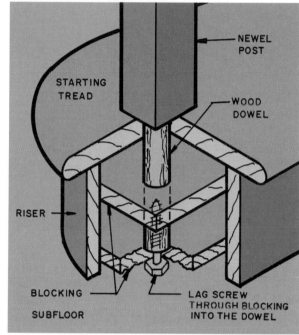

NEWEL
POST

STARTING
TREAD

WOOD
DOWEL

RISER

BLOCKING

SUBFLOOR

LAG SCREW
THROUGH BLOCKING
INTO THE DOWEL

8-56 You can secure newels that use dowels by installing a small lag screw into the bottom of the dowel.

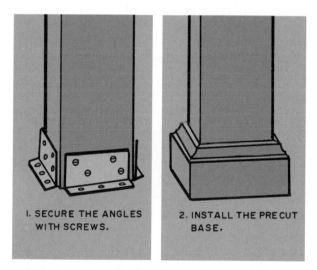

1. SECURE THE ANGLES
WITH SCREWS.

2. INSTALL THE PRECUT
BASE.

8-57 Secure the metal angles to the newel and floor with the screws provided. Then nail the premitered base around the newel. *Courtesy Universal Building Systems, 1-800-200-6770*

8-58 This kit lets you connect the newel post to the floor or to a tread with metal angles, and provides a wood base to cover them. *Courtesy Universal Building Systems, 1-800-200-6770*

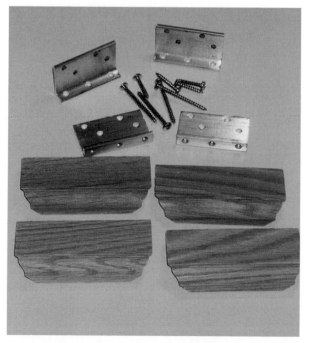

8-59 This newel post connector uses a long hanger bolt to secure the newel to the stair framing. *Courtesy Universal Building Systems, 1-800-200-6770*

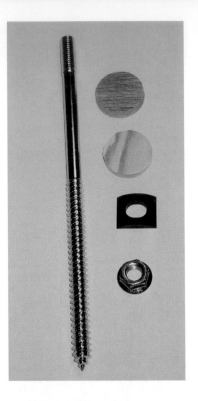

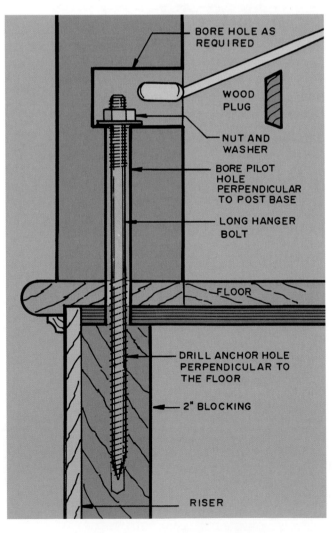

BORE HOLE AS REQUIRED

WOOD PLUG

NUT AND WASHER

BORE PILOT HOLE PERPENDICULAR TO POST BASE

LONG HANGER BOLT

FLOOR

DRILL ANCHOR HOLE PERPENDICULAR TO THE FLOOR

2" BLOCKING

RISER

Repairing Shaky Balustrades

If the balustrade has become a bit shaky, examine how it was installed. It may mean simply tightening some bolts holding the newel post (**8-54** and **8-55**). If lag screws were used and the threads in the joist are stripped, remove them and glue a wood plug in the screw hole. After the glue dries reinstall the lag screw. If possible replace the lag screws with bolts.

If the newel has been installed with a dowel, it sometimes can be pulled up by installing a lag screw in the bottom as shown in **8-56**. To remove the newel post and reglue it means the baluster will have to be pretty well taken apart. Rather than doing this consider installing metal angles which are then covered with a baseboard molding (**8-57**). These angles are sold in kit form for newel-post installation at the local building supply dealer (**8-58**).

If the newel-post has to be removed during major repairs, a good way to reinstall it is with a newel-post connector that is sold especially for this purpose (**8-59**). Installation is shown in **8-60**.

Another newel post connector kit provides a plastic plate that is screwed to the bottom of the post (**8-61**). Anchor screws are set in the stair tread or floor and the post is slid over the screw. A plastic clip snaps the post to the screw (**8-62**).

A metal plate and fasteners are available in kit form (**8-63**). The plate is screwed to the bottom of the post. The post is set in place and the plate is screwed to the tread or floor. The exposed plate is covered with wood base and a notched molding (**8-64**).

Should the handrail feel loose a simple way to tighten it is with a rail bracket (**8-65**) installed below the handrail, as shown in **8-66**. This is visible after it has been installed.

A better way to tie the handrail to the newel post is with a handrail installation kit (**8-67**). Installation is shown in **8-68**. The head of the lag screw is covered with a wood button. The kit shown in **8-69** can be used to connect the handrail to the wall, a post, or a rail-to-rail connection (**8-70**).

8-60 The long hanger bolt is screwed into the blocking set behind the riser. It provides a strong newel post connection.

Courtesy Universal Building Systems, 1-800-200-6770

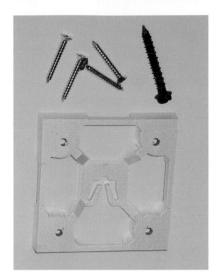

8-61 This plastic post mounting plate is typically used to secure porch posts that are braced with a strong railing. *Courtesy Universal Building Systems, 1-800-200-6770*

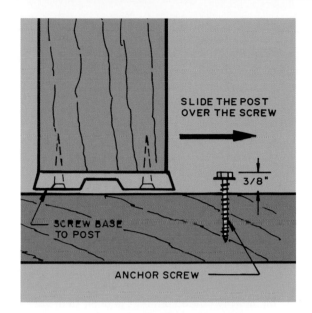

8-62 Secure the plastic base to the bottom of the post. Mount the anchor screw at the location for the center of the post. Slide the post with the notch in the base, facing the screw, over the screw until it snaps in place. *Courtesy Universal Building Systems, 1-800-200-6770*

8-63 This kit uses a steel plate connected to the bottom of the newel post and screwed to the floor. The extended plate is then covered with molding. *Courtesy Universal Building Systems, 1-800-200-6770*

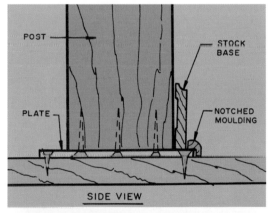

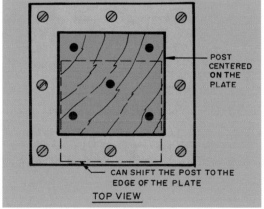

8-64 This newel post mounting plate will let you set the post centered on it or shifted to one edge. After the post is anchored with the connector you can conceal it with standard base, quarter round, or other molding. *Courtesy Universal Building Systems, 1-800-200-6770*

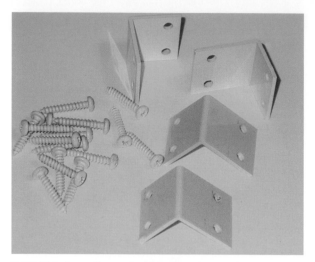

8-65 This heavily coated bracket is used to secure the handrail to the post. *Courtesy Universal Building Systems, 1-800-200-6770*

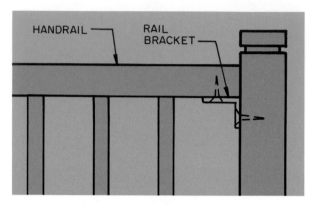

8-66 Secure the bracket below the handrail to the post. The angle bracket provides a connection from the handrail to the post and the wall.

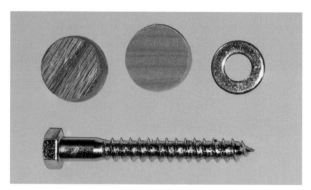

8-67 This handrail installation kit provides a connection between the newel post and handrail in post-to-post balustrades. *Courtesy Universal Building Systems, 1-800-200-6770*

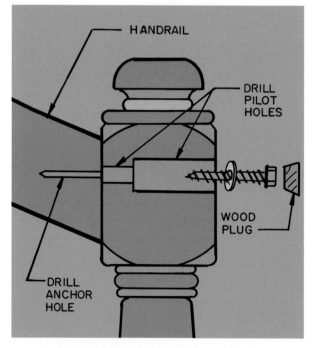

8-68 Drill pilot and anchor holes as specified by the manufacturer. Install with a socket wrench. Then glue the wood plug in place and sand smooth after the glue has dried. *Courtesy Universal Building Systems, 1-800-200-6770*

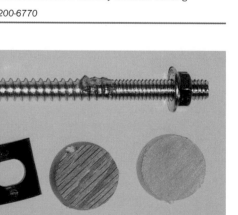

8-69 This hanger bolt can connect handrails to the newels and also connect the handrails to walls and is completely hidden. *Courtesy Universal Building Systems, 1-800-200-6770*

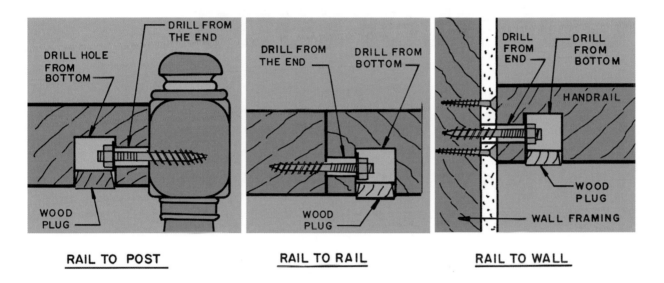

RAIL TO POST   RAIL TO RAIL   RAIL TO WALL

8-70 Install this connector by drilling a hole in the end of the handrail and in the bottom. Tighten the hanger bolt in the end of the handrail. Slide the end with machine threads in the hole and tighten the nut.

Courtesy Universal Building Systems, 1-800-200-6770

CHAPTER 9

Cabinets & Countertops

In most new construction the cabinets are selected by the homeowner and
installed by the dealer. As you examine the manufactured cabinets, be aware that there are three levels
of quality. The **economy grade** is the lowest and least expensive. Joinery and construction is simple and less
costly wood products are used. **Custom-grade** cabinets have better quality framing and joinery and are
possibly the most commonly used grade. **Premium-grade** cabinets have the best joinery and

most expensivewoods. They use considerable amounts of molding and machined panels.

If you plan to remodel a kitchen or bath you may want to build your own cabinets and install them. Detailed procedures for actually building cabinets are available in cabinetmaking books. The following information will acquaint you with how quality cabinets are constructed and with recommended installation techniques.

The decisions as to the size and type of cabinets are made as the kitchen and bath are planned. Example layouts are shown in Chapters 10 and 11. The end result of planning and study will produce a beautiful kitchen (**9-1**).

Kitchen Cabinet Types & Sizes

Manufacturers offer a good variety of cabinet sizes and combinations of drawers and doors. Typical **kitchen base cabinets** are shown in **9-2**. Standard sizes are shown in **9-3**. Manufacturers offer quite a

9-1 These high-quality cabinets with their panel construction and appropriate moldings make the kitchen very attractive. *Courtesy Wellborn Cabinet, Inc.*

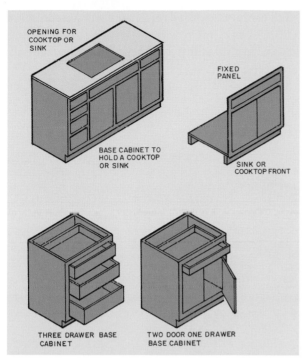

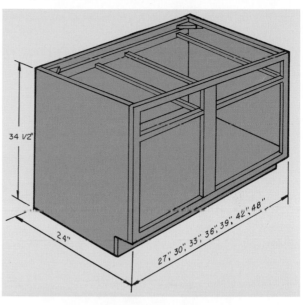

9-3 Standard sizes for kitchen base cabinets.

9-2 A few of the many designs for base cabinets for kitchens.

9-4 The entire bottom shelf rolls out, making the items accessible.

9-5 A lazy Susan in a corner makes use of an otherwise wasted area.

variety of special storage features as shown in **9-4, 9-5, 9-6,** and **9-7.**

A few of the many **wall cabinets** and the basic sizes available are in **9-8.** Recommended spacing of wall cabinets above base cabinets and appliances is shown in **9-9.** The use of wire racks expands the capacity of the wall cabinet shelving **(9-10).**

Tall cabinets are used for general storage and some hold an oven or microwave **(9-11).** The use of sliding shelves with substantial runners makes the contents very accessible **(9-12).**

Several bath variety designs and typical sizes are shown in **9-13.**

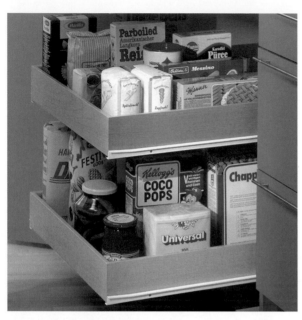

9-6 Sliding high-capacity drawers using metal drawer slides with rollers makes access to large packaged items easy. *Courtesy Blum, Inc.*

9-7 Specially designed drawer liners are used to store eating and cooking utensils.

Courtesy Blum, Inc.

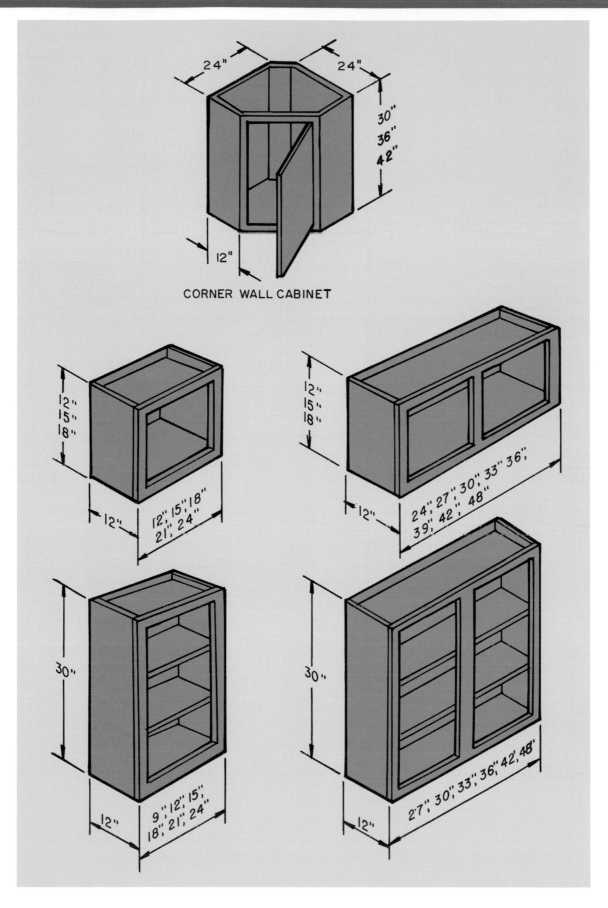

24"   24"

30"
36"
42"

12"

CORNER WALL CABINET

12"
15"
18"

12"

12", 15", 18"
21", 24"

12"
15"
18"

12"

24", 27", 30", 33" 36",
39", 42", 48"

30"

12"

9", 12", 15",
18", 21", 24"

30"

12"

27", 30", 33", 36", 42", 48"

9-8 These are some of the many types of wall cabinet available and typical stock sizes.

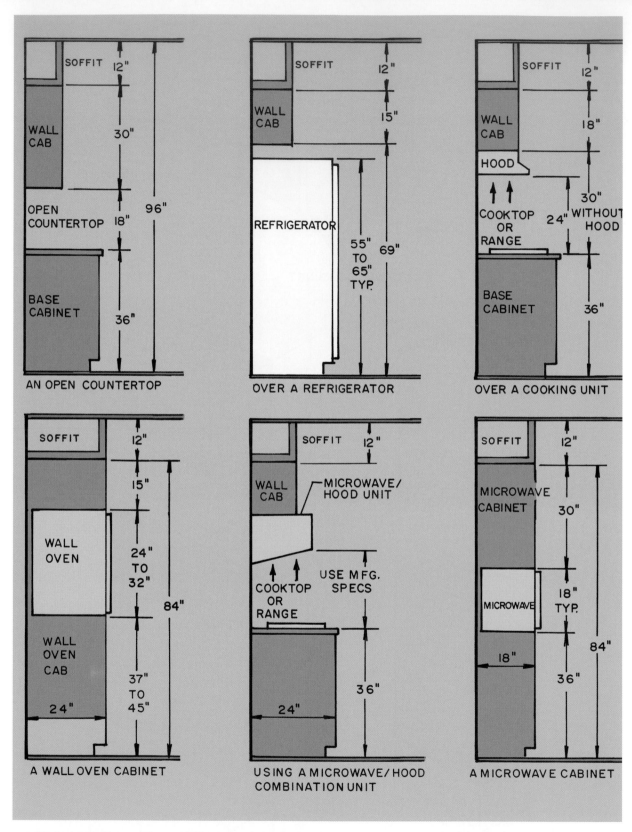

9-9 Recommended spacing of kitchen wall cabinets above countertops and appliances. *Courtesy National Kitchen & Bath Association*

9-10 Wire racks greatly increase the storage capacity of wall cabinet shelving.

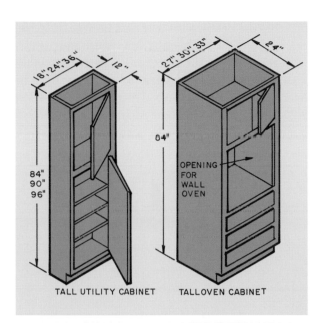

TALL UTILITY CABINET   TALLOVEN CABINET

9-11 A couple of the tall cabinet designs available with typical sizes.

9-12 Sturdy sliding shelves that run on heavy-duty roller runners expand the use of tall cabinets considerably. *Courtesy Blum, Inc.*

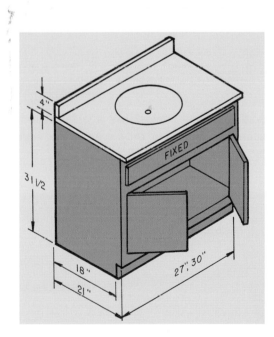

9-13 Two typical bath vanities and commonly available sizes.

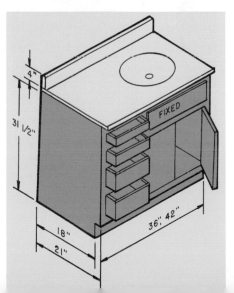

Cabinet Construction

Cabinets can be divided into two major types, based on their construction. Some are made **with a faceframe (9-14** and **9-15)**, while others are made **without a faceframe**. These are referred to as European-style **(9-16** and **9-17)**. Many of these have adjustable metal legs. They are turned on a long screw which helps to level the cabinet **(9-18)**. However, some manufacturers run the bulkheads to the floor as on cabinets with a faceframe.

CONSTRUCTION WITH A FACEFRAME

This type of construction offers several ways to relate the doors and drawer fronts to the faceframe. One way is to set them inside the frame so they are flush with the frame when closed **(9-19)**. This presents the problem of getting a perfect fit so the crack around the edges is the same width and parallel with the edges of the faceframe. The more widely used method is to allow the door or drawer front to overlap the faceframe. Several methods are shown in **9-20**. The overlap covers the edge of the faceframe so if it is a little out of line, this is not visible.

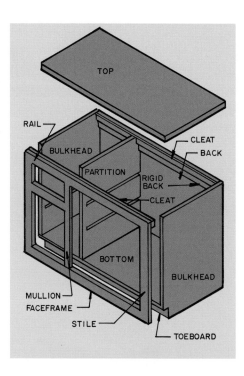

9-14 These are typical construction details for the carcass of a base cabinet with a faceframe.

9-15 The base cabinet has a wide faceframe. The doors are lipped so they overlap the edge of the faceframe.

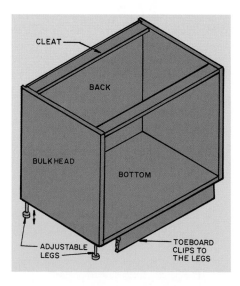

9-16 This is typical carcass construction for a European-style cabinet using frameless construction. Some use metal legs which can be adjusted by screwing them up and down.

9-17 This base cabinet uses frameless construction. The doors and drawers overlap the edges of the bulkhead, bottom and top framing, leaving a small reveal.

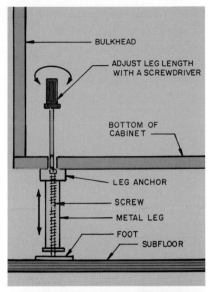

CONSTRUCTION WITHOUT A FACEFRAME

This European style of construction can have the doors set flush with the sides of the bulkheads or overlap as shown in **9-21.** They can completely overlay the bulkheads on all sides or be parted, producing a slight reveal.

There are special hinges used to hang cabinet doors on cabinets without faceframes. The manufacturer will include detailed installation instructions.

The hinge in **9-22** can be used on European-style cabinets. It permits the door to be installed and removed from the cabinet by pressing on the lock

9-18 The base cabinet can be leveled by adjusting the length of the leg. This is done by turning the internal screw with a screwdriver.

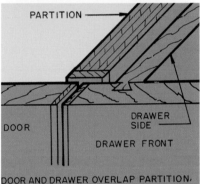

DOOR AND DRAWER OVERLAP PARTITION.

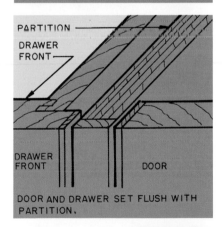

DOOR AND DRAWER SET FLUSH WITH PARTITION.

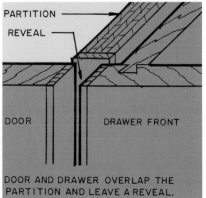

DOOR AND DRAWER OVERLAP THE PARTITION AND LEAVE A REVEAL.

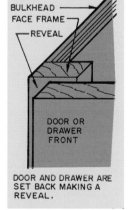

9-19 This type of cabinet construction sets the drawer fronts and doors flush with the faceframe.

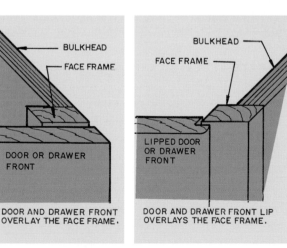

DOOR AND DRAWER FRONT OVERLAY THE FACE FRAME.

DOOR AND DRAWER FRONT LIP OVERLAYS THE FACE FRAME.

DOOR AND DRAWER ARE SET BACK MAKING A REVEAL.

9-20 Conventional cabinet construction with the drawers and doors extending over the faceframe.

9-21 Typical door and drawer installation on cabinets when a faceframe is not used.

9-22 This cabinet hinge is in two pieces. Part is installed on the inside of the bulkhead, and the other part is attached to the door.

Courtesy Melpa-Alfit, Inc.

9-23 To secure the door to the cabinet, place the lock lever over the mounting plate on the inside of the cabinet bulkhead.

Courtesy Melpa-Alfit, Inc.

9-24 To remove the door from the cabinet lift up on the lock lever.

Courtesy Melpa-Alfit, Inc.

9-25 Locate the base cabinets on the wall.

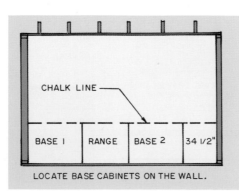

CHALK LINE

BASE 1 | RANGE | BASE 2 | 34 1/2"

LOCATE BASE CABINETS ON THE WALL.

lever on the end of the hinge. The door is secured to the cabinet by pressing down on the lock lever (9-23). It can be removed from the cabinet by lifting up on the lock lever (9-24).

Preparing the Room for Installation

When you selected the cabinets it was necessary to locate all plumbing and electrical features. The new cabinets must accommodate these. For example, the fixtures, as a sink or lavatory, must be positioned to link up with the sewer and water pipes. If during remodeling the kitchen or bath design was changed, the plumbing and electrical would also be changed to suit the new layout. Before starting the cabinet installation check this out.

Also check the condition of the walls and floor. If the walls are newly finished or repaired, check to see that they are dry. Also check to see if they are straight vertically and horizontally. A bow or out-of-plumb wall will make cabinet installation difficult.

Check the floor with a long level to see if it has any high spots or long irregularities. These must be handled before the base cabinets can be installed so the top is level.

If the walls or floor still need repair, it is best to delay cabinet installation until this is finished. The finished floor material is installed after the cabinets are in place.

The temperature in the room and the humidity should be at normal levels expected during occupancy. The floor should be clean and free of debris, shavings and dust.

LOCATE LEVELING LINES

Begin by drawing level lines on the wall to locate the top of the base cabinet without the countertop (9-25). Mark appliance locations. While the typical base cabinet is 34½ inches high, you may have to make some allowance for a bowed or damaged floor. Measure from the highest point on the floor.

Next draw level lines on the wall to locate the top of the wall cabinet (9-26). This is usually 84 inches above the floor. These lines can be located with a chalk line or laser level. If the cabinets are to butt against a soffit, this will have already been installed. (9-27).

Finally, locate the studs by marking them on the wall behind each cabinet location (9-28).

Installing the Base Cabinets

While there are a number of ways to install base cabinets, the following is typical.

1. If there are pipes or electrical services to be brought through the cabinet, mark and cut the required openings.
2. Begin with a corner unit. If the floor is not level you will have to shim the base with very thin wood wedges (9-29). Check to see that the top edge is parallel with the level line and the cabinet is perpendicular to the wall.
3. If the wall has a bow this will show when the cabinet is pushed tight against it. You may have to shim the back edge so that all the base cabinets line up when installed (9-30).

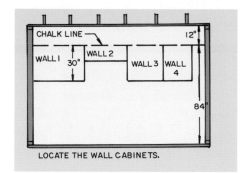

9-26 Locate the wall cabinets on the wall.

LOCATE THE WALL CABINETS.

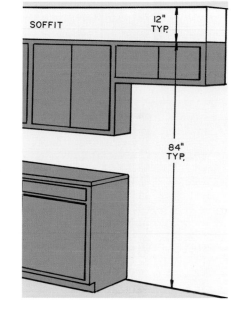

9-27 If a soffit is used, this will be installed before the cabinets. Check it for levelness.

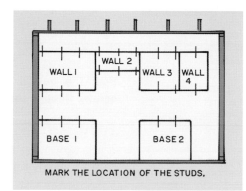

9-28 Mark the location of the studs on the cabinet outlines.

MARK THE LOCATION OF THE STUDS.

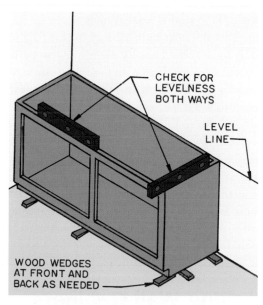

9-29 Locate the first base corner cabinet. Shim it so it is level.

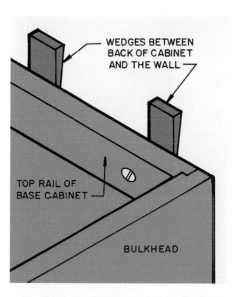

9-30 Wedges can be used to hold the base cabinets straight if the wall bows or is out of plumb.

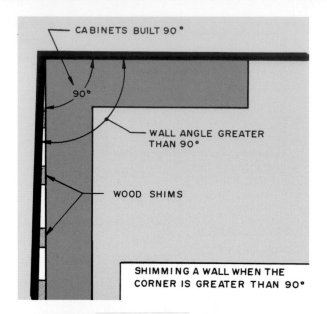

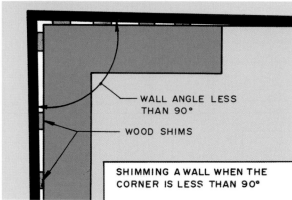

9-31 When a wall is out of square, use wood shims to hold the cabinets square.

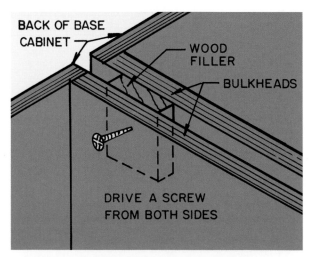

9-33 A wood filler strip is screwed in place between the bulkheads at the rear of the cabinet.

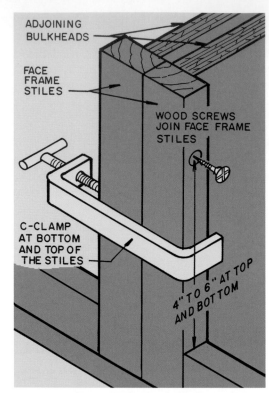

9-32 The butted cabinets are joined with wood screws through the stiles.

4. If the base cabinets turn a corner, check to see if the walls meet at 90 degrees. If they do not, shim as shown in **9-31.**

5. Secure the base to the wall studs through the back or framing if provided. Usually No. 8 or 10 flat-head wood screws will do the job. They should penetrate the stud at least ¾ inch.

6. Now butt the next base cabinet to this first one. Join it to the first base by clamping the butting stiles and securing them with flathead wood screws as shown in **9-32.** Then join the cabinets at rear as shown in **9-33.** The wood filler strip should be wide enough to hold the side bulkheads parallel. Now level the base and secure to the studs as just described.

There will be times when space is left between cabinets meeting in a corner. When this happens filler strips are installed **(9-34).** They are available prefinished to match the cabinets from the manufacturer. Two other ways to handle this space are shown in **9-35.**

Installing Wall Cabinets

If the cabinets are to be installed below a soffit, check to see if it is level (refer to **9-27**). If it is level the wall cabinets can be butted against it and secured to the wall studs. If it is a little out of level, shim the cabinets along it. After installation any crack can be covered with a small molding.

If no soffit is used (9-36) you will line up the top of the cabinets with the level line. However, first check the wall to see if it has any bows. If the wall is out of plumb, you can shim the cabinets as shown in 9-37. Also check the corners of the room to see if they are at a 90-degree angle. If not, the cabinets will have to be shimmed as shown in 9-30 and 9-31.

If there are pipes or electrical services to be brought through the cabinet, mark and cut the required openings.

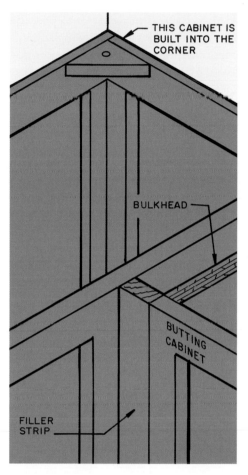

9-34 A filler strip is used to fill any gap between cabinets forming a corner or between a cabinet and a wall.

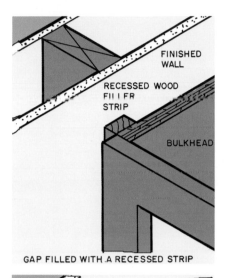

GAP FILLED WITH A RECESSED STRIP

9-35 Two other ways to conceal a gap between the cabinet and the wall.

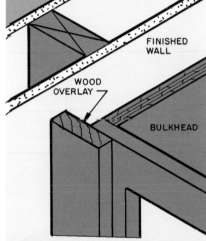

WIDE GAPS COVERED WITH AN OVERLAY.

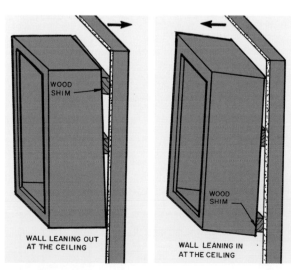

9-37 If the wall is not plumb, shim the cabinets so they are plumb.

9-36 These wall cabinets were installed in a kitchen where soffits were not used.

9-38 Wall cabinets are joined together with wood screws through the front stiles before they are raised up on the wall.

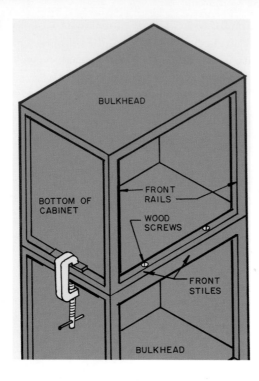

To install wall cabinets:

1. Begin by joining several cabinets together (9-38). Place them on the floor in the order in which they are to be on the wall. Clamp the butting stiles. Check to be certain the bottom rails are in alignment. Then join them with wood screws.

2. Start the installation of the wall cabinets with a unit in a corner. Raise the cabinets into position along the level line. Generally you will use some type of support that rests on a temporary plywood or particleboard top on the base cabinet. Some use boxes made to support the cabinets at the desired height (9-39).

3. Install wood screws through holes drilled in the top and bottom rear rails (9-40). The screws should penetrate the stud at least one inch. Be certain to install any needed shims before tightening the screws. Check the cabinets to be certain they are straight. A chalk line will do this for you. Check for plumb with a level.

A cabinet contractor will most likely have some type of lift that will raise a long section of assembled wall cabinets (9-41). Without a lift, it will take several people to raise the wall cabinets up on the boxes atop of the base cabinet.

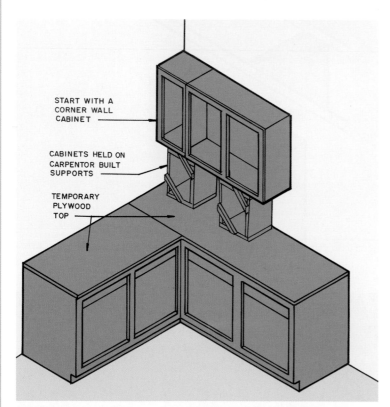

9-39 Wall cabinets can be supported while being installed by using carpenter-built supports that rest on a temporary plywood top on the base cabinets.

9-40 Wood screws holding wall cabinets to the wall should penetrate the stud at least one inch.

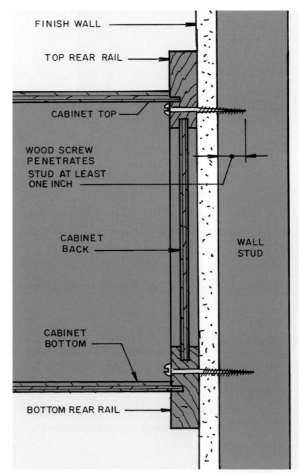

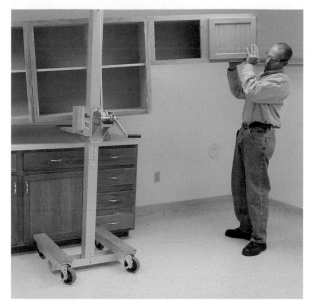

9-41 Lift the wall cabinet and roll it against the wall. Raise to the proper height. Support it on the base cabinet with the adjustable cabinet lifter and check for levelness. Screw to the studs. *Courtesy Telpro Inc.*

Renovating Old Cabinets

As cabinets age, many things can need some touching up or complete reworking. Generally, manufactured cabinets made following the guidelines of the National Kitchen Cabinet Association and certified by them will serve for many years with few problems. Following are some things that may need attention.

Drawers

If the drawers use wood drawer guides, they may start to stick over the years, mainly due to moisture. If the drawer fits tightly in the guides, sand them lightly to provide additional clearance. Then wax each so the drawer slides freely. Some type of metal drawer slide is usually used. If it is a bit hard to operate, remove the drawer and lightly oil the wheels and wipe the track clean. Quality cabinets have some type of metal drawer slide. Two types are shown in **9-42**. If you are going to continue to use the old drawers, consider mounting them with metal drawer slides. A wheel rolls in the track, giving smooth, trouble-free operation. When purchasing new cabinets make sure they have quality drawer slides.

Doors

If a door begins to sag or feel loose, examine the hinges. If they move up and down as you operate the door, probably the screws are loose in the wood stile. Try tightening them. If they seat firmly, the problem

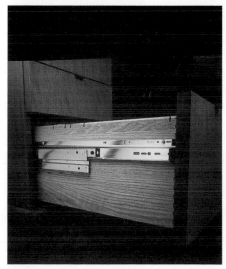

9-42 If you are replacing your cabinets or renovating the old ones, be certain to use high-quality drawer slides like these. *Courtesy Knape and Vogt Manufacturing Company*

9-43 Loose pieces of plastic laminate can be reglued.

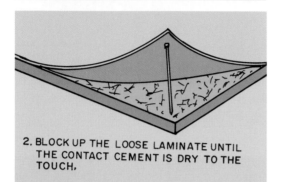

1. CAREFULLY RAISE THE LOOSE LAMINATE, BRUSH CONTACT CEMENT ON THE SUBSTRATE AND THE BACK OF THE LAMINATE.

2. BLOCK UP THE LOOSE LAMINATE UNTIL THE CONTACT CEMENT IS DRY TO THE TOUCH.

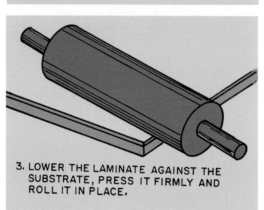

3. LOWER THE LAMINATE AGAINST THE SUBSTRATE, PRESS IT FIRMLY AND ROLL IT IN PLACE.

is corrected. If they turn in the wood and do not tighten, remove the screws, glue wood plugs in the screw holes, and when the glue is dry, reinstall the screws. Some types of crack filler compound can be used to fill the holes. Check on the label to see if they harden enough to hold the screw.

Fixing Laminate & Refinishing

The plastic laminate may start to come loose. Carefully clean as much of the old adhesive off the exposed top and the plastic laminate as possible. Try to avoid stripping the laminate back any more than is already loose. Then brush contact cement on the back of the laminate and on the wood countertop. Let the adhesive dry, lay the laminate down on the top, and press it tight. Then roll the area with a rolling pin to assure full contact (9-43).

9-44 These metal shelf standards are easy to install, offer a large number of shelf positions, and will carry a heavy load.

9-45 This exposed long barrel hinge provides an attractive feature.

If the rails, stiles, drawer fronts, and doors have the finish damaged, consider either repairing the damaged spot or refinishing the entire exposed wood. This is a big job and could require that the existing finish be removed down to the wood and a new finish be applied. Your paint dealer can recommend materials to remove the old finish and products that will make applying a new finish relatively easy. An alternative that some prefer is to sand the old finish, fill any dents and splits, and paint the wood areas with a high quality latex enamel. It is available with a flat, medium-gloss, and high-gloss finish. While the high gloss resists moisture best, the shine may be less pleasant to live with.

Still another approach is to refinish the stiles and rails and install new doors and drawer fronts. This lets you change the design of these parts and the species of wood which influences the color and darkness or lightness of the cabinet mass.

Shelves

The main function of the cabinets is storage. If the shelves are not adjustable or have a poor system, consider installing metal shelf standards. They permit a wide variety of shelf positions and will hold a heavy load **(9-44)**. Storage can be more efficient if cabinets have built-in features as shown earlier in **9-4** through **9-7.**

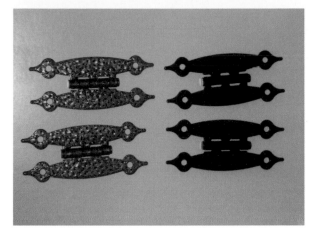

9-46 There are many very decorative, surface-mounted hinges available. They add a classic touch to cabinets and furniture.

Hardware

Another easy way to enhance old cabinets is to install new hardware. Semi-concealed, long-barreled hinges **(9-45)** and surface mounted hinges **(9-46)** are two ways to enhance cabinet appearance. Many other decorative hinges are available. Then there is the wide choice of drawer and door pulls **(9-47)**. A fresh modern look can be gotten by using wood molding pulls across the top of the drawer **(9-48)** or the bottom of the door **(9-49).**

9-47 Installing new drawer pulls and cabinet door handles will freshen up a kitchen or bath.

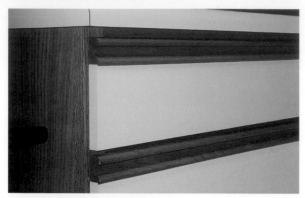

9-48 Shaped wood drawer pulls give the drawer a nice trim and finished look.

9-49 Shaped wood pulls can be installed on the bottom edge of the doors to match those used on the drawers.

9-50 A simple crown molding adds a decorative finishing touch to the wall cabinets.

Enhancing Appearance

The appearance of the kitchen can be improved by adding various moldings to the cabinets—something as simple as a single crown molding around the top of the wall cabinets (9-50); or in a large kitchen in a more expensive home, the cabinets could handle a rather elaborate trim (9-51). Applied carvings of many designs are available and make a fine decorative feature (9-52).

9-51 The wall cabinet has been finished at the ceiling with a wide center molding below a multi-piece crown and has a tall acanthus overlay on the stile. *Courtesy KraftMaid Cabinetry, Inc.*

9-52 Applied overlays add considerably to the appeal of the cabinets. This is an acanthus overlay with rosette fluted columns. *Courtesy KraftMaid Cabinetry, Inc.*

Replacing Countertops, Sinks & Lavatories

A frequently used kitchen or bathroom cabinet upgrade is to install a new countertop and sink or lavatory. This requires considerable study of the materials and fixtures that are available. For kitchens, marble, granite, ceramic tile, high-pressure plastic laminates, solid-surface materials, such as one of many acrylic and polyester polymer plastic products, and stainless steel are used. In bathrooms high-pressure plastic laminates, solid synthetic materials, and ceramic tile are used. Another approach is to replace the wood cabinet and lavatory with a pedestal lavatory (9-53).

Countertop Materials

Plastic laminates are the most widely used countertop surfacing material. They are not affected by water but will scratch if hit by something sharp. You will not want to cut foods on this surface; use a cutting board. They will also scorch if exposed to intense heat. They are available in a wide variety of colors and patterns (9-54).

Marble is expensive and requires the services of professional marble cutters. It is subject to staining and chipping (9-55). Nevertheless many homeowners prefer marble or other natural stone, such as the many granite varieties available.

Ceramic tile countertops are hard, tough, and resistant to damage from heat. While such countertops are impervious to water, the grout between them will over time stain, crack, and mildew. An epoxy grout is being used and produces better results. Ce-

9-53 A pedestal lavatory is a quick way to replace an old battered bathroom lavatory. It does limit storage so cabinets may need to be added on one of the walls.

9-54 This cabinet has a preformed plastic laminate top. The color chosen fits with the finish on the base cabinet. A wide variety of colors and patterns are available.

9-55 This island counter has a custom-made marble countertop.

9-56 This kitchen cabinet base has a ceramic tile countertop and splash. Darker tile provides a trim.

9-57 This lavatory is made from polyester resins and natural fillers. It is durable and impervious to moisture. *Courtesy Avonite, Inc.*

ramic tile can be bonded directly to a plywood or particleboard top but it will deteriorate if the grout lets moisture through. It is best applied to a cement board underlayment placed over the wood top. A wide variety of sizes, colors, and patterns are available (9-56).

There are a number of **solid-surface countertop materials** that are a cast plastic resin such as acrylic, polyester, or a combination of these plus mineral fillers (9-57). Another such material is cultured marble. The countertop and lavatory are cast as a single piece, which is made from marble chips and dust cast into a polyester resin. The more expensive units have a thicker gel coat which helps resist crazing and cracking (9-58).

Sinks & Lavatories

It is sometimes smart to replace the old sink or lavatory when you renovate or replace the old countertop. Even though the sink or lavatory works alright, over the years it begins to look dull and worn and reduces the effectiveness of the new countertop. Sinks and lavatories are available in stainless steel, cast solid-surface materials, molded cultured marble, and porcelain enamel.

Stainless steel sinks are used in the kitchen. They are tough and do not chip, but do show water marks as they age. They will scratch if cleaned with heavily abrasive cleaners. Special cleaners are available (9-59).

9-58 This lavatory is made from marble chips and dust cast into a polyester resin. It is often referred to as cultured marble.

9-59 This stainless-steel two-bowl sink has one smaller bowl. The disposal is usually connected to the drain on the bowl.

Courtesy Elkay Manufacturing Company

Porcelain enamel sinks enable you to have a choice of several colors. They are made from pressed steel and finished with a porcelain enamel which gives a glass-like surface. They will chip if hit with something sharp. Repair kits are available. They clean easily with standard nonabrasive cleaners and are difficult to scratch.

Cast solid-surface sinks and lavatories are molded of materials as described for molded countertops. They are available as separate units which are set in openings in the countertop (**9-60**) or are cast as an integral part of the countertop (**9-61**).

Replacing with New Laminate Countertops

As you consider replacing a countertop, perhaps the most economic decision is to purchase a new top with the laminate installed. One type is a postformed, self-edged top which has the top and splashboard sealed into a watertight unit, as shown in **9-62**. The curved edges provide a very attractive finish.

Another popular professionally built countertop has a wood-edge band that matches the wood in the cabinets (**9-63**). Before you actually buy a new completely finished countertop, check to see if the old top can be removed. Cabinets built in the last 20 years

9-60 This single-bowl sink is made from reinforced, modified acrylic filled with natural materials. Drill holes as needed for faucets, sprays, and soap/lotion dispensers. *Courtesy the Swan Corporation*

9-61 This double-bowl sink is made of a nonporous, homogeneous blend of a polyester resin and natural materials.

Courtesy Avonite, Inc.

9-62 This countertop has a premolded plastic laminate top that runs from the backsplash over the exposed edge with no seams. It is sold in long lengths and cut to size as needed.

9-63 This plastic laminate top is banded with hardwood molding that matches the wood used to build the cabinets.

9-64 These are commonly used methods for edge-banding a countertop.

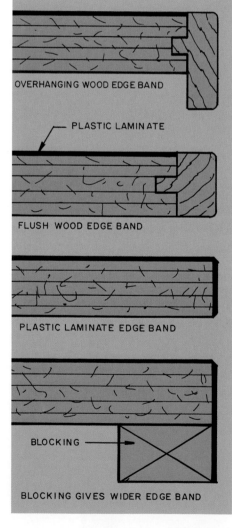

OVERHANGING WOOD EDGE BAND

PLASTIC LAMINATE

FLUSH WOOD EDGE BAND

PLASTIC LAMINATE EDGE BAND

BLOCKING

BLOCKING GIVES WIDER EDGE BAND

or so have the top installed as a separate piece, which can be removed by undoing the screws that hold it to the base. Older cabinets were made with the top permanently joined to the bulkheads and cannot be removed without damaging the remainder of the cabinet. These older tops can be refinished by installing ¼-inch medium-density fiberboard (MDF) over the old countertops. Secure it with one inch spiral underlayment nails spaced two inches along the edges and run one row down the center of the sheet. Install the new laminate to it as described in this chapter. Some try to bond new laminate directly to the old by first sanding the surface of the old to roughen it, then bonding the new to it. Not all types of plastic laminate are recommended for this procedure and the result is often disappointing. Consult the dealer supplying laminate before you try this technique.

The actual application of the plastic laminate is a difficult job and for best results you should have some special tools.

Installing Plastic Laminate

Typically a kitchen base cabinet will be 23¼ inches wide and the countertop 24 inches (refer to **9-3**). This width allows it to accommodate a standard sink and appliances, such as cooktops and dishwashers. Generally the top substrate will be ¾-inch plywood or medium-density fiberboard. A typical bathroom vanity will have a 21-inch-wide base and a 22-inch-wide top (refer to **9-13**). The backsplash is typically four inches high. In both cases a 24-inch-wide sheet of plastic laminate can be used. Wider sheets are available as needed.

Another thing to consider is how the edge of the countertop will be finished. While there are many ways to do this, those in **9-64** are commonly used.

The substrate must be smooth and free of holes or depressions. Fill as necessary with crack filler and sand. Any cracks between butting sheets of substrate should be filled.

Remove any sinks, faucets and appliances that are set into the countertop. If it has metal edging remove it. The splash must also be removed.

TYPICAL INSTALLATION STEPS

After the substrate has been prepared for the new laminate, carefully measure the top to be covered. Cut the laminate sheet to size, allowing it to be about ¼ inch longer and wider than the substrate. Cut the strips for the edge ¼ inch wider and one inch longer than each edge. This will be trimmed off after the laminate has been bonded. Try to lay out the laminate so you do not have a seam. If there is a seam try to locate it away from the major work area of the counter.

CUTTING THE LAMINATE

There are many ways to cut the plastic laminate. One way is to cut it on a **table saw** with a fine-toothed carbide-tipped blade. Cut with the good side up. If the laminate could slip under the saw fence, clamp a board to it that fits tightly against the table (9-65). Laminate can also be cut with a **saber saw** that has a fine-toothed blade. The blade will have to be replaced frequently. It is difficult to cut a straight edge this way, even if you use a guide (9-66).

A **carbide-tipped scoring tool** is also used to cut laminates (9-67). It looks a lot like a large utility knife but has a special carbide-tipped blade. To cut with this tool, place the laminate with the good side up and score it along a straightedge. Take several cuts until you cut into the core of the panel. Do not slip and scratch the face. Then hold the panel against the table top and lift one side of the sheet up until it breaks on the score mark (9-68).

APPLY TO THE EDGES

The edge of the substrate to be covered with laminate should have a strip of plywood or medium-density

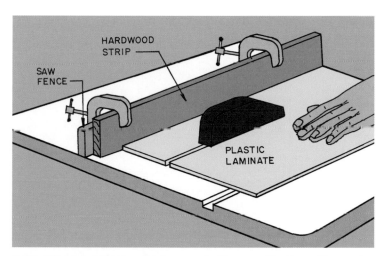

9-65 When cutting laminate on a table saw, it often slips under the fence. To prevent this, clamp a hardwood strip to the fence and fit it tightly to the table.

9-66 Plastic laminate can be cut with a saber saw using a wood straightedge as a guide. Be certain to use a fine-tooth metal cutting blade.

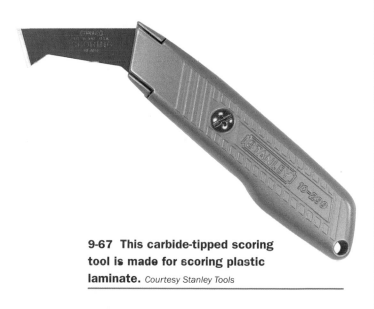

9-67 This carbide-tipped scoring tool is made for scoring plastic laminate. *Courtesy Stanley Tools*

9-68 Score the laminate with the face up along a straightedge. Score it several times. Turn the sheet over and place a wood strip parallel with the score line. Bend the sheet up to snap it in two.

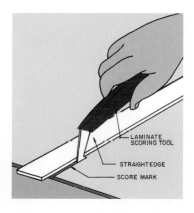

fiberboard bonded to it to provide a wider edge (9-69). This strip is typically ½ to ¾-inch thick and ¾ to 1½ inches wide. Other sizes can be used depending upon the width of the banded edge desired.

Begin by bonding the laminate to the edges first. Apply contact cement to the clean edge and the back of the laminate (9-70). Apply one coat of contact cement and let it dry. Then apply a second coat and let it dry. A disposable brush is a good tool to use.

When both surfaces are dry, test by placing a piece of paper against them. If it does not stick it is dry. If it feels tacky allow more drying time (9-71).

When dry, lay the laminate edge strip against the edge of the countertop. It should extend a little above the surface of the substrate. This will be trimmed off later. Place one end against the edge and apply light pressure along it as you apply it to the other end (9-72). Remember, once the laminate touches the edge it cannot be moved. Usually two people are needed to apply long strips. After it is hand-pressed against the edge, apply pressure along the entire edge with a roller (9-73).

Now trim the laminate flush with the top of the substrate. This is done with a router using a straight laminate-trimming bit (9-74). Check for flushness with a try square.

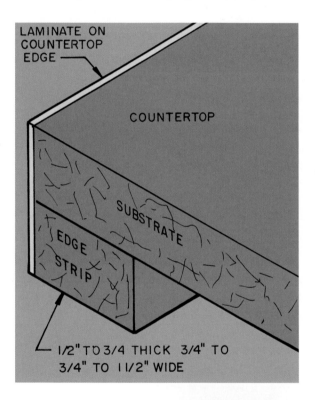

9-69 A strip is glued and nailed along the bottom edge of the countertop substrate. This establishes the width of the exposed laminate on the edge.

9-70 Apply contact cement to the edge of the substrate and the plastic laminate.

9-71 Test the contact cement with paper to see if it is dry.

9-72 Lay the laminate along the edge of the substrate. Let it extend a little above the top surface.

9-73 Firmly press the laminate against the edge with a roller.

Next give the top edge a final trim by running over it with a belt sander. Be careful to not slant it and cut down into the edge of the laminate. Position the sander so the belt rotation is inward on the top. Keep it moving and make a very light, fast stroke (9-75). Use a belt hat has a very fine abrasive. Be very careful to not overcut. This is a very brief, light touch.

If the countertop has a curved corner to be banded with laminate, the procedure is the same as first described. When the contact cement is dry, start applying the edge laminate on one end and work toward the curved corner. When you get there the lam-

inate must be heated until it is flexible enough to make the bend. You can use a heat lamp or heat gun (9-76) and keep trying to make the bend. When it bends easily wrap it around the corner and press against the edge. It will be very hot so you need heavy, heat resistant gloves. Bond the strip on down the edge until it is all in place. Press firmly as you do this so there are no gaps. Then roll the edge.

APPLY TO THE TOP SURFACE
Be certain the substrate and the bottom of the laminate are perfectly clean. Then apply a layer of con-

9-74 Remove the excess laminate with a router and a special laminate cutter.

9-75 After trimming the laminate flush with a router, lightly dress the edge with a belt sander.

I. HEAT THE LAMINATE UNTIL IT CAN BE BENT EASILY AROUND THE CORNER.

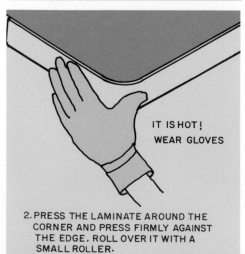

IT IS HOT! WEAR GLOVES

2. PRESS THE LAMINATE AROUND THE CORNER AND PRESS FIRMLY AGAINST THE EDGE. ROLL OVER IT WITH A SMALL ROLLER.

9-76 Heat the laminate until it is flexible enough to bend around the corner.

tact cement to each. The contact cement can be applied with a roller or brush **(9-77)**. Test the surfaces for dryness by placing a piece of paper against them **(9-78)**. If the coating appears to have areas that are dull, apply a second coat. Again test for dryness by placing a piece of paper against the surface. If it is dry place several ⅜-or ½ -inch dowel rods across the surface and put the laminate, glued side down, on top **(9-79)**. Line up the laminate so it is positioned exactly how you want it to be stuck to the substrate on all sides. Holding it steady in this position, remove an end dowel and press the laminate to the substrate. Then one by one remove the dowels, laying the plastic

9-77 Apply contact cement to the top surface and the bottom of the sheet of laminate with a roller or a brush.

9-78 Test the cement for dryness with a piece of paper. If it does not feel sticky it is dry enough to proceed.

9-79 Place wood dowels across the dry contact cement on the top and lay the laminate on top of them with the glued side down. Line it up with the edges of the substrate.

in place. Apply hand pressure against the laminate to bond it to the substrate (**9-80**). When it is in place roll it with a laminate roller (**9-81**). Start at the center and roll toward the edges.

After the laminate is in place trim the edges with a router, giving a slight bevel cut (**9-82**). Both the top edge and any corners are beveled. Do not cut so deep that you expose the substrate (**9-83**).

CLEANUP

If contact cement has gotten on the laminate, clean it with a cloth dampened with lacquer thinner. Be careful you do not wet the joints because they may come loose.

9-80 Position the laminate over the top. Then start removing the dowels and laying the sheet in place. When the sheet touches the glued top surface, it cannot be moved.

9-81 After the laminate is in place, roll it from the center toward the edges.

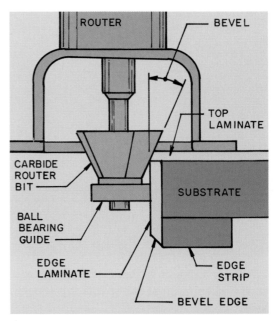

9-82 After the top laminate is secure, trim the edges with a special router bevel cutter.

9-83 The top edge and corners are beveled with a router and a special router bevel bit.

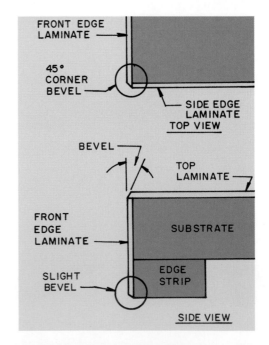

Cutouts after Installation

Once the countertop is covered it will be necessary to cut the openings for a sink or another appliance. A new top and new sink will require the location to be laid out using the template supplied with the sink. Position the template on the top and drill holes at the corners of the opening (**9-84**). Then cut to the template with a saber saw. Use a fine-toothed metal-cutting blade (**9-85**). Clean up any rough edges with a file just enough to let the sink rim fit into the opening. The installation of sinks varies, but several examples are in **9-86**. Notice that the lip of the sink in one and the sink rim in the other cover the edge of the opening, concealing any minor roughness of the laminate on the edge.

If the laminate covers an old top, the opening for the sink will be covered by the new laminate. Measure in from the edges of the top to the sides of the opening. Mark on the laminate with masking tape. Bore holes near each corner and cut the laminate with a saber saw that has a fine-toothed metal-cutting blade. Trim as necessary after the piece falls from the opening.

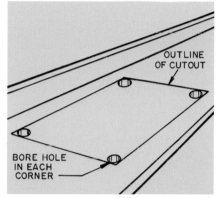

9-84 Lay the sink pattern on the countertop and bore holes in each corner.

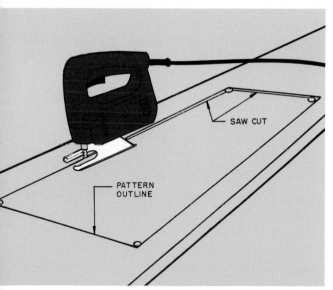

9-85 Cut the sink opening with a saber saw.

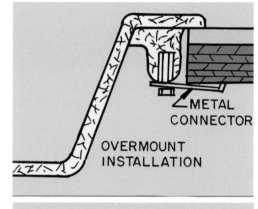

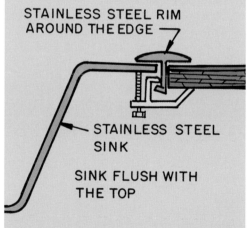

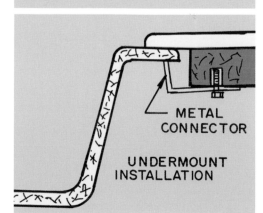

9-86 Three of the commonly used methods for installing sinks.

191

Also locate the holes needed for the faucets. Bore holes using the template that comes with new faucets. If it is a remodel job, locate and bore the laminate where the old holes are located. You can bore a small-diameter hole in the center of the holes from below and bore the larger required hole from the top.

9-87 The faucet, sink, and appliances are an integral part of a kitchen with new countertops and cabinets. *Courtesy KraftMaid Cabinetry*

Installing the Sink, Appliances & Faucets

After the opening for the sink or appliance has been cut, install as directed by the manufacturer **(9-87)**. If it is a remodel job observe how the old unit was installed before you take it out. On sinks apply a bead of caulking compound around the edge of the rim and then set the sink in place. Tighten as required and wipe away any caulk that may squeeze out.

CHAPTER 10

Kitchen Remodeling

As you look at your kitchen and wish you had something better, take time to consider the many possibilities for improvement. This could be simply refinishing the existing cabinets, buying new appliances, painting the walls, and installing a new floor. It could be more drastic, including planning a completely new layout, moving walls, upgrading the electrical and plumbing system, and buying new cabinets and appliances **(10-1).** Whatever is done, take time to plan carefully.

Visit local appliance dealers for information on new appliances and the building supply dealer for information on cabinets and flooring. Some dealers have consultants to help with the planning of a new kitchen. The National Kitchen and Bath Association has publications available that will help with the planning process. There are also books available devoted to planning and constructing kitchens. See the additional information section at the end of this chapter. (Additional information on cabinets is in Chapter 9.)

When you remodel a kitchen some plans must be made so food preparation and cleanup can continue during the remodeling. While this is difficult, a plan can be made to schedule what is to occur and the time it will take to accomplish it. For example, if new countertops are to be installed, schedule this with the contractor so you know exactly when the work will be done and how long it will take. If there is to be a total cabinet tearout and replacement, you will have to schedule not only this but also the installation of the new countertops, reinstallation of old appliances or new ones, reconnecting of the sink, dishwasher, and disposal. If the floor is to be recovered this can

10-1 New cabinets and possibly a new layout will produce a kitchen that is attractive, pleasant to use, and can increase the value of the house.
Courtesy Wellborn Cabinet, Inc.

be scheduled after all the other things have been completed. The exact things to be done will vary with the job but you can expedite things if you make a list of what needs to be done, in which order, and then schedule in the trades and suppliers. No doubt on a total remodel you may be dining on fast foods and paper plates for a few days.

Minor Repairs

Following are some easy-to-repair problems that arise as the kitchen gets older.

Loose Laminate

One problem that occurs over the years involves the plastic laminate on the countertop coming loose in a small area. To repair this, carefully lift up the loose area and spread a thin layer of contact cement on the substrate and the bottom of the laminate. Prop it up with a stick until the contact cement is almost dry, which is about five minutes. Then press the laminate against the substrate and roll it tight with a rolling pin (10-2).

Stains

Over time some substances will stain the plastic laminate. It helps if solutions which you know will stain, such as red wine, are wiped up immediately. Many stains can be removed by washing them several times with a liquid detergent. A soft brush will help. If this fails, try wiping them with a cloth dampened with household chlorine bleach or denatured alcohol. Wash with fresh water after a minute of soaking. Also available in your grocery store are pressure cans of laminate countertop cleaners. They are sprayed on and wiped down with a soft cloth (10-3).

Many stains can be avoided if you keep the plastic laminate waxed. Laminate wax is available in spray cans, so after the top is clean spray the wax, spread it with a cloth, and let it dry.

Burns

If you set something very hot on plastic laminate it will scorch. The only way to repair this is to cut out an area around the scorch and glue in a patch. This does not present the best appearance but is better than leaving the scorch. The only total solution is to replace the entire piece of laminate.

1. RAISE THE LOOSE LAMINATE AND APPLY CONTACT CEMENT TO THE SUBSTRATE AND LAMINATE.

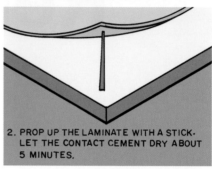

2. PROP UP THE LAMINATE WITH A STICK. LET THE CONTACT CEMENT DRY ABOUT 5 MINUTES.

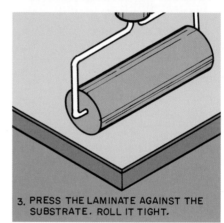

3. PRESS THE LAMINATE AGAINST THE SUBSTRATE. ROLL IT TIGHT.

10-2 To bond a loose laminate, cover the area and laminate with contact cement, let it set until almost dry, press in place, and roll flat and smooth.

Sticking Drawers

If a drawer has a wood center guide, the wood may get rough due to wear and will possibly absorb moisture. When the drawer no longer slides easily, lightly sand the sliding parts of the drawer guide. Then rub with a wax candle or paraffin wax.

If the drawer has metal drawer guides that have small wheels running in a channel (10-4 and 10-5), wipe down the channel so it is free of dust or other particles. Check to see if it has a bend somewhere along its length that could bind the wheel. Straighten out any bend. Then place a couple drops of oil on the rotating pin in the center of the wheel. Light machine oil works well. Hold a rag under the wheel to catch excess oil. Then wipe the wheel clean. Insert the drawer on the slides and see if it runs smoothly.

10-3 Laminate cleaners can remove some of the stains that appear on the countertop.

10-4 This drawer has drawer slides mounted on the bottom of each side.

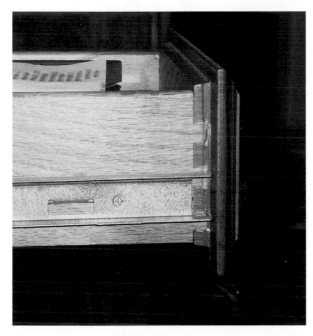

10-5 This side-mounted drawer slide is the best type to use, especially if the drawer will carry a heavy load.

10-6 Loose cabinet hinges can be secured by gluing wood plugs in the screw holes and reinstalling the screws.

Loose Cabinet Door Hinges

Sometimes the screws in the cabinet door hinges become loose. Try to tighten them. If they will not pull up tight, glue a sliver of wood in the screw hole. When the glue is dry reinstall the screw in the hole. Some types of wood caulking material can be used to fill the holes. They will hold the screw when they harden (10-6).

Walls & Floors

Over the years the walls and floors are subject to bumps and abrasions as well as spills and grease from the cooking process. Sometimes they can be repaired or cleaned but often must be re-covered. See chapters 4 and 15 for information on wall and floor materials and repairs.

Plumbing Repairs

Probably the kitchen will have more plumbing trouble than anything else. Clogs in the sink drain can usually be cleared with a plumber's plunger (10-7). If this does not work you will have to go below the sink and remove the trap (10-8). If it has deteriorated buy a new one. A very thin plastic tape is available that you wrap around the threads before assembling the parts. This reduces the chance it may leak (10-9).

Creative Ideas

If you are upgrading the kitchen but do not plan to do a complete tearout, try to find some interesting creative ways to improve the appearance. Following are a couple of ideas.

Some form of wall shelf (10-10) or interesting cabinets possibly with glass doors (10-11), can add a bit of interest. Interesting china and other objects can be tastefully displayed. A bit of refinement can be added

10-7 Drain clogs can often be removed by pushing them down into the drain pipe with water pressure developed when you pump down on a plumber's plunger.

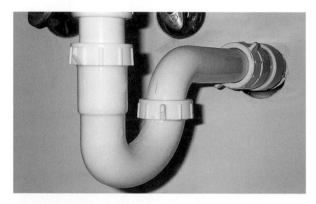

10-8 A typical P-trap used on a sink waste pipe. Unscrew the U-shaped section and remove the obstructing material. Put a bucket below the trap before you loosen it to catch the water that will flood out.

10-9 One type of tape used on pipe threads to help prevent a leak from occurring.

10-11 Cabinets with glass doors permit the display of important china and other objects which greatly enhances the appearance of the kitchen.

Courtesy KraftMaid Cabinetry, Inc.

10-10 A small wall shelf will add some interest to a plain wall.

to the existing cabinets by placing a crown molding along the top (**10-12**). This is an especially nice touch when items are displayed on top of the cabinets.

Another more expensive but quite effective thing to do is to refinish the stiles, mullions, and rails. Then install new doors. The design and color can be used to give the kitchen an entirely new look. In **10-13** the existing doors were covered with a plastic laminate, giving the kitchen a fresh, modern look.

A Complete Remodeling

If you plan a complete remodeling, first consider what you want in a new kitchen. Some of the things you do not like about the existing kitchen need to be written down. Then list what you expect from a new kitchen.

First consider the floor space. If the existing kitchen is too small, try to figure how to enlarge it. Sometimes it can be opened into an adjacent room. This means the use of this room is lost, but the kitchen gains. This happens typically when adding a dining area as part of a kitchen. Many floor plans will not permit this so consider expanding the kitchen by adding an area onto the house (**10-14**). If building codes and local appearance committees approve, this might be the best approach. If the expansion is only a few feet, you might be able to cantilever the floor out over the ground, saving the cost of the foundation (**10-15**). The distance you can extend the floor will depend on loads and building codes, so an architect or qualified construction contractor should be

10-12 A bit of crown molding along the top of wall cabinets that will not have a soffit adds to the overall appearance of the installation.

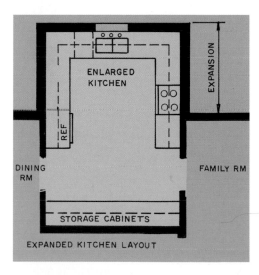

10-13 Bonding plastic laminate to the old doors provides a new look.

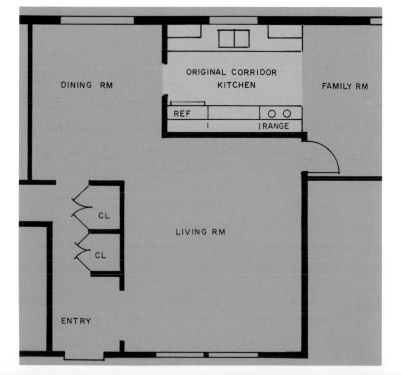

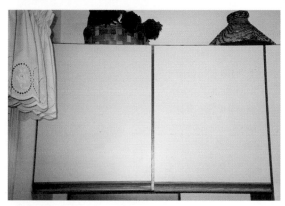

10-14 The original corridor kitchen was too small so an extension to the back of the house was built, providing the desired space and a new arrangement of appliances.

10-15 This kitchen extension was small so it was cantilevered over the ground.

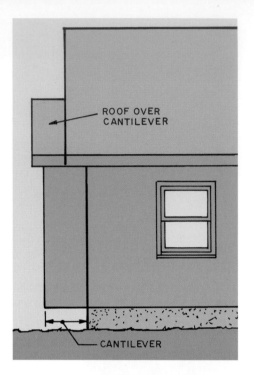

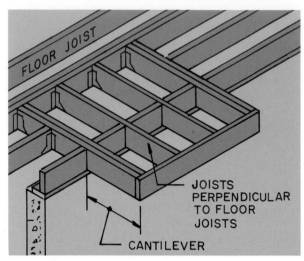

10-16 The joists on this cantilever are perpendicular to the floor joists of the house.

10-17 The joists on this cantilever are parallel with the floor joists of the house. When added to an existing house they are run back 6 to 8 feet along the house joists.

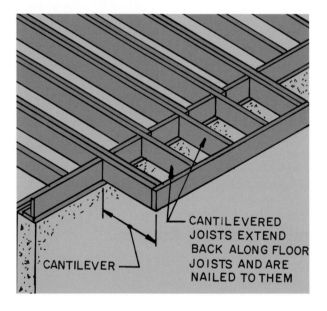

10-18 One type of roof that can be used over a cantilevered addition.

contacted. A maximum of two feet is a commonly used standard. When the floor projection runs perpendicular to the floor joists, the joists are framed as shown in **10-16.** The end joists are doubled. The use of metal joist hangers is recommended. When the cantilevered floor joists run parallel with the house floor joists, construction shown in **10-17** is used. This example is for when a cantilever is added to an existing house. In new construction the house joists would extend out the required two feet. The added joists lap back over the house joists and are nailed to them. The joists should be run back under the house floor to the

beam supporting the house joists. If you have any doubts about the load on the cantilever, consult an architect or building contractor. If a large load will be placed on the cantilever, it would be better to install a foundation.

Exterior walls and a roof are built to enclose the new space. Then the existing exterior wall can be removed. The design of the roof will vary depending upon the existing roof. One example is shown in **10-18.**

The new exterior walls and roof are built on the cantilevered floor. Construction is the same as that for typical wood frame construction **(10-19)**. Obviously, this puts a load on the cantilever before any interior cabinet or appliances are added. Remember to insulate the new floor area and seal it on the bottom with exterior plywood.

If the floor is to extend more than two feet beyond the foundation, it will be necessary to build a foundation. A typical detail is shown in **10-20**. This construction is the same as that used when the house was built. (Detailed carpentry information is available in the publications listed at the end of this chapter.)

Basic Kitchen Shapes

As you select the shape of the kitchen the location of existing doors and windows will have considerable influence, as will the essentials of good planning. The most commonly used kitchens are the **L-shaped** kitchen and the **U-shaped kitchen.** These may use an island counter to increase the efficiency and improve the traffic pattern. It also can serve as a snack area.

The **L-shaped kitchen (10-21)** has cabinets and appliances on two walls. If one wall is very long the

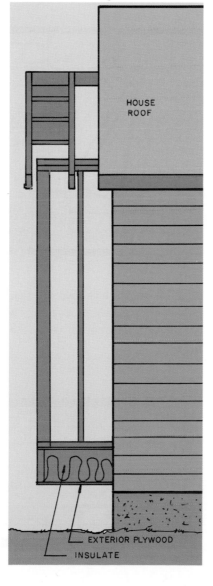

10-19 The cantilevered floor area is enclosed with walls and roof. Typical wood framing techniques are generally used.

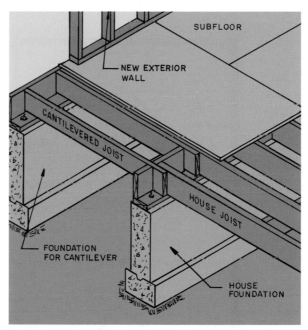

10-20 Larger room extensions are built using a full foundation.

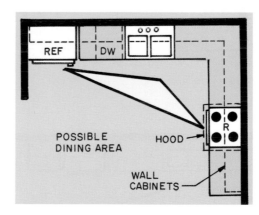

10-21 L-shaped kitchens can provide an efficient layout, but watch the length of the long wall so the work triangle does not get too large.

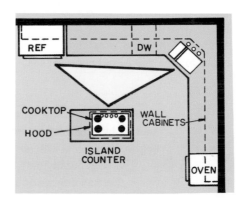

10-22 The work triangle of an L-shaped kitchen can be shortened by adding an island counter. Notice how a corner sink has been used to improve the layout.

efficiency is reduced. A small kitchen should have at least 130 lineal inches of open countertop, while a large kitchen will have at least 200 lineal inches or more.

The **efficiency** of a kitchen is determined by the distance between the stove, refrigerator, and sink. This is called **the work triangle.** The total distance between these should be not more than 22 to 26 feet. Notice in **10-22** how an island counter impacts the work triangle. It shortens the walking distances between work centers, making the kitchen more efficient.

Another efficient design is the use of a **U-shaped kitchen(10-23).** It has cabinets and appliances arranged on three walls. The size of the room available greatly influences this design. The addition of an island counter makes it easier for two people to work at the same time **(10-24).**

A more unique concept is to form a **double L-shaped kitchen(10-25).** It is especially helpful when a door, window, or other obstruction interferes with the cabinets along a wall.

If a small space is all that is available, an **I-shaped (10-26)** or **corridor kitchen (10-27)** can be used. Space is tight and there is room for only one person to work efficiently.

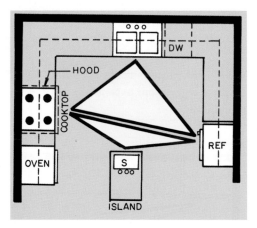

10-23 A typical layout for a U-shaped kitchen. Narrowing the U would shorten the walking distance between work centers.

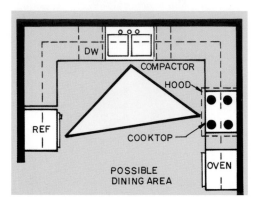

10-24 The addition of an island counter within a U-shaped kitchen provides two work triangles and makes it easier for two people to work in the kitchen at the same time.

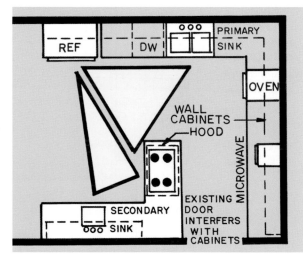

10-25 A double L-shaped kitchen can help make a useful layout when a door, window, or other obstruction interferes with the cabinets along the wall.

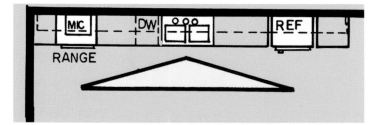

10-26 The I-shaped kitchen stretches along one wall and is suitable for use by only one person.

10-27 The corridor kitchen can be built in a long narrow space. While it is efficient to use, generally only one person can work in it.

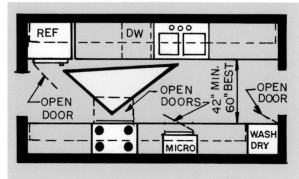

10-28 The kitchen is planned around these three major appliances and adjacent fixtures.

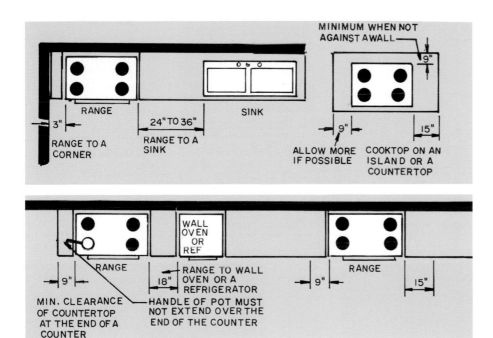

10-29 Minimum countertop allowances beside the cooking unit. Make these larger whenever possible.

Some Planning Recommendations

Consider the kitchen as having three separate but related work areas. These are planned around the three major appliances. That includes the **cooking/baking area, refrigerator/freezer area,** and the **sink/dishwasher area (10-28).**

Cooking/Baking Area

The cooking/baking area includes the surface cooking unit, a microwave, and a conventional oven. They are near the sink because this is where the food is prepared.

The conventional oven can be located some distance from the cooking unit because the prepared food in the oven does not require constant attention. The microwave is often located in the cooking area because food in it cooks rapidly so it needs frequently attention. Some prefer to locate the microwave near the refrigerator rather than the cooking unit because food generally moves from the refrigerator to the microwave.

The countertop in the cooking area should be adequate to hold a number of small cooking appliances. A hood over the cooking unit is essential. Provide storage for the cooking utensils in this area.

Minimum spacing requirements for the cooking/baking area are shown in **10-29.**

Positioning the Microwave

The microwave can be in the cooking/baking area or near the refrigerator. The bottom of the microwave should be 24 to 48 inches above the floor. Mounting recommendations are shown in **10-30**. It can be located above the cooking unit; however, there is a risk of getting burned if something is cooking on the stove **(10-31)**. One good way to handle this is to install a venting unit that holds the microwave **(10-32)**.

The Refrigerator/Freezer Area

This is where frozen foods are stored and other foods are kept cold but not frozen **(10-33)**. You should allow at least 15 inches of countertop on the handle side of the refrigerator. If it has a vertical freezer alongside, then 15 inches is needed on both sides. Some recommended spacings are shown in **10-34**. This countertop is needed to place foods going into and being removed from the refrigerator/freezer.

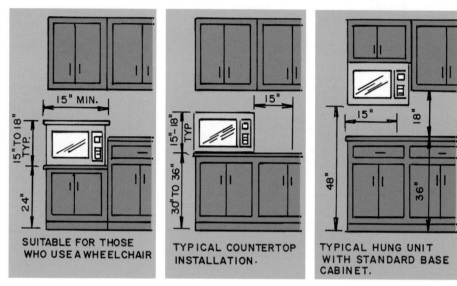

10-30 Recommended locations and minimum spacing requirements for the microwave. Remember to allow at least 15 inches of clear counter space above or below the microwave.

10-31 Microwave ovens are frequently located above the range or cooktop. *Courtesy G.E. Appliances*

10-32 This hood supports a microwave over a range and is ductless. It filters the air and returns it to the room.

10-33 A typical unit with side-by-side freezer and refrigeration compartments.

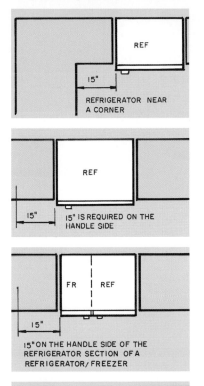

The Sink/Dishwasher Area

This area requires enough countertop to hold the pots and pans used in food preparation, as well as the foods being processed. It also must hold the soiled cooking utensils and dishes that go into the dishwasher. Remember, raw food preparation takes space and baking preparations often require even more **(10-35)**.

The dishwasher is located next to the sink **(10-36)** because this is where the items are rinsed before going into the dishwasher. If you want a trash compactor

10-34 Minimum countertop allowances beside the refrigerator/freezer.

10-35 This sink serves as the center of the food-preparation area for cooking and serving. The countertop is clear on both sides, providing needed work space.

10-36 The sink is the center of the cleanup area. Counter space on both sides is necessary. The dishwasher is next to the sink. A trash compactor could go on the other side.

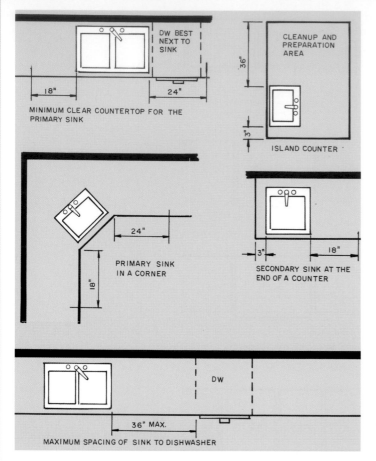

10-38 This open wicker basket is used to store food items such as potatoes and onions. *Courtesy KraftMaid Cabinetry, Inc.*

10-37 Other minimum spacing recommendations for the primary and secondary kitchen sinks.

it is usually placed on the opposite side of the sink as the dishwasher.

If you plan to use a garbage disposal unit in one of your sinks, first check the local codes. They are banned in some areas. Place the switch far enough away so you cannot reach the disposal when turning it on.

Recommendations for locating the primary sink in relation to the dishwasher and providing the required minimum countertop are in **10-37.** It is very helpful in large kitchens to locate a second sink that will cut down the traffic pattern and reduce crowding on the primary sink. This is often located on an island counter.

Storage

Storage is required for packaged and fresh foods, frozen foods, kitchen accessories, utensils, dishes, sil-

10-39 There are a wide variety of wire racks you can place on the shelves of a cabinet to increase the storage capacity.

Table 10-1 Recommended Minimum Kitchen Storage Capacity.

	150 sq ft or less	larger than 150 sq ft
Base cabinet frontage	156 inches	192 inches
Wall cabinet frontage	144 inches	186 inches
Drawers (Individual total frontage)	120 inches	165 inches

Courtesy National Kitchen and Bath Association

verware, electrical appliances, linens, cleaning supplies, brooms, mops, and other such items. A large house will require more storage space than a small house and will typically have a larger kitchen and dining area. Typical minimum storage recommendations are in **Table 10-1**.

As you plan your storage space, you want to locate items such as utensils and food products as near as possible to the work center where they will be used. Take advantage of the many features offered by cabinet manufacturers to utilize the space available most efficiently.

For example, in **10-38** is an open wicker basket that is great for storing potatoes and other vegetables. It slide out for easy access and permit air to circulate around the vegetables. Wire racks are available that permit greater use of shelving while permitting easy removal of each item stored (**10-39**). Base cabinets with sliding drawer shelves provide access to items at the back of the shelf (**10-40**).

Tiered storage shelves make it easy to reach the items on the back of the shelf. This is especially useful for storing small items such as spices (**10-41**).

Recycling

Consider some means of storing materials to be recycled. This can be in a cabinet or in an area just outside the exterior kitchen door.

Waste Disposal

This is a constant problem with no easy solution. As mentioned earlier a disposal can be used to remove biodegradable materials if local codes permit. A trash compactor will compress other materials in a plastic

10-40 This base cabinet has sliding drawer/shelves which make the rear of the shelf easily accessible.

Courtesy KraftMaid Cabinetry, Inc.

10-41 Tiered storage shelves make it easy to access small items stored at the rear of the shelf.

Courtesy KraftMaid Cabinetry, Inc.

10-42 This is a handy top-mount waste disposal basket system. It can be lined with plastic garbage bags for messy waste or used unlined for recycleable materials.

Courtesy KraftMaid Cabinetry, Inc.

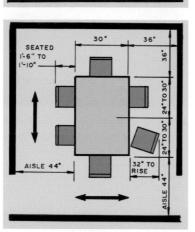

10-43 These are minimum spacing recommendations for seating at a dining table.

SQUARE / RECTANGULAR
2'-6" x 2'-6" SEAT 2
2'-6" x 3'-2" SEAT 4
2'-6" x 5'-0" SEAT 6

ROUND
3'-0" DIA SEAT 3
3'-6" DIA SEAT 4
4'-6" DIA SEAT 6

bag, providing easy disposal to an exterior trash can. A plastic bin lined with a plastic trash bag is often used and frequently located below the sink (**10-42**).

A Dining Area

If you plan a dining area at the end of the kitchen, the space recommendations in **10-43** will be helpful. You need to allow space on all sides so those seated can leave without disturbing the others. An island counter provides a good place for quick snacks without requiring a lot of room (**10-44** and **10-45**). A booth

10-44 The height of eating counters can be varied. This depends upon the height of the chair or stool to be used.

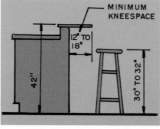
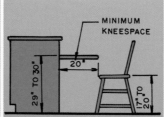
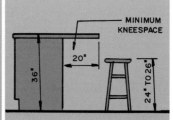

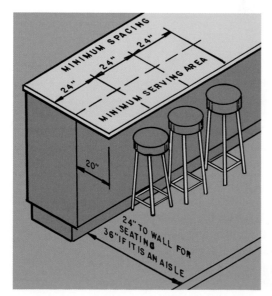

10-45 An eating counter should provide at least 24 inches of clear countertop for each person and adequate knee space, depending on the chair or stool selection.

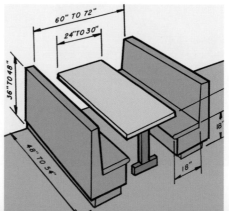

10-46 Booths are popular in the kitchen eating area. These are typical sizes for a booth to seat four people.

is an efficient way to provide for dining in the kitchen. It occupies less space than a table and chairs. Recommended sizes and spacing are shown in **10-46**.

Traffic Patterns within the Kitchen

How traffic flows through a kitchen greatly affects the efficiency. A flow through the center, as shown in **10-47**, can disrupt food preparation. Be certain traffic flow is not blocked if an appliance door is open (**10-48**).

A traffic aisle alongside a counter where no appliance will be used should be 36 inches wide. When aisles meet at a 90-degree angle, one should be at least 42 inches wide. This will also allow a person in a wheelchair to make the turn (**10-49**).

Ventilation

Cooking and baking produce moisture, fumes, and grease in the air and odors that eventually cling to the cabinets, walls, curtains, and other exposed items.

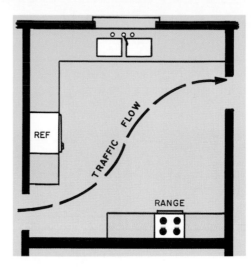

10-47 When the primary traffic flow is through the center of the kitchen, it can disrupt food preparation activities.

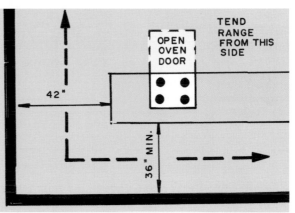

10-49 A 36-inch-wide aisle is recommended for traffic flow past a counter where no appliance is to be used. If aisles intersect, one should be at least 42 inches wide.

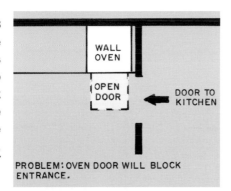

10-48 Position the appliances so they do not block the entrance into the kitchen.

PROBLEM: OVEN DOOR WILL BLOCK ENTRANCE.

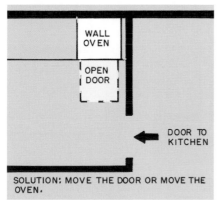

SOLUTION: MOVE THE DOOR OR MOVE THE OVEN.

10-50 Hoods are essential over ranges and cooktops.

Range hoods are used to remove the pollutants from cooking and exhaust them to the outdoors. Usually they have a light to illuminate the cooking surface (10-50). They have pipes that are placed to vent to the outside through the attic, the soffit over the cabinets, or directly through the wall (10-51). A range hood should have a fan rated at 150 cubic feet per minute.

Some surface cooking units use a downdraft venting system (10-52) that is part of the cooking unit. A fan pulls the fumes into openings beside the burners and exhausts them outside through a duct.

The hood shown earlier in 10-32 pulls the fumes from the stove or cooktop, circulates them through a filter, and discharges the air back into the kitchen. This unit also supports a microwave.

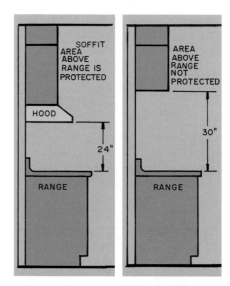

10-51 You can run the hood vent pipes through the attic joists, inside the wall cabinet soffit, or if on an outside wall, directly through the wall.

10-53 The hood should be 24 inches above the range. If no hood is used, raise any cabinets above the range to 30 inches minimum.

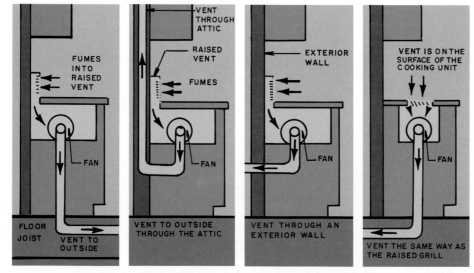

10-52 A downdraft cooking venting system pulls the fumes from the cooking surface and exhausts them outdoors.

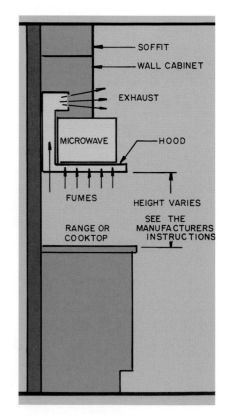

10-54 The clearance needed by ductless hoods and those carrying microwaves should be placed as directed by the manufacturer.

The distance between the surface of the cooking unit and the ventilation hood should be 24 inches. If a hood is not used, raise any cabinets over it to 30 inches (10-53). If the hood supports a microwave, adjust the height as recommended by the hood manufacturer (10-54).

Plumbing Considerations

If your improved kitchen leaves the sink, dishwasher, and ice maker on the refrigerator in the same locations, little needs to be done to connect to the new sinks and appliances. New pipe should be used to make the connections from the pipes in the wall to those providing water and waste disposal. Generally, the hot and cold water will enter the kitchen from pipes run through the wall. A valve is installed on the end and a flexible pipe is run to the faucets which are mounted on the countertop (10-55). Single-lever faucets are possibly the most commonly used kitchen faucet (10-56). They have tubes extending below, to which the hot and cold water are attached, and one for the spray hose (10-57).

If the kitchen is more than 10 years old, it would be a good time to replace the faucets. If the revised kitchen moves the sink and other appliances, a plumber should be brought in to move the potable water and waste pipes to the new locations. This job

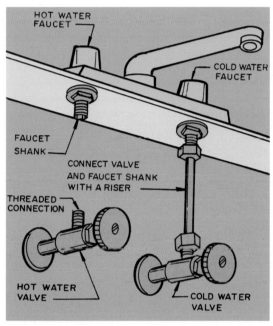

10-55 A flexible riser tube connects the valve at the wall to the faucet shank.

10-56 This single-lever faucet regulates the water flow and temperature and has a spray which is very useful for food preparation as well as cleanup.
Courtesy Franke Consumer Products, Inc./Kitchen Systems Division

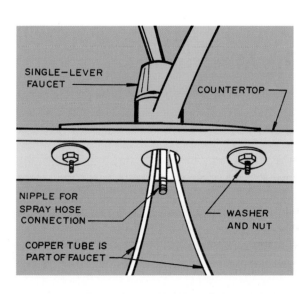

10-57 This single-lever faucet is supplied with copper water tubing. It is bolted to the countertop.

10-58 This two-bowl sink has bowls of the same size and matching faucets, spray, and soap/lotion dispenser. *Courtesy Elkay Manufacturing Company*

10-59
This single-bowl sink is made from a reinforced, modified acrylic filled with natural materials. Drill holes for faucets, sprays, and soap/lotion dispensers.
Courtesy The Swan Corporation

must meet local building codes and may be inspected by a building inspector.

Sinks

Since the kitchen sink gets heavy use, accompanied often by stains and damage, replacing it with a new sink is a good move. There are a number of excellent sinks made from materials not available years ago. Also consider changing the size and configuration of the sink. If a new countertop is to be installed, it is easy to install a new sink. Kitchen sinks are available in various sizes of single, double, and triple bowls and have multiple openings for sprays, faucets, and soap/lotion dispensers (**10-58**). Before you select a large sink, verify the length of the countertop needed for its installation.

Single-Bowl Sinks

Single-bowl sinks are usually used as a second sink. Try to get one that is big enough to hold the larger cooking utensils (**10-59**). A good size is 22 inches wide and 25 inches long; however, smaller sizes are available. A single-bowl sink usually requires a 30-inch cabinet.

Two-Bowl Sinks

Two-bowl sinks are widely used for the main sink. A disposal is often mounted on the smaller bowl (**10-60**). Some have the bowls the same size (refer to **10-58**). If you have a long counter, the sink in **10-61** is very handy. It has a drain board on one side. Sometimes a two-bowl corner sink will fit into the plan better (**10-62**).

10-60 This two-bowl stainless steel sink has one smaller bowl. The disposal is usually mounted below the drain on this bowl. *Courtesy Elkay Manufacturing Company*

Sink Materials

Sinks are available made from stainless steel, porcelain enamel on cast iron or steel, and man-made composites.

Stainless steel sinks are made in a number of thicknesses of the steel. The heavier the gauge of steel, the more durable the sink. They do not chip but will show water marks, and some chemicals may cause them to discolor. Stainless steel sinks are shown in **10-60** and **10-61**.

Porcelain enamel sinks have been used for many years. These are made from pressed steel and cost less

10-61 This stainless steel sink has a drainboard on one side. One bowl is large enough to hold a dish drying rack.

Courtesy Elkay Manufacturing Company

10-62 A corner sink fits diagonally in the corner formed by the intersection of the base cabinets.

Courtesy Elkay Manufacturing Company

10-63 This water purification system uses a membrane filter cartridge to remove particles and sediments, heavy metals such as lead, compounds such as nitrates and cysts, viruses, bacteria, and other impurities. *Courtesy Kinetico Incorporated*

than those using cast iron, but are not as durable. The porcelain enamel will chip if hit by a heavy object. They are durable and easy to clean. Repair kits are available.

Molded kitchen sinks are becoming widely used and are made from composite materials such as quartz composites, granite composites, or reinforced modified acrylic or polyester resins. The sink shown earlier in **10-59** is made from natural minerals and a modified acrylic. The color runs completely through the material so even if the surface is worn the color is still visible.

Improving Water Quality

While you are renovating the kitchen is a good time to consider the quality of the potable water delivered to it. If you are connected to a municipal water system it will have to supply water that meets government specifications. If you have a private well the quality can vary. You should have the water tested at least once a year to guarantee it is safe. The best time to take samples is in the spring or summer following a rainy period. Also test after you replace a pump, have work done on the well, or replace pipes. If the house is old it may have lead pipes or copper pipes soldered with lead solder. These will permit lead to get into the water system, which presents a serious health problem. All these pipes should be replaced. Well water even if tested to be safe can have an undesirable smell caused by hydrogen sulfide, which is referred to as rotten egg odor. If there is iron in the water it will cause rust stains on clothes, fixtures and other items. Impurities can be removed by a water filtration system (10-63). Hard water can be handled with a water softener. Mud, silt, and other sediments in water can be filtered out. Water with a high carbon dioxide content is called acid water and reacts on brass and copper pipes and fittings. High sodium or magnesium content produce a salty, brackish flavor. Systems are available to reduce these distasteful conditions.

The kind of treatment needed should be decided by a representative of an accredited water treatment company.

Water Treatment Systems

Information is available from the National Sanitation Foundation (NSF). The address is in Additional Information at the end of this chapter. Following are some of the in-home water treatment systems available. Check to see if the one you are considering will remove the contaminants you're concerned about. Also observe how much water they treat per minute. If you are only treating water to the kitchen sink, ice maker, or chilled drinking water, a lower-volume unit can be chosen. Also observe the cost of installation and operation. How often will you have to replace the filtering material?

Activated Carbon Filters

This system uses normal water pressure to force the water through canisters filled with activated carbon. The carbon filters may be granular, powdered, powder-coated paper, or pressed carbon black (10-64). The carbon filters trap the contaminating substances. This type of filter will remove from the water some organic contaminants that may cause undesirable tastes, odors, and colors. Generally it will not remove inorganic chemicals such as salts and metals. Some types of filters will, however, remove lead.

Activated carbon filters deliver from a half-gallon to three gallons per minute. The filter will normally be replaced after filtering 1,000 gallons. Never fail to replace the cartridge when it is time. A loaded cartridge will allow bacteria to collect and multiply.

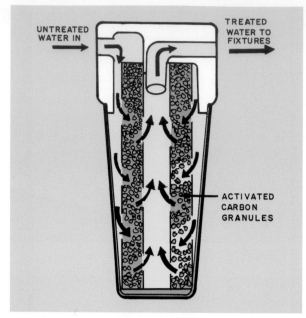

10-64 An activated carbon water purification system filters the incoming water through a granular carbon cartridge from which it flows to the faucet.

10-65 This water filtration system has a faucet with three handles: hot water, cold water, and pure filtered water. The filtration system is below the sink. It uses a replaceable ceramic cartridge.

Courtesy Franke Consumer Products, Inc., Kitchen Systems Division

Ceramic Filters

This system uses a replaceable ceramic cartridge that is mounted below the sink. The sink in **10-65** has a faucet with three handles: hot water, cold water, and pure filtered water. The third handle lets you pour filtered water when you need it. You use unfiltered water to rinse dishes and cooking utensils. The cartridge removes microscopic particles, cyptosporidium, lead, and chlorine. It will not treat microbiologically unsafe water.

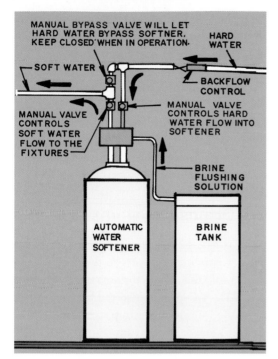

10-66 This water softener system removes the high mineral content from the water. This makes it more pleasant to use for bathing, cleaning everything in the house becomes easier, it leaves laundry clean and soft, and dishes clean and spot free.

10-67 This twin-tank water softener automatically adjusts to meet the needs of an entire household. It has a high flow rate to meet the needs of laundry, showers, and kitchen. *Courtesy Kinetico Incorporated*

10-68 This compact twin-tank water softener removes calcium and magnesium ions, producing soft water. It uses block salt as shown in this illustration. *Courtesy Kinetico Incorporated*

10-69 The compact water softener can be mounted below the sink and/or dishwasher. Dishes come out without mineral deposits on them. *Courtesy Kinetico Incorporated*

Ultraviolet Disinfection Units

These units destroy bacteria, deactivate viruses, and leave no odor or taste in the water. They are not effective in removing chemical pollutants. The unit must be carefully maintained.

Water Softeners

Often water taken from a private well is classified as hard water because it has a high mineral content. Minerals in well water will stain plumbing fixtures a brown color and build up on the electrodes of electric water heaters and the inside of the water pipes.

A typical water softener system is shown in 10-66. The untreated water enters through a canister containing a synthetic resin such as zeolite. The hard calcium and magnesium ions dissolved in the water are exchanged for soft sodium ions bonded to the resin. Once the resin is saturated with calcium and magnesium, it automatically flushes the salt water (called brine) from the adjoining tank. This regenerates the resin and the process repeats. The brine tank must be reloaded with salt on a regular schedule. The system

213

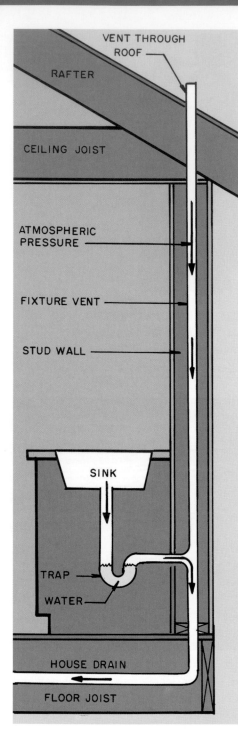

10-70 The sink trap retains water, blocking the entrance of sewer gas into the kitchen.

shown in **10-67** requires no electricity and operates completely automatically. It will serve the needs of an entire house.

A compact water softener (**10-68**) can be installed below the kitchen counter (**10-69**) to serve the needs of that sink and related appliances such as a dishwasher. It uses a bed of resin beads which holds sodium ions. The beads hold the calcium and magnesium ions and release sodium ions. Removing the calcium and magnesium makes the water soft.

Waste Disposal Systems

A basic waste disposal system is illustrated in **10-70**. The sink waste line connects to the building waste system at a stub that comes through the wall below the sink (**10-71**). A P-trap is inserted between the sink drain and the wall stub. The trap remains full of water which keeps sewer gases from running back up the waste pipe into the house through the sink drain. The

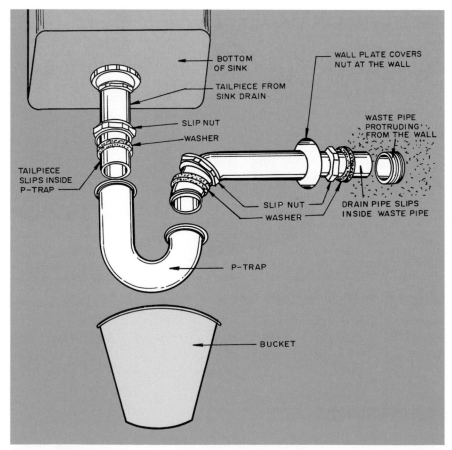

10-71 The sink drain is connected to a trap which is connected to the waste pipe that extends through the wall below the cabinet top. If you remove the trap, put a bucket under it before you loosen it.

waste line called a vent pipe, runs through the roof —which keeps the waste disposal system operating at atmospheric pressure.

Other appliances such as the dishwasher and disposal are connected into the waste line below the sink, along with the sink waste line (10-72).

Some Electrical Considerations

If the kitchen is rather old it is possible that the electrical system needs major upgrading. This could include adding more electrical outlets, improving the lighting, and providing power to a new dishwasher, disposal, stove, and oven. The old wiring may no longer meet building codes and could be a fire hazard.

10-72 This shows disposal and dishwasher waste pipes connected to the kitchen sink waste pipe, which then connects to the waste pipe in the wall.

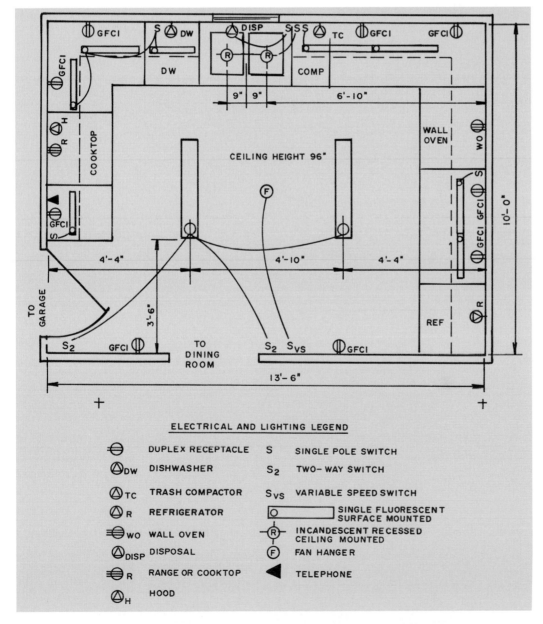

10-73 A typical architectural drawing showing the plan for a kitchen, including electrical and lighting requirements.

ELECTRICAL AND LIGHTING LEGEND

DUPLEX RECEPTACLE	S	SINGLE POLE SWITCH	
DW DISHWASHER	S_2	TWO-WAY SWITCH	
TC TRASH COMPACTOR	S_{VS}	VARIABLE SPEED SWITCH	
R REFRIGERATOR		SINGLE FLUORESCENT SURFACE MOUNTED	
WO WALL OVEN	R	INCANDESCENT RECESSED CEILING MOUNTED	
DISP DISPOSAL	F	FAN HANGER	
R RANGE OR COOKTOP		TELEPHONE	
H HOOD			

If you plan an entirely new kitchen, the new electrical plan will be part of the drawings showing the cabinets, appliances, and plumbing. The design and installation must meet local building codes and be installed by a licensed electrician. If you plan to do the electrical installation, check with the local building inspection department to see what requirements they will specify.

A typical electrical plan with lighting and electrical requirements indicated is shown in **10-73.** This plan provides general illumination for the entire room plus lighting below the cabinets to increase the illumination on the countertop and appliances. An electric stove or oven requires 220-volt current. Gas appliances require a piping diagram to the stove and oven. Outlets along the wall above the countertop are required for operating small appliances. As a rule of thumb place a duplex outlet every four feet along the countertop. Remember to add some outlets in the walls, to operate the vacuum cleaner and other portable appliances.

Additional Information

National Sanitation Foundation, P.O. Box 1468, Ann Arbor, MI 48106.

National Kitchen and Bath Association, 687 Willow Grove St., Hackettstown, NJ 07840.

Other Sterling Publishing Co. Inc. books by William P. Spence, including *Carpentry and Building Construction, Constructing Kitchens, Residential Framing,* and *Encyclopedia of Home Maintenance and Repair.*

CHAPTER 11

Remodeling Bathrooms

In today's homes the design and furnishing of the bathrooms receive major consideration. Bathrooms are larger, handle a wider range of services than older bathrooms, and are decorated and color coordinated with the same flair as the living room, dining room, and other rooms in the house (11-1).The bathroom gets heavy use and the fixtures, walls, floor, and ceiling are exposed to heat and considerable moisture. The air is often exposed to high humidity for many hours if the room is not properly ventilated. Bathrooms in older houses are often poorly planned, including the spacing and arrangement of fixtures, the types of fixtures, and storage space. For example, are towels and other supplies stored in a hall closet rather than in the bathroom? Older bathrooms are often small, resulting in an area that is not comfortable to use and which actually is very inconvenient. If remodeled and possibly enlarged, a much better facility can be developed. The lighting in old bathrooms is typically very poor and makes using wall mirrors difficult. Finally, ventilation is usually absent, resulting in high humidity and odors that are retained.

11-1 This contemporary bathroom has color-coordinated fixtures and a creative decorating scheme, providing a cheerful, attractive atmosphere.

Courtesy Crane Plumbing/Fiat Products/Universal Rundle

Maintenance of Existing Bathrooms

If the bathroom is old enough so that one of the fixtures is stained or does not function properly, consider replacing all of them. This will enable you to coordinate the colors, have newly designed fixtures, and the latest, most efficiently functioning units. This is a good investment, especially if you are considering selling the house in the near future. An old, smelly, deteriorated bathroom is a major hindrance when selling a house.

Many older houses will have three or four bedrooms and one bathroom. Today this is simply not acceptable. As you plan the remodeling, consider adding one or two additional bathrooms. If you have a large living area and enjoy entertaining, a powder room near this area is a great help. It only needs a lavatory and a toilet. It can be rather small. The size of a large closet is typical. If you have a recreation area, a small bathroom there is helpful, especially if it has easy access to the yard where children play. They can use it and not track through the house. In the bedroom area a full, large bathroom off the master bedroom is very much appreciated. Build one or more smaller full-size bathrooms for use by those in the other bedrooms.

Minor Repairs

Since the bathroom gets heavy use it may require repairs more often than the rest of the house. Typical examples include loose plastic laminate on the lavatory top, stains on the lavatory, shower, tub and floor, sticking drawers, and cabinet doors with loose hinges. These repairs are discussed in Chapter 10.

Walls & Floors

The floors are regularly exposed to water and the high humidity in the air, which can damage the walls and ceiling. If the flooring is physically damaged, a repair can be made by removing the damaged area and replacing it; however, if the floor covering is old, consider installing all new material. Since water is a problem, roll vinyl floor covering or ceramic tile is a good choices.

If the structure of the floor is weak and the floor is sagging, reinforce the joists as discussed in Chapter 15. Since the bathroom may have a whirlpool or hot tub, the floor has to carry a load much greater than

normal. This is also an important consideration if one of these is added to an existing bathroom.

Most walls are covered with gypsum wallboard, which if painted with a paint that will resist humidity, or if covered with a wallpaper that has a vinyl coating, will ensure damage be minimal. Information on repairing walls and ceilings is in Chapter 4.

Clogged Drains

The lavatory drain will most likely slow down or completely stop flowing after a few years. The major cause is hair in the trap. Liquid drain cleaning solutions do help if used regularly, but sooner or later other action will have to be taken. The easiest step is to try to force the clog out with a plunger (11-2). If this does not work the lavatory trap will have to be removed and cleaned as shown in Chapter 10. If the trap is old and possibly the threads are worn, replace it.

When replacing the trap, installing faucets or making other threaded connects, wrap the threads with a special tape made for this purpose. This reduces the possibility the connection may leak (11-3).

If the trap is clean the obstruction will most likely be in the waste pipe in the wall or under the floor. Remove the pipe to the waste pipe in the wall and

11-2 Bathroom sink, tub, and toilet traps can sometimes be unclogged with a plumber's plunger.

work an auger down the pipe (**11-4**). Rotate it and probe back and forth. When it hits the clog slide it into the clog and rotate the auger until it breaks it up (**11-5**). Then run lots of water down the pipe to flush away the debris.

BATHTUB DRAIN CLEARING

When the bathtub begins to drain slowly try to clear the clog by running an auger down the overflow tube. Remove the popup from the drain and flow-control linkage that is in the overflow tube (**11-6**). Then feed

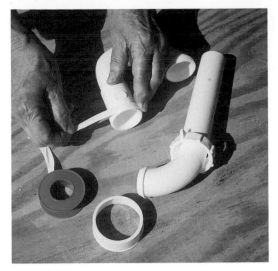

11-3 This sink trap is being assembled using special tape wrapped around the threads on each connection.

11-4 If the trap is clear the clog will be somewhere in the waste pipe. Remove the trap and connecting pipe and run an auger down the waste pipe that is in the wall.

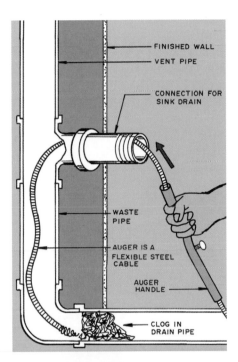

11-5 As you move the auger down the waste pipe, rotate it with the crank handle and move it back and forth to help it slide down the pipe. Rotate and butt it into the clog to break it up.

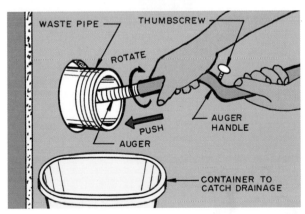

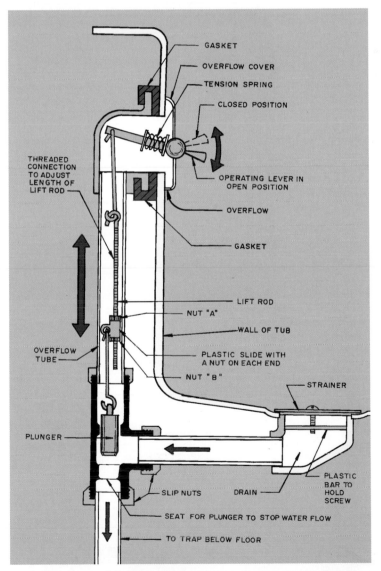

11-6 This type of bathtub drain uses a plunger controlled by a lift rod to regulate the flow of water out of the bathtub.

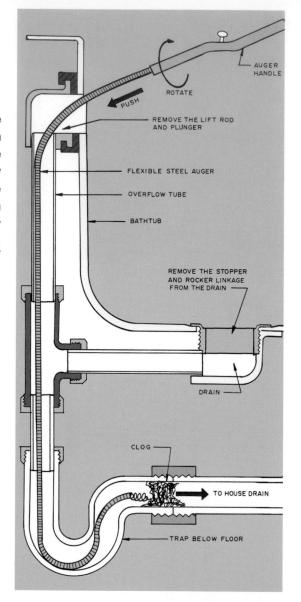

11-7 The bathtub drain can be cleared by running the auger down the overflow tube.

the auger down the overflow tube, rotating it with the handle to reach and break up the clog (**11-7**). Flush the pipe with water.

ANOTHER DRAIN-CLEARING TECHNIQUE

It is possible to unclog most drains using water pressure from a garden hose. The lavatory or bathtub requires that the drain popup be removed. Then wrap cloth tightly around the hose and insert it in the drain. Turn on the water (**11-8**). If the pressure is enough to move the clog all is well. If it is not, you will begin to get water squirting out around the cloth. Have someone ready to turn off the water if this occurs. This will also work well on clogged floor drains.

If the lavatory has an overflow opening, hold a cloth over it so the water does not back up out of it.

11-9 A clogged toilet can sometimes be unclogged by pumping water into the drain with a plumber's plunger.

11-8 Some clogged drains can be cleared using water pressure from a garden hose. Be certain to have someone ready to turn off the water if the clog does not clear.

Clogged Toilets

Sometimes a clogged toilet can be cleared by using a plumbing plunger. Place the plunger directly over the opening inside the bowl and push down on it with a pumping motion (**11-9**). If this does not do the job you can run an auger through the trap and drain as shown in **11-10**.

Some Other Toilet Problems

After many years the toilet may begin to show wetness around the bowl at the floor. This indicates the wax seal between the floor flange and the waste pipe at the bottom of the toilet needs replacing (**11-11**). While you can shut off the water to the toilet, drain the tank, dip out the water in the bowl, disconnect the water tank from the bowl, unbolt the bowl from the floor, lift it off the wax seal, replace the wax seal, and reassemble the fixture and check for leaks, it is a difficult job and best handled by a plumber. Some local codes may prohibit you from doing this job.

Also, after many years toilets with separate water tanks will begin to leak where they bolt to the toilet bowl (**11-12**). You might first try tightening the connecting bolts; however, be very careful. If you tighten too much you could crack the ceramic tank or the bowl. Generally it requires you to shut off the water and drain the tank. Disconnect the water line and unbolt the tank from the bowl. Replace the washers on the bolts that hold the tank to the bowl, the spud washer, and the washer that sits in the hole at the bottom of the tank (**11-13**).

Possibly the most common problem occurs when the **flapper** wears and no longer sits on the drain pipe at the bottom of the tank. The flapper releases the flow of water providing the flushing of the bowl (**11-14**). When worn it lets water leak into the bowl, causing the filler valve to turn on, replacing the lost water. This can increase your water bill. The best solution is to replace the flapper.

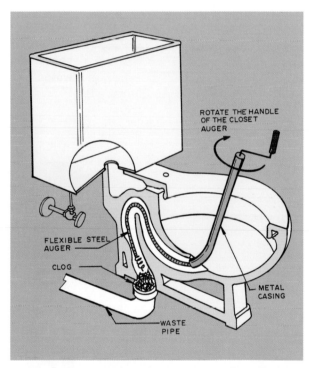

11-10 A closet auger can clear toilet clogs that the plunger will not clear. Hold the metal casing in one hand and rotate the hands as you carefully push the auger through the trap.

11-11 A typical detail of the toilet connection to the waste pipe in the floor.

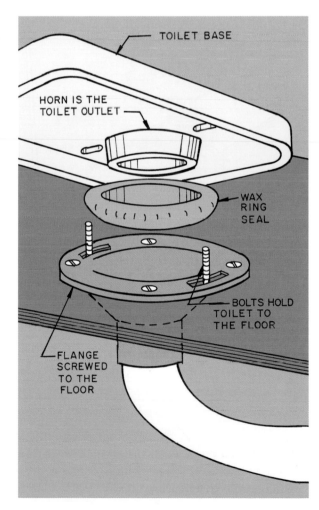

11-12 This toilet has a separate water tank that bolts to the back top surface of the bowl. As the gasket ages, leakage can occur at this connection.

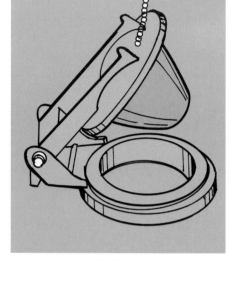

11-14 An old, worn flapper will let the water in the tank leak into the bowl, wasting considerable water.

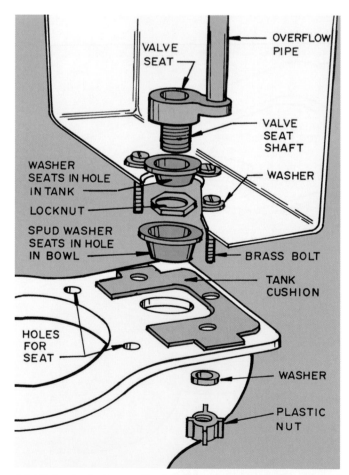

VALVE SEAT

OVERFLOW PIPE

VALVE SEAT SHAFT

WASHER SEATS IN HOLE IN TANK

WASHER

LOCKNUT

SPUD WASHER SEATS IN HOLE IN BOWL

BRASS BOLT

TANK CUSHION

HOLES FOR SEAT

WASHER

PLASTIC NUT

11-13 Leaks between the tank and bowl could be caused by defective bolt washers, spud washers, or the washer that seats in the hole at the bottom of the tank.

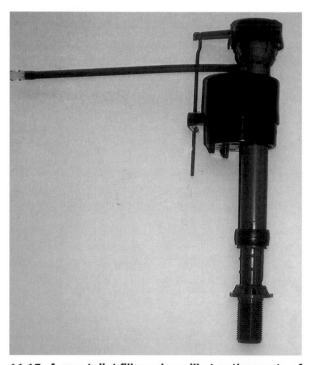

11-15 A new toilet-filler valve will stop the waste of water when the old valve fails to completely stop the flow of water into the tank.

Another common problem that occurs with toilets is when the valve that controls the flow of water to the tank no longer keeps it shut off. The easiest solution is to buy a new valve (**11-15**). It is inexpensive and easy to replace. A typical installation is shown in **11-16**. Follow the installation instructions that came with the valve.

Faucet Problems

The faucets may start to drip after a few years. There are several designs and the correction depends upon the faucet in place. Following are some commonly found examples.

Repairing a Compression Faucet

A typical two-handle **compression faucet** used in bathrooms is shown in **11-17**. It will have a threaded stem with a **seat washer** on the end or two **O-rings**.

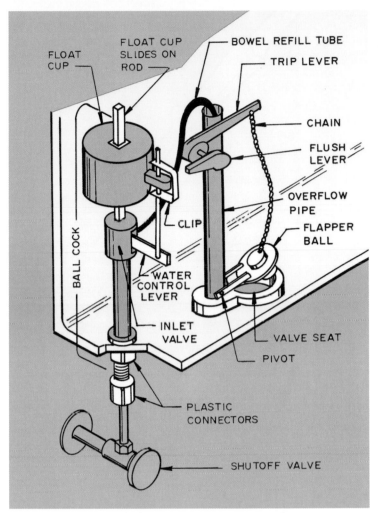

11-16 A typical mechanism for filling and flushing the toilet. The float controls the height of the water in the tank. The flapper is raised to flush the bowl.

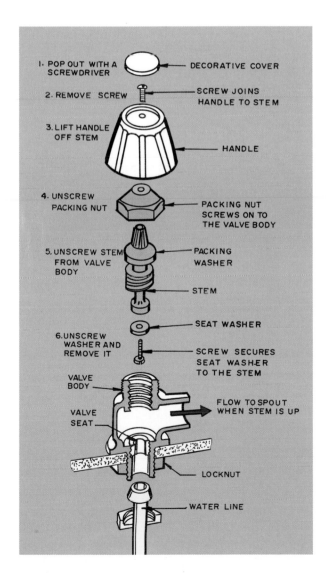

11-17 A typical two-handle compression faucet that uses a washer at the end of the stem to control the flow of water.

11-18 The parts of a typical compression faucet. The seat washer stops the flow when it is lowered against the valve seat.

223

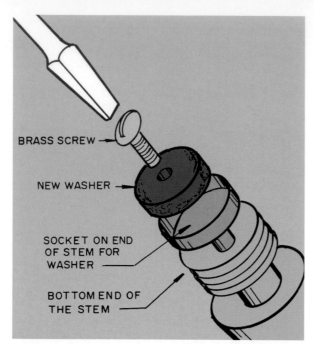

BRASS SCREW

NEW WASHER

SOCKET ON END
OF STEM FOR
WASHER

BOTTOM END OF
THE STEM

11-19 Install the new washer in the socket on the bottom of the stem.

When it is turned closed, the seat washer is pressed against the **seat**. This prevents the water from flowing. When the stem is turned raising the washer, the water will flow. If the washer gets old and worn it will let water drip into the lavatory. To correct this replace the seat washer. The steps to do this are shown in **11-18. Do not forget to turn off the water to the faucet.**

After the stem has been removed from the valve body, remove the brass screw, holding the seat washer to the end of the stem. Install a new washer (**11-19**) and reassemble the faucet.

If the faucet is very old the valve seat may be worn and the new washer will not fit tightly against it. If this occurs you will have to resurface the valve seat using a **valve-seat dresser (11-20).**

Older faucets will sometimes start leaking around the stem below the handle. When new they have a preformed **packing washer** placed beneath the cap nut (**11-21**). Replace this washer to stop the leak. You can also buy a thread-like **self-forming packing**. This

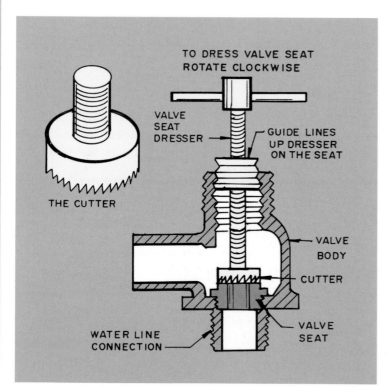

TO DRESS VALVE SEAT
ROTATE CLOCKWISE

VALVE
SEAT
DRESSER

GUIDE LINES
UP DRESSER
ON THE SEAT

THE CUTTER

VALVE
BODY

CUTTER

WATER LINE
CONNECTION

VALVE
SEAT

11-20 Worn valve-seat surfaces can be resurfaced by lightly dressing them with a valve seat dresser.

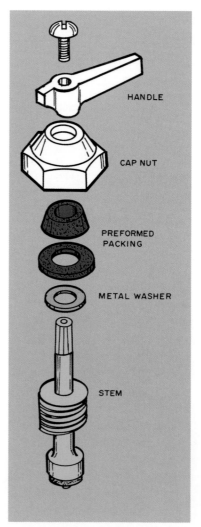

HANDLE

CAP NUT

PREFORMED
PACKING

METAL WASHER

STEM

11-21 If the faucet is leaking at the stem and has a packing washer, replace the washer.

is a round, pliable material about 1/16 inch in diameter and several inches long. Wrap it around the stem. Then tighten down the cap to force it to fill the space under the cap, sealing any openings around the stem (**11-22**).

Cartridge Faucets

Newer faucets use a **cartridge** with two O-rings to control the flow of water (**11-23**). The O-rings are placed on the cartridge of the faucet (**11-24**). They are available from the dealer handling plumbing supplies. Hopefully the dealer will handle the brand of faucet you have.

The disassembled faucet is shown in **11-25**. To replace the O-rings remove the handle, unscrew the

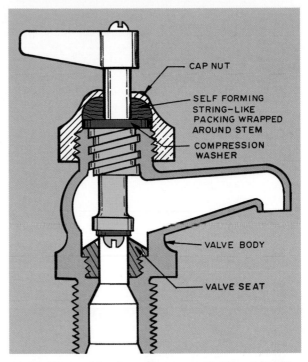

CAP NUT

SELF FORMING STRING-LIKE PACKING WRAPPED AROUND STEM

COMPRESSION WASHER

VALVE BODY

VALVE SEAT

11-22 Some older faucets use a pliable, string-like packing material wrapped around the stem under the cap nut.

11-23 Cartridge faucets are an improvement over the compression faucet. They use O-ring seals to control the flow of water. The actual design will vary with the brand of faucet. *Courtesy Moen Incorporated*

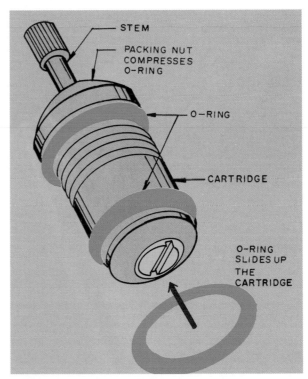

STEM

PACKING NUT COMPRESSES O-RING

O-RING

CARTRIDGE

O-RING SLIDES UP THE CARTRIDGE

11-24 The cartridge faucet uses O-rings to seal the stem.

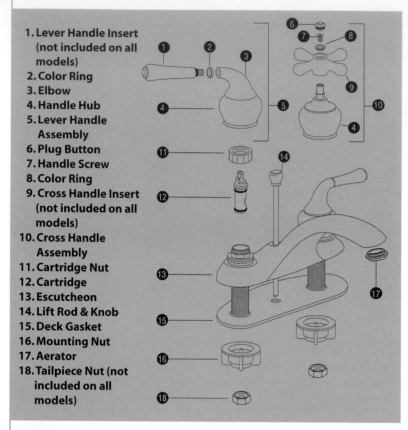

1. Lever Handle Insert (not included on all models)
2. Color Ring
3. Elbow
4. Handle Hub
5. Lever Handle Assembly
6. Plug Button
7. Handle Screw
8. Color Ring
9. Cross Handle Insert (not included on all models)
10. Cross Handle Assembly
11. Cartridge Nut
12. Cartridge
13. Escutcheon
14. Lift Rod & Knob
15. Deck Gasket
16. Mounting Nut
17. Aerator
18. Tailpiece Nut (not included on all models)

11-25 This exploded drawing shows the parts of a two-handle cartridge lavatory faucet. *Courtesy Moen Incorporated*

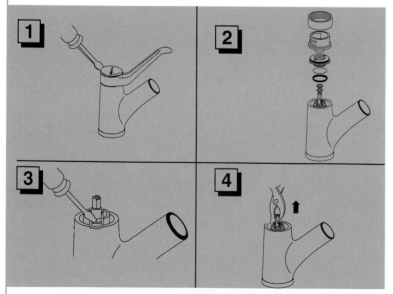

11-26 Should you need to replace the O-rings on a cartridge, remove the handle and cartridge nut and lift out the cartridge.

Courtesy Moen Incorporated

cartridge nut, and lift out the cartridge **(11-26).** Install new O-rings and replace the cartridge. Be certain to get the exact replacement O-rings.

Considering a Partial Renovation

If you are satisfied with the size of the bathroom and the placement of the fixtures, you just need to replace those that are old, stained, and no longer function properly. Visit the local cabinet dealers and plumbing fixture dealers and become acquainted with the new fixtures available. Lavatories are available in a variety of materials. Bathtubs and showers come in many designs and materials. Toilets have new features including water-saving designs. Do not forget to get the costs of installing these. New plumbing may be needed, so this could be a major expense.

The renovation will most likely include refinishing the walls and installing new flooring. The material and colors chosen should be coordinated with those of the new fixtures.

In the following pages detailed information will be given concerning bathroom layout and fixtures. Study these before making a final decision.

Going Ahead with a Complete Remodeling

Most older bathrooms are poorly laid out and many are very small. To get an efficient and attractive bathroom, consider a complete remodeling, including moving fixtures and possibly enlarging it. First consider what you do not like about it and then list the things you would like to do to improve it.

Start with the floor area that's available. If it is too small to accommodate your new plans, try to find a way to enlarge it. It may be expanded into part of an adjoining room or have a segment added if it is on an outside wall **(11-27).** (See expansion suggestions detailed in Chapter 10.)

Basic Bathroom Designs

As you consider the placement of the fixtures, remember to observe the location of existing doors and windows. While these can be changed, they are part of the planning process. Then consider the use the bathroom will be expected to serve. The following examples are minimum layouts and more room is recommended if it is at all possible. Notice the minimum layout dimensions for each fixture.

HALF BATH

A half bath, also referred to as a powder room, is usually located near the area where groups will be entertained or where children playing outside can use it without having to run through the house. The layout in **11-28** is typical. Notice you need at least 36 inches open space in front of the toilet. The lavatory can be very small and even a wall hanging fixture or pedestal fixture could be used to save space (**11-29**). If space is really tight a layout like that in **11-30** will serve; however, the door must open out.

THREE-QUARTERS BATH

Often when a third or fourth bathroom is added it has a shower instead of a bathtub (**11-31**). This is often referred to as a three-quarters bathroom. It is very effective and can serve an area with three or four bedrooms quite adequately. Actually some prefer a

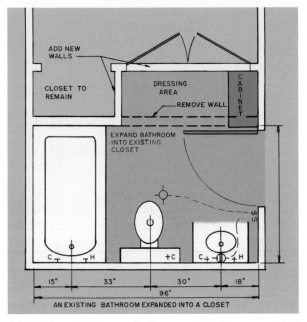

11-27 Sometimes a bathroom can be expanded into a closet or space in the next room.

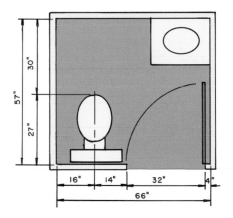

11-28 This is a typical layout for a half bath. The space for the occupant is at absolute minimum, so an additional foot or two would be welcome.

11-29 A pedestal lavatory occupies little space and since it has no cabinet, provides additional open floor space.

11-30 This is a very tight half bath. Notice the door must swing out or it cannot be used.

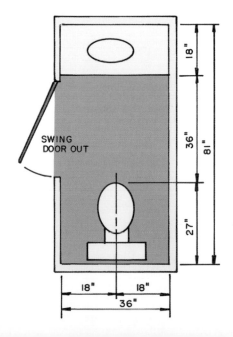

11-31 One layout for a bathroom with a shower instead of a bathtub. Notice the minimum spacing requirements for each fixture. A larger shower than that shown should be used if possible. Remember that you need floor area to leave the shower.

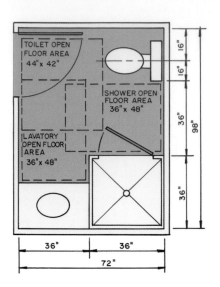

shower instead of a bathtub even if the tub has a shower head off the wall. Do not overlook the value and pleasure of a large, comfortable shower.

FULL BATHROOM

A typical full bathroom has a lavatory, toilet, and bathtub. It plays a major role as the main bathroom in most houses. While some build a master bathroom off the master bedroom, the full bathroom is the type most often used. It can vary from a minimum layout as shown in **11-32** to a larger design with lots of open floor space (**11-33**). Some place the toilet in a compartment and add double lavatories in an attempt to allow two persons to use the facility at the same time, but some do not like this multiple-use feature (**11-34**).

MASTER BATHROOM

The master bathroom is placed off the master bedroom and is not conveniently available to other members of the family. It is the place where people can splurge and include features not in the other bath-

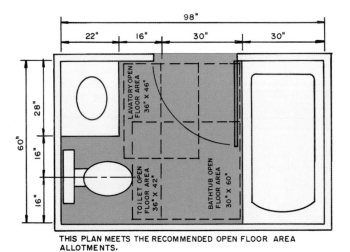

THIS PLAN MEETS THE RECOMMENDED OPEN FLOOR AREA ALLOTMENTS.

11-32 This is a minimum full-bath layout. It is rather tight so allow a few feet of extra room if possible. Notice the minimum spacing dimensions for the fixtures.

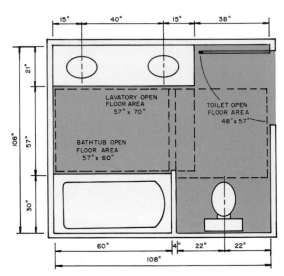

11-33 This is a larger full bath that provides more open floor area to make it easier to dry off and to dress after a bath. Notice it has the toilet in a semi-private compartment with two lavatories.

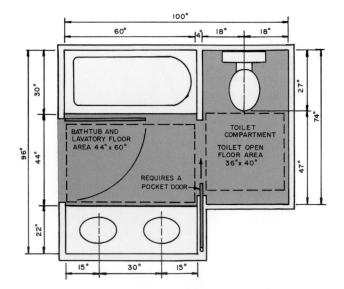

11-34 This full bath places the toilet in a compartment and provides two lavatories so two people can use the facility at the same time.

rooms (**11-35**). A nice but minimal master bathroom is shown in **11-36**. It features double lavatories, a toilet, a shower and a bathtub or whirlpool (**11-37**).

The master bathroom layout in **11-38** has a huge whirlpool set against a frosted glass window. The toilet and bidet are in a large compartment. Some even add a large closet and dressing area to the master bathroom. Often this area will have carpet on the floor, while the wet areas will have ceramic tile (**11-39**).

Storage

Every bathroom, regardless of size, needs storage for towels, soap, toiletries, and other supplies. A large bathroom can have a shallow closet as shown in the bathroom in **11-35**. This should be deep enough to hold folded towels and tissue. Other storage can be in the lavatory vanity cabinet (**11-40**) or some type of wall-hung cabinet (**11-41**). These are good for medicines and other small articles. A mirrored cabinet

11-35 A beautiful master bathroom with a vanity, storage cabinet, pedestal lavatory, and durable, colorful, sheet-vinyl floor covering. *Courtesy Congoleum Corporation*

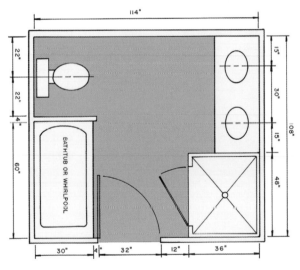

11-36 This is a comfortable master bathroom featuring both a shower and a bathtub or whirlpool.

11-38 The comfortable master bathroom provides a lot of open space and multiple fixtures.

11-37 A whirlpool bathtub adds an attractive feature to the bathroom. Notice it has been raised above the floor and the platform is covered with ceramic tile.

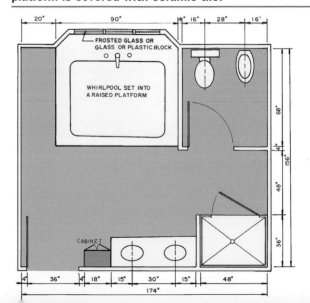

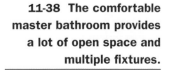

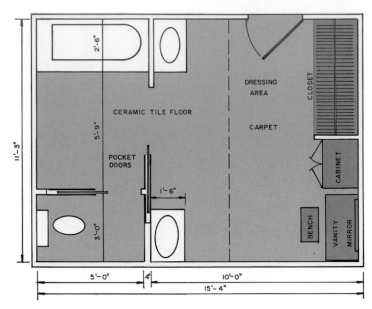

11-39 This master bathroom has a private compartment plus a cabinet, closet storage, and a vanity/dressing area. Closet doors are often omitted.

11-40 Lavatory cabinets can provide considerable storage space including some drawers.

11-41 This typical wall-hung cabinet is deep enough to store tissue and many other necessities.

11-43 The cabinetry in this bathroom provides considerable storage and reduces the "bathroom" look. *Courtesy KraftMaid Cabinetry, Inc.*

11-42 A cabinet with a mirror is probably the most used storage unit.

above the lavatory will hold medicines and small items (**11-42**).

If you want to lose the bathroom look, consider using some of the attractive cabinetry available from cabinet manufacturers. This also provide a lot of storage. The bathroom in **11-43** is an excellent example of the use of quality cabinetry.

Ventilation

Showering and bathing produce a lot of moisture in the air. The high humidity glosses over the mirrors and dampens cabinets, walls, and floors. This leads to deterioration and the formation of mold. If the bathroom has a window, natural ventilation is a help, but much of the year you do not want an open window, so mechanical ventilation is required. Building codes require a bathroom to have an operable window or mechanical ventilation. A typical ventilating fan should be able to provide eight air changes per hour. To find the size fan required, multiply the cubic feet inside the room by eight and divide by 60 minutes. For example: bathroom 720 cubic feet x 8 air changes = 5,760 cubic feet to move, per hour. Dividing 5,760 by 60 minutes = 96 cubic feet per minute. A fan that will move 96 cubic feet per minute is required.

Ceiling fans remove the air and discharge it outside (**11-44**). Others not only remove the air but have a ceiling light as part of the fixture (**11-45**). Some also contain electric heating coils which provide instant heat to keep you comfortable as you dry off after a shower. The fan exhausts better when about one inch of air space is left below the door.

Plumbing Considerations

If the remodeling or a complete renovation leaves the new fixtures in the same place as those being replaced, major plumbing work will not usually be necessary. This is a good time to replace the toilet floor connection if it is in poor shape. Hot and cold water pipes in the walls and floor can be replaced if they are clogged, reducing water flow, or leaking.

If the fixtures are moved to a new location, the plumber will have to remove and discard the old pipes and fittings and install new ones to the changed location. This work must meet local building codes.

Generally the hot and cold water will be run to a lavatory through the wall with the pipes stubbed through the drywall. Flexible lines are run from these to the lavatory faucets. Always place a valve on the

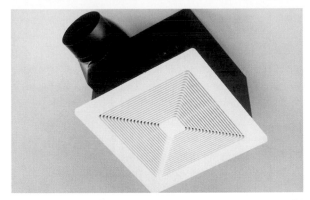

11-44 This ceiling ventilation fan removes humid air from the bathroom and discharges it outside the house. *Courtesy Broan-Nutone LLC*

11-45 This beautiful ceiling light provides general illumination and contains a ventilating fan that pulls air in through the grill openings next to the ceiling.

Courtesy Broan-Nutone LLC

11-46 Valves are mounted on the water line below the lavatory so the water to the faucets can be shut off when necessary for repairs.

stub so you can shut off the water to the lavatory (11-46). Water to the bathtub and shower is run up the wall and the faucets are mounted on the wall. Usually the wall will have ceramic tile (11-47) or will have the faucet enter through the fiberglass wall (11-48). Steps to remove these faucets for repair are shown in 11-49. Both compression and cartridge type faucets are available. A diagram of a typical potable water distribution system is shown in 11-50.

The waste disposal pipe for the lavatories usually extends through the wall under the fixture. Toilet,

11-47 This faucet is mounted on a wall finished with ceramic tile.

11-48 Some bathtubs and showers have a fiberglass surround through which the faucet and spout enter the shower.

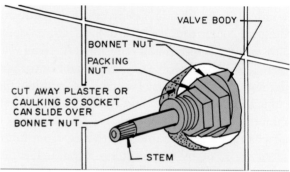

SOME HAVE AN ESCUTCHEON. PRY OFF WITH A SCREWDRIVER

STEM COVER – TURN COUNTERCLOCKWISE TO REMOVE

HANDLE

STEM

I. REMOVE THE HANDLE AND UNSCREW THE STEM COVER TO EXPOSE THE VALVE.

VALVE BODY

BONNET NUT

PACKING NUT

CUT AWAY PLASTER OR CAULKING SO SOCKET CAN SLIDE OVER BONNET NUT

STEM

2. THE PACKING NUT IS SCREWED INTO THE BONNET NUT SO WHEN YOU UNSCREW THE BONNET THE PACKING NUT IS REMOVED WITH IT.

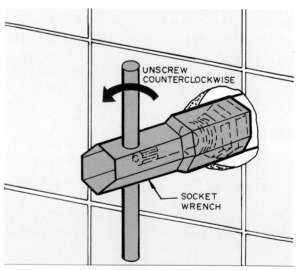

UNSCREW COUNTERCLOCKWISE

SOCKET WRENCH

3. USE A PLUMBER'S DEEP-SOCKET WRENCH TO REMOVE THE BONNET NUT AND VALVE STEM. ANY OTHER THAT WILL FIT CAN BE USED.

11-49 Wall-mounted compression faucets are opened for repairs by unscrewing the bonnet nut with a deep-socket wrench. The washer is replaced as mentioned earlier for compression faucets.

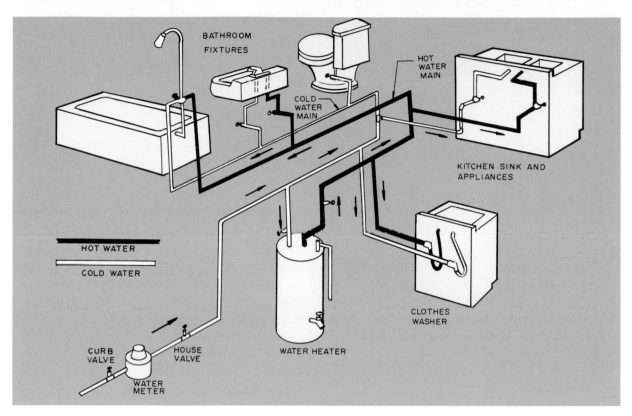

11-50 A typical piping system for supplying potable cold and hot water to bath and kitchen fixtures.

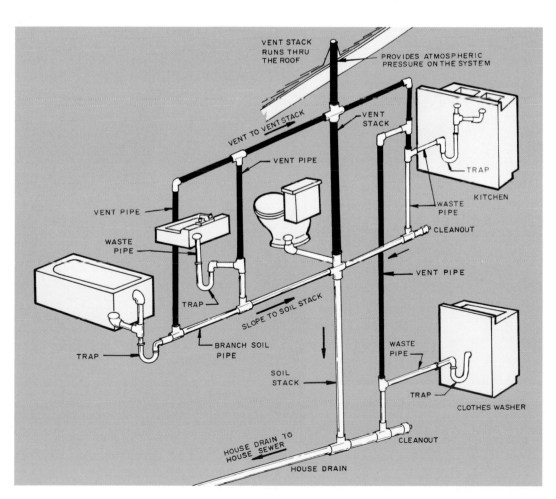

11-51 A typical waste disposal piping system for bathrooms and kitchens.

11-52 This vitreous china lavatory has a glass-like finish and is impervious to moisture. This unit is installed in a cutout in a plastic-laminate-covered countertop.

shower, and bathtub waste lines penetrate the floor beneath the fixture (11-51).

See Chapter 10 for more information on plumbing systems.

Lavatories

Lavatories are available made from vitreous china, enameled steel, and solid synthetic composite materials such as polyester and acrylics, combined with natural materials. Cultured marble is a composite of marble chips, dust, and a polyester resin. Cultured onyx is a composite of calcareous (porous limestone) chips and a polyester resin. A vitreous china lavatory is shown in 11-52. It has a glass-like finish, is impervious to moisture and easy to clean. Both vitreous china and enameled steel lavatories are set into openings cut in the countertop. A typical cultured marble lavatory is shown in 11-53. The lavatory and countertop are cast as a single unit. There are no cracks along the edge which could allow water to seep into the plywood substrate as can occur on tops with plastic laminates and a drop-in lavatory. Lavatories come in

11-53 Cultured marble lavatory and countertop are cast as a single unit, so there are no cracks between them—making it easier to clean.

11-54 This one-piece toilet is often called a low-profile toilet. It uses 1.6 gallons of water per flush and is made from vitreous china.

a wide range of sizes and shapes, so a visit to the local plumbing fixture dealer would be well worth the trip.

Toilets

There have been many changes in the design and operation of toilets over the past few years. A visit to the local plumbing fixture dealer will let you see the new designs and operating features. One big development is a standard, set by the U.S. Government, which requires the use of toilets that use only 1.6 gallons of water per flush. Older toilets use 3.5 gallons per flush.

One-piece toilets are very popular (**11-54** and **11-55**); however, the two-piece unit with a separate tank bolted to the back of the bowl is widely used (**11-56**). Gravity-fed and pressure flushing systems are available.

Bathtubs

Most bathtubs currently available are enameled steel, acrylic, fiberglass with a gel, or an acrylic coating, acrylonitrile-butadiene-styrene (ABS). They are available in a wide range of sizes and shapes, with rectan-

11-55 This is a one-piece toilet using a siphon-jet pressure flushing system.

11-56 This two-piece toilet has the tank bolted to the back of the bowl. It is a gravity flushing type.

11-57 This recessed acrylic bathtub has the exposed side finished. The wall will be finished with ceramic tile or some other waterproof wall covering.

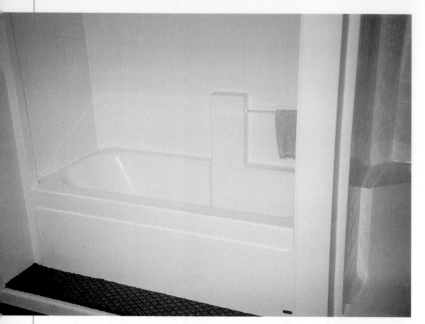

11-58 This recessed acrylic bathtub is set in a recess between walls. The acrylic material covering the wall is an integral part of the tub.

11-59 A prefabricated shower stall is a completely assembled unit consisting of the floor and walls. A glass door is mounted on the open side.

gular and square shapes most popular. The colors available match the toilets, showers, and lavatories made by the same manufacturer.

Some bathtubs are made to fit between two walls with only the exposed side finished **(11-57)**. A variation of this design has one end and the long side finished; the other end butts a wall. A very popular design uses a recessed tube set between walls, but it has the acrylic or fiberglass material running up over the wall, much like a shower stall **(11-58)**. This gives a seamless, totally waterproof enclosure that is easy to clean and will be free from mold.

Whirlpool bathtubs provide pressurized water jets which help relax tired muscles. They are often placed on a raised platform as shown earlier in **11-37** or are recessed in the floor.

Showers

If a bathtub is recessed between walls it can serve as a shower by mounting a shower head on one wall and installing sliding glass doors.

A prefabricated shower stall is typically made from acrylic or fiberglass formed into a completely assembled unit with a glass door on one side **(11-59)**. While 32-inch square units are available, consider using a

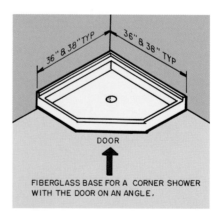

FIBERGLASS BASE FOR A CORNER SHOWER WITH THE DOOR ON AN ANGLE.

11-60 Typical one-piece fiberglass shower pans. The glass enclosure mounts on them.

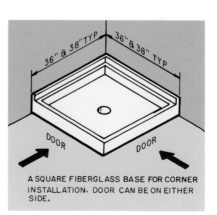

A SQUARE FIBERGLASS BASE FOR CORNER INSTALLATION. DOOR CAN BE ON EITHER SIDE.

36- or 48-inch size. The extra inches make it much easier to use.

If you want something better, use a unit that has a shower pan **(11-60)** with a metal-framed glass enclosure **(11-61).** Shower stall pans are available made of acrylic, ceramics, or enameled steel. Remember to coordinate the size of the pan and the selection of the glass enclosure. The walls can be covered with ceramic tile or any other suitable material **(11-62).**

Cabinets

Cabinet manufacturers have available a wide range of very attractive and functional cabinets for use in a bathroom. They provide the major design features of the room and much needed storage **(11-63).** See Chapter 9 for more information on cabinets.

11-61 This corner shower uses a fiberglass shower pan and is enclosed on three sides by a metal-framed glass enclosure assuring a watertight unit. *Courtesy Alumax Bath Enclosures*

11-62 Ceramic tile is widely used on the walls of showers with pans that are built-in or custom-built.

Courtesy American Olean Tile Company

11-63 This cherry cabinet adds a great deal to the appearance and ambience of the room. *Courtesy KraftMaid Cabinetry, Inc.*

237

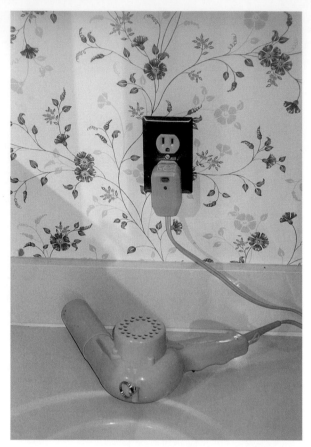

11-64 It is important to have at least one wall outlet above the lavatory. It must be monitored by a ground-fault circuit-interrupter.

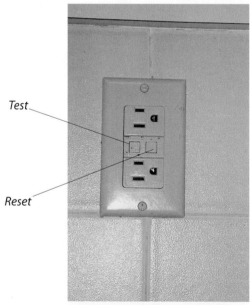

Test

Reset

11-65 This is a wall-mounted duplex outlet with ground-fault circuit-interruption protection.

Electrical Considerations

Whether you are remodeling or building an entirely new bathroom, consider where the electrical outlets should be placed. One duplex outlet should be mounted on the wall by the lavatory **(11-64)**. Under no conditions should these be installed face up on the lavatory countertop. This outlet is used for small appliances, such as a hair dryer, electric toothbrush, or shaver. Place additional outlets along the wall for the vacuum cleaner or other appliances. Since this is a damp area, the current should be monitored by a **ground-fault circuit interrupter** (GFCI) **(11-65).**

To brighten the entire room, the bathroom will need **general illumination,** such as some type of ceiling light. You will also need **task lighting** that focuses on a limited area such as the lavatory and mirror.

Sound Control

It is very important to construct the bathroom walls to control the flow of sound from it to the abutting rooms. The sound control effectiveness of a wall assembly is specified as its **sound-transmission class** (STC). STC rates a partitions resistance to airborne sound transfer at speech frequencies; the higher the number, the better the isolation. Bathroom walls should have an STC of 52 or higher. In **11-66** are two wall assemblies that are suitable. The assembly shown in **11-67** is also quite good and a little less expensive to build. Your building supply dealer will have sound reducing insulation batts which greatly reduce sound transmission through interior walls. Note that STC ratings favor materials that absorb high frequencies and not those that are more effective at absorbing lower frequencies.

Floor Loads

The floor should be reinforced to carry the load of a bathtub or whirlpool. Some whirlpools are very large and hold a lot of water. Water weighs 8.3 pounds per gallon, so multiply the capacity of the tub or whirlpool to get the maximum load that could be placed on the floor joists. Typically the floor under the bathtub will have double joists under the outside edge as shown in **11-68.** A whirlpool will require an architect to determine how to frame the floor to carry it.

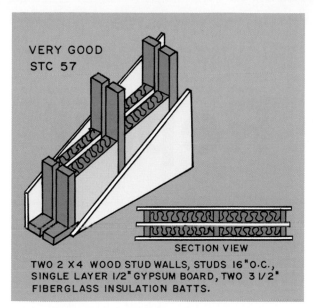

VERY GOOD
STC 57

SECTION VIEW

TWO 2 X 4 WOOD STUD WALLS, STUDS 16"O.C.,
SINGLE LAYER 1/2" GYPSUM BOARD, TWO 3 1/2"
FIBERGLASS INSULATION BATTS.

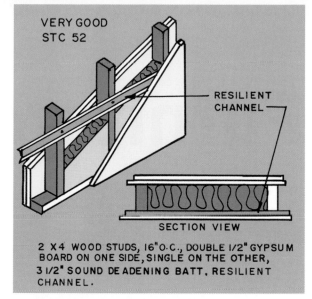

VERY GOOD
STC 52

RESILIENT
CHANNEL

SECTION VIEW

2 X 4 WOOD STUDS, 16"O.C., DOUBLE 1/2" GYPSUM
BOARD ON ONE SIDE, SINGLE ON THE OTHER,
3 1/2" SOUND DEADENING BATT. RESILIENT
CHANNEL.

11-66 These two wall assemblies equal or exceed the minimum STC rating for bathrooms.

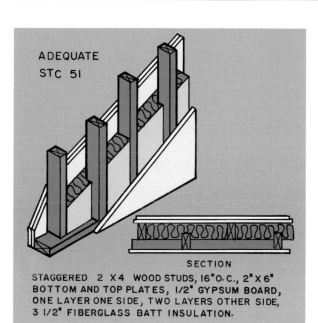

ADEQUATE
STC 51

SECTION

STAGGERED 2 X 4 WOOD STUDS, 16"O.C., 2"X 6"
BOTTOM AND TOP PLATES, 1/2" GYPSUM BOARD,
ONE LAYER ONE SIDE, TWO LAYERS OTHER SIDE,
3 1/2" FIBERGLASS BATT INSULATION.

11-67 This wall
assembly is very
good but doesn't
quite meet the
STC rating for
bathroom wall
assemblies.

11-68 A
typical way to
reinforce the
floor under a
bathtub.

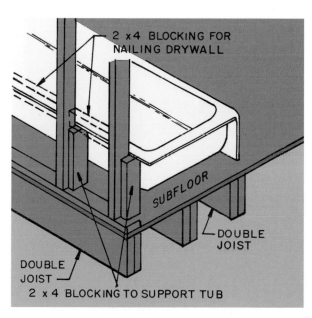

2 x 4 BLOCKING FOR
NAILING DRYWALL

SUBFLOOR

DOUBLE
JOIST

DOUBLE
JOIST

2 x 4 BLOCKING TO SUPPORT TUB

CHAPTER 12

Furniture Repair

12-1 This beautiful carved chest is typical of furniture over 100 years old that is passed down through generations.

Over the years furniture will begin to need repairs, many of which can be made by the owner. Families have pieces they inherited from deceased family members or which were given to them by parents (12-1). These can be in need of renovation before they are ready for everyday use.

Should a Piece Be Restored or Simply Repaired?

Some people enjoy touring antique shops and yard sales and often buy a nice piece of furniture that needs some restoring. If a piece is a genuine antique that has considerable value, it is best to consult a professional antiques dealer to see if your restoring it will greatly reduce its value. In these cases a professional antique restoration shop should do the job.

As you examine a piece being considered for restoration, note the species of wood and type of finish, and try to find the type of joints used. Each of these will influence what you do to perform the restoration (12-2).

Examining Furniture Joints

Joints on very old furniture were often cut with hand tools like saws and chisels. They were worked on until they fit, so each joint may be a slightly different size. For many years now furniture factories have used power tools that accurately cut each joint the same size so the pieces can fit any furniture piece of that type. Computers now control some of the manufacturing processes.

The following joints are those commonly found in old and new furniture construction.

The **mortise-and-tenon** gives a strong joint. The **blind mortise-and-tenon** produces a hidden joint. The joint with a **through tenon** lets it be seen and is used for its design feature. These are mainly used to connect table and chair rails to the legs **(12-3)**.

The **dowel joint** is widely used for many applications. Some typical places it is used are shown in **12-4.**

12-2 This old table was discovered in a wet basement where it had been for many years. The wood had almost no finish and was a light gray.

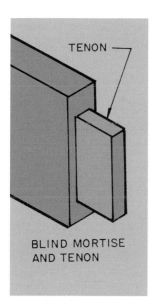
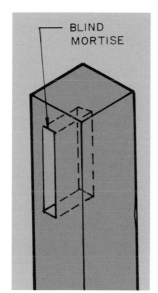

BLIND MORTISE AND TENON

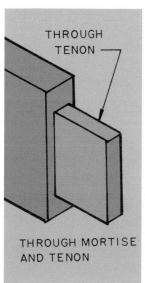
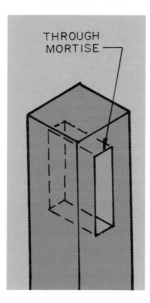

THROUGH MORTISE AND TENON

12-3 Typical mortise-and-tenon joints.

241

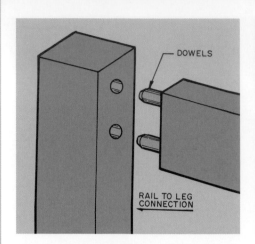
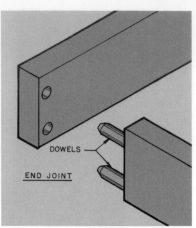
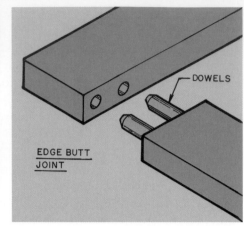

12-4 Several applications of dowel pins used to secure wood pieces.

Lap joints are used to secure butting or crossing members. The two pieces are notched so when they are joined the surfaces are usually flush (12-5).

The simplest joint is the **butt joint.** Edge joints can be glued, nailed, or screwed together. End butt joints are joined with dowels (12-6).

Rabbet joints are widely used to secure members along the edges or ends. They are not strong enough to resist heavy strain (12-7). They are widely used to install the backs on cabinets and furniture.

Dadoes and **grooves** are used when you want to recess a member into one it butts. A **dado** is a rectangular slot cut across the grain of a board (12-8).

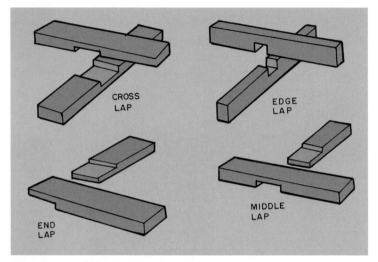

12-5 Some of the commonly used lap joints.

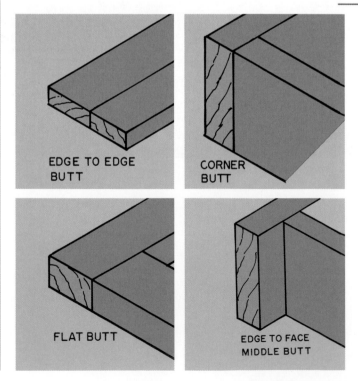

12-6 Butt joints are the simplest to construct and are secured with adhesives, nails, screws, or dowels.

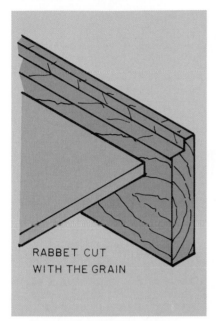
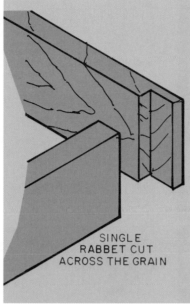
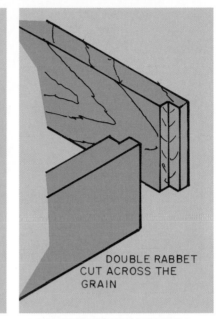

RABBET CUT
WITH THE GRAIN

SINGLE
RABBET CUT
ACROSS THE GRAIN

DOUBLE RABBET
CUT ACROSS THE
GRAIN

12-7 Rabbet joints are commonly used in furniture construction.

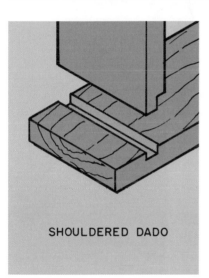

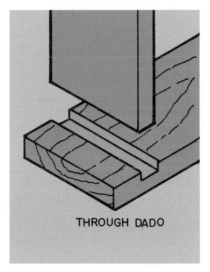

SHOULDERED DADO

STOPPED DADO AND A
SQUARE SHOULDER

THROUGH DADO

12-8 Dadoes are frequently used in furniture construction. They are cut across the grain of a board.

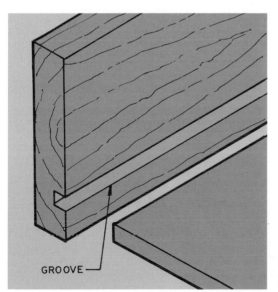

GROOVE

12-9 Grooves are rectangular slots cut with the grain of the wood.

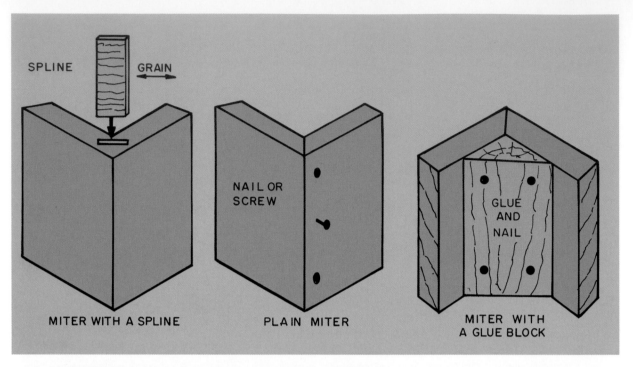

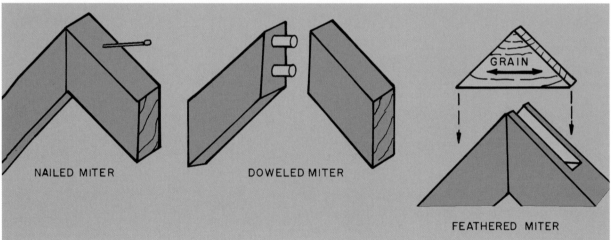

12-10 These are some of the ways wide and flat miters are secured.

A **groove** is a rectangular slot cut with the grain of a board (**12-9**). Dadoes are often used to hold shelves and grooves are often used to hold a panel such as a drawer bottom.

Miter joints are widely used because they hide the end grain of butting pieces and are decorative. While widely used to connect flat pieces, such as moldings, they can join wider pieces as shown in **12-10**.

Dovetail joints are strong and are used in situations like connecting a drawer front to the side (**12-11**). They indicate quality construction. While they can be cut with a saw and chisel, the use of a power router and template produces a better fitting joint, and much faster.

Box joints are an alternative to the dovetail but since the pins do not lock it is not quite as strong (**12-12**).

A more recent connection device is a **plate**, also called a **biscuit**. It is used to join butt, edge, and miter joints. The plate is a thin wood member set into curved slots that have been cut in the wood with a special saw. The plates are glued into the slots holding the wood pieces together (**12-13**).

As you examine the furniture be aware that there are a number of mechanical fasteners used to secure the wood pieces together (**12-14**). One type of fastener is a connector that is used to secure the joint by drilling the proper diameter holes in each piece.

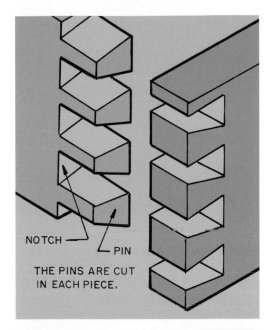

NOTCH
PIN

THE PINS ARE CUT
IN EACH PIECE.

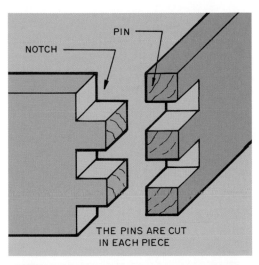

NOTCH
PIN

THE PINS ARE CUT
IN EACH PIECE

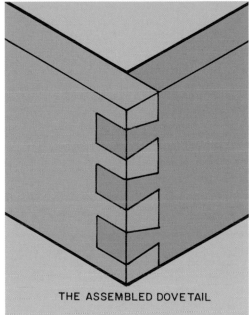

THE ASSEMBLED DOVETAIL

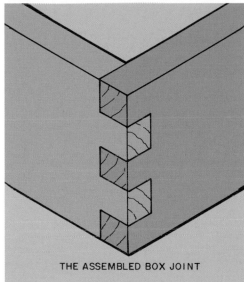

THE ASSEMBLED BOX JOINT

12-12 Box joints are a good way to secure corners. They are also decorative.

12-11 Dovetail joints are strong and used where the parts tend to be pulled apart, such as securing a drawer front to the drawer sides.

12-13 Edge joints and miters can be securely joined. Use pressed wood plates (biscuits) set in slots machined in the butting edges.

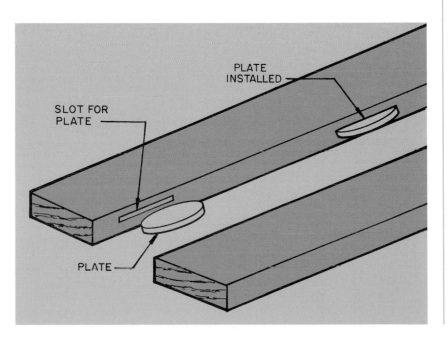

PLATE
INSTALLED

SLOT FOR
PLATE

PLATE

245

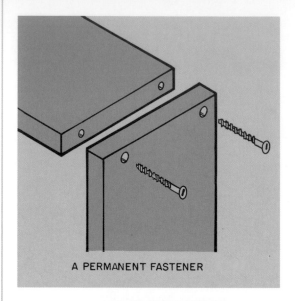

A PERMANENT FASTENER

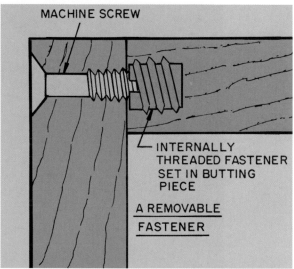

MACHINE SCREW

INTERNALLY THREADED FASTENER SET IN BUTTING PIECE

A REMOVABLE FASTENER

12-14 Various mechanical fasteners can be used to secure wood members. Some permit the assembly to be taken apart.

KEYHOLE FITTING

SCREW

12-15 A typical fitting used for assembling furniture sold in pieces disassembled. It permits rapid assembly and disassembly.

Another type receives the threaded fastener, which is stronger than using wood screws. A third type places a metal insert in one of the pieces. A special threaded fastener screws into this, making a tight connection. This connection can be unscrewed and reconnected as desired. It is used on furniture sold unassembled. The threaded fastener permits fast and easy assembly when you get the furniture home. It gives a strong and solid connection. Other metal connectors are available such as the one in **12-15.** This permits rapid assembly and disassembly but is not as rigid as the screw-type connectors.

Glues, Adhesives & Cements

One major task when working with old furniture needing repair is the securing of loose joints. Key to this is the selection of a good glue or adhesive. **Glues** are made from natural materials—vegetable and animal products such as hides and bones. **Adhesives** are made from synthetic materials. There are two types, thermosets and thermoplastics. They are the most commonly used bonding agents. Examples include epoxy and polyvinyl adhesives. **Cements** are made from synthetic rubber suspended in a liquid. An example is contact cement used to bond plastic laminates to the substrate.

The choice depends upon the material to be bonded, its condition, the difficulty of holding the bonded parts, together and the specifications of the bonding agent. **Table 12-1** shows a summary of the most frequently used bonding agents. Visit the local hardware store and read the specifications on the label to be certain the bonding agent being considered will do the job.

Repairs by the Home Woodworker

After you have examined the piece, and determined that repairs can be made without damaging the piece or greatly reducing its value, proceed with your repairs. Typical repairs that the home woodworker might feel comfortable making include repairs to joints, split wood, and veneer as well attention to warping, hardware, or upholstery.

Repairing Mortise-and-Tenon Joints

The most secure repair for a mortise-and-tenon joint is to take it apart, clean off the glue, and reglue it. It

Table 12-1 Characteristics of Bonding Agents.

Type	Form	Solvent	Mixing Procedure	Applications	Service Durability	Clamping Time
Acrylic	Liquid and powder	Acetone	Mix liquid w/powder	Wood, glass, metal	Water-resistant	Sets in 5 minutes, cures in 12 hours
Aliphatic resin	Liquid (yellow glue)	Warm water	None	Edge- and face-gluing, laminating	Interior use	1 hour
Casein glue	Powder	Soap and warm water before it hardens	Mix w/water	Furniture, laminated timbers, doors, edge-gluing	Water-resistant	2 hours
Cellulose cement	Clear liquid	Acetone	None	Wood, glass, metal	Water-resistant	2 hours
Contact cement	Liquid	Acetone	None	Bonding to plastic countertops	Water-resistant	No clamping time
Cyanoacrylate (Superglue)	Liquid or gel	None	None	Liquid-metal, plastic, rubber, ceramics	Water-resistant	Sets in 30 seconds, cures in 30 to 60 minutes
Epoxy	Liquid and a catalyst	Acetone	Mix liquid and catalyst	Wood, glass, metal, ceramics	Interior use	Sets in 5 minutes, cures in 2 hours
Hot-melts	Solid	Acetone	Melt w/electric glue gun	Molding, overlays	Interior use	None
Liquid hide glue	Liquid	Warm water	None	Furniture, edge-gluing, laminating	Interior use	2 hours
Polyvinyl acetate (white glue)	Liquid	Soap and warm water	None	Edge- and face-gluing, laminating, paper	Interior use	Sets in 8 hours, cures in 12 hours
Polyvinyl chloride	Liquid	Acetone	None	Wood, ceramics, glass, metal, plastic	Water-resistant	Sets in 5 minutes, cures in 12 hours
Resorcinal resin	Liquid and powder	Water before it hardens	Mix liquid w/catalyst	Laminated timbers, sandwich panels, general bonding, boats	Waterproof	16 hours
Urea-formaldehyde	Powder	Soap and warm water before it hardens	Mix w/water	Lumber and hardwood plywood, assembly gluing	Water-resistant	16 hours
Urethane	Liquid	Alcohol before hardening	None	Gel-wood, porous materials, wood to wood, glass, and metal	Interior use	1 hour

12-16 The mortise-and-tenon joints on this old table have become loose. To make repairs, first remove the top. Notice it was made without corner blocks.

is often difficult to take the joint apart to clean it unless some of the other joints are also loose and the legs and rails can be disassembled. You have to consider how the piece of furniture is built. For example, a small table will most likely require that the top be removed (**12-16**). If all the rail and leg joints are loose they can be separated by carefully tapping them with a mallet. If only one joint is loose it will have to be repaired without taking the solid joints apart. Protect the legs with a piece of carpet (**12-17**). Once the rails are free of the mortise, scrape the old glue off the tenon and inside the mortise (**12-18**). Once they are clean insert the tenon in the mortise. Hopefully they are still their original size. They should make a tight fit. If the scraping has reduced the size of the tenon or enlarged the mortise you can glue a piece of wood veneer (**12-19**) on the tenon and trim it with

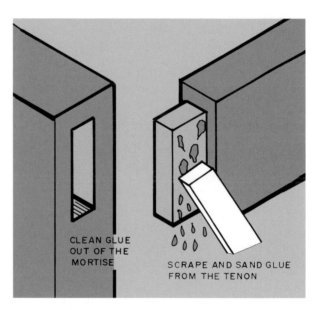

12-18 Before regluing a mortise-and-tenon joint, scrape and sand off all the old glue on the tenon and in the mortise. Try to maintain their original sizes.

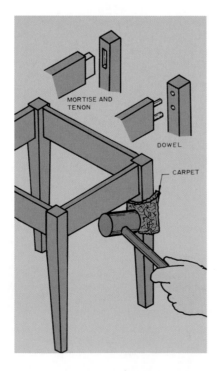

12-17 Protecting the legs so they are not damaged, carefully tap the legs loose on one end, then the others. This technique is used with mortise-and-tenon and dowel joints.

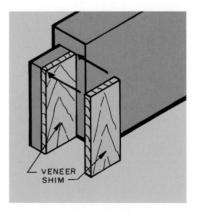

12-19 If the fit is loose after cleaning off the glue, you can glue a piece of veneer on one side of the tenon and trim it to fit snuggly into the mortise.

a chisel until a tight fit is developed. Then apply an adhesive to the tenon and inside the mortise. Insert the tenon in the mortise and clamp until the adhesive dries.

If it is not possible to disassemble the joint, a fairly good repair can be made by clamping the joint closed and installing a dowel through the leg and tenon (12-20). Bore a hole through them the size of the dowel. Apply glue inside the hole and to the dowel and tap it in place. If you plan to refinish the leg, let the dowel stick out a little so it can be sanded flush. If a refinished job is not planned, sand the end of the dowel before it is installed. Tap it smooth and stain to match the leg. You might prefer to make a dowel of the same wood as the leg.

Loose leg-to-rail connections can be strengthened with wood or metal corner blocks (12-21). Other types of corner connector are available at the hardware store.

Repairing Dowel Joints

Loose table and chair rails, legs, and bracing are common in older furniture. These are often held together with dowels. To make an effective repair, the joint must be taken apart. This will possibly mean some disassembly as shown earlier in **12-17.** Sometimes some of the dowels will be loose and one or more may be stuck. If you cannot break them loose by tapping carefully with a mallet **(12-17),** try working

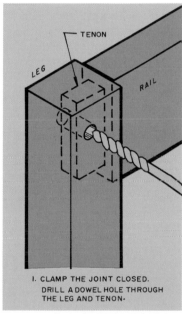

12-20 A loose mortise-and-tenon joint can be secured by clamping it closed and installing a dowel through it and the leg.

I. CLAMP THE JOINT CLOSED. DRILL A DOWEL HOLE THROUGH THE LEG AND TENON.

2. GLUE A DOWEL IN THE HOLE.

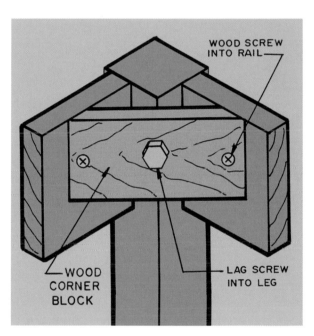

12-21 The leg-to-rail connection can be strengthened by adding wood or metal corner blocks. Many manufacturers install these as the furniture is made.

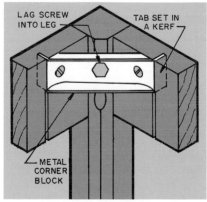

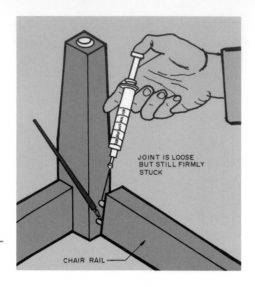

12-22 A joint that has one part loose and one part stuck can sometimes be loosened by injecting various solvents to soften the adhesive. This helps separate the parts without damage to them.

a solvent into the joint with a small brush or syringe (12-22). Try various solvents such as acetone, lacquer thinner, and paint thinner. Allow the joint to soak. It may take several applications of solvent.

Once the joint is separated, scrape and sand off the old glue on the dowels (12-23). Be careful not to reduce the size of the dowel. Also sand out the glue in the dowel hole. Wrap the sandpaper around a small dowel or pencil and work inside the hole. Try to avoid enlarging the hole (12-24).

Now attempt a dry assembly of the joint. If it does not fit tightly mix some sawdust with the adhesive to give it gap-filling capabilities. If the fit is very loose replace the dowels with larger ones. If the fit is not too loose you can enlarge the dowel by covering it with adhesive and wrapping it with thread. Then insert it in the holes (12-25).

Another way to enlarge a dowel that is a bit too small is to cut a kerf in the end. Make a thin wedge a little shorter than the depth of the saw kerf. Place the wedge in the kerf, apply adhesive to the dowel, and

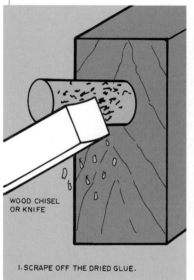

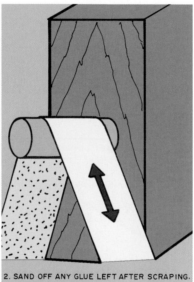

12-23 Scrape the glue off the dowel with a knife or chisel and clean up any left with fine-grit sandpaper.

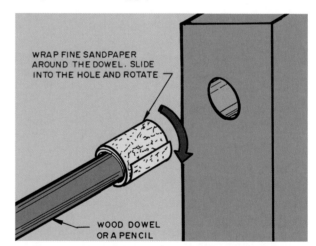

12-24 You can remove dried glue from the dowel hole by sanding it with fine-grit sandpaper.

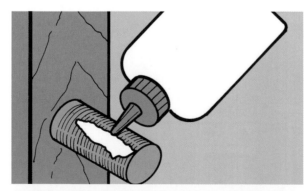

12-25 Sometimes you can enlarge a dowel enough to get a tight fit by wrapping thread around it and coating it with adhesive before sliding it into the dowel hole.

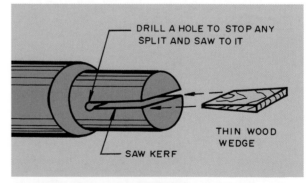

12-26 Put glue on the wedge and the dowel and place the wedge in the kerf. Drive the dowel into the dowel hole. The wedge expands the dowel, making a tight fit.

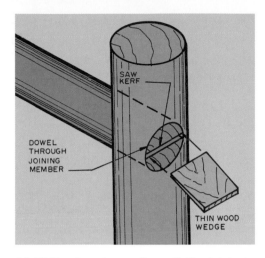

12-27 If a dowel runs through the joining member and it is loose, cut a kerf in the end of the dowel, apply glue to a wedge, and drive it in place from the outside.

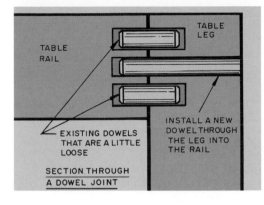

12-28 If a rail connected to a leg with dowels is a little loose, the joint can be tightened by installing a new dowel through the leg into the rail.

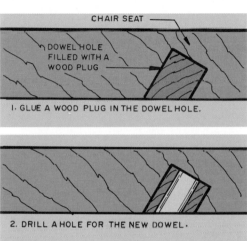

12-29 If a dowel on a chair or table leg breaks off, cut the end smooth and install a smaller dowel in it. Fill the dowel hole in the seat or tabletop and bore a hole in it for the new dowel.

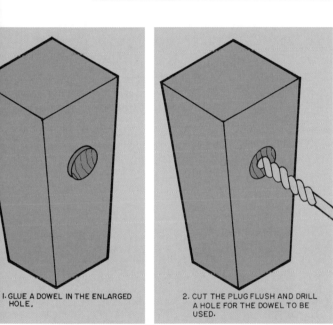

12-30 If a dowel hole becomes enlarged, glue a tight-fitting dowel plug in it. When the glue dries bore the hole for the new dowel.

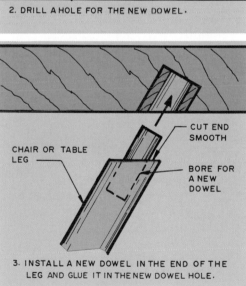

tap it into the hole. The wedge should be thick enough to cause the end of the dowel to expand and bind to the sides of the hole (12-26). If the dowel runs through the butting member the wedge can be tapped in from the exposed end (12-27).

A loose dowel joint can also be tightened by clamping it closed and installing a dowel through the leg into the rail as shown in 12-28.

Sometimes the dowel will have broken off. This can be repaired by cutting it off flush with the end of the member, such as a chair leg, and installing a new dowel in it (12-29). If the new dowel is smaller than the dowel hole, plug it with a piece of dowel the right size and after the adhesive dries drill the correct size for the new dowel (12-30).

Once the dowel fits tightly in the hole, coat it with adhesive (12-31) and coat the inside walls of the hole (12-32). Do not fill the hole with glue or you will not be able to get the dowel to go all the way in. Press the dowels in the holes and clamp the assembly. While

12-31 When assembling a dowel joint, coat the dowel with the glue before you slide it in the hole.

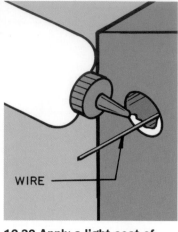

WIRE

12-32 Apply a light coat of glue to the inside of the dowel hole. Do not fill the hole because this could prevent the dowel from fully entering.

bar clamps are the best way to do this **(12-33)**, you can improvise by using a rope as a clamp **(12-34)**.

Repairing Split-Wood Parts

If a wood piece has split with the grain, open the split a little with a tool, such as a screwdriver, and force glue into the split. Clamp the split closed until the glue has dried. Wipe off any glue that squeezes out as the clamp is tightened **(12-35)**.

If a round member has a split with the grain, open the split enough to force in some glue and clamp it as shown in **12-36**.

Sometimes a small piece will split off the edge of a board. If you saved the piece, glue it back and hold it in place with masking tape **(12-37)**.

Repairing Loose Wood Veneer

If a piece of wood veneer becomes loose, carefully lift it and be certain the exposed surfaces are clean. Lightly sand if necessary. Then work glue under the lifted veneer with a wire or thin piece of wood or plastic. Try to get the glue on the bottom of the veneer and on the substrate. Follow the directions given on the glue container. Then lower the veneer against the substrate and clamp it. Protect the surface with wax paper and a board **(12-38)**.

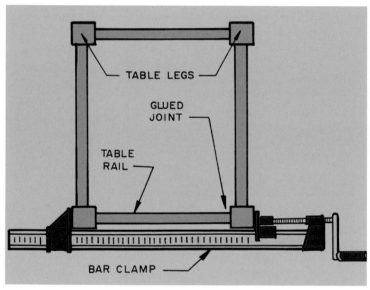

TABLE LEGS

GLUED JOINT

TABLE RAIL

BAR CLAMP

12-33 The best way to clamp a glued joint is with a bar clamp.

12-34 You can use a rope or length of rubber as an emergency clamp to hold a glued joint as it dries.

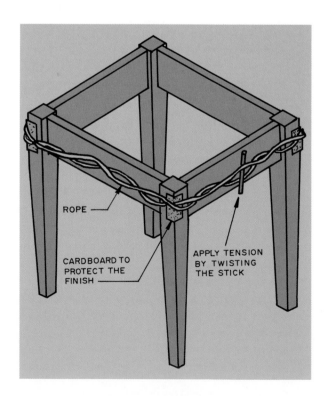

ROPE

CARDBOARD TO PROTECT THE FINISH

APPLY TENSION BY TWISTING THE STICK

Repairing Loose Hinges

Over time the screws holding hinges may work loose. First try to tighten them. If they twist in the hole remove the screws, glue wood pins in the holes, and after the glue dries reinstall the screws (12-39). Some types of wood filler will hold screws. Force the plastic filler into the holes until they are full, and let set until hard. Then reinstall the screws.

If the surface behind the hinge has deteriorated and cannot hold a screw, cut the recess deeper and glue in a block of wood. Then install the hinge in the normal manner (12-40).

Repairing Warped Tabletops

If the tabletop was not given a sealer coat on the bottom, it could, over a period of years, absorb moisture from the air. This can cause the wood to swell and the top to warp.

1. HOLD THE SPLIT OPEN AN FLOW IN THE ADHESIVE

2. REMOVE THE TOOL AND CLAMP THE JOINT. WIPE OFF ADHESIVE THAT SQUEEZES OUT OF THE JOINT.

12-35 A split with the grain of a board can often be repaired by opening it up a little, filling it with glue, and clamping it until it is dry.

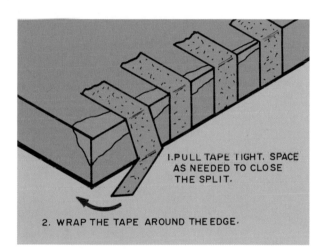

1. PULL TAPE TIGHT. SPACE AS NEEDED TO CLOSE THE SPLIT.

2. WRAP THE TAPE AROUND THE EDGE.

12-37 Split pieces off of edges can be repaired by regluing the split-off piece. Clean off as much glue on the surface as possible before taping it. Some surface-finish repair will usually be necessary.

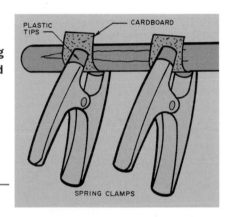

PLASTIC TIPS   CARDBOARD

SPRING CLAMPS

12-36 Splits along the sides of round members can be glued and clamped. Spring clamps or C-clamps are generally used.

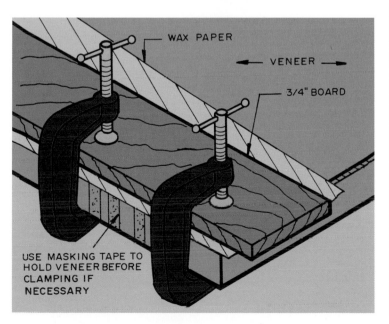

WAX PAPER

VENEER

3/4" BOARD

USE MASKING TAPE TO HOLD VENEER BEFORE CLAMPING IF NECESSARY

12-38 Loose veneers on tabletops can be reglued and then clamped until the glue dries. Be certain to protect the finish on the veneer.

253

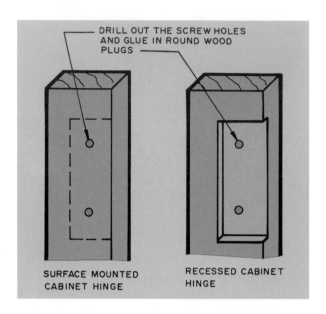

12-39 When hinge screws become loose and cannot tighten, drill out the screw holes, glue in wood plugs, and reinstall the screws.

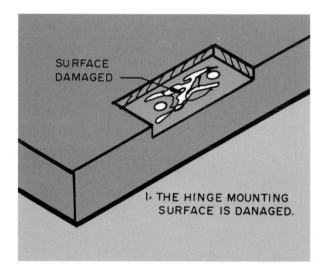

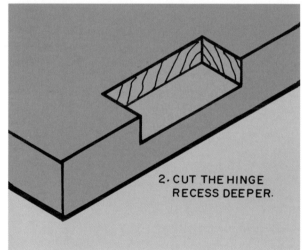

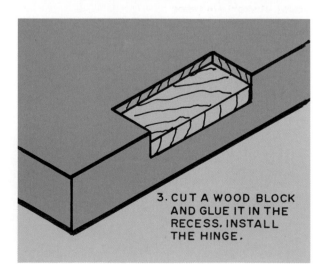

12-40 If the wood in the hinge recess is badly damaged, cut it deeper and glue in a solid-wood block. Then reinstall the hinge.

First place the top in a location where the humidity in the air is low so it can lose moisture and flatten naturally. When it becomes flat, seal the bottom with a standard sealer, such as polyurethane.

If there is the possibility it may not stay flat or does not return completely, clamp the top flat for 24 hours. Then install cleats on the bottom, placed perpendicular to the grain (12-41). Usually 1 x 2-inch cleats will hold it flat. Use thicker cleats if necessary. Be certain the screws do not break through the top.

Repairing Scratches

Scratches in the finish can be hidden somewhat by using some of the various touch-up products available at the local building supply or hardware dealer.

Scratches in the stain can be touched up with a marker containing a stain. The stain marker is available in a range of colors so you can select the one as close to the stain on the furniture as possible. Apply the stain by brushing it on with the tip (12-42). Carefully wipe off excess stain on the surface with a soft cloth.

One way to repair deeper scratches and fill nail holes is with a crayon-like material in the shape of a pencil. This is available in a range of colors. Work the material into the scratch or hole by rubbing the pencil across it (12-43). Then carefully wipe off the excess material and smooth the material in the damaged area.

Replacing Webbing

Webbing has been used for many years to support the seat and back cushions in chairs. It may be a heavy woven fabric or a rubber strip about two inches wide. After many years the webbing stretches and sags and must be replaced. This requires that the upholstery fabric and padding of both cushions be removed and replaced.

Begin by removing the old webbing. It is held by a series of upholstery tacks. The webbing can be pried loose with an old chisel or screwdriver. Before installing new webbing check the chair frame to see if the rails are securely joined to the legs. If any part

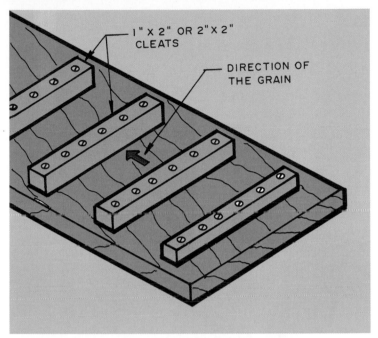

12-41 Wood cleats will help hold a top flat. Remember to apply a coat of sealer to the bottom after the top loses its excess moisture.

12-42 You can touch up very small scratches with stains available in tubes having a felt tip.

12-43 Deeper scratches and nail holes can be filled with a crayon-like pencil. A wide choice of colors is available.

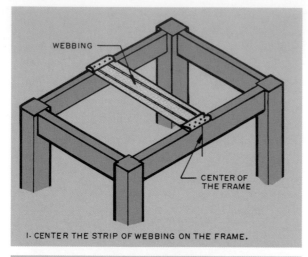

I. CENTER THE STRIP OF WEBBING ON THE FRAME.

WEBBING

CENTER OF THE FRAME

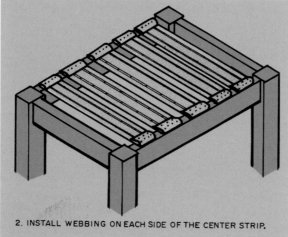

2. INSTALL WEBBING ON EACH SIDE OF THE CENTER STRIP.

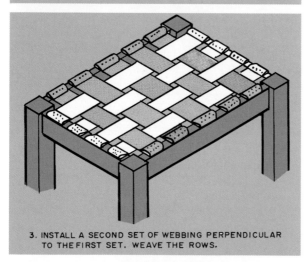

3. INSTALL A SECOND SET OF WEBBING PERPENDICULAR TO THE FIRST SET. WEAVE THE ROWS.

12-44 The steps for installing webbing.

12-45 When you nail the first end of the webbing to the chair frame, fold it over and drive seven 16-ounce upholstery tacks through it into the frame.

has been damaged, replace it or repair it. A damaged rail can be strengthened by gluing a new piece of wood to it and then securing it with screws.

Install the first piece of webbing in the center of the seat frame (12-44). Fold over one end about one inch and nail it to the frame with seven 16-ounce upholsterer's tacks as shown in 12-45. Then pull it tight to the other side with a webbing stretcher and tack it to the frame with three tacks. Cut off the webbing about one inch beyond the frame, fold it over the frame, and nail four tacks between the first three as shown in 12-46. The rest of the webbing is spaced about ½ inch apart. Adjust the spacing so a full width is on each end. Now run the webbing across the frame in the other direction, weaving it in and out of the first rows and nail on each end as just described (12-44).

Now you can lay a one-inch-thick rubber foam pad over the webbing and tack it to the frame. Then lay a ½-inch-thick cotton felt pad over it and wrap it over the side of the frame. Now lay the fabric over these and tack it to the bottom of the frame (12-47).

Repairing Chairs or Sofas with Sagless Springs

Sagless springs are curved metal wire that is very strong and has the ability to return to a flat condition when bent. As a chair ages, these sometimes come loose from the frame. The **clips** holding them to the

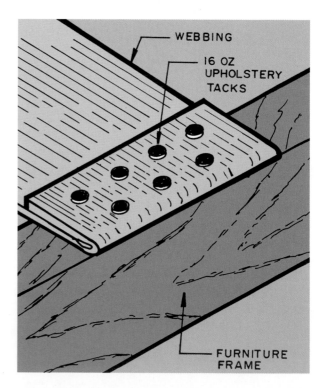

WEBBING

16 OZ UPHOLSTERY TACKS

FURNITURE FRAME

frame are usually nailed and if the chair is abused or the frame deteriorates, the clips may pull loose. Excessive overloads may bend the clips. If this is the problem, replace the clips. If the springs are pulled loose from the frame you will have to repair the damaged surface of the frame and reinstall the clips. You may want to use wood screws instead of nails (**12-48**).

To get at the clips it is usually necessary to remove the upholstery. Typically it laps over the frame and

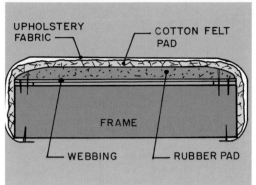

12-47 After the webbing has been replaced, cover it with a rubber pad over which cotton felt is laid. The upholstery fabric is laid over the entire thing.

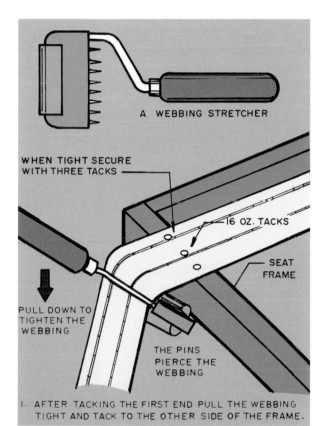

1. AFTER TACKING THE FIRST END PULL THE WEBBING TIGHT AND TACK TO THE OTHER SIDE OF THE FRAME.

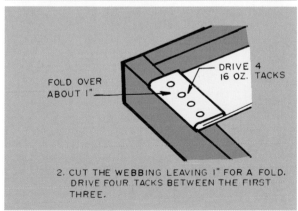

2. CUT THE WEBBING LEAVING 1" FOR A FOLD. DRIVE FOUR TACKS BETWEEN THE FIRST THREE.

12-46 Pull the webbing tight with a webbing stretcher and secure it to the frame with three 16-ounce upholsterer's tacks. Cut it one inch long, fold it over the frame, and drive four tacks, one to each side of the first three.

12-48 These are typical of the clips, plates, helical springs, and links used to install sagless springs.

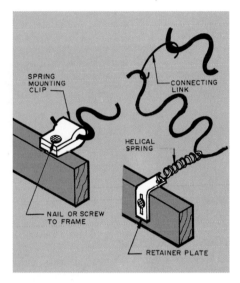

12-49 A typical sagless spring installation.

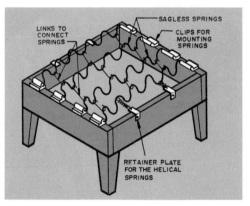

12-50 After the springs are installed, cover them with a sheet of burlap and tack it to the bottom edge of the frame.

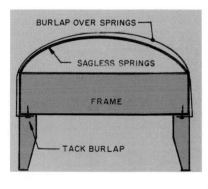

12-51 Lay the foam pad over the burlap and wrap the upholstery fabric tightly around it. Tack the fabric to the bottom edge of the frame.

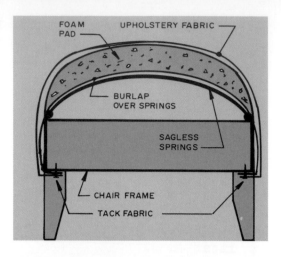

is tacked on the bottom. When the upholstery cover is removed and the cushion is taken away, the frame with springs will appear like that in **12-49**. Notice the springs are nailed on each end of the frame with mounting clips. They are tied together in the center with connecting links and connected to the side frame with **helical springs** connected to a **retainer plate** (refer to **12-48**).

Once the springs are back in place cover them with a heavy burlap that is pulled tight and tacked to the bottom of the frame (**12-50**). Then place the foam rubber pad over the burlap, pull the upholstery fabric tightly around, and tack it to the bottom of the frame as shown in **12-51**. If the fabric is old this would be a good time to replace it. Use the old piece as a pattern.

Repairing Padded Chair Seats

Chairs such as the straight chair in **12-52** have a plywood seat base covered with a foam rubber pad which is then covered with an upholstery fabric. The plywood is secured to the chair frame with wood screws

12-52 This straight chair has a foam pad and fabric installed over a plywood base.

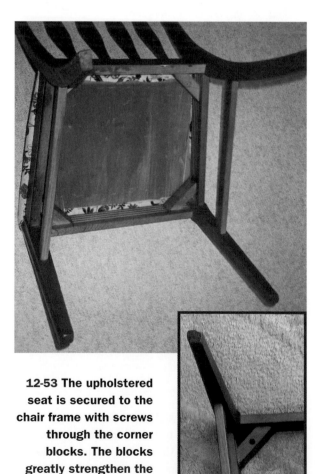

12-53 The upholstered seat is secured to the chair frame with screws through the corner blocks. The blocks greatly strengthen the leg-rail connection.

(12-53). Sometimes webbing is used instead of plywood. Over the years the padding will deteriorate and the seat will have a low spot in the center. It is then time to replace the pad. If webbing has been used it will have to be replaced as well. If the fabric is faded consider replacing it.

Remove the old padded seat from the frame. It will usually be held with several screws. The upholstery fabric will be tacked or stapled to the bottom of the plywood. Pull all the staples or upholstery tacks and remove the fabric and pad (12-54).

Get a new foam pad that is 1 to 1½ inches thick from an upholstery supply shop. Cut it about ½ inch wider and longer than the plywood (12-55). This will provide some extra material that is squeezed by the fabric, forming a nicely rounded edge.

Cut the upholstery fabric to size. If possible, use the old piece as a pattern so you get a piece the right size. To be safe cut the new fabric an inch or so larger than the pattern. Lay this over the pad and turn the seat with the plywood facing up. Tack it to the plywood in the center of the front and back sides (12-56), stretching the material until it forms the curved edge

12-54 The upholstery fabric will be tacked or stapled to the plywood base.

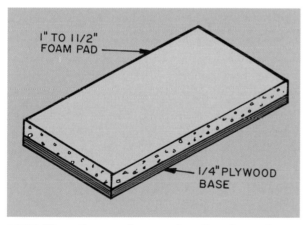

12-55 Place the new foam pad on the plywood base.

12-56 Turn the seat over on the fabric and tack the fabric on the center on opposite sides. Pull it tight to form a rounded edge.

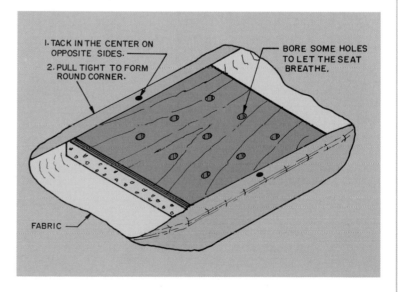

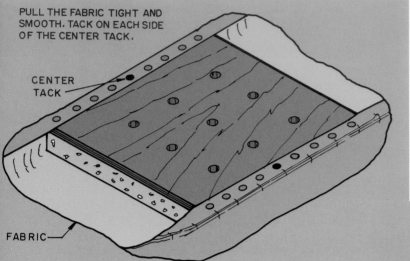

PULL THE FABRIC TIGHT AND SMOOTH. TACK ON EACH SIDE OF THE CENTER TACK.

CENTER TACK

FABRIC

of the seat. Now keeping the material tight, drive No. 6 or 8 upholstery tacks to the right and left of the center tack, keeping the fabric pulled tight, being careful to form a uniform curved edge. Space the upholstery tacks about 2 inches apart (12-57).

Now repeat this for the other two sides (12-58). Finally it is necessary to fold and tack the corners to the plywood base as shown in 12-59.

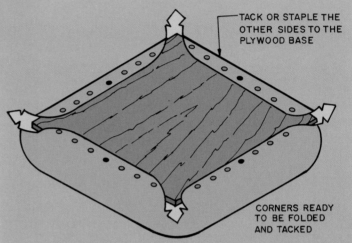

TACK OR STAPLE THE OTHER SIDES TO THE PLYWOOD BASE

CORNERS READY TO BE FOLDED AND TACKED

12-57 Tack on each side of the center tack. Pull the fabric tight, forming a smooth and uniform corner.

12-58 Tack the other opposite sides.

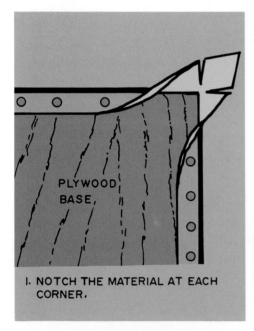

PLYWOOD BASE.

I. NOTCH THE MATERIAL AT EACH CORNER.

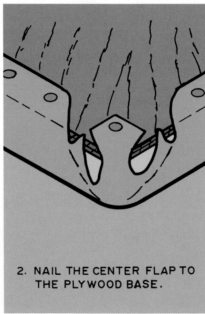

2. NAIL THE CENTER FLAP TO THE PLYWOOD BASE.

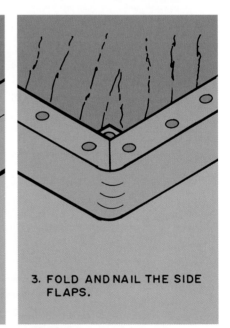

3. FOLD AND NAIL THE SIDE FLAPS.

12-59 This is how to make the corners so they form a smooth transition as they are nailed to the plywood base.

CHAPTER 13

Adding a Closet

Older homes often lack adequate closet space and the space provided is often not used to best advantage. This means you need to decide if revising the interior of the existing closet will meet your needs, or if it would be best to expand the existing closet or to add additional closets to the room. Efficient use of the space can keep the size of the closet down. This is especially important if space is at a premium (13-1).

Planning Considerations

As you make preliminary plans for a new closet, first consider the sizes commonly used. A standard closet should be at least two feet wide, with no obstructions, inside. A couple of extra inches would be wise if space is available. The length depends on your needs and the space that can be made available. A typical person can use 5 to 6 lineal feet of closet space in a bedroom. In addition, other closet space is needed in areas close to the bedrooms such as those opening onto a hall or into a bathroom.

Consider other needs in a closet such as a hanging rod, shelves, and even drawers. The hanging rod usually has one or more shelves above it. Very popular are the wide variety of metal shelves. These have wire framing, allowing the air to filter through the closet. In areas of high humidity these help keep air circulating in the closet, reducing the chances of mold building up.

Also remember to consider the types of doors to be used and the sizes available.

Types of Closet

The **bedroom closet** is used to hang clothing and store all sorts of other objects. While it is tempting to add

13-1 This closet makes good use of the hanging space and provides storage above for large items.

13-2 Recommended placement of closets.

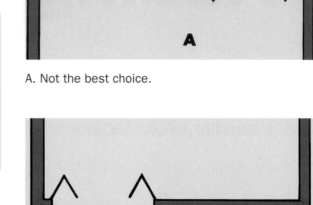

A. Not the best choice.

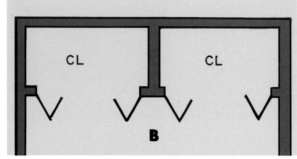

B. Closets can extend across the end of a room.

C. Closets can open into adjoining rooms.

a closet in a corner, this is not necessarily the best arrangement for using the rest of the wall (**13-2**). It leaves an area that is difficult to use; however, in an older house you may have no other choice. Ideally the new closet should extend the entire length of a wall. An alternate solution is to build the closet along the entire wall but open part of it in the next room, thus improving storage in the adjoining room. Bedroom closets should be a minimum of two feet deep, wall to wall (**13-3**).

13-3 Sizes for a typical bedroom closet.

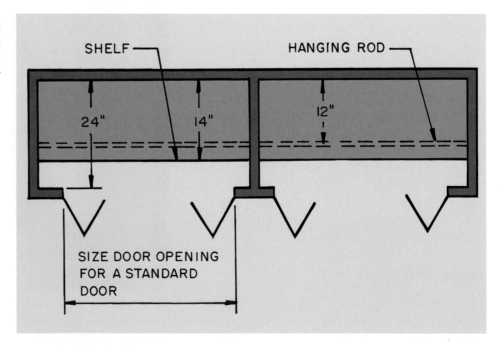

SHELF

HANGING ROD

24"

14"

12"

SIZE DOOR OPENING FOR A STANDARD DOOR

Walk-in closets are popular but require a large floor area (13-4). They are usually built to open off the bedroom but may open off a bathroom or dressing area. Some open on a hall in the bedroom area of the house. A small walk-in closet will have hanging space on one side. If it is a little larger a cabinet could be put on one side. A larger closet will have hanging space on both sides. Plan for a minimum three-foot-wide aisle; however, adding an extra six inches is a good idea if you have the space.

Another type of bedroom storage unit is a **wardrobe (13-5)**. It has hanging space like a closet but can have drawers and sections with shelving. Wardrobes are built using the same materials and techniques as cabinets.

Another type of closet to consider is a **linen closet.** It is located near the bedroom area and often opens off the hall. A linen closet is typically 12 to 18 inches deep and as wide as you wish. Three to four feet wide is typical. A linen closet usually has shelves from top to bottom. Sometimes a small linen closet is built off a bathroom.

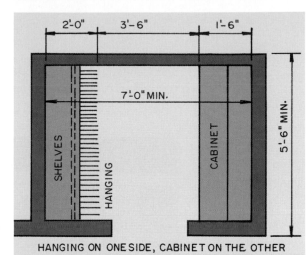

13-4 Typical walk-in closet layouts.

HANGING ON ONE SIDE, CABINET ON THE OTHER

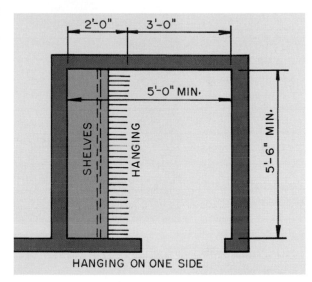

HANGING ON ONE SIDE

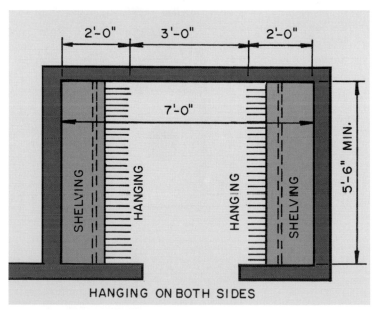

HANGING ON BOTH SIDES

13-5 This expensive, high-quality hardwood wardrobe is an attractive addition to the bedroom and provides additional hanging and drawer storage.

In a major remodeling of the house you may want to install a washer and dryer (13-6) or put the new water heater or furnace in a closet. If you do this be certain to check the local building codes so fire and other requirements are met. When remodeling the kitchen consider adding a pantry closet or a tall cab-inet (13-7) to hold the many small items needing storage.

Another place to find closet space is under a stair (13-8). The hanging space is limited but it does not require major carpentry. Shelving can be installed under part of the lower end of the stair.

Some Design Sizes

As you plan the interior of the closet the location of the hanging rod and shelves is important. Typical height locations are shown in **13-9**. It is helpful if some of the shelves can be adjustable so objects of varying heights can be accommodated. Typical shelf spacing of 12 to 14 inches is common. The goal for the hanging rod is to set the height for the clothes most likely to be hung.

If you are planning a children's bedroom, consider arranging for a hanger rod and shelves that can have their height adjusted as the children get older. If the children are young you might start as shown in **13-10**.

13-6 A washer and dryer can be installed in a closet, saving considerable floor space. It should be near the bedrooms. Notice the washer that is front loaded is installed on top of a platform finished with ceramic tile. This makes it easier to load. *Courtesy GE Consumer Products*

13-7 A pantry in a kitchen requires little floor space but holds a considerable amount of food items, cooking utensils, and linens. *Courtesy KraftMaid Cabinetry*

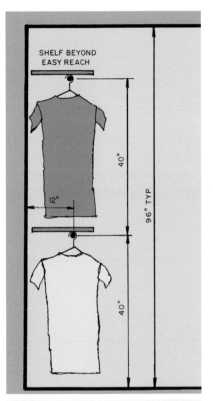

SHELF BEYOND EASY REACH

40"

12"

96" TYP

40"

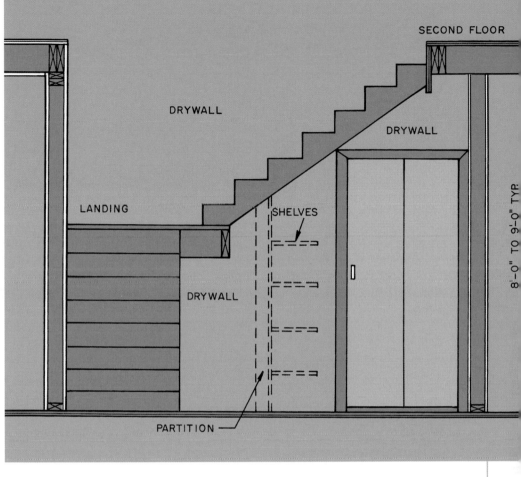

SECOND FLOOR

DRYWALL

DRYWALL

LANDING

SHELVES

DRYWALL

PARTITION

8'-0" TO 9'-0" TYP.

13-8 The area under stairs can provide hanging space plus room for shelving in some instances.

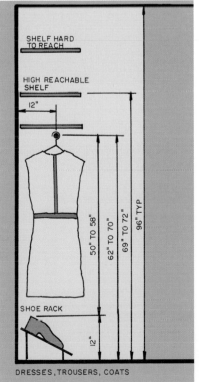

SHELF HARD TO REACH

HIGH REACHABLE SHELF

12"

50" TO 58"

62" TO 70"

69" TO 72"

96" TYP

SHOE RACK

12"

DRESSES, TROUSERS, COATS

13-9 Typical heights for hanging rods and shelves for adult bedroom closets. Select a height suitable for your reach above the floor.

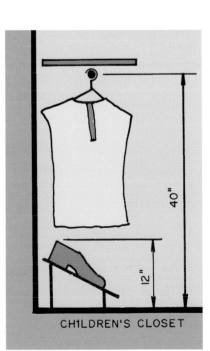

40"

12"

CHILDREN'S CLOSET

13-10 For very young children start the hanging rod about 40 inches above the floor.

Interior Closet Layout

Whether you build a new closet or remodel the interior of an existing one, efficient use of the space available is essential. This can be accomplished by building in wood shelves, drawers, and hanging rods or by selecting items from commercial closet systems which use wire shelves and carry various designs. In both cases the layout choices are the same.

In **13-11, 13-12,** and **13-13** are examples that were planned using wood shelves and dividers. The total cost would be small and the woodworking skills within the ability of most home craftsmen or craftswomen.

Wire shelving is also a popular solution to making efficient use of closet space. It is strong and, when installed with the proper fastening devices, can carry heavy loads **(13-14)**. A few of the many layout possibilities are in **13-15.**

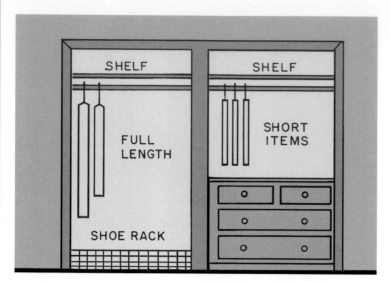

13-11 This arrangement provides hanging long and short items and drawer storage plus a shoe rack and shelving above the hanging rod.

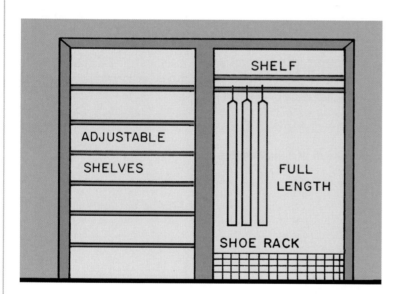

13-12 This arrangement provides generous adjustable shelving storage, a shoe rack, hanging space for full-length clothing, plus shelving above the hanging rod.

13-13 An arrangement providing hanging space for long and short clothing plus adjustable shelving.

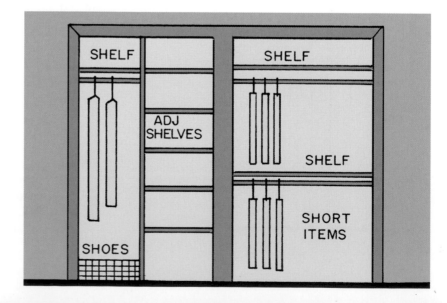

13-14 Wire closet shelving provides efficient storage and can carry heavy loads.

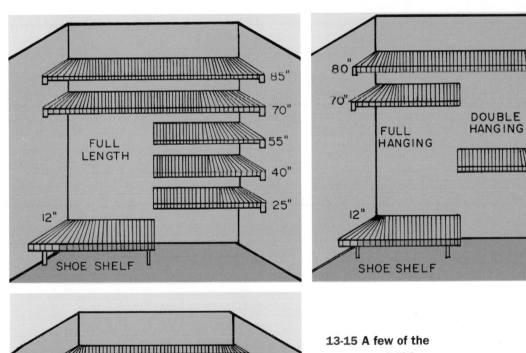

85"
70"
55"
40"
25"

FULL
LENGTH

12"

SHOE SHELF

80"
70"

FULL
HANGING

DOUBLE
HANGING

40"

12"

SHOE SHELF

85"
75"

FULL HANGING

12"

SHOE SHELF

13-15 A few of the many possibilities for using wire closet shelving.

Door Selection

The types of doors typically used on closets include swinging, bi-fold, pocket, and bypass (13-16). The **swinging door** is usually a poor choice because it requires a large open floor area in front of the closet and is difficult to use (13-17). The **bypass door** is satisfactory but has the disadvantage of opening only half the closet at one time (13-18). If you see you need something in the covered half you have to slide both doors to open it. **Pocket doors** open the entire closet, providing easy access (13-19). They require special hardware and wall construction to allow the door to slide into the wall. The **bifold door** seems to be the best. They are easy to install and allow the entire closet to be opened (13-20).

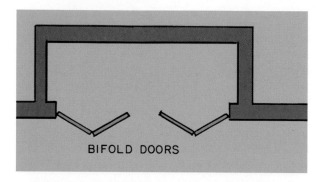

BIFOLD DOORS

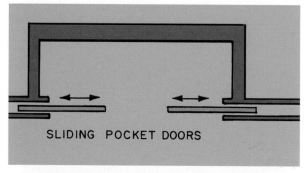

SLIDING POCKET DOORS

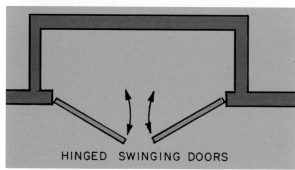

HINGED SWINGING DOORS

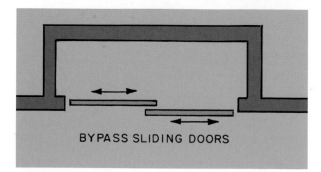

BYPASS SLIDING DOORS

13-16 These are the doors commonly used on closets.

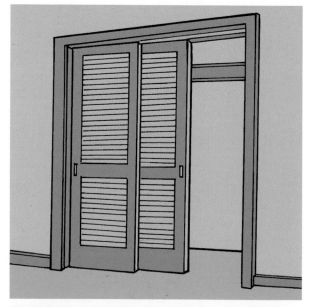

13-17 Swinging doors get in your way as you open them and require a large open floor area in front of the closet.

13-18 Bypass doors always leave half the closet covered.

13-19 Pocket doors expose the entire closet but require special hardware and need a wall into which they can slide.

13-20 Bifold doors expose the entire closet, are easy to install, and require little open floor area in front of the closets.

Building a Closet

Typically a new closet is built in the corner of a room. While this does not give the best appearance to the room, you only have to build two walls. These walls are framed in the same manner as the partitions in the house. A typical framed closet is shown in **13-21**. As you plan the construction, think of it as two separate walls which will be assembled and brought together to form the closet, as shown in **13-22** and **13-23**.

Begin by removing the baseboard in the closet area. You may have to pull it from the entire wall and cut and reinstall it after the closet is finished. If it is not damaged during removal, parts of it can be used around the closet.

Next mark the outline of the closet on the floor. As explained earlier in this chapter, it should be at least 24 inches wide, unobstructed on the inside. Remember to allow for the thickness of the drywall used to finish the inside of the closet and the studs forming the front wall **(13-24)**. Now mark the location of the top plate on the ceiling.

Plan the placement of studs, the corner framing, and the door location. (A typical layout is shown in **13-22** and **13-23**.) Notice how the corner is assembled **(13-25)**. This permits the walls to be firmly nailed together and provides a nailing surface for the drywall.

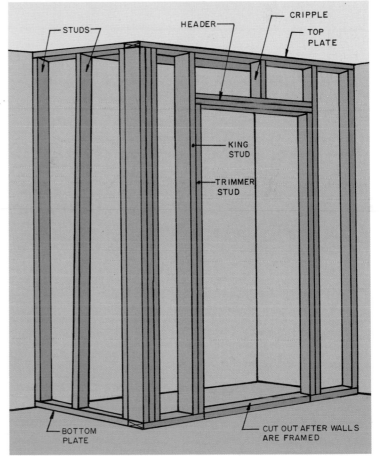

13-21 Typical framing for a closet. The corner stud has a spacer block. A single 2 x 4 header can be used though some prefer to double it.

269

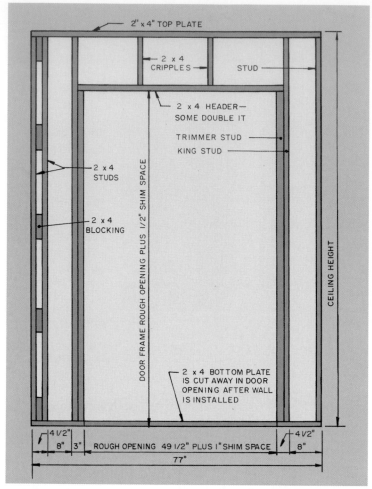

13-22 A framing plan for the front wall with the door.

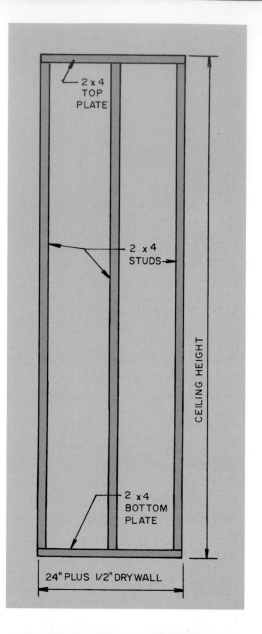

13-23 A framing plan for the side wall.

13-24 Lay out the location of the bottom plate on the floor and the top plate on the ceiling.

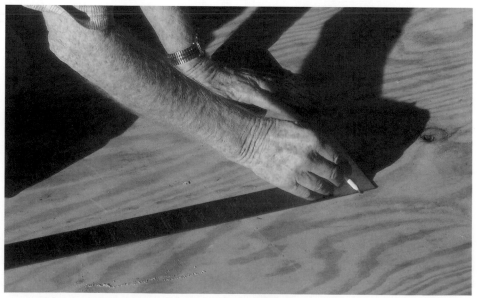

Lay out the location of the studs on the bottom and top plates. Symbols used may vary but those shown in the following examples are typical. In **13-26** is the layout for the front and side walls of the closet detailed in **13-21**.

Each wall is assembled according to the plan. The studs and plates are laid on the floor and nailed together with two 12d or 16d common or box nails. Box nails are thinner and less likely to split the wood **(13-27).**

Raise each assembled wall and place it on the marks on the floor and ceiling. Check it for plumb with a level. Then nail the bottom plate to the subfloor every two feet. Nail the top plate to the ceiling joists **(13-28).**

When all is secure cut out the bottom plate that is across the door opening **(13-29).**

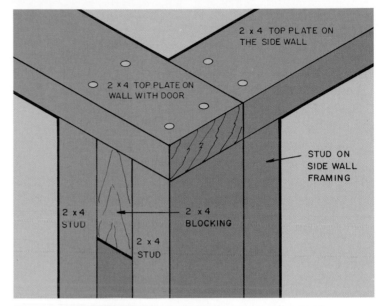

13-25 Framing for the corner where the two walls meet.

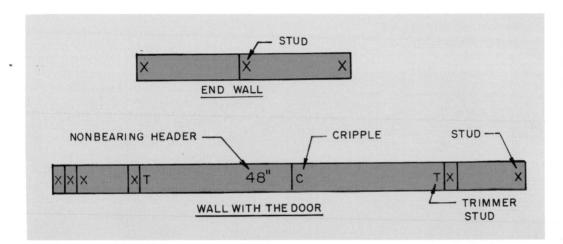

13-26 Mark the location of the studs, trimmer, and cripple on the top and bottom plates.

13-27 Assemble each wall on the floor.

271

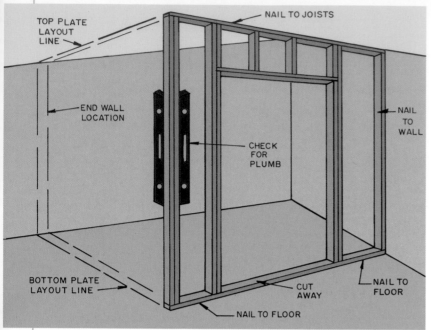

13-28 Raise each wall in place. Position the wall on the marks on the floor and ceiling. Nail to the floor and ceiling joists through the plates. Check it for plumb with a level.

Now install the drywall on the inside and outside of the closet. Review Chapter 4 for detailed information. Install the baseboard and casing around the door opening. See Chapter 7 for more details. Finally, install the hanging rod and shelving. Remember, if you want a light in the closet the electrician will want to install it and the wall switch before the drywall is installed (13-30).

Installing a Bi-fold Door

Begin by installing the door jamb. Following is a typical way to do this. Consult the instructions that come with the door.

1. Measure the required height above the floor. This includes the height of the door plus clearance needed at the floor (13-31). If the finished floor is already installed measure from it, allowing a ¾-inch space between it and the bottom of the door. If the finished floor has not been installed, find out its thickness and allow for it. Allow for

13-29 When the walls are securely nailed in place, cut out the bottom plate that runs across the door opening.

13-30 The light switch is usually installed on the wall next to the door casing.

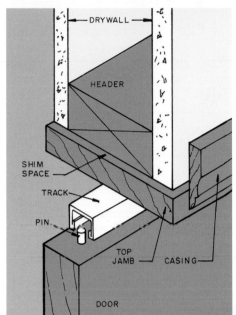

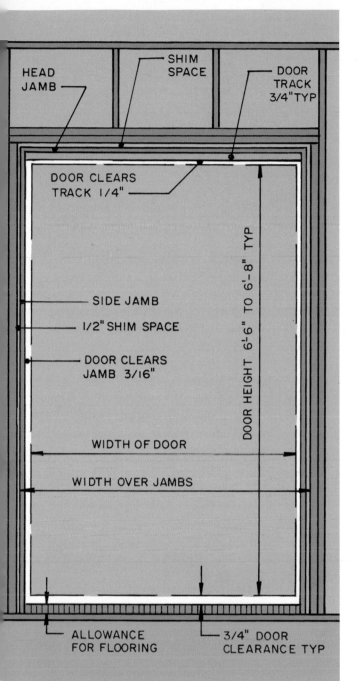

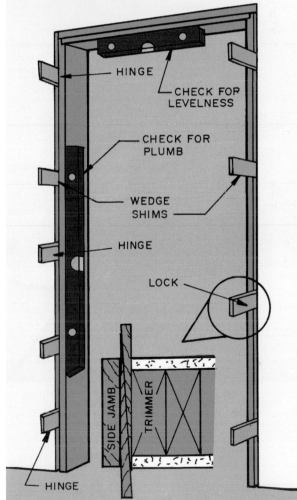

13-32 Typical construction at the head jamb.

13-31 When sizing the door frame consider the type of flooring, the door height and width, the size of the track on the head jamb, plus clearance spaces at the floor and track.

13-33 Place the frame in the rough opening and shim to get it level and plumb. Place shims as needed to get the head jambs level and the side jambs plumb.

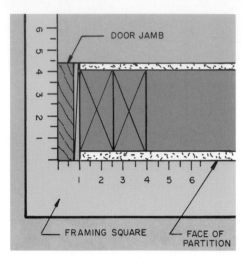

13-34 Check to be certain the jamb is at right angles to the face of the wall.

track at the head jamb. The door slides on a pivot in the track. Also allow about ¼-inch clear space between the door and the track (**13-32**). This total distance is typically one inch. You may have to shorten the door a little. The frame provided should allow a ³⁄₁₆-inch space on the sides of the door. Check this measurement before installing the frame.

2. Set the jamb in the rough opening. Position it so the head is in line with the mark on the stud, locating the inside face of the head jamb. Place shims on each side as shown in **13-33**. Start at the top and place them down to the floor. Push in so they are just tight.

3. Check the head jamb for levelness with a long level. Adjust this if needed by placing a shim under the side jamb at the floor.

4. Check the side jambs for plumb with a long level. Lightly tap the shims together until they hold the frame in place and it is level and plumb. Do not overtighten the wedges because this can cause the frame to bow. Check to see that the jamb is at a right angle to the face of the partition (**13-34**).

5. Now nail through the jamb and wedges into the trimmer stud (**13-35**). Use finish nails long enough to penetrate the stud at least 1 to 1½ inches. Use two nails per shim (**13-36**). Nail through the shims on the head jamb into the header. Use a sharp utility knife to trim off any of the wedges that pro-

13-35 Nail through the jamb and shims into the trimmer stud. Pneumatic-driven finish-nailers are the fastest and easiest way to drive the nails. *Courtesy Senco Fastening Systems*

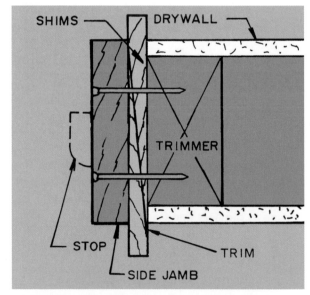

13-36 When the door frame is level and plumb, drive two finish nails through each pair of shims into the trimmer stud. The part of the shims sticking out is cut off flush with the edge of the door jamb.

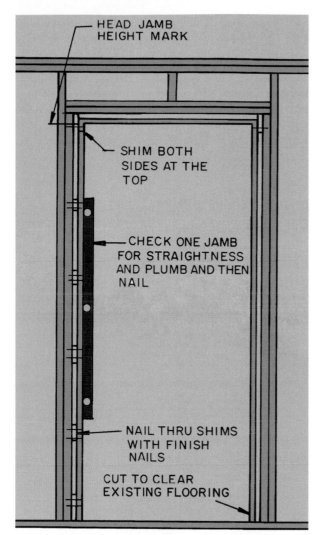

HEAD JAMB
HEIGHT MARK

SHIM BOTH
SIDES AT THE
TOP

CHECK ONE JAMB
FOR STRAIGHTNESS
AND PLUMB AND THEN
NAIL

NAIL THRU SHIMS
WITH FINISH
NAILS

CUT TO CLEAR
EXISTING FLOORING

I. INSERT THE FRAME IN THE ROUGH
OPENING. INSERT SHIMS AS NEEDED
TO PLUMB ONE SIDE JAMB. NAIL
WHEN PLUMB AND SHIMS ARE
TIGHT. BE CERTAIN SIDE JAMB IS
SQUARE WITH THE FACE OF THE
WALL.

13-37 The basic steps for installing the door frame for a bi-fold door.

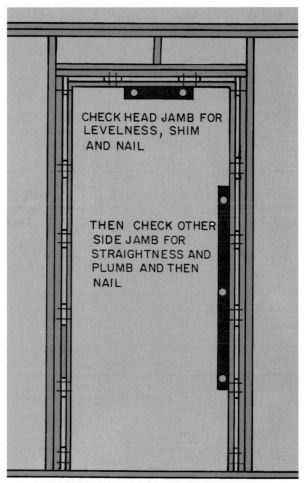

CHECK HEAD JAMB FOR
LEVELNESS, SHIM
AND NAIL

THEN CHECK OTHER
SIDE JAMB FOR
STRAIGHTNESS AND
PLUMB AND THEN
NAIL

2. LEVEL THE HEAD JAMB, SHIM
AND NAIL. PLUMB, SHIM AND
NAIL THE OTHER SIDE JAMB.

trude out from the frame. They will be covered with the casing. Cut into the wedge a little and snap it off. A summary of these steps is shown in **13-37**.

Install the casing as explained in Chapter 7. Now that the frame is in place and the casing installed, review the manufacturer's instructions for installing the bi-fold door. Following is a typical procedure.

Begin by installing the track on the head jamb. First cut it to length so it fits between the side jambs. Cut it ⅛ inch shorter than this width. Use a hacksaw with a fine-tooth blade **(13-38)**. Then install the track on the head jamb **(13-39)**. Put the center of the track on

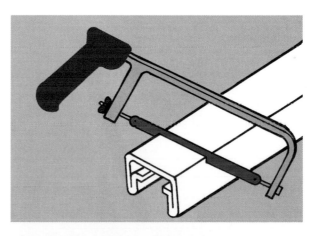

13-38 Cut the track to length. Cut it about 1/8 inch shorter than the overall length.

13-39 Screw the track to the head jamb.

the centerline of the head jamb (13-40). It is predrilled and the screws are supplied with the hardware.

Locate and bore holes for the pivots on the top of the door (13-41). They are generally located one inch in from the end of the door and are on the centerline (13-42). Bore the hole the depth recommended by the manufacturer. Tap the pivot pins into the holes, leaving the specified amount above the top of the door. This is typically one inch. Smooth, round pivot

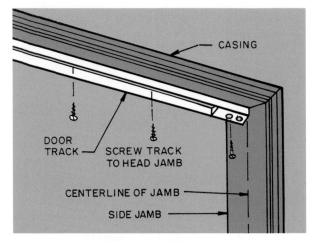

13-40 The centerline of the track is located on the center- line of the jamb.

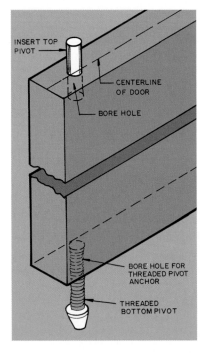

13-41 Round plastic pivot pins are placed at the ends of the two doors. A threaded pivot pin is placed in the bottom of the door edge that is next to the side jamb.

13-42 Insert the pivot pins in holes in the top outside corners of the door. A threaded bottom pivot is located at the edge next to the side jamb. Locate door as instructed by the manufacturer's instructions.

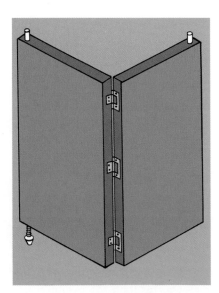

pins are installed in the top corners of the doors. Install the threaded pivot pin in the bottom of the door on the end that will be next to the side jamb. This pin will fit into the floor bracket.

Next install the jamb bracket at the floor on the side jamb (13-43). The center of the bracket must be on the centerline of the side jamb. Secure it to the jamb with screws. If the floor has a carpet it will be placed just above the top of the carpet (13-44).

A spring-loaded bracket is installed in the very center of the track where the two doors meet. The top pivot presses against this, which holds the doors closed (13-45). Install a pivot bracket next to the jamb at each end of the track (13-46). The top pivot pin on the side of the door fits into the hole in this bracket.

To install the door, insert the top pivot pin that's on the end of the door next to the jamb into the pivot bracket, and the top pin on the end of the adjoining

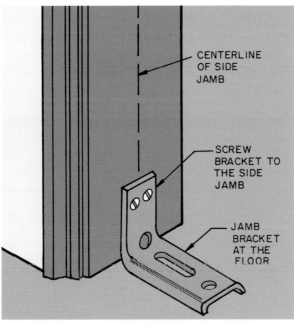

13-43 The jamb bracket is installed on the centerline of the side jamb. It must line up vertically with the centerline of the track.

13-44 The jamb bracket is installed on top of a carpeted floor. This provides the clearance needed for the doors to swing open.

13-45 A spring-loaded bracket is installed in the center of the track. The butting doors press against the spring when they are closed.

13-46 A pivot bracket is installed in the track next to each side jamb. The pivot pins in the ends of the doors fit into these brackets.

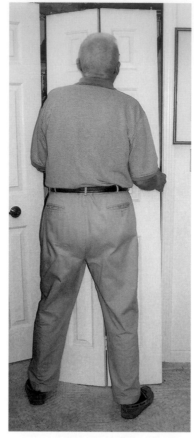

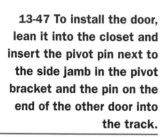

13-47 To install the door, lean it into the closet and insert the pivot pin next to the side jamb in the pivot bracket and the pin on the end of the other door into the track.

door into the track **(13-47)**. Move the bottom of the door over the jamb bracket at the floor. Lift the door as high as possible and set the bottom pivot pin into the bracket. Then lower the door. Adjust the height by screwing the threaded pivot pin until the door moves easily **(13-48)**. If the door is not plumb **(13-49)** lift it out of the floor bracket and move it toward or away from the side jamb as needed **(13-50)**. When it is plumb, lower the threaded pin down into the floor bracket.

Once the doors are plumb, install the door alignment tabs on the inside of each of the doors that meet when the doors are closed. Adjust these until they hold the butting doors closed **(13-51)**. Now you are finished. Admire the final results of your work

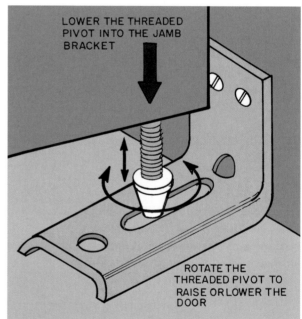

13-48 After the pins on the top of the door are in the track, place the pin in the side of the door by the side jamb over the jamb bracket. Lower the pin into the bracket. Screw the pin, raising or lowering the height of the door so it moves freely.

13-49 This door has been installed but is not plumb with the side jamb.

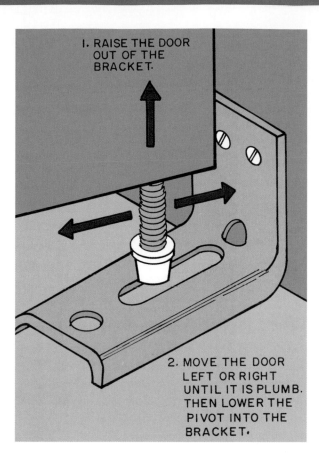

1. RAISE THE DOOR OUT OF THE BRACKET.

2. MOVE THE DOOR LEFT OR RIGHT UNTIL IT IS PLUMB. THEN LOWER THE PIVOT INTO THE BRACKET.

13-50 To plumb the door, lift it out of the jamb bracket and move it right or left as needed and set it back on the bracket.

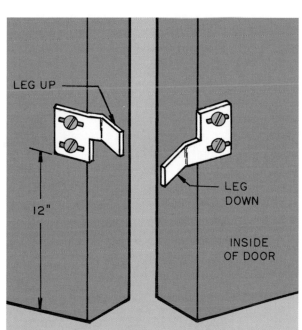

LEG UP

12"

LEG DOWN

INSIDE OF DOOR

13-51 Install the door aligners on the inside of the doors where they meet in the center of the closet. Adjust the brackets until the doors close tightly.

13-52 The closet is finished. The casing is in place and the bifold doors have been installed properly.

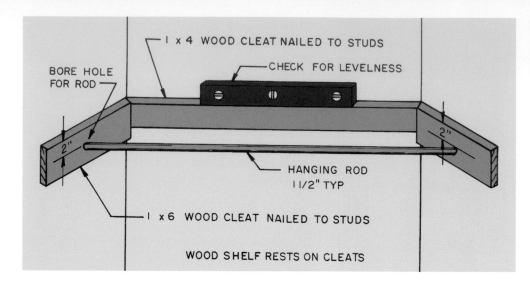

BORE HOLE FOR ROD

1 x 4 WOOD CLEAT NAILED TO STUDS

CHECK FOR LEVELNESS

2"

2"

HANGING ROD 1 1/2" TYP

1 x 6 WOOD CLEAT NAILED TO STUDS

WOOD SHELF RESTS ON CLEATS

13-53 A typical installation for a wood hanging rod with a wood shelf above.

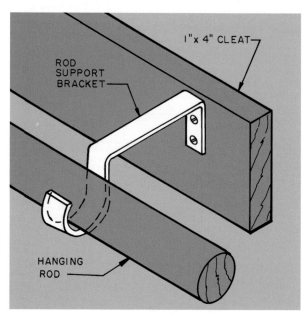

1"x 4" CLEAT

ROD SUPPORT BRACKET

HANGING ROD

13-54 A metal rod support bracket adds support to long hanging rods.

(13-52). You can now install the shelves and hanging rod.

Installing Wood Shelves & Hanger Rod

A common way to install wood shelves and the hanger rod is to use wood cleats on the wall to hold the shelf and a wider support on the side wall into which the hanging rod is inserted (13-53). The rod is typically 1¼ or 1½ inches in diameter. A metal support bracket is available that is installed in the center of a long rod and shelf to help support the load (13-54).

Wire Closet Shelves & Hangers

As you plan for the storage facilities in the closet it is a good idea to consider the use of wire shelving. The manufacturers provide planning and installation instructions and copies are often available for examination at the local building supply dealer. The choice of shelving and hangers makes it easy to use the closet space efficiently. The various installation braces and connectors make it easy to produce a strong installation (13-55). Wire shelving has the added advantage of letting light down through the closet, making it easier to see the items in it. Also it allows air to circulate through the closet, reducing the chance of mold developing, as often happens in hot, humid climates.

As you make the plan review the spacing recommended earlier in this chapter. Then make a simple drawing showing the type of shelf and the required sizes (13-56). You can find out the sizes and types of shelving available at the local building supply dealer and use this information as you prepare your layout. A frequently used plan is to use part of the closet for long, hanging items and part for shorter items (13-57).

The 16-inch-wide shelving shown in **13-58** provides exceptional storage for large items and has a clothes-hanging rod below. A 12-inch-wide shelf is available that has the hanging rod on the outside edge of the shelf (13-59). For general storage where a hanging rod is not needed, shelving as shown in **13-60** is available; it comes in a number of standard lengths or can be cut to the required length with a bolt cutter or hacksaw. This leaves a rather sharp end on each wire which must be covered with a plastic cap (**13-58** and **13-62**).

The shelves are secured by support brackets placed every three feet and at the ends of shelves that are not

13-55 This closet in a home office has metal shelves to use all of the space available and carry the heavy weight of the books.

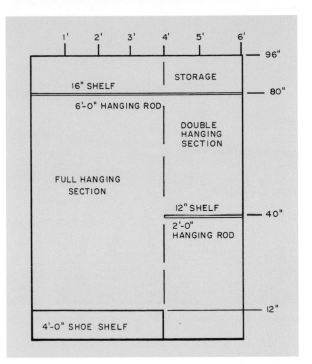

13-56 A simple plan for layout and type of shelving wanted. It helps as you purchase the items and make the installation.

13-57 Using some of the closet space for long hanging items and some for short ones increases the efficiency of the layout.

13-58 This wide-wire shelving provides considerable storage and a clothes-hanging rod.

13-59 The narrower clothes-hanging shelf has the hanger on the outside edge.

13-60 This type of shelving is used for general storage.

secured to a wall (13-61). When the shelf butts a wall it is supported with a wall bracket (13-62). Cut the shelf a little short so it does not rub on the wall when it fits down into the wall bracket. The back edge is secured to the wall with special clips (13-63). They are placed on each end of the shelf and every 12 inches across the edge. A popular accessory is a shoe shelf (13-64), which is convenient to use and keeps things orderly. It is secured to the wall along the back edge with clips.

SECURING THE SUPPORT BRACKETS & CLIPS

Since the load on the shelving can be considerable the supports must be properly secured to the wall. The support bracket can be secured to wood studs with wood screws that penetrate the stud 1¼ inches. If the wall is faced with ½-inch drywall, a 1¾-inch wood screw is needed. Usually the spacing required is such

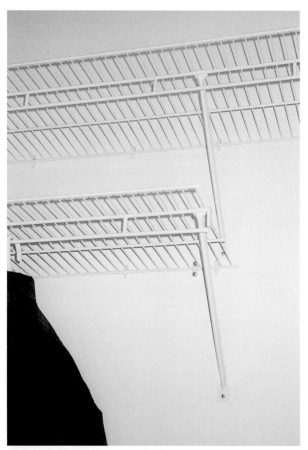

13-61 The top shelf is supported every three feet. Notice the end of the lower shelf also has a support.

13-62 The front edge of the shelf is supported on the ends next to the wall with a special wall bracket.

13-63 These small clips are placed along the back edge of the shelf about every 12 inches.

13-64 A shoe shelf is a nice addition to the closet layout.

13-65 Toggle bolts can be used to secure the support brackets to the drywall.

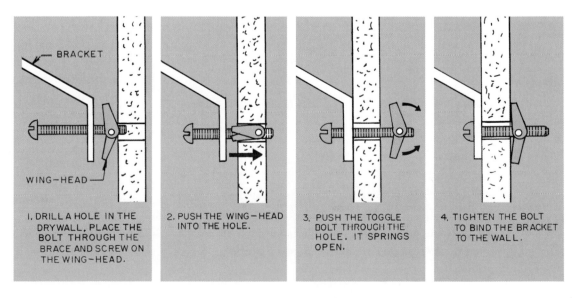

BRACKET

WING-HEAD

1. DRILL A HOLE IN THE DRYWALL. PLACE THE BOLT THROUGH THE BRACE AND SCREW ON THE WING-HEAD.

2. PUSH THE WING-HEAD INTO THE HOLE.

3. PUSH THE TOGGLE BOLT THROUGH THE HOLE. IT SPRINGS OPEN.

4. TIGHTEN THE BOLT TO BIND THE BRACKET TO THE WALL.

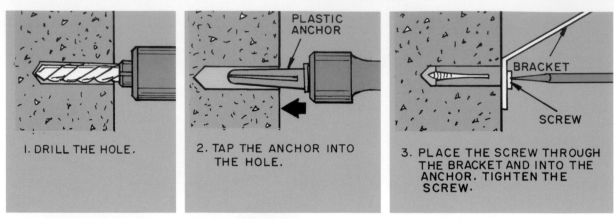

I. DRILL THE HOLE.

2. TAP THE ANCHOR INTO THE HOLE.

PLASTIC ANCHOR

BRACKET

SCREW

3. PLACE THE SCREW THROUGH THE BRACKET AND INTO THE ANCHOR. TIGHTEN THE SCREW.

13-66 Plastic anchors set in holes drilled in masonry walls will hold the support bracket.

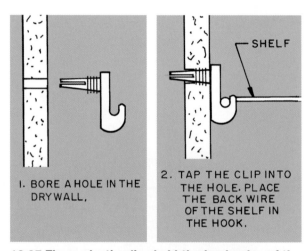

SHELF

I. BORE A HOLE IN THE DRYWALL,

2. TAP THE CLIP INTO THE HOLE. PLACE THE BACK WIRE OF THE SHELF IN THE HOOK.

13-67 These plastic clips hold the back edge of the shelf to the wall. The shelf actually hangs in the hooked opening.

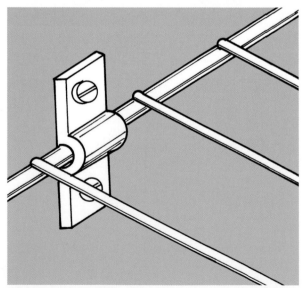

13-68 These plastic clips hold the back edge of the shelf to the wall and are screwed into the wall studs.

that the wood screws do not hit a stud, so are secured with a toggle bolt (13-65). If the shelves are to be secured to a masonry wall a plastic screw anchor can be used (13-66). Other fasteners can also be used on masonry walls.

There are several different clips available for holding the back edge of the shelf to the drywall. The clips in 13-67 are installed by boring the specified size hole in the wall and tapping the pointed end into it. These are installed about 12 inches apart. The back wire edge of the shelf is placed in the hooked end. Then a second type of clip is placed over this wire edge and screwed to each stud in the back wall (13-68). This increases the holding capacity and keeps the shelf from lifting out of the first clip.

Installing Low-Cost Storage Cabinets

Sometimes you do not want to go through the trouble of framing up a closet with studs and drywall, or you only want to add a little additional closet-type storage. One economical way to do this is to install a storage unit like the one in 13-69. It contains shelving behind each door (13-70). This unit is sold disassembled in a kit. It is assembled with special fastening devices and requires only a screwdriver to put it together. One type of connector uses a specially threaded fastener that goes through a hole in one of the joining pieces and cuts threads in a hole drilled in the other piece (13-71). Another fastener uses a threaded insert in one piece and a special machine screw that passes through a hole in the adjoining member (13-72). The joint can be taken apart and reconnected as needed. A third type uses a cylinder-head machine screw that screws into a cross-nut that is inserted in a hole bored

13-69 This storage unit was built from a kit that is assembled at home. This particular unit is 16 inches deep and contains adjustable shelving behind each door.

13-70 The shelving in this storage cabinet is adjustable, enabling it to store items of widely varying sizes.

13-71 This corner butt joint is secured with a special one-piece fastener, forming a rigid but releasable connection.

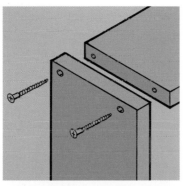

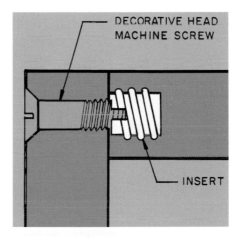

DECORATIVE HEAD MACHINE SCREW

INSERT

13-72 This two-piece fastener has a decorative countersunk-head machine screw that screws into a threaded insert installed in the adjoining member.

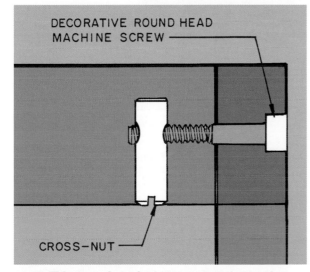

DECORATIVE ROUND HEAD MACHINE SCREW

CROSS-NUT

13-73 This two-piece fastener uses a decorative cylinder-head machine screw that screws into a cross-nut inserted into the adjoining member.

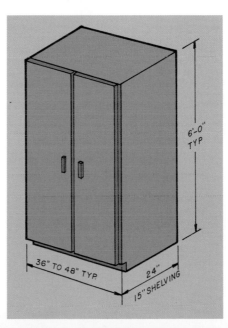

6'-0" TYP

36" TO 48" TYP.

24"

15" SHELVING

13-74 This inexpensive wardrobe can be easily built in the home workshop. It can be sized to suit individual needs.

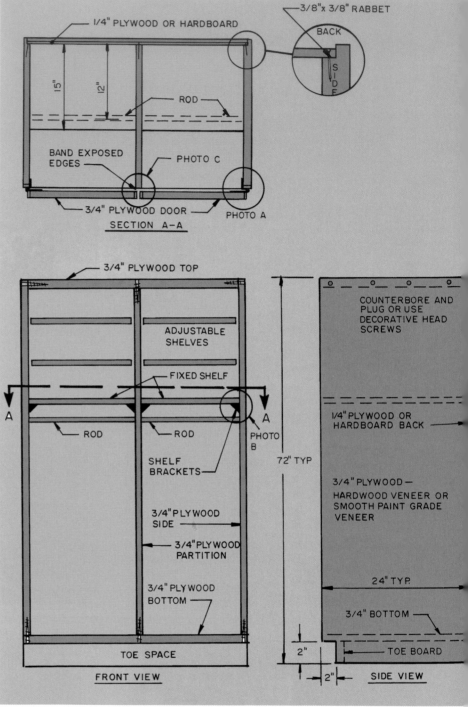

SECTION A–A

1/4" PLYWOOD OR HARDBOARD

3/8"x 3/8" RABBET

BACK

SIDE

15"

12"

ROD

BAND EXPOSED EDGES

PHOTO C

3/4" PLYWOOD DOOR

PHOTO A

3/4" PLYWOOD TOP

ADJUSTABLE SHELVES

FIXED SHELF

A

ROD

ROD

PHOTO B

SHELF BRACKETS

3/4" PLYWOOD SIDE

3/4" PLYWOOD PARTITION

3/4" PLYWOOD BOTTOM

TOE SPACE

FRONT VIEW

A

72" TYP

COUNTERBORE AND PLUG OR USE DECORATIVE HEAD SCREWS

1/4" PLYWOOD OR HARDBOARD BACK

3/4" PLYWOOD— HARDWOOD VENEER OR SMOOTH PAINT GRADE VENEER

24" TYP.

3/4" BOTTOM

TOE BOARD

2"

2"

SIDE VIEW

13-75 Typical construction details for a simple frameless wardrobe.

in the adjoining member (13-73). It also is a releasable type connection.

A more expensive, higher-quality and decorative unit is a factory-built wardrobe made from hardwoods, which is given the same finish as that used on the other bedroom furniture (refer to 13-5). They are rather expensive but make an impressive appearance.

You can also build rather low-cost wardrobes using the same construction details as are used to build kitchen cabinets. These are free-standing units and can be moved about the room. Typical overall dimensions are shown in 13-74. They can be sized to suit your needs. Typical construction details are shown in 13-75. Books on furniture and cabinet construction will give additional fabrication information.

It adds to the appearance if the plywood used for the sides and doors has a hardwood veneer such as oak or birch. If the exposed edges of the door, sides, top, and bottom are covered with solid-wood strips or wood-veneer edge banding (13-76). This material is available in several wood species. It is glued to the edge of the plywood. (The wardrobe can be built with a faceframe as shown below in 13-82 and 13-84.)

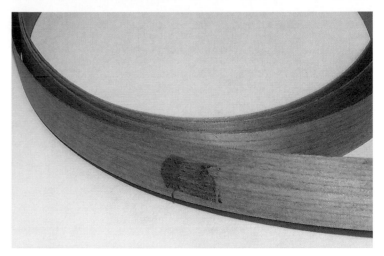

13-76 This veneer edge-banding material is available in several wood species. It is glued to the exposed plywood edges.

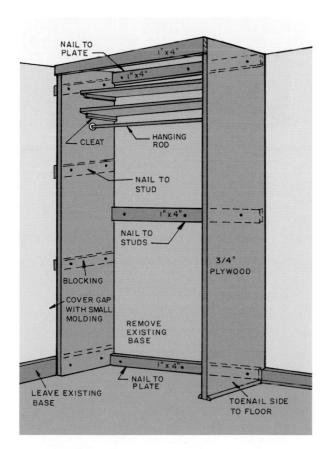

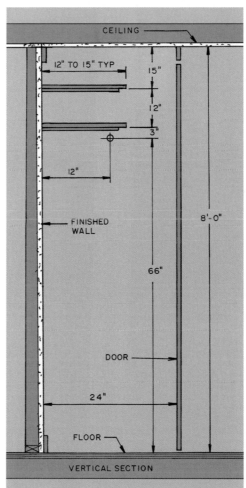

13-77 This is a simple, easily constructed, cabinet-type closet. It does not need a top or back but must be securely nailed to the walls.

13-78 Typical construction details for a cabinet-type closet. If it gets wider than 4 feet, install a partition.

Building a Cabinet-Type Closet

A very simple solution to adding a closet is to build it using ¾-inch plywood sides and doors and secure it to the wall. A design for a simple unit is shown in **13-77** and **13-78.** While the method of construction can vary, the one shown is quite adequate. It could be improved by adding a ¼-inch plywood back. This would add to the rigidity of the assembled unit. The sides run to the ceiling so no top is needed. A small molding could be added at the ceiling to trim it out.

The interior arrangement is up to you. Two door units are most frequently used; however, a third section can easily be included. Some typical layouts are shown in **13-79.**

The closet can be built with or without a faceframe. In **13-77** no faceframe was used. This requires the use of the special hinge shown in **13-80.** The section view in **13-81** shows that the door overlaps only half of the plywood side.

If a faceframe is used there are a number of hinges that could be used. One could be mounted on the exposed face while the other could be mounted on the edges of the door and faceframe **(13-82).** The faceframe adds considerably to the appearance of the closet. The unit shown in **13-83** also has a crown molding along the top. A completed closet is shown in **13-84.**

The doors can be ¾-inch plywood. They can be made stiffer and less likely to warp if 1 x 2 or 1 x 3 wood molding is glued and nailed around the edges **(13-85).** This also improves the appearance. Another decorative feature to consider is moldings on the door, creating the appearance of a paneled door **(13-86).**

The easiest way to install permanent shelving is with plastic shelving brackets **(13-87).**

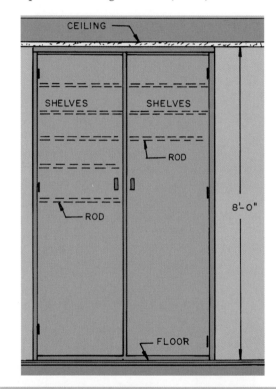

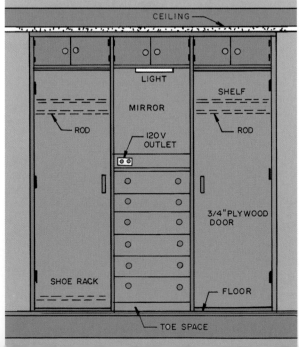

13-79 Examples of interior arrangements used with cabinet-type closets.

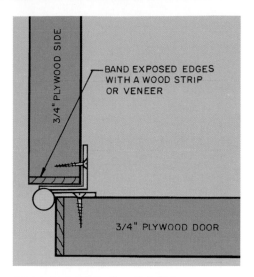

13-81 An installation detail for mounting the door hinges when a faceframe is not used.

13-80 The doors can be mounted without a faceframe by using this special hinge, which is mounted on the inside of the side and the door—exposing the attractive hinge barrel to view.

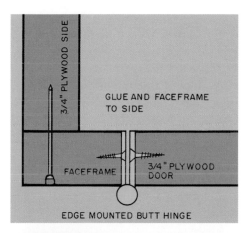

13-83 Moldings can be added at the ceiling to give the unit a finished appearance and cover any crack that may appear.

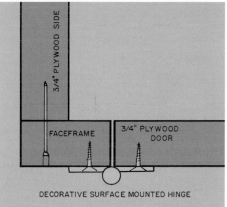

13-82 Two ways to install doors when the cabinet-closet has a faceframe.

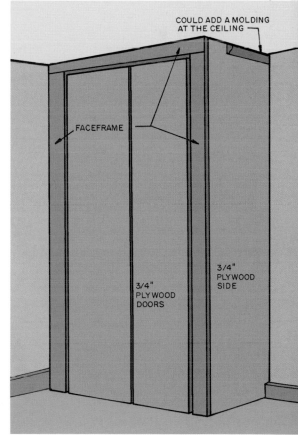

13-84 A completed cabinet-type closet.

13-85 A narrow, solid-wood strip or a molding can be added around the edge of the door to give additional stiffness and improve the appearance.

INSTALL A WOOD STRIP
OR MOLDING AROUND
THE EDGE OF THE DOOR
1 1/4" TO 1 1/2" WIDE TYP

3/4" PLYWOOD
DOOR

13-86 Molding can be applied to the door to improve the appearance and give a panel-like impression.

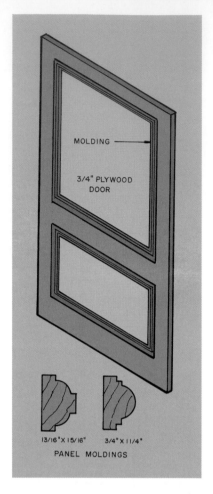

MOLDING

3/4" PLYWOOD
DOOR

13/16" X 15/16"   3/4" X 1 1/4"
PANEL MOLDINGS

13-87 Fixed shelves are easily installed with shelf brackets.

CHAPTER 14

Remodeling a Basement

A basement contains a large number of square feet which if properly handled can greatly increase the living area in a house **(14-1).** There are many things to be considered as you develop plans for this remodeling job. It pays to work out exactly what is to be done with the space and list the factors that must be considered before deciding to go ahead with the project.

Preparing a Remodeling Plan

Begin by making a floor plan of the area and show where the rooms will be located **(14-2).** Is a recreation room the main project for this remodeling? Do you want a workshop or laundry room? Extra bedrooms can be located which will require a bathroom. In almost any situation a half bath or full bath is a good addition. This will increase the cost because of the need to tear into the existing concrete floor. These and other needs should be carefully planned and their cost estimated.

Assessing the Dampness of the Basement

Before you get into any basement remodeling, check the area for problems with water infiltration or dampness on the inside walls. Addressing these problems is essential and it is important to be able to determine whether the expense of a contractor will need to be figured into the estimated budget.

14-1 This basement recreation room has a pool table and comfortable chairs in a TV viewing area. Notice the suspended ceiling with flush mounted fluorescent lights. The walls are finished with gypsum drywall and the floor with vinyl flooring.

Courtesy Mr. and Mrs. Sherman Creson

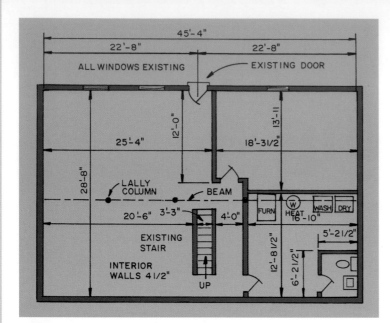

14-2 Make a floor plan showing the exact location and size of the rooms to be in the remodeled basement.

14-3 Small cracks in the foundation should be opened up a little, cleaned out, and filled with an epoxy-compound joint filler.

CHECKING THE FOUNDATION

If the house is old there may have been some settling resulting in cracks in the foundation. This allows water to seep into the basement, creating an unhealthy and unlivable situation. If the foundation is in bad shape it will require major repairs and a foundation contractor should be consulted. There are cases where the house was set on blocks and the old foundation was removed and replaced. Generally, if there are a few small cracks these are repaired by cutting along the crack with a chisel to enlarge the opening at the surface (14-3). It should be cut about ¼-inch deep and wide. Then fill it with a joint sealer such as an **epoxy-cement compound.** Follow the instructions on the container. The wall usually must be dry.

If the crack is larger and possibly allows moisture to penetrate to the inside of the foundation, open up the crack about ¾-inch wide and one inch deep (14-4). Be certain to remove all loose chips and dust (14-5). Some use a vacuum cleaner after brushing the crack clean. Now wet the crack with a garden hose, which helps clean it and dampens the area. Let the crack dry until the surface moisture is gone.

Fill the crack with hydraulic-type patching cement. **Hydraulic-type cement** is sold in dry form and is mixed with water to form a soft, clay-like material. It hardens rapidly so do not mix more than can be used in a few minutes. This type of cement is also useful when repairing a place where a large piece of the concrete block or poured concrete wall has been knocked out (14-6).

Work the hydraulic cement into the crack, overfilling the area. Let it cure for about 30 minutes and then strike it smooth with a trowel (14-7). Cover with a heavy cloth and keep moist for several days. Cracks between the concrete floor and foundation can be

14-4 Larger cracks should be widened and undercut to hold in place the hydraulic-type patching cement.

14-5 Remove all dust and chips before sealing the crack.

filled with an **elastomeric joint sealant**, which comes in a tube and is applied with a caulking gun (**14-8**).

If the foundation was properly waterproofed when the house was built, the control of cracks on the inside of the wall may be enough to provide a dry foundation wall; but, if the inside of the foundation feels damp much of the time, it should be waterproofed before installing finished walls over it. You can check for moisture by first drying a small area with a hair dryer. Then tape a layer of plastic sheet over the wall (**14-9**). After 48 hours remove the sheet. If the surface of the sheet is wet and the wall behind

14-8 If moisture enters the basement where the concrete floor butts the foundation wall, seal the crack with an elastomeric joint sealer.

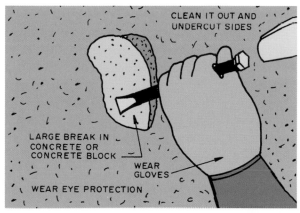

14-6 Large breaks in concrete can be repaired by cleaning the surface and undercutting the edges before applying the hydraulic patching cement.

14-9 Select an area on the wall and dry it with a hair dryer. Tape a piece of plastic sheet over the area.

14-7 Work the hydraulic cement into the crack, overfill a little, and strike smooth after it cures for a short while.

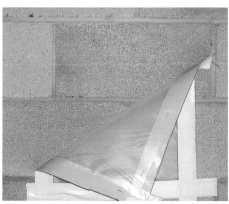

14-10 After 48 hours remove the sheet. If the wall is dry it is not transmitting moisture. Any moisture on the plastic is due to condensation from the trapped air.

14-11 If the wall is damp it indicates that moisture is penetrating the wall, and exterior waterproofing is necessary.

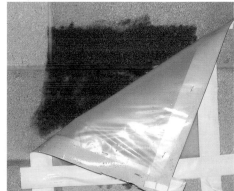

14-12 Gutters are a great help because they keep the water coming from the roof from soaking into the soil around the perimeter of the foundation.

14-13 The downspout directs the water from the gutter to some means on the ground that causes it to flow away from the foundation.

14-14 If the land slopes away from the foundation a splash block will direct it away from the house.

it is dry, the moisture is from condensation in the air (14-10). If the wall is wet it indicates moisture is seeping through the wall (14-11). If the moisture is from condensation in the air you know you will have to provide mechanical ventilation and install a dehumidifier. If the wall is wet you will have to waterproof the wall.

EXTERIOR WATER CONTROL

Dampness on the inside of basement walls can be reduced somewhat by installing **gutters** along the edge of the roof (14-12) and using **downspouts** to move the water to the ground (14-13). At the ground add an **elbow** to direct the water onto a **splash block** (14-14), which is sloped away from the foundation. If the area has frequent rains consider placing a length of downspout on the elbow and running the water six feet or so away from the foundation as shown in 14-15. Extra protection can be had by laying a piece of plastic sheet over the area from the foundation out several feet. Bury it several inches below the ground or cover it with topsoil, gravel, or wood chips.

If you do not have gutters, slope the soil away from the foundation and install a plastic sheet as shown in 14-16. An alternate plan would be to cover the area about six feet away from the foundation with precast

concrete paver blocks. Put a plastic sheet below the blocks (14-17). The pavers keep the water from the roof from digging into the soil and causing erosion.

In a situation where the site is flat and rain is frequent, consider building a sealed pit below each downspout, filling it with gravel, and running a pipe from it to a **drywell** (14-18). If there is enough slope the drain could run to open, above ground outlet.

A drywell is a large opening dug in the ground and lined with loosely laid concrete blocks set without mortar; or it is a large metal drum that has holes punched in the sides. The opening is filled with gravel. Rain water flows into the drywell and percolates out into the soil (14-19).

If the house is situated where the surrounding area is higher than the building site, surface water will flood the area around the house every time it rains. This can be controlled by building a **berm** (14-20) or **swale** along the side of the house that faces the down-hill water flow (14-21). A berm is a built-up earthen bank around the area much like a dike. It directs the water away from the house (14-22).

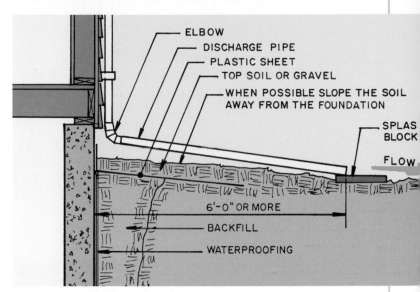

14-15 On relatively level lots use a section of downspout off the elbow at the ground to discharge the water some distance from the foundation.

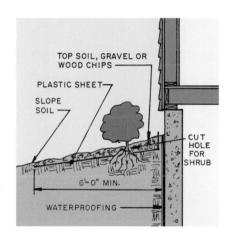

14-16 If you do not want to use gutters, you can direct rain runoff from the roof by placing a plastic sheet a few inches below the surface of the sloped soil.

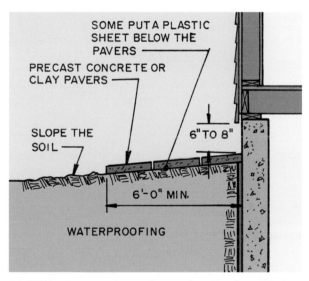

14-17 Cast concrete or clay paving blocks will also direct roof runoff rain away from the foundation. It is important to slope the soil to assist in the runoff.

14-18 If the lot has a high water table, is level, and adequate slope cannot be provided, the downspouts can discharge water into a recessed, watertight enclosure and then to a drywell or daylight, if possible.

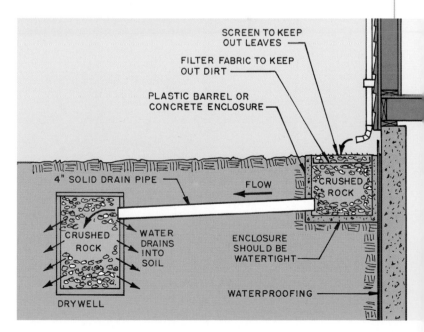

If the water flow is not too severe the berm can be sodded. If erosion is likely to occur cover the side facing the slope with a plastic sheet and lay crushed rock over this. One problem to this solution is where are you directing the water? Are you flooding the street or a neighbor's lot? Local governments have surface flow regulations so you might have to get permission to redirect the surface water.

Another technique is to dig a shallow trench, referred to as a swale, around the house instead of a berm. A swale is a recessed area in the soil (14-23).

The swale could be lined with a plastic membrane and covered with gravel, or, if the flow is not too heavy, be sodded with a tough grass.

An ever bigger problem is if the house is in an area that has considerable subsurface water flow or springs. This would require a swale that has the bottom dug several feet deep and filled with large crushed rock placed over a perforated plastic drainpipe in the bottom. The top is covered with filter fabric to keep leaves and dirt from clogging the rock. The side of the drain pit facing the house is covered with a plastic

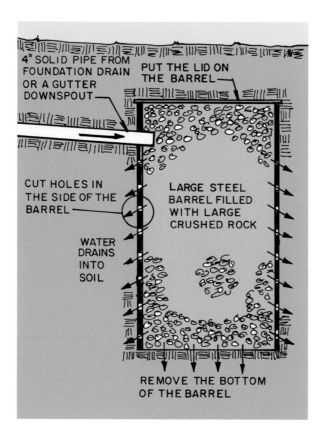

14-21 Surface water that flows toward the house can be redirected by building a berm or digging a swale.

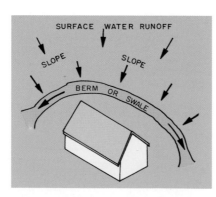

14-19 A drywell receives water from a foundation drain or downspouts and percolates it into the surrounding soil. If the flow is large, several drywells can be connected with 4-inch solid pipe.

14-20 This small berm was installed along a concrete driveway to turn the water coming down the driveway away from the yard and house. It was made with compacted sand and clay and covered with pine straw.

sheet or waterproof material to keep the water collected in the drain from moving on toward the foundation (14-24).

WATERPROOFING THE FOUNDATION

If the basement is dry and you know the foundation was given a waterproof membrane when the house was built, perhaps all you need to do is to divert surface water away from the foundation. If the inside of the foundation is damp or has leaks, the exterior wall must be waterproofed. Since this is an expensive operation, be certain to discuss it with an experienced, reputable waterproofing contractor.

There are waterproofing products available that can be applied to the inside of the basement wall. This is the easiest and least expensive way to handle the situation. Try to secure information about effectiveness from those who have used each product. All cracks should be sealed before the coating is applied.

The frequently used waterproofing materials include liquid membranes and sheet membranes.

A **liquid membrane** is some form of sprayed polymer membrane that is covered with an insulation and protection board (14-25) or a heavy-bodied cutback asphalt coating covered with drainage board (see 14-30).

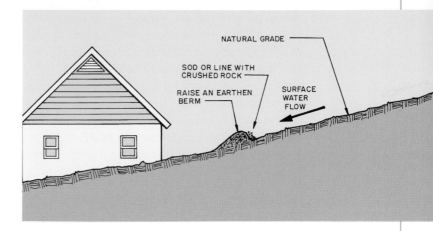

14-22 A berm is effective for diverting heavy surface water flows.

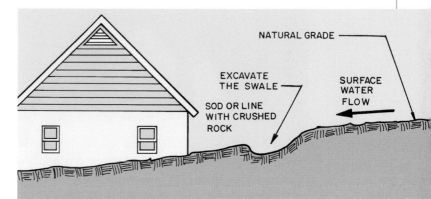

14-23 A swale will redirect low-volume surface water flows.

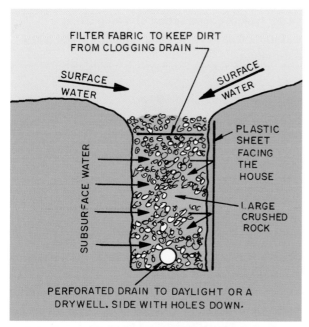

14-24 Those with high levels of subsurface water can direct the water away from the foundation by building a swale with a drain dug in the bottom that is filled with crushed rock. A perforated drainpipe is laid in the bottom and run to daylight or a drywell.

14-25 This polymer waterproofing membrane can be spray-applied over almost any wall of poured concrete or masonry. It will be covered with an insulation board or drainage board. *Courtesy Mar-Flex Systems, Inc.*

Sheet membranes are bonded to the clean foundation wall and are made of self-adhesive polyethylene. They are covered with a protective thermal insulation board or a drainage board (**14-26**).

Often the drainage board is covered with a filter fabric which keeps the soil from clogging the drainage channels. Some boards have the fabric already bonded to them. A typical installation is in **14-27**.

Following is a generalized example of the waterproofing process. Remember to read and follow the manufacturer's instructions that accompany the materials you plan to use.

Begin by excavating the area around the foundation. If this is a deep excavation the contractor needs to pay attention to providing temporary support to the exposed earth wall so a cave-in does not occur, burying the workers (**14-28**). Then clean the exposed foundation area.

Prepare the cleaned surface to receive the waterproofing compound or membrane as instructed by

14-26 This polyethylene sheet membrane is bonded to the exterior of the foundation—forming a large unbroken, impervious, waterproof layer. The sheets are overlapped and installed as directed by the manufacturer. *Courtesy Carlisle Coatings and Waterproofing, Inc.*

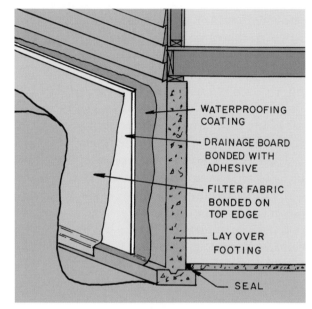

14-27 Typical construction of a foundation waterproofing system using drainage board and a filter fabric.

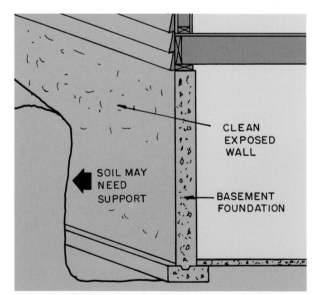

14-28 To waterproof an old basement foundation, excavate around the house and thoroughly clean the exposed wall.

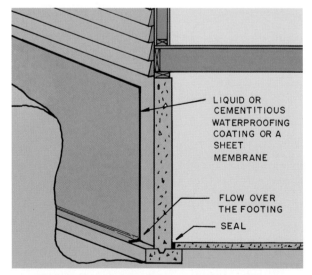

14-29 Apply the chosen waterproofing material to the clean foundation. Lay it over the footing. Seal the basement floor to the foundation on the inside of the wall.

the manufacturer. Liquid coats may be sprayed, brushed, or troweled on the surface. A key to success is to get a uniform coating of at least the minimum specified thickness. Sheet products are unrolled and pressed against the surface. Be careful to cover all coves or irregularities in the surface. Pieces can be cut and laid over the original sheet in difficult areas or where it may have been damaged (14-29).

After the waterproof coating has set as required, the soil can be replaced along the foundation (14-30).

REMOVE THE SUBSURFACE WATER

After the foundation has been waterproofed, lay 4-inch perforated pipe around the perimeter to drain away subsurface water that is next to the foundation (14-31). The exact plan will vary depending upon the shape of the house and the slope of the land. Locate

cleanouts at each corner so any clogs that occur here over time can be routed out. Plan the slope of the pipe so it drains from the high side of the foundation to the low side. Slope the bottom of the excavation and the rock fill to provide this slope. The water should run from the drainage system to daylight or to a drywell.

Lay a large piece of filter fabric over the footing and up the foundation and excavation as shown in 14-32. Then lay the 4-inch perforated drain pipe along the side (14-32) or on top (14-33) of the footing over a bed of large crushed rock. Place the perforated side down. Now lay a bed of the rock over the pipe, filling the excavated area above it about 10 to 12 inches. Then wrap the filter fabric over the rock and backfill the excavated area. Some will fill in 10 to 12 inches of coarse sand on top of the filter fabric and then

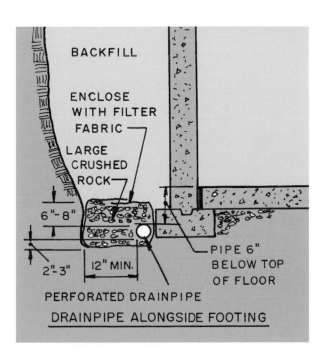

14-30 The opening around the foundation is backfilled after the liquid waterproofing coating has been applied and cured. Notice the use of drainage board below grade.

PERFORATED DRAINPIPE
DRAINPIPE ALONGSIDE FOOTING

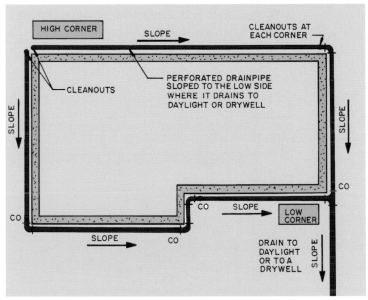

14-31 Water around a foundation can be drained by installing a 4-inch perforated plastic drainpipe around the perimeter and running it to daylight on a drywell.

14-32 The perforated plastic drainpipe can be placed alongside the footing in a bed of large crushed rock enclosed in a filter fabric.

299

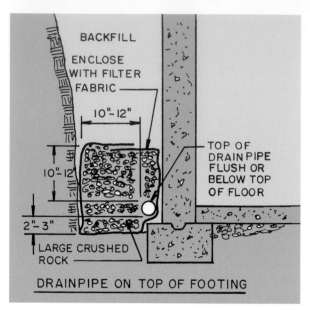

BACKFILL
ENCLOSE WITH FILTER FABRIC
10"-12"
10"-12"
2"-3"
LARGE CRUSHED ROCK
TOP OF DRAINPIPE FLUSH OR BELOW TOP OF FLOOR

DRAINPIPE ON TOP OF FOOTING

14-33 The perforated plastic drainpipe can be placed on top of the footing and next to the foundation wall. It is surrounded by a bed of large crushed rock enclosed in a filter fabric.

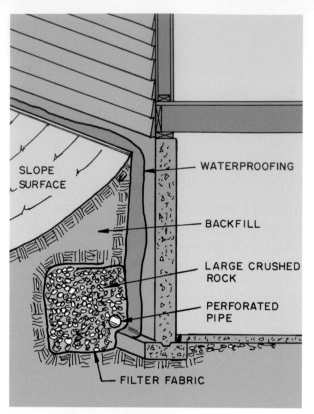

SLOPE SURFACE
WATERPROOFING
BACKFILL
LARGE CRUSHED ROCK
PERFORATED PIPE
FILTER FABRIC

14-34 A typical installation for removing subsurface water along a foundation.

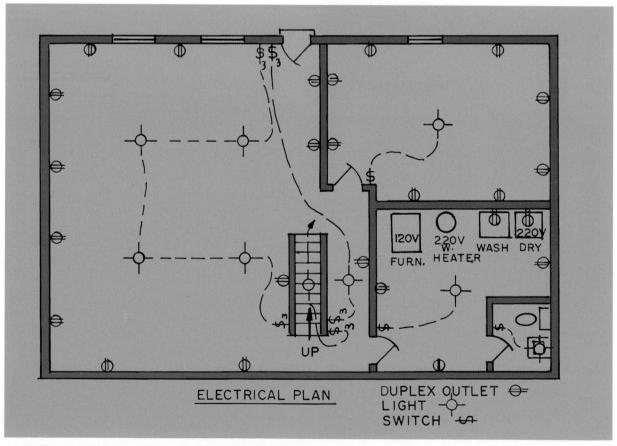

ELECTRICAL PLAN

120V FURN.
220V W. HEATER
WASH
220V DRY
UP

DUPLEX OUTLET
LIGHT
SWITCH

14-35 The electrical requirements are located on the floor plan.

finish filling the excavation with the soil that was removed. Remember to slope the backfill away from the foundation to assist in moving surface water away from the foundation (**14-34**).

Checking the Overhead Floor Joists

Before getting into any basement remodeling, check the overhead floor joists to see if they are in good condition. Since they will support the ceiling they must be straight and strong. Review Chapter 15 for additional information.

Ceiling Planning

Generally the ceiling will be gypsum drywall or a suspended ceiling or fiber ceiling tiles. Review Chapter 4 for additional information.

Electrical Planning

A remodeled basement will require quite a few duplex outlets in the walls and a number of ceiling lights with switches. After you develop the floor plan locate these on the drawing as shown in **14-35**.

This will be a big project which must meet electrical codes and be properly installed. The building inspector will check the installation before it is covered up.

Following are some typical planning considerations. Duplex outlets are located near the corners of the rooms because furniture is typically grouped in these areas. Other outlets are along the walls spaced so no usable wall space is more than six feet from an outlet.

If a laundry is to be in the basement it will require special-purpose outlets for high-current appliances such as an electric washer/dryer.

General illumination in a room is provided by one or more ceiling lights (refer to **14-1**) controlled from one or more locations by wall switches. Consider other lighting sources such as cornice lighting or sconces (**14-36**). Both fluorescent and incandescent lighting fixtures are widely used.

Wall switches are located on the latch side of the door and placed around the room so they are available on normal traffic patterns. They are normally 48 inches above the floor.

After the electrical plan has been developed the electrician will design the various items into circuits to meet codes. It will also be necessary to add several more circuit breakers in the electric panel. Sometimes reaching the existing panel is difficult so a smaller panel is installed in the basement with a single line feeding it from the main panel on the first floor. Remember, all installations must meet local codes and usually must be checked by a licensed electrician if you do the work.

Plumbing Considerations

If you plan to add a bathroom or laundry to the remodeled basement there are some big problems to face. Running hot and cold water is usually not a major problem. The lines can run through and between the floor joists and down inside the new partitions. The waste disposal system, however, can present major problems. If the basement floor is high enough above the city sewer or septic tank that the line can flow to it with adequate natural slope, the big task is cutting a channel through the concrete floor. But if the city sewer is above the basement floor you will have to install a lift pump to raise and discharge the waste material.

If it is not possible to run the water and waste pipes between the floor joists they can be hung below and cased with a box made from solid wood or plywood. This can be painted to match the ceiling or stained so

14-36 A sconce is very decorative and provides a soft spot illumination.

it appears to be a wood beam (14-37). Vertical pipes such as the soil and vent stack for a toilet can be run up the wall and enclosed in the same way. Remember to cover the cold water pipes with an antisweat insulation so condensation will not drip down on the finished covering material.

Small heat ducts can be boxed in with wood or plywood. Larger or multiple ducts can be framed with 2 x 2 stock and covered with drywall or ceiling tile to match the ceiling, or with ⅜-inch plywood (14-38).

Before finalizing your remodeling plans get a preliminary bid on the plumbing requirements. Be certain to follow local codes.

Heat & Air-Conditioning

In handling the heating and air-conditioning needs, it must be determined if the existing system has the capacity to handle this large increased area. If you are adding just one room in the basement, generally the existing system can handle it. If adding the entire basement, chances are the existing system will be inadequate. Another factor is that the air temperature in the basement rooms will be controlled by a thermostat that is on the floor above. Another factor is whether it is physically possible to extend ducts or hot water heating lines to the basement rooms. One certain solution is to add a separate heating system for the basement rooms.

Since basements, even if watertight, tend to develop high humidity, the system should provide mechanical ventilation and be able to dehumidify the air. A licensed heating and air-conditioning contractor should be consulted.

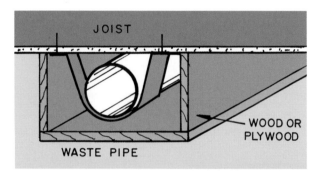
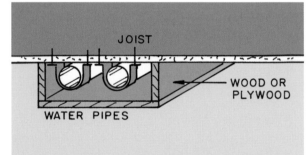

14-37 Exposed water and waste pipes can be concealed by covering them with a wood or plywood box.

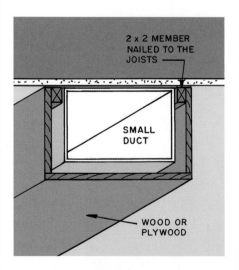

14-38 Heat ducts require rather large enclosures. Large and multiple pipes require that a wood framework be built, over which some type of covering is installed.

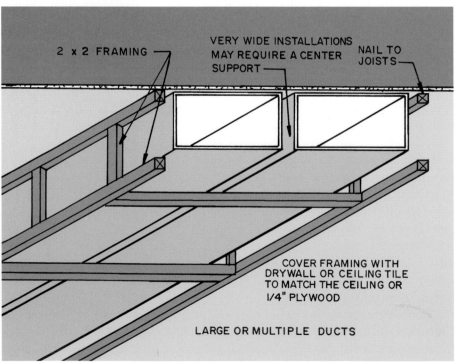

Floor Drains

As you make remodeling plans, locate the floor drain and consider how it works into the overall plan. Hopefully it will be located in the new laundry room or bathroom. If located in a living area, remember it has a trap below the concrete floor which when full of water blocks the flow of sewer gas into the basement. One thing that can be done is to prepare a cover over it that can be removed occasionally so water can be poured in to keep the trap full. Some advocate filling the drain with concrete to permanently seal it. While this will give total blockage of sewer gas you have given up any chance of draining away water that may leak into the basement. In either case something must be done to give some concealment of the floor drain to enhance the appearance of the room. You could even put a small rug over the cover. The floor will slope a little toward the drain. If you permanently close the drain you can lay a top coating to fill the slope, thus leveling the floor around it. The topping can be a concrete mix or some other coating material. Consult the local building supply dealer for products currently available.

If you plan to put vinyl tile on the floor, the drain will cause no problem as described above. If you plan to install wood or laminated wood flooring, a different situation faces you. You may have to topcoat the entire basement.

In older homes it was common practice to give a considerable slope to the floor toward a floor drain. In newer homes this slope is usually less and in some cases the floor drain is omitted and the floor is level. If the slope is excessive it will make it difficult for furniture to sit properly and even make walking difficult as you approach the floor drain. In extreme cases consider laying a concrete topping over much of the basement floor. Toppings two inches thick or more may be required. This is a big job.

Possible Foundation Changes

If at least half the foundation is above grade, consider installing windows and an outside door. If the basement floor is on grade on one side of the house, a walkout door can be cut through the foundation (14-39). If the floor is three or four feet below grade you can excavate the area outside the door opening,

14-39 This design makes full use of the basement, including double access doors and standard double-hung windows.

frame it with concrete block or cast-in-place concrete walls, and build concrete steps up to grade (14-40). Consider putting a drain in the floor and connecting it to the perimeter drain that runs around the foundation. Or run a pipe to an above-ground outlet. Prefabricated, precast concrete stairwells are available that provide the enclosure and the steps leading to grade (14-41). After the unit is set in place and secured to the basement foundation, a watertight metal enclosure is installed over it (14-42). Metal doors form a watertight enclosure. They swing up, providing easy access to the basement door (14-43).

Adding a door or windows to a foundation requires cutting openings in it. If it is built with concrete blocks this is considered reasonable. Trying to cut through a cast concrete foundation is generally too difficult and expensive.

If you are adding an addition to the house which will have a walkout basement, consider framing the exposed exterior walls with wood or steel studs and sheathing, and finishing as is done on normal wall framing (14-44). This will make it possible to frame the door and window openings as is done on the other

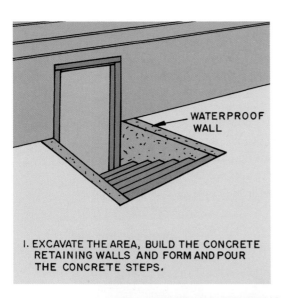

WATERPROOF WALL

I. EXCAVATE THE AREA, BUILD THE CONCRETE RETAINING WALLS AND FORM AND POUR THE CONCRETE STEPS.

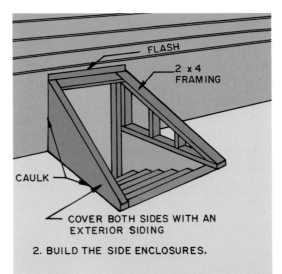

FLASH

2 x 4 FRAMING

CAULK

COVER BOTH SIDES WITH AN EXTERIOR SIDING

2. BUILD THE SIDE ENCLOSURES.

14-40 If the basement floor is below grade, an outside door can still be installed by excavating an area and building a concrete retaining wall and steps.

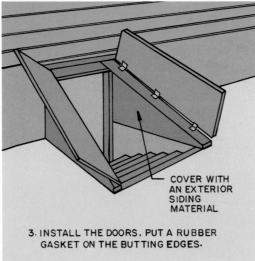

COVER WITH AN EXTERIOR SIDING MATERIAL

3. INSTALL THE DOORS. PUT A RUBBER GASKET ON THE BUTTING EDGES.

14-41 The precast concrete stairwell is lowered into the excavated area outside the door and secured to the foundation. *Courtesy The Bilco Company*

walls of the house. The basement door in **14-45** was framed in a concrete block foundation.

Basically to prepare an opening in a concrete block foundation you first have to support the overhead load, which includes the floor, exterior wall, and roof (**14-46**). Then you can remove the concrete blocks, forming the rough opening for the doors or windows (**14-47** and **14-48**).

One way to install the doors and windows in the masonry opening is to secure 2-inch-thick framing around the opening. This frame is referred to as a **buck**. It is held to the concrete blocks with metal

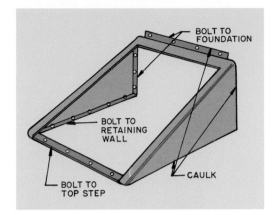

14-42 A prefabricated metal enclosure is bolted to the stairwell and foundation wall.

14-43 The metal doors form a watertight enclosure yet provide easy access to the basement door.

Courtesy The Bilco Company

14-44 These interior and exterior photos show an exposed basement that has a wood framed wall with door and window openings framed as in the above-grade walls.

14-45 This basement door has been framed in a concrete block foundation.

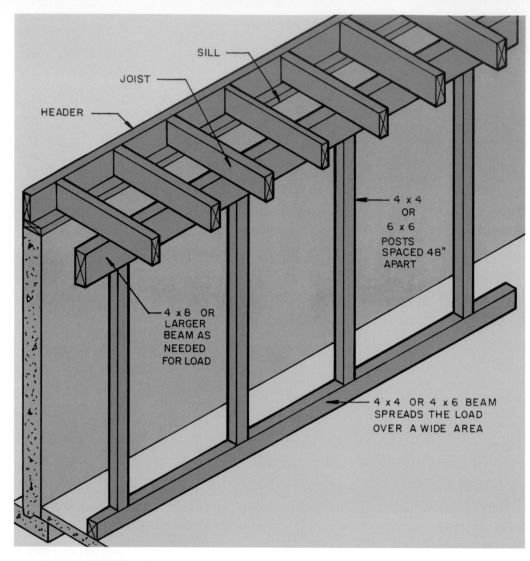

SILL

JOIST

HEADER

4 x 4
OR
6 x 6
POSTS
SPACED 48"
APART

4 x 8 OR
LARGER
BEAM AS
NEEDED
FOR LOAD

4 x 4 OR 4 x 6 BEAM
SPREADS THE LOAD
OVER A WIDE AREA

14-46 If you plan to cut a door or window opening through an existing concrete block wall, first build a support to hold the floor joists and other loads placed upon them.

14-47 The concrete blocks were cut and knocked out, providing an opening for a door.

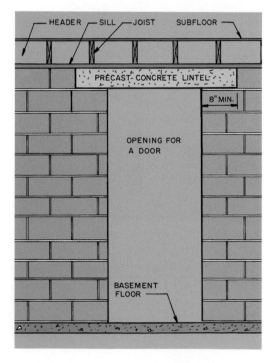

HEADER — SILL — JOIST — SUBFLOOR

PRECAST CONCRETE LINTEL

8" MIN.

OPENING FOR
A DOOR

BASEMENT
FLOOR

anchors set into the mortar joint (**14-49**). The clear distance inside the bucks must be the same as the manufacturer-specified rough opening. A generalized installation detail when a buck is used is shown in **14-50**. Always consult the installation instructions provided by the manufacturer.

Some windows can be installed directly to the concrete opening (**14-51**). Before you size the opening be certain how you are going to install the doors and windows so the rough opening required will be accurate. Consult the manufacturer's installation guide.

BUILDING WINDOW WELLS

If the grade is near the windowsill it is best to build a window well to carry off water from rain and surface drainage. Window wells are used on windows with the sill above or below grade. Prefabricated metal well enclosures are available; however, they can be built using concrete blocks or cast-in-place concrete.

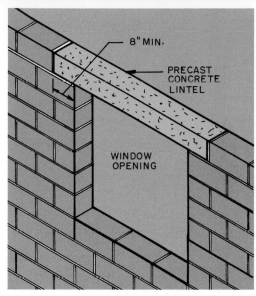

14-48 A typical window opening in a concrete block foundation wall.

8" MIN.

PRECAST CONCRETE LINTEL

WINDOW OPENING

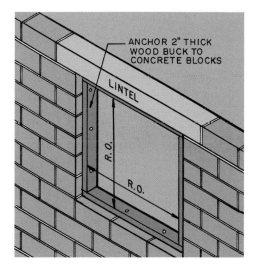

ANCHOR 2" THICK WOOD BUCK TO CONCRETE BLOCKS

LINTEL

R.O.

R.O.

14-49 Some windows require a wood buck be installed inside the opening in the concrete block wall.

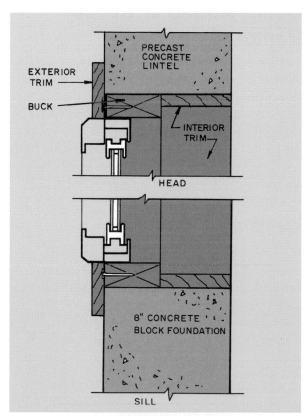

PRECAST CONCRETE LINTEL

EXTERIOR TRIM

BUCK

INTERIOR TRIM

HEAD

8" CONCRETE BLOCK FOUNDATION

SILL

14-50 A generalized installation illustration for inserting a vinyl window in a concrete block wall.

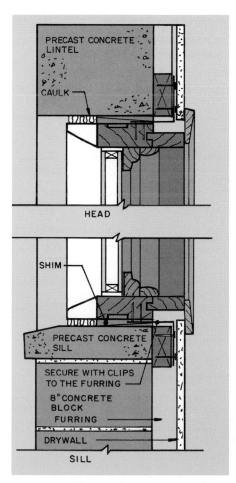

PRECAST CONCRETE LINTEL

CAULK

HEAD

SHIM

PRECAST CONCRETE SILL

SECURE WITH CLIPS TO THE FURRING

8" CONCRETE BLOCK FURRING

DRYWALL

SILL

14-51 A generalized detail for a wood window installed in a concrete block foundation opening. It is shimmed on each side to get it plumb and level, and is secured to furring with manufacturer-supplied clips.

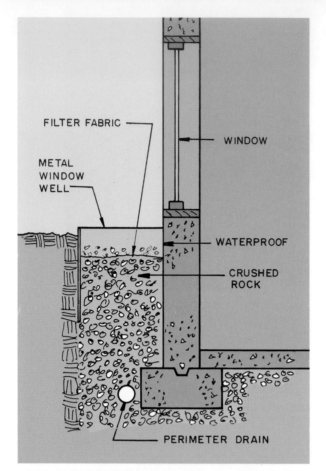

14-52 A window well is used to remove rain and surface water from windows at or below grade. This installation has the water flow down through the crushed rock and be removed by the perimeter drain.

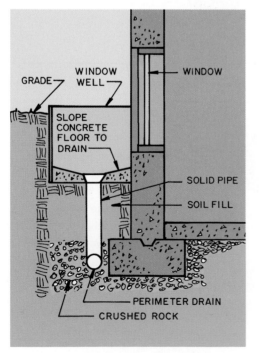

14-53 This window well has a concrete floor sloped to a drain which is connected directly to the perimeter drain. This drain provides immediate discharge of the water.

There are several ways window wells can be built. The choice depends upon local conditions and climate. If the area has moderate rainfall and the soil is sandy, the window well in **14-52** should be able to handle the water. The water flows down through the crushed rock and is carried away by the perimeter drain. If the rainfall is heavier you may want to direct the water falling into the window well with a floor drain that is connected directly to the perimeter drain (**14-53**).

Another approach to using a window well is to use perforated plastic drainpipe to carry the water away from the well to a location 8 or 10 feet away from the foundation and dispense the water in a small drywell. Slope the pipe so it drains away from the window well (**14-54**).

Window wells can also be made larger than needed and serve as small greenhouses. If they are built something like the one in **14-55** they will allow natural light to enter the basement and permit plants to be grown. This provides an attractive feature when most of the window is below grade. The transparent cover also keeps the well fairly dry since it sheds the rain.

Going Ahead with the Remodeling

You have addressed the basic issue of the soundness of the basement space for conversion to a living space. Your plans for what is to be done with the space, as well as what the ventilation, electrical, and plumbing needs will be, are completed. Once whatever foundation work you determined was needed has been done, you are ready to go ahead with building out the space and finishing the walls and floors.

Locating the Partitions

Mark the location of the partitions on the floor by measuring carefully and snapping a chalk line, locating both sides of the partition (**14-56**). These are used to set the assembled wall partitions in place. If the floor plan was developed by an architect be aware that they usually locate interior partitions by their centerline; therefore, you will have to allow for this as you lay out both sides of the partition on the floor.

Framing the Partitions

The partitions forming the walls of the basement rooms are usually nonloadbearing. They consist of a bottom plate, studs, and a top plate (**14-57**). A single

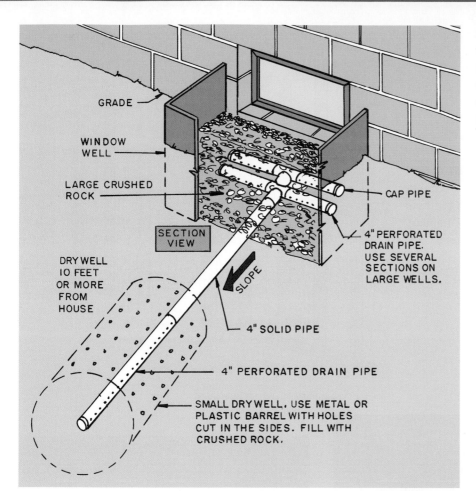

GRADE

WINDOW WELL

LARGE CRUSHED ROCK

SECTION VIEW

DRY WELL 10 FEET OR MORE FROM HOUSE

SLOPE

4" SOLID PIPE

4" PERFORATED DRAIN PIPE

SMALL DRY WELL. USE METAL OR PLASTIC BARREL WITH HOLES CUT IN THE SIDES. FILL WITH CRUSHED ROCK.

CAP PIPE

4" PERFORATED DRAIN PIPE. USE SEVERAL SECTIONS ON LARGE WELLS.

14-54 The water falling in the window well can be collected by perforated drain pipes in the bottom and directed to a small drywell.

14-55 A window well can be enclosed. The glazed covering directs the water outside the well, and it allows natural light to enter the basement; it is a great place to grow plants.

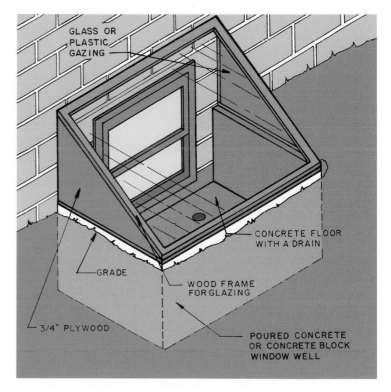

GLASS OR PLASTIC GAZING

GRADE

3/4" PLYWOOD

CONCRETE FLOOR WITH A DRAIN

WOOD FRAME FOR GLAZING

POURED CONCRETE OR CONCRETE BLOCK WINDOW WELL

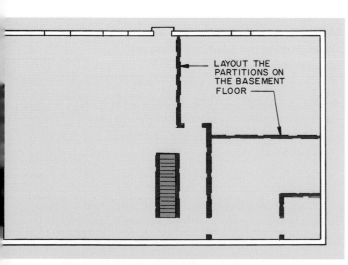

LAYOUT THE PARTITIONS ON THE BASEMENT FLOOR

14-56 Locate the partitions on the floor.

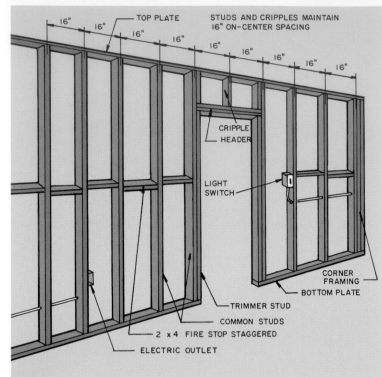

14-57 Typical interior partition framing.

14-58 A typical interior partition framed on a concrete floor.

14-59 This wide header is over the doors to a closet. Notice the overhead joists are perpendicular to the partition.

top plate can be used. If the partition carries any floor load the top plate should be doubled. Openings over doors will require a double 2 x 4 header and one or more cripples (short, vertical blocks that typically continue the stud spacing). The studs and cripples are placed 16 inches on center, although 24 inches is sometimes used. A typical installation is shown in **14-58.** The header above a door or closet is framed as shown in **14-59.**

If a partition runs perpendicular to the joists it will be nailed to them **(14-60)**. If it runs parallel with the joists a nailer and blocking is installed **(14-61)**. The partition is nailed to this and the ceiling material is nailed to the blocking. If one end of the partition butts the concrete or concrete block basement wall, it can be nailed to the wall **(14-62)**. Some partitions simply butt it and are not nailed to it.

Determine the height of the partition. Since this is in a basement, the floor-to-ceiling height will probably be different than on the above-ground floors **(14-63)**. Also the height may vary some if the floor has been sloped toward a drain. If this is very small

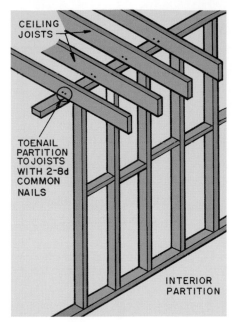

14-60 Partitions that run perpendicular to the joists are nailed to the joists.

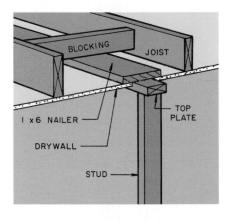

14-61 Partitions that run parallel with the joists are nailed to blocking and a nailer run between the joists.

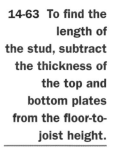

14-63 To find the length of the stud, subtract the thickness of the top and bottom plates from the floor-to-joist height.

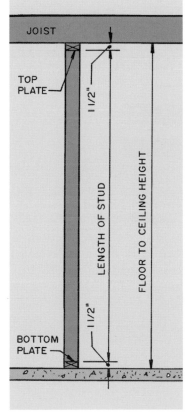

14-62 This partition butts the concrete block basement wall. It can be nailed to the wall.

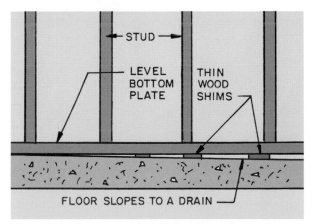

14-64 If the concrete floor has a slight slope the partition can be leveled with a series of thin shims.

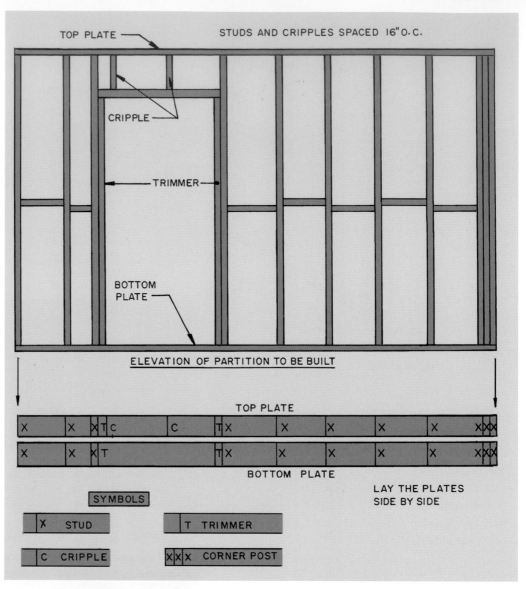

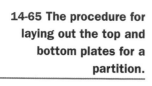

14-65 The procedure for laying out the top and bottom plates for a partition.

STUDS AND CRIPPLES SPACED 16"O.C.

TOP PLATE

CRIPPLE

TRIMMER

BOTTOM PLATE

ELEVATION OF PARTITION TO BE BUILT

TOP PLATE

BOTTOM PLATE

LAY THE PLATES SIDE BY SIDE

SYMBOLS

X STUD   T TRIMMER

C CRIPPLE   XXX CORNER POST

14-66 Assemble the partition flat on the floor. Nail the plates to each stud or cripple with two 16d common nails.

the top plate can be shimmed at the joists and the bottom plate shimmed at the floor (14-64). If the slope is small the gap at the floor can be covered with the baseboard and shoe molding. If the floor has considerable slope you may have to lay a layer of leveling compound to bring it up flush with the area near the wall.

Begin by cutting to length the top and bottom plates. Accurately mark the location of each stud and corner member as shown in 14-65 on the top and bottom plates. They can be laid out at the same time by placing them side by side.

Cut the studs to the required length. Then assemble the partitions flat on the floor (14-66). Nail the top and bottom plates to the studs with two 16d common or box nails.

Place the bottom plate of an assembled partition along its layout line on the floor and raise the wall

14-67 Raise the assembled partition. Place it on the layout line on the floor. Check for plumb and squareness with butting walls.

(14-67), placing the top plate against the overhead floor joists. Check to be certain that it is aligned with the floor marks, it is perpendicular to the wall it will butt, and it is plumb. Nail the bottom plate to the concrete floor with high-carbon masonry nails (14-68 and 14-69) and the top plate to the floor joists or blocking with common nails. When a partition butts another partition along its length the studs are prepared to nail the butting partition to it as shown in 14-70. If the partitions form a corner (14-71) the framing is made using three studs on one of the par-

14-68 When the partition is in place, fasten the bottom plate to the concrete floor with high-carbon steel nails that have been heat-treated and tempered.

14-69 These are flat, hard-cut masonry nails. Other types are available such as a round, fluted nail. When nailing, wear safety glasses.

14-70 Here are two ways to prepare a partition so a butting partition can be nailed to it and provide a nailing surface for the drywall.

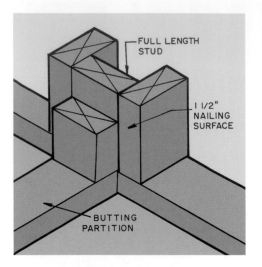

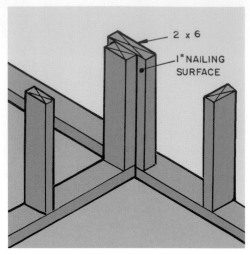

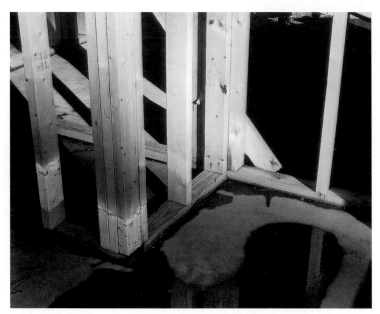

14-71 These interior partitions meet, forming a corner.

titions (**14-72**). If the end of the partition butts a concrete or concrete block wall, nail it to the wall as shown earlier in **14-62.**

After all the partitions are in place install the electrical outlets, light switches, and plumbing and heating that may run in the walls (**14-73**). Finally, finish the walls with paneling, drywall, or some other finish material.

Finishing the Masonry Basement Wall

How you decide to finish the masonry walls depends upon how nice you want the finished room to be, how much you want to spend, and, in some cases, the climate. In warm climates the walls can be painted using a masonry paint designed for this purpose. This is typically a polyvinyl acetate or acrylic emulsion; a

14-72 If interior partitions meet, forming a corner, a three-stud corner post can be used.

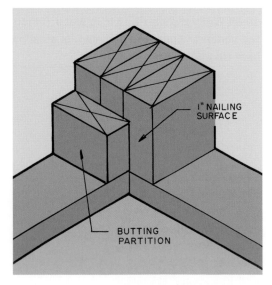

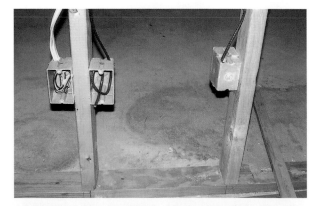

14-73 Once the partitions are in place it is time to install the electrical, plumbing, and other mechanical systems.

form of latex paint. It forms a tough film that resists the alkali of the masonry, and also helps seal out moisture. The wall should be primed with a primer recommended for the product you select. There are other formulations of masonry paint available. Consult the local paint dealer for recommendations.

If a lot of the foundation is above grade and the area has some freezing weather, it is a good idea to insulate the foundation wall at least two feet below grade. Be certain the wall is watertight and no moisture penetrates through it. It helps if you paint a coat of waterproofing masonry paint over the wall. This minimizes moisture penetration. Any moisture getting on the insulation reduces its effectiveness and provides the opportunity for health-damaging mold to develop.

One way to insulate the foundation is to nail 1 x 2 furring strips spaced 16 inches O.C. and to place rigid foam insulation boards between them. Nail the furring with 1½-inch masonry nails and bond the insulation board with a recommended adhesive. Not all adhesives can be used with insulation board (**14-74**). The insulation sheets should fit very tight against the wood strips. If you hit an area where it is difficult to fit it in place, buy a can of foam insulation and carefully spray the area or crack full (**14-75**). It comes out like shaving cream, and expands a great deal, so only a small amount is needed for most gaps. Experiment to get the feel of this product.

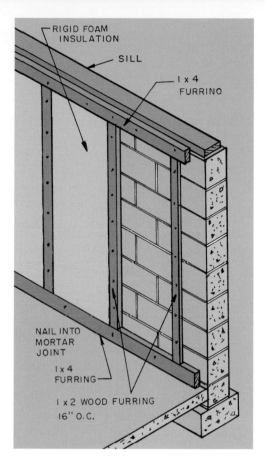

14-74 Furring strips can be nailed to the masonry wall and panels of rigid foam insulation are placed between them.

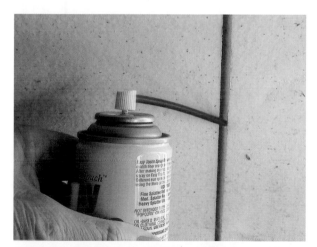

14-75 Canned foam insulation can be sprayed into openings to seal them, making the insulation completely cover the area.

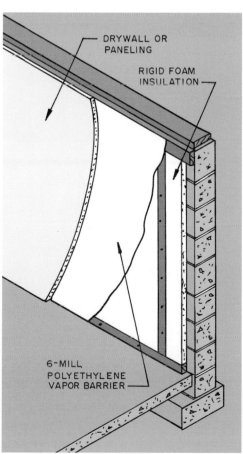

14-76 Cover the insulated wall with a polyethylene vapor barrier and the finished wall material.

14-77 The basement foundation wall can be finished by installing a 2 x 4 stud wall over it. This allows for 3½ inches of insulation for electrical outlets and switches.

INSULATION STAPLED TO THE SIDE OF THE STUD

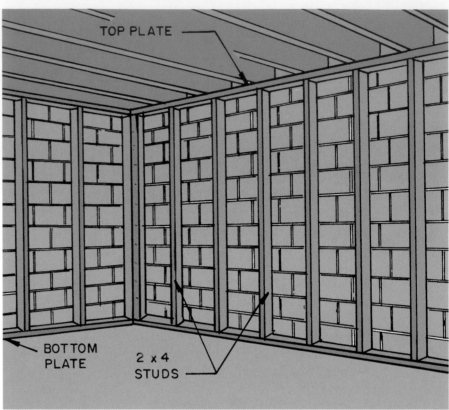

TOP PLATE

BOTTOM PLATE

2 x 4 STUDS

14-78 Staple the fiberglass insulation blankets to the inside edge of the studs.

Cover the furring and insulation with a 6-mil polyethylene sheet vapor barrier and then nail the gypsum wallboard or paneling to the furring strips (14-76).

An even better way, after the wall has been sealed with a waterproof masonry paint, is to build a wall using 2 x 4 studs in the same manner as was used for interior partitions (14-77). Space the studs 16 or 24 inches on center. Nail the top plate to the overhead

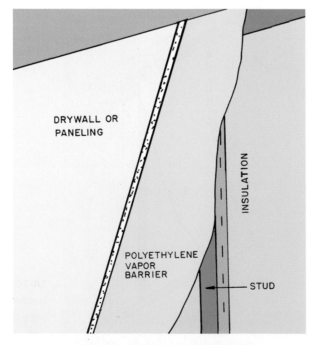

DRYWALL OR PANELING

INSULATION

POLYETHYLENE VAPOR BARRIER

STUD

14-79 Cover the fiberglass insulation with a polyethylene vapor barrier, and then the finish wall material.

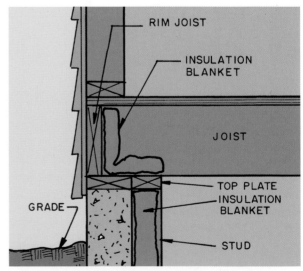

RIM JOIST

INSULATION BLANKET

JOIST

TOP PLATE INSULATION BLANKET

GRADE

STUD

14-80 Insulate the rim joist before you install the ceiling material.

joists and the bottom plate to the concrete floor. Then install fiberglass batts between the studs (14-78) and cover this with a 6-mil polyethylene sheet vapor barrier. Next install the drywall or paneling over it (14-79). Since the wall has a 3½-inch depth, electrical outlets can be installed in it.

In all cases be certain to insulate the rim joist before the ceiling is installed (14-80).

Finishing Basement Floors

As mentioned earlier in the chapter, basement floors in older homes were typically sloped rather steeply toward a floor drain. New homes generally have the floor level and often omit the floor drain. This is due to the use of very effective exterior waterproofing coatings on the foundation.

The floor with quite a bit of slope will have to be leveled. One way to do this is to plug the floor drain with concrete and lay a concrete topping over the area, bringing it up to level with the surrounding area.

Most types of finished flooring material can be installed over a concrete floor; the most commonly used are resilient floor coverings: ceramic tile, laminate floating flooring, and solid wood.

REPAIRING CRACKS
If the slab is sound but there are a few cracks, enlarge and undercut them with a chisel, and fill with an epoxy patching material. Trowel it flush with the surrounding surface.

LEVELING DEPRESSIONS
Depressions should be brought up to level by troweling a layer of epoxy patching material, working it flush with the surrounding surface. It can be leveled by screeding with a straightedge.

ELIMINATING HIGH AREAS
Any high areas will cause great difficulty as you try to lay the flooring. They can be ground down some with a concrete grinding machine, which can be rented.

REMOVING SEALED SURFACES
In order for the flooring to bond to the concrete floor, there must be a clean, sealer-free surface. Sealed surfaces are slick and new flooring will not bond to them properly. One way to clear this is to rent a floor sander to sand the surface. If this does not do it, rent a machine that will lightly scarify the surface. Remember to wear eye protection and a dust mask.

CHECKING THE FLOOR FOR MOISTURE
All flooring laid over concrete floors requires the slab to remain dry at all times. To check the surface for moisture transmission, tape a two-or three-foot-square piece of plastic over an area. If moisture collects under it after 48 hours, you have a major problem and possibly should not bond any flooring material to it. This process is shown by photos in 14-9 through 14-11.

Installing Floor Covering

If the concrete floor is dry all year and has been smoothed so there are no depressions or high spots, you can bond the resilient floor covering directly to it. Any irregularities in the surface will show because the floor covering will mold to their shape. Other types of flooring can also be used as described below.

RESILIENT SHEET FLOORING
Resilient sheet floor covering can be installed in several ways. Probably the **perimeter adhesive method** would be the easiest. Use the adhesive recommended by the manufacturer. Adhesive used on nonporous surfaces, such as concrete, is usually different from that used on porous surfaces, such as wood.

The procedure is shown in 14-81. Trowel an 8-inch-wide band of adhesive along one wall. Let it set a short time to permit the moisture to dissipate. Lay the flooring back on the adhesive along this wall and roll it with a 100-pound floor covering roller (14-82). Notice that the end of this adhesive was kept 12 inches from the side walls. This helps when you adhere the next edge. Now travel to one corner and down the next wall. Lay the floor covering in it and roll. Continue around the room.

When choosing a trowel select one that has the proper size notch on the edge (14-83). The manufacturer will specify the correct size notch to use. Be careful not to use too much adhesive because it could make it difficult to roll the floor covering flat.

VINYL COMPOSITION TILE
The concrete floor should be prepared in the same way as described for sheet flooring earlier in this chapter. While there are several ways to lay out the tile, the square layout procedure is commonly used. Begin by snapping a chalk line, locating the centerlines of each wall. Be certain they meet at 90 degrees. Check this with a carpenter's framing square or the Pythagorean 3-4-5 technique as shown in 14-84.

Now lay a row of tiles without adhesive from the center to an end and side wall. This will show how

14-81 One plan often used to lay sheet resilient flooring.

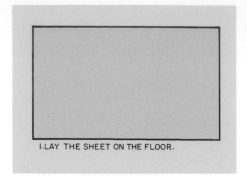

1. LAY THE SHEET ON THE FLOOR.

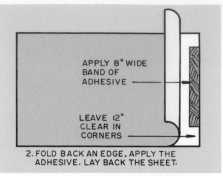

APPLY 8" WIDE BAND OF ADHESIVE

LEAVE 12" CLEAR IN CORNERS

2. FOLD BACK AN EDGE, APPLY THE ADHESIVE. LAY BACK THE SHEET.

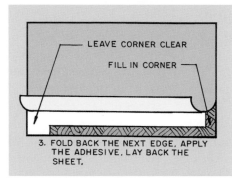

LEAVE CORNER CLEAR

FILL IN CORNER

3. FOLD BACK THE NEXT EDGE, APPLY THE ADHESIVE. LAY BACK THE SHEET.

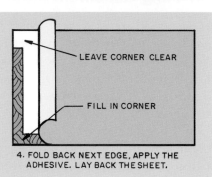

LEAVE CORNER CLEAR

FILL IN CORNER

4. FOLD BACK NEXT EDGE, APPLY THE ADHESIVE. LAY BACK THE SHEET.

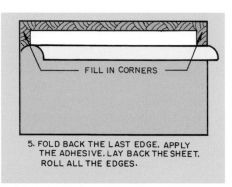

FILL IN CORNERS

5. FOLD BACK THE LAST EDGE. APPLY THE ADHESIVE. LAY BACK THE SHEET. ROLL ALL THE EDGES.

14-82 After the flooring is laid over the adhesive, roll it with a 100-pound floor covering roller.

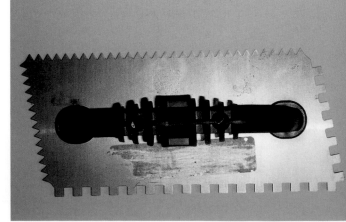

14-83 A trowel used to spread the adhesive has notches in the edges to control the depth of the adhesive.

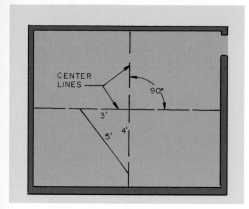

14-84 Locate the centers of the walls and snap chalk lines to locate the center of the room. They must cross at 90 degrees. Use the Pythagorean "3-4-5 technique" or a carpenter's square to check this angle. Adjust the lines until they are perpendicular to each other.

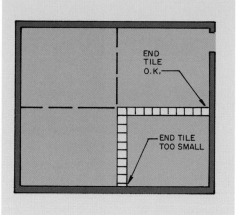

14-85 Make a trial layout of the tiles from the center to each wall. Notice whether the tiles at the wall are less than half a tile wide.

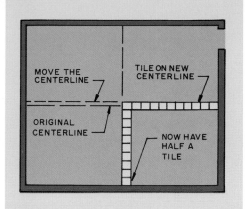

14-86 Move the centerline off center, as necessary, to increase the width of the tile at the wall to at least half a tile.

14-87 Spread the adhesive with a notched trowel.

the tiles will work out at the walls (14-85). If the last tile is less than half a tile, move the centerline until the end tiles are at least equal to half the tile width (14-86).

Lay the adhesive over the first quarter with a notched trowel (14-87). After it has had time to set begin laying the tiles from the center toward each wall. Fill in the first quadrant (14-88). If there are partial tiles at the wall leave them off and install after all four quadrants have been laid. Do not trowel adhesive over this edge area until you are ready to lay the partial tiles.

One way to measure the tiles to be cut along the wall is shown in 14-89. Make certain the grain and pattern of the tile are running the same direction as

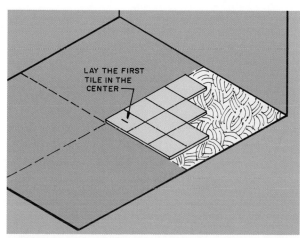

14-88 Lay the adhesive over the first quarter. Start laying the tile in the center and work to the walls.

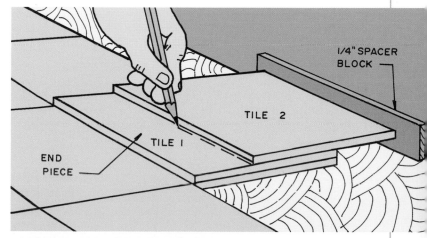

14-89 Mark and cut the tiles to fill the border along the wall. Keep them ⅛-inch from the wall.

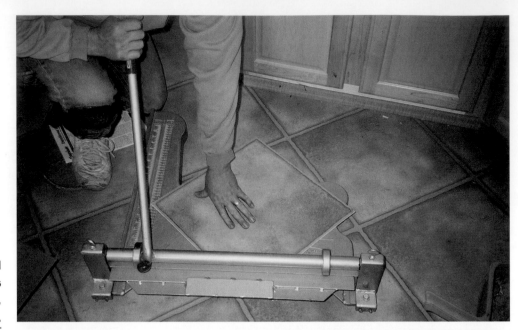

14-90 The vinyl tile cutter makes long, clean, straight cuts.

14-91 Before cutting thicker vinyl composition tiles with a utility knife, heat the back with a propane torch to soften it.

14-92 After heating the vinyl composition tile cut it with a utility knife.

the floor tile. Cut the tiles ¼ inch shorter than the space so they can expand when necessary.

Long, straight cuts in vinyl composition tiles can be best made with a resilient tile cutter (14-90). It makes a sharp, clean, straight cut. Round or notched cuts can be made with a utility knife. Some of the thicker, harder composition tiles must be softened by heating on the back with a propane torch (14-91) and then cut with a utility knife (14-92).

CERAMIC TILE FLOOR COVERING

As with other types of floor covering, the concrete floor must be dry and free of depressions and high spots. As you consider this, go to a tile dealer for help selecting the tile. Generally it will be a glazed ceramic floor tile or a paver tile. Then choose the best adhesive for the existing conditions. These include non-polymer thin-set mortars, latex portland cement thin-set mortars, epoxy adhesives, epoxy mortars, and several mastics. You also have to select a grout to fill the spaces between the tiles. Among these are sanded polymer-modified grouts, unsanded polymer-modified grouts (for spaces ⅛ inch wide), and epoxy-modified grouts (for spaces up to 1 inch wide). The grouts are available in a variety of colors—

14-93 The dark grout makes the joints become the dominant feature of the floor; the lighter color tiles recede into the background.

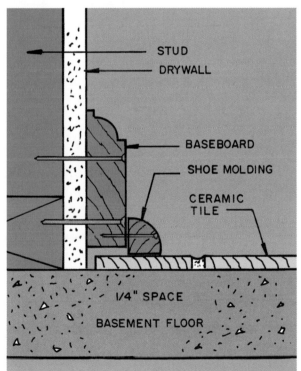

14-94 The floor can be laid to within ¹/₄ inch of the wall, and the space can be covered with a baseboard and shoe molding.

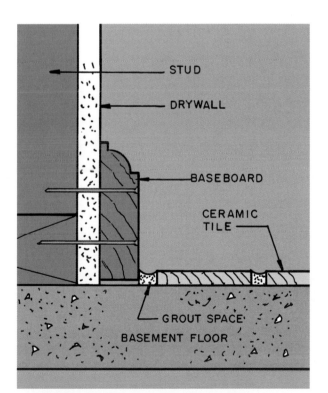

14-95 Some prefer to lay the tile up to the baseboard, leaving a grout space between them and not using a shoe molding.

14-96 One way to make a layout is to establish the centerlines of the room, then lay out the tile without mortar to see how they work out. Adjust as necessary, then run chalk lines locating the perimeter of the field of tile.

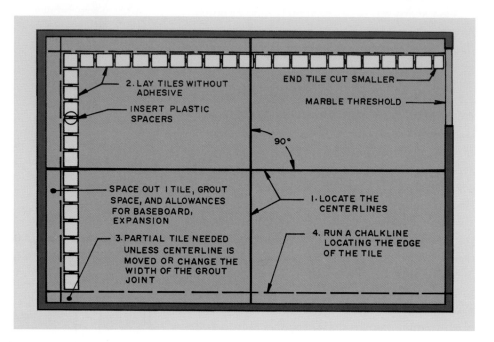

2. LAY TILES WITHOUT ADHESIVE

INSERT PLASTIC SPACERS

END TILE CUT SMALLER

MARBLE THRESHOLD

90°

SPACE OUT 1 TILE, GROUT SPACE, AND ALLOWANCES FOR BASEBOARD, EXPANSION

1. LOCATE THE CENTERLINES

4. RUN A CHALKLINE LOCATING THE EDGE OF THE TILE

3. PARTIAL TILE NEEDED UNLESS CENTERLINE IS MOVED OR CHANGE THE WIDTH OF THE GROUT JOINT

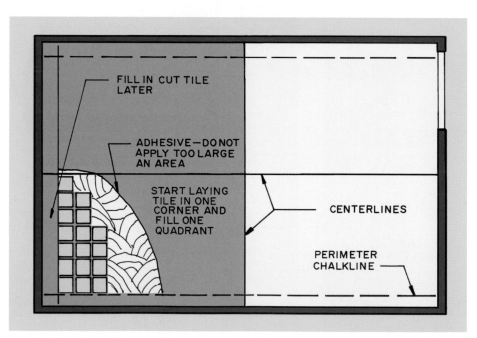

FILL IN CUT TILE LATER

ADHESIVE—DO NOT APPLY TOO LARGE AN AREA

START LAYING TILE IN ONE CORNER AND FILL ONE QUADRANT

CENTERLINES

PERIMETER CHALKLINE

14-97 Many tile setters prefer to start laying in one corner using the perimeter chalk line marks as a guide. Set one quadrant at a time. Do not trowel on more adhesive than you can cover with tile before the adhesive begins to set.

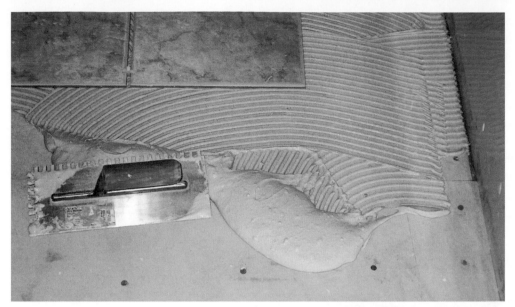

14-98 Spread the adhesive over a small area and lay the tiles as you progress. Use a notched trowel. Different adhesives require different size notches. Check the manufacturer's recommendations on the container.

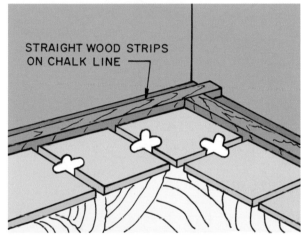

STRAIGHT WOOD STRIPS ON CHALK LINE

14-99 Straight wood strips laid along the chalk line can help line up the first row of tiles.

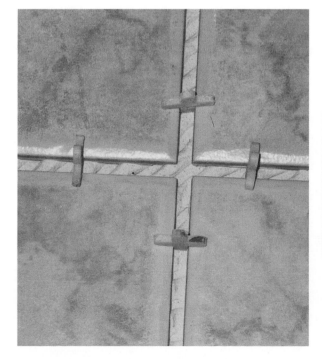

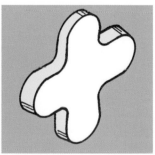

14-100 Plastic spacers are used to space the tiles in both directions.

the color(s) selected greatly influence the final appearance of the room (14-93).

Decide how you plan to finish the floor at the wall. The two commonly used methods are shown in **14-94** and **14-95.**

Begin by laying out the centerlines of the room as shown in **14-96.** Be certain they are at 90 degress. Often the walls are not at right angles, so work the measurements from the centerlines. Then along the two most prominent walls, measure in a distance equal

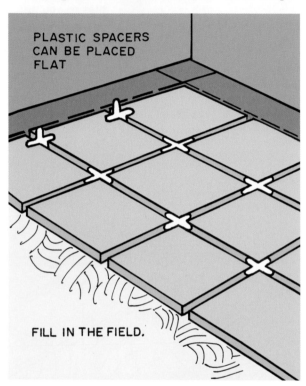

PLASTIC SPACERS CAN BE PLACED FLAT

FILL IN THE FIELD.

to one tile, and one grout space, plus ¼ inch. Snap a chalk line along this measurement. Then lay tiles without mortar along each line to see how they work out. The end tiles on the other wall will generally have to be cut to fit the remaining space. If they are very small pieces, move the centerline a little and cut the tiles on both walls.

Again check to be certain the layout lines along the wall are perpendicular to the centerlines. As you space out the tiles without mortar you have to decide on the width of the mortar joint. Plastic spacers are available and it is a good idea to use them to establish the grout space.

Once the trial layout is set, many tile layers prefer to start in a corner (**14-97**). Trowel the adhesive over several square feet. Use a trowel like the one shown earlier in **14-83** that has the correct size notch. Spread the adhesive so there are no voids or lumps (**14-98**).

Long, straight wood boards can be laid along the edge next to the wall to help line up the first row (**14-99**). All the other rows depend upon getting this first row straight. Lay several tiles along the layout

14-101 After the tiles are laid along the layout line by the wall, fill in the field tiles in that area.

14-102 After you have laid a section of ceramic tiles, seat them into the adhesive by tapping on a bedding block.

14-103 The manually operated tile-snap cutter holds the tile in place as it is scored by a cutting wheel on the handle of the cutter. The tile is broken along this scored line by pulling on the handle.

line, inserting **plastic spacers** to establish the width of the grout space (14-100). Then lay down more tiles, filling the field (14-101). Continue laying the tiles in small sections across the end of the room. Then tap them lightly with a **bedding block,** which can be made with a piece of 2 x 4 lumber covered with carpet (14-102). This gently sets the tiles into the adhesive and helps to level them. Remove any adhesive that may have filled the grout spaces. The grout space should be open the full thickness of the tile. Check the surface for levelness with a long carpenter's level. Now lay another series of tiles across the room and set. Continue until the room is covered.

Do not walk or kneel on freshly laid tile. If something happens to make this necessary, lay a large piece of plywood over the tile and kneel on it.

After the adhesive has set the length of time specified on the container, you can cut and install the tiles around the perimeter. Straight cuts can be made with a tile-snap cutter (14-103). Irregular cuts can be made with a nipper (14-104 and 14-105). If you have a lot of tile to cut, consider renting a power tile-cutting saw (14-106). Be certain to wear eye protection, or better still, a full face shield.

When the entire floor has cured it is time to grout the spaces between the tiles. First remove the plastic spacers. Even though the adhesive has set, work off a sheet of plywood.

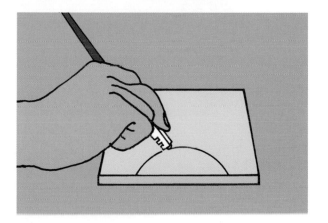

14-104 To make irregular cuts, score the line of cut on the tile by cutting through the glaze with a glass cutter.

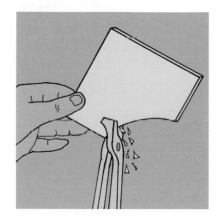

14-105 After the curve is scored, begin breaking out small pieces with a nipper. Work carefully and do not try to take big bites.

Prepare the grout as directed. It may be in dry form and require water to be added and thoroughly mixed or be premixed. Pour some on the face of the tile and spread it into the joints with a rubber float. Wipe off the excess with the edge of the float (14-107). Wipe on an angle across the spaces so you do not remove the grout from them. Pack the grout tightly into the spaces. Wipe off the excess grout with a squeegee (14-108).

After the grout sets about 15 to 30 minutes, wipe the tile surface clean with a soft cloth or sponge (14-109). Finally, tool the joints so the grout is smooth and slightly rounded. Tile setters use a tool called a **jointer**. The back side of a plastic spoon works pretty well (14-110). Block off the room and let the grout cure the required time.

ENGINEERED WOOD, LAMINATE & SOLID-WOOD FLOORING

These products can be installed over the concrete basement floor. The concrete must be dry and stay

14-106 This power tile-cutting saw uses a diamond blade. Since tile is hard and brittle, work carefully and do not let the tile twist while cutting. Wear eye protection, or better still, a full face shield.

14-107 Spread the grout with a rubber float.

14-108 Some tile setters use a squeegee to finish wiping away the excess grout.

14-109 After the grout has set for about 30 minutes, carefully wipe the surface of the tile clean. A soft sponge does a good job. Rinse it frequently.

dry all the time. There are a number of ways these can be installed. This is explained in detail in Chapter 16.

Additional Information

For additional detailed information consult the following Sterling Publishing Co., Inc. Publications by William P. Spence: *Carpentry & Building Construction, Installing & Finishing Flooring, Residential Framing, Finish Carpentry, Encyclopedia of Home Maintenance & Repair,* and *Installing & Finishing Drywall.*

14-110 After the tile has been cleaned, go over each grout joint with a rounded, smooth object. A tile jointer is commonly used but a plastic spoon can be used as a substitute.

CHAPTER 15

Renovating Old Floor Structures

The floors are a major feature when you enter a house. A well-maintained, good-quality finished floor sets the tone for the room. When you enter a house the floor is one of the first things you notice (15-1); however, in older homes this is one area that often needs restoration. An important part of the renovation of a house is careful examination of the **floor structure**. There are many things that could have happened to the floor structure that would require some minor repairs or even a major one. Perhaps the first thing noticed is a bit of sagging, which quickly alerts you to the need for careful examination.

Determining Whether Floors Are Sagging

A sagging floor may cause doors to stick and cause distortion in the finished flooring. For example, wood flooring may buckle and in some places the joints open. Check for levelness by running a chalk line across the room in several places in both directions (15-2). The sag may be caused by some defective joists or the supporting girders may have some sag. The girder can be checked as shown in 15-3.

Before correcting the sag it is often necessary to correct the feature that caused it. For example, if the crawl space or basement is damp, the years of high humidity can cause the joists to decay. To check for decay punch the joists with an ice pick or carpenter's awl. If it penetrates the joist it must be replaced. The damage is often referred to as **dry rot**, which is a

15-1 The beautiful southern pine flooring sets the tone for the room. *Courtesy Southern Pines Council*

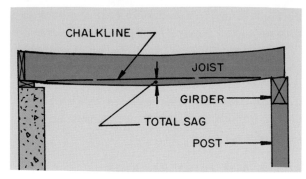

15-2 Use a chalk line to check to see if the floor joists have developed a sag.

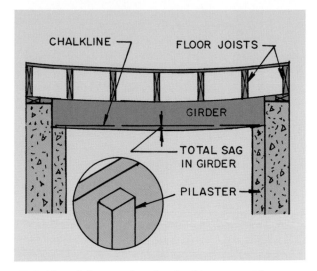

15-3 The girder can be checked for sag with a chalk line.

fungus that is nourished by long-term exposure to moisture in areas with little or no ventilation.

To control moisture in the basement, you will need to have the foundation walls waterproofed to control leakage and the area permanently ventilated. Ventilating fans may be necessary. A crawl space needs the foundation to be waterproofed to stop leakage, and the ground must be covered with 6-mil polyethylene plastic sheet material. Glue the sheets to the foundation and overlap the glued sheets within the crawl space. Foundation vents are typically built when the house is constructed (15-4). In periods when the outside air is very dry, open the vents. When the air is humid, close them. In some areas they can remain closed all the time if the area under the floor is carefully protected as just mentioned.

If the joists appear rotted and you punch holes in them with an ice pick or awl and notice little bugs inside, you have termites. These joists will have to be replaced and the area treated for termites. Do not try to treat it yourself. Only a registered termite company can do the job.

Correcting Sagging by Jacking Floors & Girders

Sagging and deteriorated joists and beams must be straightened or replaced to get the floor flat and level. This requires jacking them up to make the necessary repair. Since jacking the floor affects everything above it, the process must be handled carefully so you do not do damage to other parts of the house. It can cause the drywall to crack, doors and windows get out of plumb and jamb, and even may cause roof problems. It is wise to seek professional advice and help before attempting to raise the house even a little.

While there are a number of ways to do this, typically the house jacks are supported on large wood

15-4 This is an inside view of a foundation vent with the metal-hinged door closed. A more airtight and insulated cover can be had by cutting rigid plastic insulation pieces to fit tightly inside the masonry opening.

members either on the basement floor or in the crawl space. Wood posts run from the jacks to a heavy girder overhead which will raise the joists (15-5). After you get the members in position, tighten the jacks until the girder is firmly against the joists. Wait 12 to 24 hours and then raise the floor a little. Wait again and raise the floor a little more. Continue until it reaches the desired result.

After repairs are made, lower it slowly. Lower it a little each day until the weight is fully on the new girder and posts or piers.

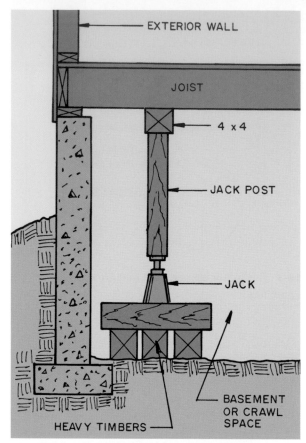

15-5 A typical system for jacking a series of sagging joists.

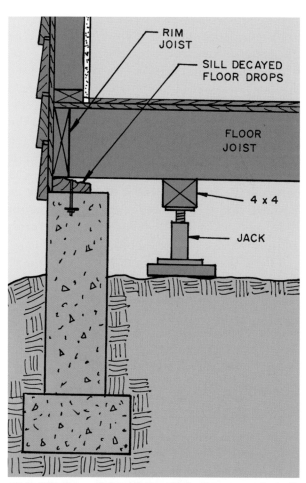

Before making the repair, check the local building code on what it requires for a permanent support of the new girder. Some permit the use of adjustable metal jack posts, while others require the use of a **lally column**. A lally column is a round steel column that is bolted to the floor and the girder (15-6). It is a permanent installation. The jack post can be unscrewed and removed.

SILL & RIM JOIST REPAIRS

Decay may also be in the **sill** bolted to the top of the foundation (**15-7**). Since sills are on top of the foundation they are vulnerable to decay. They must be replaced with pressure-treated lumber. This requires that the joists in the area to be removed must be raised

15-6 Lally columns are made for holding wood and steel girders. They are installed permanently.

15-7 Lift the weight off the sill in the area where it is to be removed. Cut it up and break it out.

enough to remove the rotted sill. Use enough jacks so the lift is spread over the entire length.

The sill is most likely bolted to the foundation. It may have to be split with a chisel to free it. If you raise the floor high enough to lift it off the bolts you could damage the exterior wall, doors, and windows. The new pressure-treated sill will have to be notched to slide up against the bolts. Most likely you will not be able to loosen or tighten the nuts.

Then there is the possibility that the **rim joist** is also decayed or infected with termites. This can indicate the termites are up in the wall studs and a major rebuilding must occur. If the infestation or decay is limited to the sill and rim joist and the house has wood, vinyl, or some other siding material that can be removed, you will have to support the floor and ceiling joists in the area where the rim joist is to be removed, as shown in **15-8.** Use several supports so the weight is spread out over a long distance. Remove the siding, cut back the sheathing, and pry off the rim joist. In some cases it helps if you cut the nails holding it to the floor joists. Be careful not to split the ends of the floor joists. The rim joist and sill can be cut into shorter pieces, making removal a bit easier. Before replacing them clean the foundation, treat for termites as a precaution, and brush on a wood preservative on the ends of the joists. Then install the new rim joist.

In some very old houses the joists were set directly upon the foundation **(15-9).** This contact will cause the bottom to decay, and the floor and actually the entire exterior wall to sag. This can affect the doors and windows in the exterior wall.

If the damage is limited it is possible to make a repair by jacking up the joists until the sag has been removed, placing an aluminum or copper flashing sheet over the top of the foundation; brush a coating of wood preservative over the damaged joist and then nail a six-foot length of pressure-treated stock the same size as the old joist to the side of the old joist **(15-10).** If the ends were infected with termites, remove the infected wood and treat the remaining material to prevent reinfestation.

If the damage due to rot or termites is more than a small deteriorated area, it is best to replace the entire joist.

If it is necessary to replace a number of floor joists, often a new straight joist can be installed alongside the warped or defective joist. Raise the floor in the area to be repaired until it is level and install the new joists **(15-11).** Be certain the new joists are at the proper level at the foundation and girder. Be certain

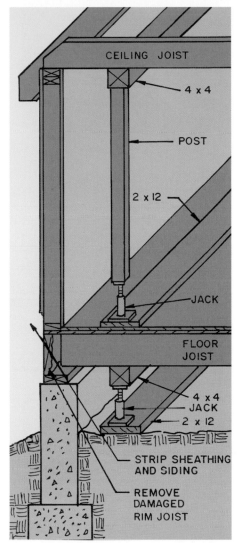

15-8 Support the floor and exterior wall loads in the area where the rim joist must be removed. Remove the siding, cut and remove the sheathing, and pry off the damaged section of rim joist.

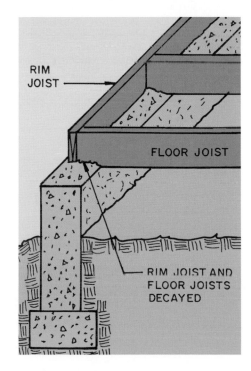

15-9 If floor joists have been installed directly on the foundation they will, over time, decay.

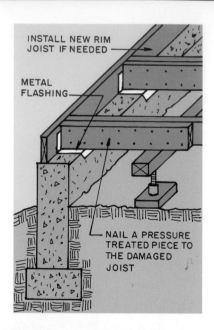

15-10 Joists that have decayed on the foundation can be repaired by jacking the floor level and nailing pressure-treated sections of joists on each end.

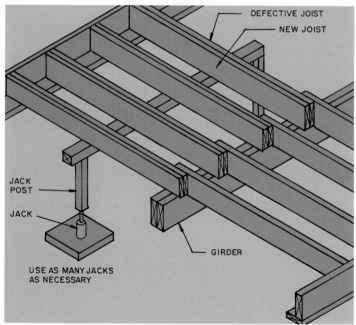

15-11 Joists that need replacing can often be left in place and a new joist installed alongside them.

the sill is in good condition and the girder is straight and sound. This is not an easy project and will require several people to help. It will require the sagging floor to be jacked level in the area where the replacement joists are to be installed.

Check the new joist to see if it has a slight crown (15-12). If it does, place the crown up. If the new joist is warped or has very much crown, discard it.

After the new joists are in place gradually lower the jacks, letting the weight slowly begin to be transferred to the joists.

Minor Sagging or Springy Floors

Often the sill and girder are sound and straight but there is some minor sagging of floor joists in an area of the house. This means over the years the joists, while sound, have deflected under normal use or by an overload such as the addition of a hot tub or a grand piano. The floor can be **leveled** by jacking the joists to the level position and installing a girder which rests on piers or posts. The following procedures can also be used to stiffen floors that are level but **springy.**

IMPROPER JOIST PENETRATIONS

One common cause of sagging solid-wood floor joists is that they are notched or have holes bored through them to permit the passage of pipes and electric wires. If these are in the wrong place or cut too large the joist could sag over time. Notching and boring holes in joists should be avoided whenever possible.

Notching and holes in I-joists, laminated veneer, parallel strand, or other manufactured beams and joists should be located and sized according to the instructions of the manufacturer. Recommended notch and hole sizes and locations are shown in **15-13.**

Notches can be only in the end third of the joist. They should not exceed one sixth of the depth of the joist and never be in the center third. They should

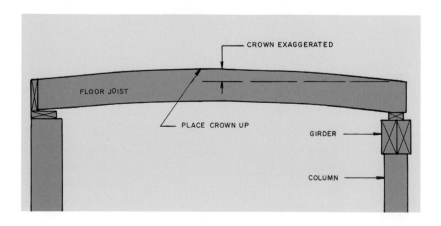

15-12 If the new joist to be installed has a slight crown, place the crown up. If the crown is excessive, do not use the joist.

never be longer than one-third the joist depth. It is recommended that square-cut notches not be used. Instead cut the sides on a 45-degree angle. Joists notched on the very end should never be over one-fourth the depth of the joist and should be cut on a 45-degree angle.

Holes bored in joists should not be within two inches of the top or bottom. The diameter of the hole should not exceed one-third of the depth of the joist. Square holes are not recommended.

If it appears that notching or holes in the joist contribute to its sagging, glue and nail two-inch stock on both sides of the notched section **(15-14).** This occurs in floors over crawl spaces and floors over basements.

FLOORS OVER CRAWL SPACES

If the floor is over a crawl space, generally **concrete block piers** are used to support a girder below the sagging joists (15-15). Typically 8 x 16 concrete blocks make an adequate pier for this type of support. For heavier loads 12 x 16 or 16 x 16 piers will be used (15-16).

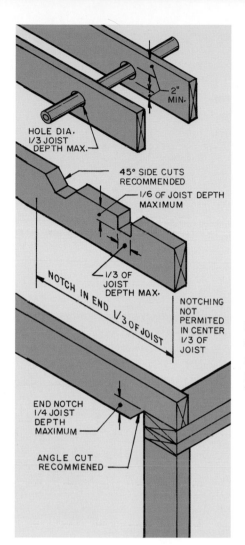

15-13 Recommended allowances for notching or boring holes in joists.

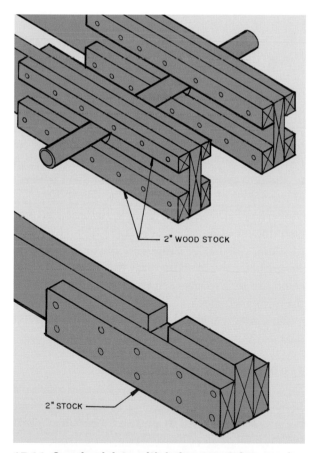

15-14 Sagging joists with holes or notches can be strengthened by adding 2-inch wood strips on both sides.

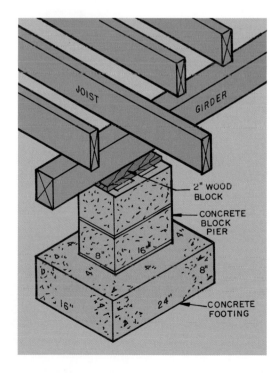

15-15 A typical concrete block pier supporting a girder.

333

15-16 Frequently used sizes of concrete block piers and footings.

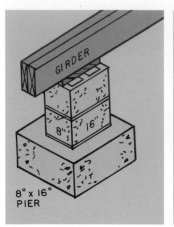
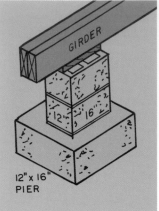
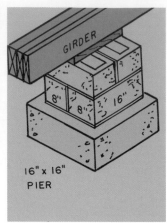

15-17 Wood blocking is placed on top of the pier.

15-18 Fill the core of the top concrete block with concrete. Even better, fill all the cores to the footing.

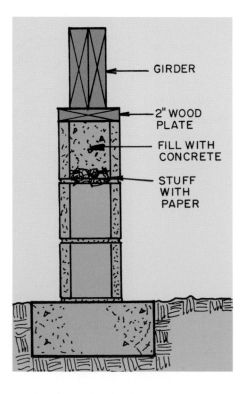

Pour a concrete footing for each pier as shown in **15-15** and **15-16**. The footing is usually eight inches thick and extends four inches beyond each side of the pier. Typically a 4 x 6 girder can be used with piers spaced every four to six feet. If the girder is larger the piers can be spaced farther apart. If a heavy concentrated load exists on the floor, the girder and pier sizes will need to be calculated.

Jack up the sagging joists until they are level or just a little bit higher than needed as shown earlier in **15-7**. Lay up the concrete block pier and let the mortar harden until it has reached bearing strength. The number of days of curing time will vary depending upon the type of mortar used. After the mortar has cured use solid-wood blocks to shim the joists to the pier as shown in **15-17**. You may want to jack up the joist a little higher than needed so the blocking can be slid into place. It is recommended that the cores of the top block be filled with concrete. To do this, stuff crumpled newspaper down in the core and pour eight inches of concrete over it **(15-18)**. The wood blocking gets better support if this is done. Some simply place the blocking over the open cores. You could also lay a solid 4-inch-thick concrete block over the top 8-inch block.

Now slowly lower the jacks, placing the joists on the girder.

FLOORS OVER BASEMENTS

Sagging floors over basements are handled much the same way as just described for crawl spaces except the girder is supported by posts. One problem is that the addition of more posts produces obstructions to the use of the open space in the basement.

Begin by jacking the sagging joists level. Then install the new girder and **jack posts.** Jack posts are

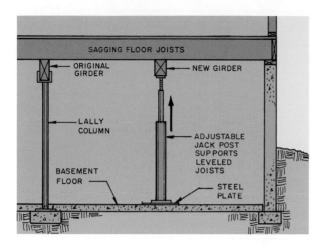

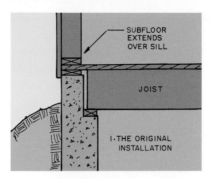

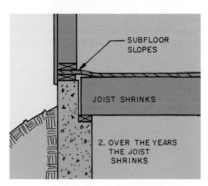

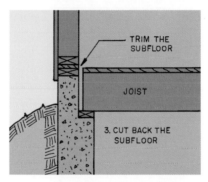

15-19 After the floor joists are jacked straight, they are held in position with a girder and several jack posts. The jack post is screwed against the girder after the joists have been jacked straight. If the load is heavy a steel plate is placed on the floor below each jack post.

metal posts that can be adjusted after the top bearing plate is raised near the girder; the top section screws up, pressing the bearing plate against the girder (15-19). The bottom plate should rest on a ⅜-inch-thick steel plate about two feet square. This spreads the load over the concrete basement floor so it will not crack. If the installation is used to just reduce bounce in the floor, the post bottom plate can rest directly on the concrete floor.

The jack post can be used to reduce the amount of spring in a springy floor, or if the amount of lift needed is very light; the use of heavy jacks is not necessary. If the lift area is large or there is considerable bow, the joists must be lifted with a hydraulic jack. Then the jack posts are installed to hold them straight.

Evaluating Floor Slope along Exterior Walls

Another problem that sometimes is noticed is that the floor along an outside wall develops a slope. This may occur if the house was framed at the foundation, allowing the subfloor to extend over the sill. The joists are set on a foundation ledge which has been set so the subfloor is level with the top of the sill (15-20). Over time the joists may shrink, causing the subfloor to slope down from the sill. Correct this by cutting the subfloor along the wall to remove the slope and eliminate any chance of additional slope developing.

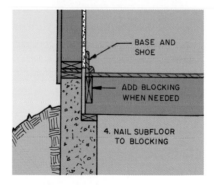

15-20 Sloping floors along the exterior wall can be corrected by cutting the subfloor back so it does not extend over the sill.

335

The gap at the floor is covered by the baseboard and shoe molding.

If the floor was framed with blocking between the joists, nail the subfloor to it. If it was not, install 2 x 4 blocking so the edge of the subfloor can be nailed securely.

If the Problem Is Sagging Girders

If the girder supporting the floor joists has developed a sag, it can sometimes be jacked straight, which will also pull a slope out of the floor. If the amount of sag is small it can be jacked straight and held there by installing one or more posts. If the amount of sag is great, usually it is not possible to jack it straight. Remove the girder. The jacking and supporting process in basements is the same as shown in **15-19**. If in a crawl space, a footing is poured, and a jack post or pier needs to be built as is shown earlier in **15-15**. This is a big job and requires careful planning so the entire floor area is fully supported.

If the house is old it will likely have wood posts supporting the girder. If these appear to be deteriorated they must be strengthened or replaced. Since the wood posts rest directly on the concrete footing, moisture in that area will cause them to deteriorate at their base. The top and main part of the post will usually be sound. You can replace the post with an adjustable jack post as just described or fasten new wood posts to it as shown in **15-21**. If it is a 4 x 4 post, screw a new one to the old one. If it is a 6 x 6 post, secure one that size. Use pressure treated posts and cap the end on the footing with aluminum or copper. Posts could have 4 x 4 posts secured on each side. This gives a balanced support.

Consider supporting the girder temporarily, by removing the old post and replacing it with a lally column or an adjustable jack post.

Repairing Squeaky Floors

Probably the most common cause for floor squeaks is that the nails holding the subfloor work loose and the subfloor slides up and down, rubbing on the nail. If the subfloor had been glued to the joist and then nailed with ring shank nails, or better still, secured with wood screws, this problem is not likely to occur. Another cause is when the subflooring or wood flooring loosens up a little and the pieces rub against each other. To correct the problem it is necessary to tighten the subfloor to the joists and secure wood flooring to the subfloor.

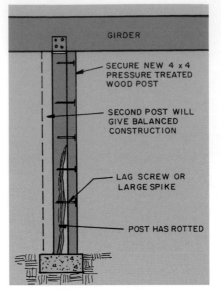

15-21
Rotting wood posts can be strengthened by securing pressure-treated posts to one or both sides.

Repairing Loose Subfloor

When working from below, the squeak can be located by having someone walk as you listen. Then mark the location of each squeak.

If a couple of joists have bowed, leaving a gap between them and the subfloor, you can glue and nail 2 x 4 blocks to the joist and glue the top edge so it bonds to the subfloor as shown in **15-22**. You can place a piece of wood against the block and the floor and drive it tight against the block to set it firmly against the subfloor before nailing it in place. If the joist is straight but there is a squeak it can be silenced by carefully opening the joint between the subfloor and joist with a screwdriver and filling it with polyurethane caulk or carpenter's cement **(15-23)**. Work the caulk or glue into the opening with a thin stick or piece of metal. Wear gloves as you do this to keep these materials off your hands.

Another procedure after you locate the exact place where it squeaks is to drive very thin wood shims between the joist and subfloor. Coat them with glue **(15-24)**. Again, do not use thick shims or you will cause a bulge in the floor.

Another repair that can be made if you can get under the floor is to glue and nail or screw 2 x 2 cleats along each side of the joist in the area of the squeak **(15-25)**. It helps if a weight, such as someone standing on the floor, is used to push the floor tight to the joist as the cleats are installed. The cleats will stop the up-and-down movement of the subfloor.

If the floor does not have bridging or it is loose or not installed in the area of the squeak, sometimes

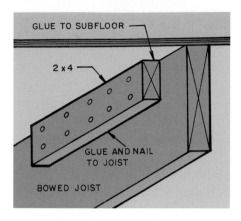

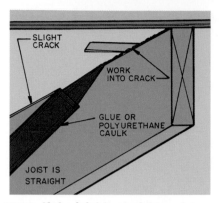

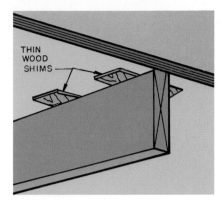

15-22 If you have one or two joists that have become bowed, stabilize the subfloor by adding blocking alongside the joist.

15-23 If the joist is straight and a very small crack develops between it and the subfloor, work a carpenter's glue or polyurethane caulk into the crack.

15-24 Thin wood shims can be covered with glue and driven into the crack between the joist and subfloor. Be careful they do not cause a bulging in the floor.

15-25 Wood cleats can be used to pull the subfloor and joist together. Glue and screw them in place.

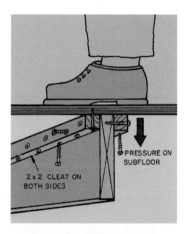

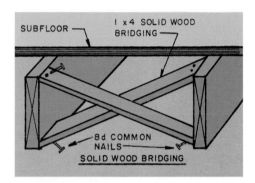

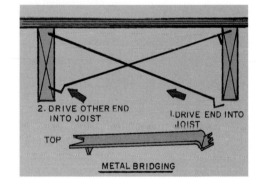

15-26 Metal and wood cross bridging helps stabilize the floor joists.

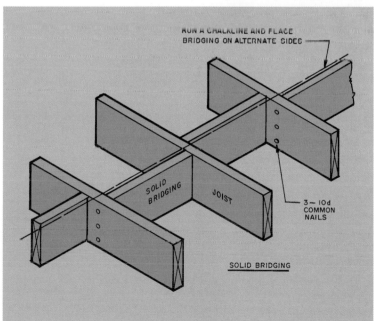

15-27 Bridging can be made using solid 2-inch-thick members nailed between the joists.

installing it will help reduce movement. Cross bridging can be manufactured steel members or cut from 1 x 4 wood (15-26). Solid bridging can also be used (15-27). If the bridging touches as it crosses, this could squeak. Make a saw cut between the pieces to separate them, or else they may cause a squeak.

Sometimes a wood floor will buckle. It can be pulled flat by driving wood screws through the subfloor into the flooring (15-28). Be certain the screws do not penetrate the flooring. Place a weight on the buckled area as you drive the screws.

If you cannot get under the house you can make a fairly good repair by nailing through the floor into the joist. The big problem is finding the joist. If you are nailing through carpet you can walk on the squeak and by the sound get close to the joist. Then drive a finish nail through the carpet. If it punches through the subfloor but misses the joist, remove it and move over and try again until you hit it (15-29). Even though the finish nail is slender, work carefully so you do not damage the carpet. Open the pile so the nail enters the backing and goes through the pad. On some carpets the pile is very fine and sometimes this

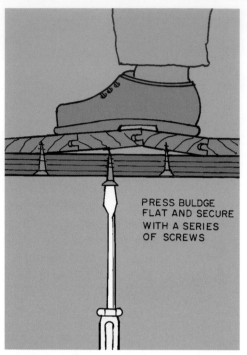

PRESS BULDGE FLAT AND SECURE WITH A SERIES OF SCREWS

15-28 If you have a buckle or a squeak in wood flooring, you can pull the flooring tight against the subfloor with a series of wood screws.

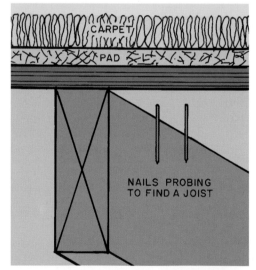

NAILS PROBING TO FIND A JOIST

15-29 Open the pile and carefully drive a finishing nail to find the joist. Several tries will usually be necessary.

15-30 Once a joist is found, drive the nail into it until the head is just above the pile. Typically, several nails along the joist will be necessary.

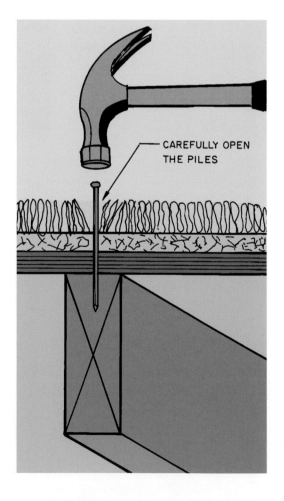

CAREFULLY OPEN THE PILES

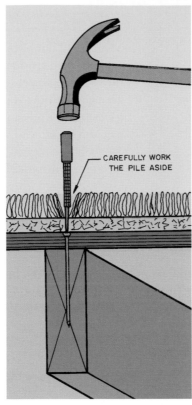

15-31 Finish driving the nail with a slender, fine-tip nail set. Work carefully so you do not damage the pile.

CAREFULLY WORK THE PILE ASIDE

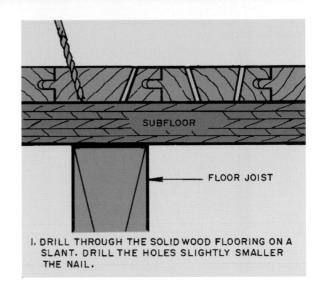

SUBFLOOR

FLOOR JOIST

1. DRILL THROUGH THE SOLID WOOD FLOORING ON A SLANT. DRILL THE HOLES SLIGHTLY SMALLER THE NAIL.

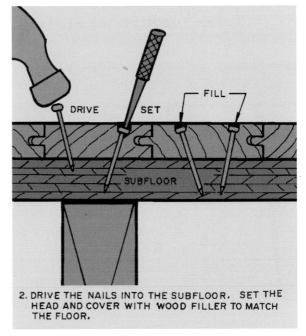

DRIVE   SET   FILL

SUBFLOOR

2. DRIVE THE NAILS INTO THE SUBFLOOR. SET THE HEAD AND COVER WITH WOOD FILLER TO MATCH THE FLOOR.

15-32 Squeaks can be silenced in wood floors by nailing through the flooring into the subfloor. Use galvanized finishing nails.

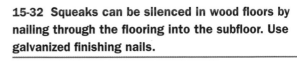

COUNTERBORE

SHANK

PILOT

COUNTERBORE TOOL

COUNTERBORE

SHANK HOLE   WOOD FLOOR

PILOT HOLE

SUBFLOOR

1. DRILL COUNTERBORE, SHANK AND PILOT HOLE.

2. INSTALL THE SCREW.

WOOD PLUG

3. GLUE IN THE PLUG AND SAND FLUSH.

15-33 Wood screws set through the squeaky wood floor into the subfloor are covered with wood plugs.

15-34 These wood plugs are of the same material as the floor. Typical counterbored holes are shown.

15-35 Wood plugs can be cut from a different species of wood and provide a decorative appearance.

15-36 The plug cutter will bore into a piece of wood, producing a plug.

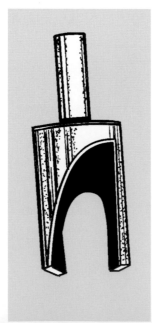

technique leaves a small blemish. Make a sample test in a closet or other area where a small blemish, if it occurs, will not be noticeable.

Once you find the joist, drive the finish nail through the carpet into the joist (15-30). When the head gets near the surface of the carpet, finish driving it through the pile and pad with a slender, fine-point nail set (15-31). Wood floors can be secured from above by nailing through them and the subfloor into the joist (15-32). Use galvanized finishing nails. Set the nails and fill with a wood filler that matches the color of the floor. Another technique is to drive flathead wood screws through the floor into the joist (15-33). The hole for the screw head can be counterbored, and a wood plug cut from the same material as the floor can be glued in place (15-34). Carefully sand and finish. Plugs of a different species of wood can be used to get a decorative appearance (15-35). The plugs are cut from scrap flooring with a **plug cutter** (15-36).

Reducing Sound Transmission

The transmission of sound through floors can be annoying. While you are renovating the floor structure and installing new finish flooring is a good time to reduce sound transmission. Floor assemblies are rated by their ability to reduce sound transmission. This reduction is measured in decibels. Floor-ceiling assemblies are rated by their **sound-transmission class** (STC) and **impact insulation class** (IIC). STC is a single-number rating that indicates the effectiveness of a material or an assembly of materials to reduce the transmission of airborne sound through the unit. The larger the STC number the more effective the material is as a sound transmission barrier.

IIC is also one number giving an approximate measure of the effectiveness of floor construction to provide insulation against the sound transmission from impacts such as walking or dropping things on the floor.

Floor-ceiling assemblies separating a living space from another living space should have an STC of 50 and an IIC of 50. Several floor-ceiling assemblies and their ratings are shown in **15-37.** An easy way to improve the STC and IIC is to use sound-deadening board over the subfloor. Sound-deadening board is a lightweight panel that is strong and impact resistant. It can be used under carpet, wood, or laminate flooring. Install as recommended by the manufacturer.

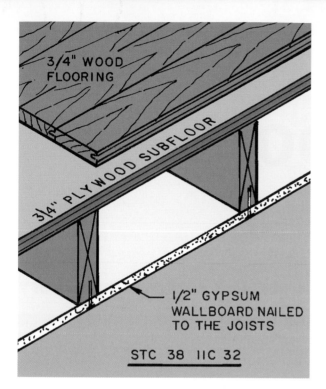

3/4" WOOD FLOORING

3/4" PLYWOOD SUBFLOOR

1/2" GYPSUM WALLBOARD NAILED TO THE JOISTS

STC 38 IIC 32

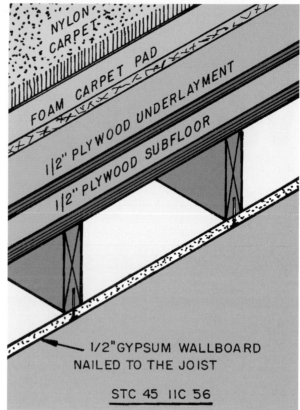

NYLON CARPET

FOAM CARPET PAD

1/2" PLYWOOD UNDERLAYMENT

1/2" PLYWOOD SUBFLOOR

1/2" GYPSUM WALLBOARD NAILED TO THE JOIST

STC 45 IIC 56

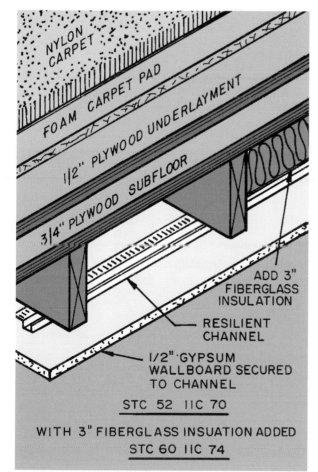

NYLON CARPET

FOAM CARPET PAD

1/2" PLYWOOD UNDERLAYMENT

3/4" PLYWOOD SUBFLOOR

ADD 3" FIBERGLASS INSULATION

RESILIENT CHANNEL

1/2" GYPSUM WALLBOARD SECURED TO CHANNEL

STC 52 IIC 70

WITH 3" FIBERGLASS INSUATION ADDED
STC 60 IIC 74

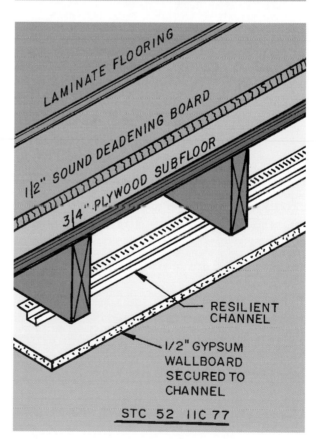

LAMINATE FLOORING

1/2" SOUND DEADENING BOARD

3/4" PLYWOOD SUBFLOOR

RESILIENT CHANNEL

1/2" GYPSUM WALLBOARD SECURED TO CHANNEL

STC 52 IIC 77

15-37 Several typical floor-ceiling assemblies are shown with their sound transmission class (STC) and impact insulation class IIC ratings. Actual ratings will depend on the brand of materials used and the method of assembly.

CHAPTER 16

Wood Floors

If you are remodeling a very old house, the floors may be in such a condition that you are not certain of the species of wood or the type of finish. Often years of use have darkened them and they have become scratched, nicked, checked, or otherwise damaged. It is likely they may have been refinished

16-1 This Southern pine floor has been stained a light honey color. *Courtesy Southern Pines Council*

16-2 This maple wood floor is the dominant feature of the room and sets the tone when choices are made for furniture and upholstery. *Courtesy Harris Tarkett, Inc.*

several times and the finishing materials used may not have been compatible. The finish may be soft and uneven. If you finally decide to remove the old finish it is very possible you will find beautiful wide pine flooring which, if it can be salvaged, will produce a beautiful floor that is in keeping with the style, design, and period when the house was built (16-1). Also maple flooring has been used for many years and is very durable and refinishes beautifully (16-2). Other woods such as oak and birch are hard and refinish nicely.

Assessing Old Flooring

As you consider what to do begin by examining the **floor structure.** The joists must be strong and straight. The girders must be sound and level. (Review Chapter 15 for detailed information on checking and restoring deteriorated floor structures.)

As you assess the condition of the old flooring, if it appears that repairing damage and trying to refinish are more difficult than you want to try, consider installing laminated or engineered flooring over it. This is discussed later in this chapter.

It is also possible to tear up the old wood flooring, renail and repair the old subfloor, and install new solid-wood flooring. This produces a floor similar to the original that was part of the house in early years. You could secure the old wood flooring securely to the subfloor and lay a new solid-wood floor over it. This will require some unusual actions such as cutting off the doors, changing the baseboard and shoe molding, and providing a transition piece between this new floor and where it meets some of the old floors that remain unchanged. In each case review the end results and the cost of each procedure. Part of the final solution is to determine what you want and part is to see if it is worth the cost and effort. Working with flooring is hard work.

Maintaining Wood Floors That Are in Good Condition

Wood floors, if properly maintained, will last many years and can be reconditioned. The major action to be taken is to prevent sand and other dirt from entering the house (16-3). These grind into the wax and finish and can, over a few years, wear away the finish, exposing the bare wood at each exterior door. A doormat on the outside (16-4) and a throw rug on the inside significantly reduce the flow of granular material into the house (16-5). This, followed by

16-3 This is one reason doormats and throw rugs at each entrance are important.

16-4 Doormats outside the door encourage visitors and family to wipe their shoes.

16-5 Throw rugs inside of each exterior door are essential to keeping abrasive materials off the floor.

16-6 Frequent vacuuming is essential to keeping the floor free of abrasive particles that damage the floor finish.

frequent cleaning of the mat and rug, as well as frequent vacuuming of the entire room, is important **(16-6)**.

The finish will last longer if you wax it once or twice a year. Apply and polish as directed on the container. Paste waxes give the best protection but are more difficult to apply than a liquid buffing wax. Do not use liquid waxes that have a water base. After applying the wax, buff with a power buffer, which can be rented, and the proper buffing pad is sold with it.

Before waxing and buffing, be certain the floor has all blemishes, such as black shoe marks, tar, wax, and tree resin, removed as explained later in this chapter.

Never wet-mop or wash wood floors with soapy water. Not only is it hard on the finish, it could get between the boards on the edges and over time possibly cause them to warp.

The use of pads on the feet of chairs and larger cups below the legs of heavier furniture will provide signif-

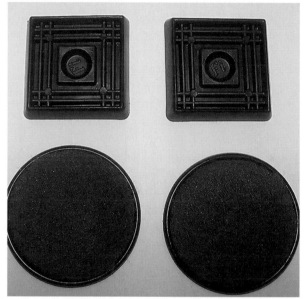

16-7 Furniture leg pads and coasters protect the wood floor while the furniture is in use and when it is being moved.

icant protection (16-7). This is especially helpful with items that are occasionally moved, such as a dining room table or living room sofa.

Minor scratches on floors that have been stained can be hidden somewhat by coloring with a stain marker as shown in **16-8.** You can also wipe into the scratch stain that's used to finish furniture and carefully clean off any on the surface of the flooring (16-9).

It is important to remove all surface marks such as those caused by shoes, tar, tree resin, and other substances tracked into the house. These can be removed by wiping the area with a soft cloth dampened with mineral spirits, turpentine, or kerosene. Some can be removed with a soft pencil eraser. Before doing this, wipe a spot in a closet to see how the floor finish reacts. If this damages it, the damage will not be noticed here (16-10).

The entire floor can be renewed by coating with a top dressing designed especially for maintaining wood flooring. Usually it is applied to the entire floor every 6 to 12 months. The more the room is used every day, the more frequently it should be applied. Various products are recommended for different floor finishes. Apply it to the entire floor so that when it is finished the sheen on the surface is uniform across the room.

Another product is a floor-care kit which contains the cleaners and mop required to remove dirt, food spots, marks as from shoe heels, gum, asphalt, crayon, or other materials on the surface of hardwood floors (16-11).

16-9 Small scratches can be colored by wiping a diluted wood stain over them and then wiping the surface clean.

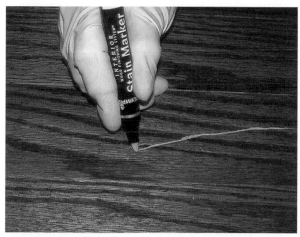

16-8 Small scratches can be colored and concealed with a stain marker. It contains a liquid stain that is applied to the scratch with a felt tip.

16-10 Surface scuffs, tar, wax, and other blemishing material should be removed regularly—and always before waxing the floor.

16-11 Special top dressing materials are available that are spread evenly over the entire floor to renew the finish.

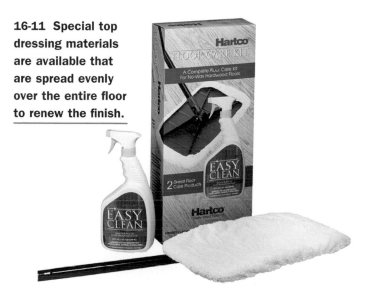

Determining the Finish on the Old Floor

Whether you plan to try to touch up and restore the old finish or sand it off and completely refinish it, you should find out what finish is on it. Typically it will be polyurethane, varnish, shellac, wax, a penetrating oil, or floor paint.

FINISH TESTS

To test to see if the finish is **shellac,** apply a little denatured alcohol on a small area. If the finish dissolves, it is shellac. To test for **varnish,** apply a coating of lacquer thinner or a commercial paint remover. If it **cracks** and raises from the surface it is varnish. To test for **polyurethane,** coat an area with paint remover. If the finish forms **bubbles** it is polyurethane. To test for **penetrating oil,** apply the above materials and if nothing happens, it is an oil finish. To test for a **wax** finish place several drops of water on the surface. If it develops a **white spot,** it is a wax finish.

LIMITATIONS OF VARIOUS FINISHING MATERIAL

Floors finished with a **wax finish** are difficult to refinish with one of the other finishes. The wax penetrates the pores of the wood and goes down into the cracks between the strips. It is best to refinish an old wax floor with a wax refinish. **Penetrating oil** finishes flow into the pores of the wood, and sanding to completely remove them would require removal of a lot of the wood flooring. It is best to refinish the floor with a penetrating oil finish. Both **water-** and **oil-based polyurethane** can be touched up with either type of finish and can also be used to refinish the floor after it has been sanded. **Shellac** and **varnish** can be recoated with the same material after the floor has been sanded. If a polyurethane finish is to be applied, shellac and varnish should be completely sanded off. **Painted** floors can be repainted with a good-quality floor paint after the original paint has been completely sanded away. If a paint dealer has a clear product that could be applied over it, the job would require deep sanding to get the paint out of the pores, and some careful work to clean the cracks. Consider repainting a floor that has already been painted.

Renovating the Existing Floor Finish

One of the first things to do when studying an old finish on a floor is to see if the old finish can be restored without sanding. Sometimes removing the layers of old wax which hold dirt will leave the floor with a clearer, brighter surface. Wax can be removed by wiping the floor with mineral spirits. Since fumes will develop, provide ventilation.

One way to clean up an old floor where the finish is intact but dirty is to gently rub the floor with a combination liquid cleaner and wax on a No. 00 or 000 steel wool pad. Then wipe off the excess material, which will be dirty, with a soft cloth. Let the surface dry as recommended on the cleaner-wax container. If there are dark spots remaining apply a second coat of cleaner-wax. When the overall appearance is reasonably uniform, buff it with a power buffer. You can rent one and buy the proper buffing pad.

Sometimes there is only a small, dark area that needs cleaning. Working carefully, rub it with No. 00 or 000 steel wool dipped in floor wax. Be careful not to cut through the finish. Sometimes a little mineral spirits on the steel pad will help. Some report success cleaning varnished floor areas with a soft cloth dampened with cold tea.

If the floor clears up by cleaning and you do not have the job of sanding and refinishing, it is quite a relief.

WILL IT STAND SANDING?

As you continue to assess the condition of the old floor, examine it to see if it is solid wood or a veneer. Some veneers can be sanded once, but if it is an old floor it may have already been sanded years ago. Try to examine the edge of the veneer flooring by removing the cover on a heat register in the floor or possibly removing a threshold at a door. You might remove a piece of baseboard in an area where perhaps damage will not be noticeable, and check the edge of the flooring.

Solid-wood flooring also has limits on how much it can be sanded. You will generally need at least ⅛ inch clear of the top lip of the groove left after sanding **(16-12).** If you go deeper you begin to get splinters and the flooring must be replaced. Typically, ¾-inch-thick solid-wood flooring can be sanded twice and sometimes three times, if the first sandings were light. Check the original thickness of the flooring because ½ and ⅜-inch thicknesses are used.

One way to try to find out how much material is left is to slip a thin piece of metal between two pieces of flooring until it hits the tongue. Mark the top of the flooring on the metal, remove it, and measure the distance **(16-13).** Do this in several places around the room, especially in areas where the floor was subject to the most use, such as near a door, and in areas serving as walkways across the room.

Removing the Old Finish

If the old wood floor has to be refinished, the old finish coating has to be removed and the exposed wood surface sanded in preparation for receiving the new finish. While it is possible to remove the old finish using a chemical stripping agent, as used when refinishing furniture, generally this is not recommended. The chemical must be completely removed from the surface along with the old finish and this is a big mess on an area as large as the floor of a room. It then has to be neutralized so it does not interfere with the new finish. It is also very difficult to get it out of the cracks between the boards which, if not completely clean, will hinder the curing of the new finish. The best procedure is to sand the floor, removing the old finish, and dress the bare wood.

PREPARING FOR SANDING

Since this is an old floor it can, over the years, have had nails driven into it to stop squeaks, and staples to hold vinyl floor covering. Set the nails below the surface deep enough so the sanding will not touch them. They can be filled after the sanding is finished. Remove old staples or any other items that may be secured to the floor. The sanding pad will be torn to pieces if it hits a protruding nail, staple, or other such item. You will know it because sparks will fly. The sanding pad will have to be replaced.

If a nail is flush with the surface of the flooring, it must also be set below the surface. If one is hit you will see a shiny spot. This alerts you that it has stripped the abrasive from the sanding pad so the area covered by this part of the pad is not sanded the same as the rest of the floor. When you apply a finish these strips will appear as tiny raised ridges.

Now clean off any other materials that have been stuck to the floor such as candle wax, old glue, and tar. These get in the sanding pad and clog it and will be smeared across the surface of the flooring. If they are not completely removed, including in the pores of the wood, the new finish will most likely not stick in that spot and even be a different color than the surrounding wood.

Check for loose or damaged boards. Secure loose boards and replace damaged boards. Finally, thoroughly sweep the floor so it is free of all dirt, wood chips, or any other loose material. Some prefer to vacuum after sweeping to get a better job. (Review

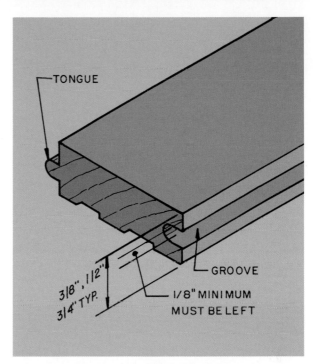

16-12 Do not sand closer than ⅛ inch from the top lip of the groove or the piece will lift off the tongue and splinter when hit with the sander.

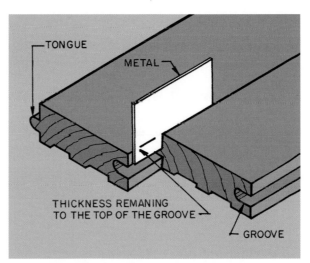

16-13 Measure the amount of solid wood left, to see if there is enough to permit the floor to be sanded.

16-14 This split section must be removed.

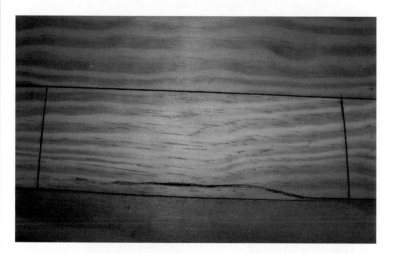

16-15 Mark the ends of the damaged section to be removed.

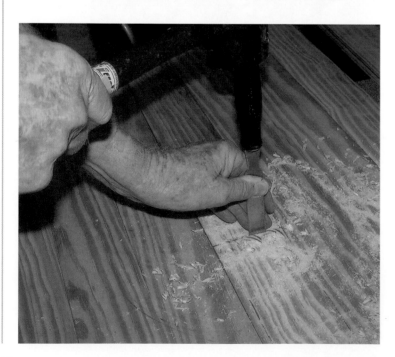

Chapter 15 for additional tips on preparing the old flooring for refinishing.)

Replacing Damaged Flooring

Before starting to sand the floor, examine it for places where a strip may have split or received heavy surface damage (16-14). This section must be removed and replaced. The replacement strip must be the same species of wood as the old floor. The following process can be used to fix short, damaged pieces in the range of 18 to 24 inches. Longer pieces begin to tear up too large a section and may require removing and replacing several strips.

Begin by marking the ends of the piece to be removed (16-15). Bore holes on each end with a large diameter bit such as a spade bit (16-16). Do not bore through the subfloor. Cut each end with a wide-blade wood chisel (16-17). Cut to the line, keeping the cut as straight as possible until the piece is free on both ends (16-18). A wedge cut will help free the piece (16-19). Now split the piece into three or four strips (16-20) and carefully pry them out of the floor. Be careful you not to damage the edges of the flooring on each side of the opening. After the damaged piece is

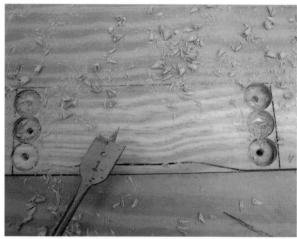

16-16 Bore large-diameter holes on each end. A spade drill or large auger bit will do the job.

16-17 Cut along the mark at each end of the damaged section. Be careful to make a sharp, straight cut on the surface of the flooring.

16-18 Take repeated cuts straight down and then on an angle.

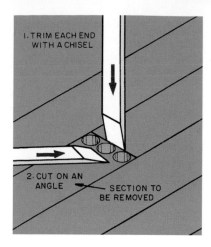

1. TRIM EACH END WITH A CHISEL

2. CUT ON AN ANGLE — SECTION TO BE REMOVED

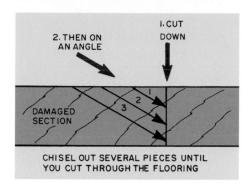

2. THEN ON AN ANGLE

1. CUT DOWN

DAMAGED SECTION

CHISEL OUT SEVERAL PIECES UNTIL YOU CUT THROUGH THE FLOORING

16-19 The angle cut will remove small pieces and then cut straight down again and repeat the angle cut.

16-20 Once the ends are free, split the damaged piece into several pieces. Pry out each piece but do not damage the edges of the butting flooring.

16-21 Measure and cut the replacement piece to length. This must be very accurate.

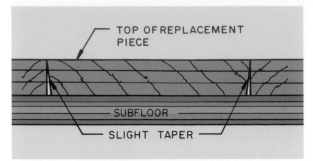

16-22 If the replacement piece ends are cut on a slight angle, it helps make it fit into the opening and produces a tighter end joint.

16-23 After the replacement piece has been cut to length, trim off the bottom edge of the groove.

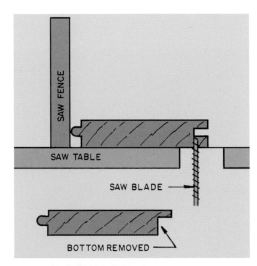

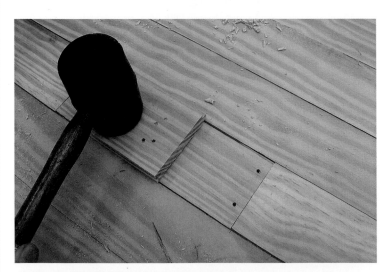

16-25 Once the piece is in place carefully tap it if necessary to get it set flush with the surrounding flooring.

16-26 Drill pilot holes for nails in each end of the piece, then install and set the nails.

removed, the edges of the opening cut by the chisel will need a little trimming to get them straight. Next measure and cut to length the replacement piece (16-21). It helps get a tight fit on the ends if each end has a slight taper to the bottom (16-22). You have to remove the bottom of the groove so when the piece is inserted it lays over the tongue of the adjoining piece. This is easily done by trimming it on a table saw (16-23). However, it could be trimmed off with a chisel. Apply carpenter's glue to each edge of the new piece and insert it in the opening, putting the tongue in the groove on one side and laying the piece down over the tongue on the other side (16-24). Tap it with a mallet to get it flush with the floor (16-25). Then drill holes in each end slightly smaller than the diameter of the finishing nails used to secure it to the subfloor. Set the nails below the surface of the flooring (16-26).

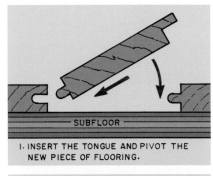

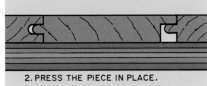

16-24 Put carpenter's glue on each edge and lay in the new piece of flooring.

Sanding Wood Strip Flooring

After the damaged pieces of flooring have been replaced and all nails, screws, tar, and other contaminating materials have been removed, sweep and vacuum the floor clean. It is now ready to be sanded.

GET THE ROOM READY

Since sanding creates a lot of dust which not only falls on the floor but gets into the air, you should cover heat and air-conditioning outlets, cold-air returns, cabinets, plumbing fixtures, smoke and carbon-monoxide alarms, and any other items that could be damaged by the dust. Close the doors to adjoining rooms and seal with duct tape. Open doorways should be covered with a plastic sheet. When possible run some fans to exhaust the air through the windows. The dust bag on the sander will collect a lot of the dust, but much of it escapes.

Finally, remove the shoe molding and label each piece so it can be replaced (16-27). This makes it easier to sand close to the wall.

PERSONAL PROTECTION

Always wear a disposable dust mask (16-28) or better still, a mask with a replaceable filter (16-29). The disposable dust mask is helpful but it should be replaced frequently, as should the filter in the better mask. It is recommended that you wear a hat and sturdy shoes with soft soles that will not mar the floor. It is essential to wear total eye protection and remove and wash it occasionally. Some also wear ear plugs because the sound is intense and could over time harm hearing.

A COUPLE OF DANGERS

If you have to sand a floor that has been painted, you need to realize the paint may contain **lead**. Lead was used in paints until 1976. Congress passed the Lead-Based Paint Poisoning Prevention Act which limited the use of lead in paints in residential construction to 0.06 percent. Most manufacturers stopped adding it to their paints completely. If the house is old and you think the floor was painted before 1976 you need to check samples of the paint for lead. **Lead dust when ingested even in minute amounts can have serious health effects.** Lead could be in paints on interior trim and exterior wood siding.

It is difficult to handle the removal of lead-based paints. Check with the local building inspection department for local testing sources and recommendations for removing it. Paint samples can be checked for lead content with a portable X-ray fluorescence

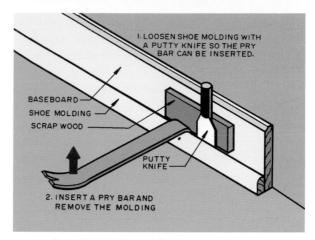

16-27 Pull the shoe molding loose with a pry bar. Open a crack with a putty knife and work in the pry bar. Pull it loose at each nail. Protect the baseboard with a scrap block.

16-28 This low-cost, disposable dust mask will give some protection but should be replaced frequently.

16-29 This is a better dust mask with a thick replaceable filter pad. It will do a better filtration job and the pad does not have to be replaced quite as often.

(XRF) analyzer. This must be used by a qualified tester. Kits are available at the local building supply dealer for checking for the presence of lead (16-30). Follow the instructions carefully. If they meet U.S. Government Consumer Product Safety Commission standards, they will detect the presence of leachable lead, which must be removed. These kits do not propose to replace inspection by a licensed lead inspector or testing laboratory. They will alert you to the fact that lead may be present. Other tests will confirm this and indicate if special lead-removal procedures will be necessary. Also remember there may be several coats of paint. While the top coat may be free of lead, one or more previous coats may be lead based, so test each layer of paint.

The U.S. Department of Health and Urban Development has guidelines for the removal of lead-based paints. Since these change from time to time, check with them for the current list of procedures. You can also get up-to-date recommendations by contacting the National Lead Information Center hotline at 1-800-424-LEAD.

Generally, lead-based paint will not present a health hazard as long as the paint is not chipping or flaking; chips or flakes, if ingested by children, are a serious hazard. Also consider areas where children may chew on things such as window sills, cribs, and toys.

Sanding lead-based paint raises dust in the air, which when inhaled is dangerous. The dust also settles on everything in the room and gets on your hands and possibly in food. If you use a heat gun or blow lamp

to heat and vaporize the paint, hazardous lead fumes fill the air. These procedures should not be used. Any paint chips or fragments already loose should be carefully swept up and put in a sealed container and taken to a hazardous waste disposal site. Wear a respirator and head and body protection so particles do not get on you.

The hazards present make it wise to contract with a professional lead-based paint contractor to do the actual removal.

If you find your home has lead-based paint and are concerned about health problems that you or your family may have already developed, you can have your family physician run a blood test to find out how much lead has already entered your system.

Another danger you may run into is the presence of **asbestos**, which for years was used in patching and joint filling compounds and adhesives to bond floor tile and linoleum, which also contained asbestos. Contact your local building inspection department for advice on how they require this to be handled.

SANDING GRITS

When you are refinishing **old floors** the main job is to remove the old finish while removing as little of the wood as possible. Since the floor may have been sanded before, the wood remaining may be minimal and you do not want to cut through to the tongue and groove. When sanding **new flooring** the goal is to produce a flat, clean surface ready for finishing.

The first consideration is the choice of abrasive paper. It is specified by grit numbers. A 12-grit paper has 12 pieces of grit in one square inch whereas a 100-grit paper will have 100 pieces of grit per square inch. The larger the grit number, the finer the abrasive grains.

The choice of the grit to use will vary depending upon the condition of the floor. A coarse grit is used for the first sanding, followed by sanding with two or more grits, each a bit finer than the one before. The choice will also vary some depending upon the species of the wood. For example, a floor with **new hardwood-strip flooring** might make the first sanding with a 40-grit paper. Next re-sand with 60-grit and then an 80-grit. The final sanding could be with a 100-grit screen. New softwood floors might be started with a 50-grit paper for the first pass and move to an 80-grit and finally to 100-grit paper. Always finish with the grit size recommended by the manufacturer of the finish to be used.

If the floor has a finish and probably lots of wax over the years, you will have to experiment to get the

16-30 This is typical of the lead detection kits available at your building supply dealer. Carefully observe the directions and check to see if it has been certified.

best starting grit. Always start with the finest grit possible; however, when you try one and it quickly gums up, move one grit coarser. Start with a coarse grit, such as 36-grit, which is referred to as an open-coat abrasive because the coarse particles are spread widely apart so it does not gum up as easily as finer grit papers.

Once the finish has been removed the exposed surface can be flattened and smoothed with a series of finer grit papers. Typically two to four additional sandings are needed each with a finer grit paper. Do not make large differences in the grit between sandings. For example, if you started with 40-grit the next will be a 50-grit or not more than 60-grit. This continues until you are sanding with the final grit recommended by the manufacturer of the finish to be used. If you skip too fast to finer grits it will most likely not remove scratches from the previous coarser grit sanding. These will show when the floor is finished. Some finishers recommend moving up only one grit for each sanding.

Since the sander does not get close to the walls, this unsanded area is sanded with a disc sander designed for this purpose, or it is hand sanded. It can also be cleaned with a hand scraper.

Finally, the floor is buffed with an 80- or 100-grit screen. This blends the scratched surface into a more uniform finished surface. Usually the buffer screen grit is one level finer than the grit of the last sanding. For example, if the last sanding was 80-grit, a 100-grit screen is used for buffing.

Cleaning the Floor

The floor should be vacuumed **after each sanding and buffing.** Use a powerful industrial vacuum. Also consider going over the windows, cabinets, and other itmes in the room if the sander permits quite a bit of dust to escape.

After the **last sanding,** the floor, windows, baseboard, doors, sills, cabinets, fixtures, and other items in the room should be thoroughly cleaned. Do the walls have a light film of dust? They may need cleaning as well. After the room is thoroughly clean, let it sit for several hours or better still, overnight so the dust in the air can settle. Then re-vacuum the floor and check the doors, windows, and other dust collecting items in the room. Before applying the finish, wipe down the floor with a **tack rag.** A tack rag is a cloth impregnated with a slow-drying or non-drying varnish or resin. It picks up the dust as it is wiped across the surface. It can also be used on walls, cabinets, and other surfaces. Any dust not removed could easily be disturbed during the finishing process and settle in the finish coats.

Dust & Fire

Always empty the sander dust-collection bag frequently **during the day** and **always** have it empty when you **leave the job** at the end of the day. If you do not, by morning the bag could be on fire and damage or destroy the house. This is caused by spontaneous combustion. The bag contains natural decomposed sawdust which is a fuel. The enclosed fuel will begin to generate heat in the bag as it continues to decompose and since oxygen is present, all the ingredients for a fire are present—fuel, heat, and oxygen.

Using the Drum Sander

The old finish is removed and the floor is made flat and ready to finish with a **drum sander (16-31). The** sander has a dust bag that collects much of the dust;

16-31 A power floor sander has a bag that collects the dust; however, some dust remains on the floor and must be removed with a vacuum. Some dust also filters into the air.

16-32 Old floors with heavy finishes are often first sanded on a 30- to 45-degree angle with an open-coat abrasive. Sand from the center of the room to each wall and sand back to the center by pulling the sander back over the sanded area. Overlap the paths in the center about 2 feet.

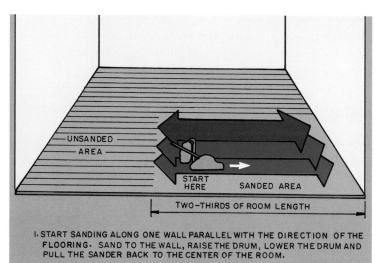

1. START SANDING ALONG ONE WALL PARALLEL WITH THE DIRECTION OF THE FLOORING. SAND TO THE WALL, RAISE THE DRUM, LOWER THE DRUM AND PULL THE SANDER BACK TO THE CENTER OF THE ROOM.

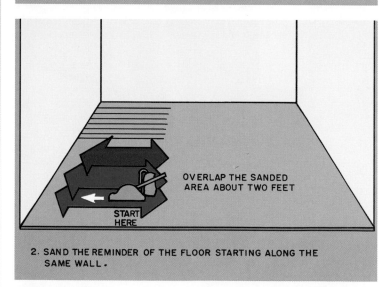

2. SAND THE REMINDER OF THE FLOOR STARTING ALONG THE SAME WALL.

16-33 Starting at a side wall, sand about two-thirds of the width of the room; then turn around and, starting along the same wall, sand the remaining one-third of the floor. Overlap the end passes about 2 feet and the sides of the passes 3 or 4 inches.

however, some does escape into the room on the floor and in the air. The operator must wear ear and eye protection and a dust mask. The abrasive paper is on a drum at the front of the sander. The abrasive action between the drum and the floor moves the sander forward so the operator must stay in firm control.

When removing an **old finish,** sand the first time on a diagonal of 30 to 45 degrees (**16-32**). This will remove the old, possibly gummy material faster than if the first run is parallel with the flooring. It will also help remove any slight cup that has developed in the old flooring. This sanding will leave unsanded flooring along the walls that will have to be sanded with a disc edge sander and some hand-scraping. Next make two to four sanding passes parallel with the run of the flooring as shown in **16-33.**

When sanding **new flooring,** sand in the direction that the flooring is laid. In other words, sand with the grain of the wood. If there is a small area that is particularly uneven, sand it **lightly** on a 45-degree angle before sanding with the grain. Be very careful not to remove too much wood or you may create a depression.

As you sand keep the sander moving whenever the drum is touching the floor. If it sits still for even a few seconds, it will cut a concave depression in the floor. To keep this from happening when you stop for a second, press down on the handle. This raises the drum off the floor.

Start sanding next to a side wall until about two-thirds of the way across the room (**16-33**). Tilt the handle down and start the motor. When it reaches full speed, slowly lower the drum to the floor and immediately move to the wall. Failure to move ahead will produce a groove where the drum hits the floor. Just before the sander reaches the wall, gradually raise the drum off the floor. This produces a tapered or feathered cut at the wall. Now lower the drum and pull the sander back over the same area of the floor to the beginning of the cut at the center of the room.

At the center, lift the drum off the floor and move the sander over for the next pass. It should overlap the first pass about half the width of the drum, which is typically 4 inches. Repeat the sweep to the wall and back to the middle of the room. Repeat these sweeps until the width of the room has been covered. Then turn around and, starting in the middle of the room, sand the remaining third. Let each pass overlap the end of the first pass about two feet.

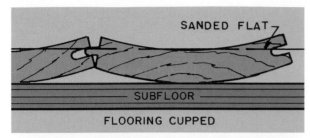

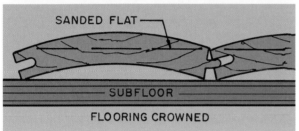

16-34 If the floor has considerable cup or crown, it is not possible to sand it flat without cutting into the tongues and grooves.

SANDING CUPPED FLOORING

Very old flooring such as the 1 x 6 and 1 x 8 pine boards that were widely used are often cupped. Some are convex and others concave (16-34).

If the cup is very slight the floor can be flattened by sanding it on an angle of 15 to 30 degrees, and then sanding parallel with the flooring with the same grit paper (refer to 16-32). However, if there is much cup you will most likely cut into the tongue and groove, which will ruin the floor. While you might sand it on a diagonal a little, the usual solution is to hand-scrape the old finish and the raised area a little, but try not to get it flat.

SANDING THE EDGES OF THE FLOOR

Once the floor has had the first sanding, sand the area along the edges which the drum sander could not reach. This is sanded with a disc sander (16-35). Usually this sander uses discs that have the same grit as that used on the floor area. Each time the floor is sanded, sand the edges with the same grit paper on the disc. This continues until the finest-grit abrasive has been used.

To use the disc edge sander, hold the sander by the handles and turn on the power (16-36). When it reaches full speed, lower the disc to the floor and begin moving it along the edge of the wall. Move it back and forth in a slow, sweeping motion across about 15 to 18 inches (16-37).

Most prefer to begin sanding in a left-hand corner and move to the right. Remember to keep the sander moving while the disc is in contact with the floor. If

16-35 This is a typical disc sander designed to sand the area along the walls that the large drum sander cannot reach.

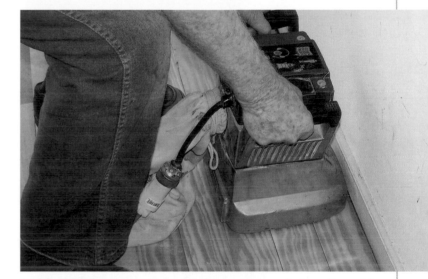

16-36 Hold the edge sander by the handles and place it next to the wall. Turn on the power and lower the disc to the floor. Keep it moving.

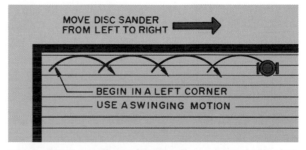

16-37 Start sanding with the disc sander in the left corner and move to the right; move the sander back and forth in a slow, sweeping motion.

16-38 A small finishing sander can be used to final dress the area along the wall where the disc sander may have left some surface scratches.

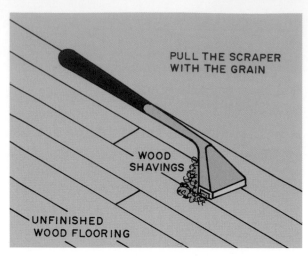

16-39 A scraper is used to remove very thin wood shavings where scratches from the disc sander remain.

16-40 A buffer can be used with a screen to dress the sanded floor. It can also be used to smooth coats of finishing materials. *Courtesy Southern Pines Council*

it remains still for a couple of seconds, it will sand a depression in the floor. Do not press down on the sander to try to get it to cut faster. Let the weight of the sander apply the pressure on the disc.

Remember to empty the dust-collection bag frequently.

There will be areas around the wall where the disc sander cannot reach. These are areas close to the baseboard, corners where the round disc cannot reach, and under various items such as a **hot water heating system radiator**. The disc sander may leave marks where it overlaps the area sanded by the drum sander. These can sometimes be removed by using a small, handheld random-orbital or straight-line motion power sander **(16-38)**. In areas where these small sanders cannot reach, a scraper can be used **(16-39)**. It removes thin wood shavings and leaves a smooth surface.

Finally, a buffer with an abrasive nylon pad and screen can be used after all the sanding has been finished, to blend the areas that are somewhat different in finish due to the use of different sanding and smoothing techniques **(16-40)**. A fine screen such as 100- or 120-grit is commonly used. Generally it is finer than the grit of the last sanding paper.

It is especially important to buff along the walls because this is where most differences in the finish will occur. Work from one wall parallel with the flooring over to the center of the room. Then buff the other half in the same way you did the sanding.

Finishing Wood Floors

Even if the persons sanding the floor did a good job of cleaning away the dust, before you begin to apply any finish, clean it again. Thoroughly vacuum with an industrial vacuum cleaner (16-41). If the floor is old and the old finish has been removed, it is likely there will be cracks between some of the boards. Special effort must be made to clear the dust from between the boards. If the cracks are very wide you probably will want to fill them with a crack filler. The local paint dealer will have a choice of products and can advise you. If you do not plan to stain the floor the filler should be a color as close to the wood as possible. If the floor will be stained the filler should be a color much like the finished stained floor.

Finally, wipe the windows, doors, cabinets, baseboard, and other features with a tack rag. Then wipe the entire floor. From this point on only the floor finishers should be in the room.

The Finishing Process

The steps followed to finish a wood floor will vary depending upon the situation. They can include staining, filling the grain, sealing, and applying a topcoat.

FILLING THE GRAIN

Some hardwoods, like oak and walnut, have an open grain and require filling. Closed-grain hardwoods, such as maple and birch, do not require filling. If open-grain woods are not filled, the top coatings will settle into the open pores and the surface will not be smooth. Filling will also fill hairline cracks between the flooring strips.

If the floor will not be stained, select a filler that is neutral or will match the color of the floor after the topcoat has been applied.

Generally the filler is applied to the floor before light color stains are applied, and after dark color stains have been applied. Always try staining and filling some sample pieces of the flooring to check the colors.

Some types of stain also provide some sealing of the bare wood. This provides a colored matte finish over which the durable topcoatings are applied.

When the flooring is filled before it is stained, or if it is not going to be stained, apply the filler to the floor after it has been sanded. Then go over the filled surface again with a very fine paper after it has hardened.

Commonly used commercial fillers are lacquer-based or latex-based and dry in about one hour.

APPLYING THE FILLER

When applying the filler before staining, be certain to vacuum the sanding dust from the pores and hairline cracks between the strips. Then pour a small amount of filler on the floor and spread it with a trowel. Wipe it across a section of flooring, remove all excess filler from the surface, and finish by wiping the surface clean with a coarse cloth like a burlap pad (16-42). Wipe across the grain. Let it dry and lightly sand with a fine abrasive.

If the floor has been stained, you might scratch the stain with a trowel, so work the filler into the pores and cracks with a cloth, wiping across the grain. If

16-41 This hardwood floor has been sanded, filled, and cleaned; it is ready for stain and the finish coat.

16-42 A coarse rag can be used to work filler into the pores and cracks. Wipe across the grain to remove excess filler.

16-43 This southern pine flooring has been stained a dark color to give the room the appearance desired. *Courtesy Southern Pine Council*

you wipe with the grain, much of the filler will be wiped out of the pores and cracks. After the excess filler has been wiped off the floor, carefully wipe it again across the grain with a clean cloth. Some dampen the cloth with turpentine.

SELECTING A STAIN

Stains are applied to the floor to change their color. A darker color than that of the natural wood (16-43) may be desired. Many floors are finished without staining.

Stains are available using either a dye or pigment coloring agent suspended in some type of vehicle (liquid). Stains are typically designated as water-based, oil-based, and fast-dry. **Water-based stains** set faster than oil-based and have little odor; however, they do raise the grain and require a light sanding after they dry. **Oil-based stains** set slower than water-based, giving more time during application to work with them. These are the most commonly used stains for floors. **Fast-dry stains** have an oil vehicle with driers added. This speeds up the setting time. Some types have a sealer providing a protective coating over the stain.

Once a stain has been selected, try it out on a few pieces of scrap flooring to be certain it is what you want. The stain selected should be compatible with the type of final finish. The selection of the type of stain as well as the type of final finish should be made before proceeding with stain application.

APPLYING A STAIN

After the floor has been thoroughly cleaned with a tack rag, the stain is **wiped** on the floor, allowed to set a while, and **wiped off.** The longer it sets, the darker the color. Place some stain on scrap flooring and time it at several intervals before it is wiped off. This will let you know how long to let it set before wiping.

Plan where to begin application. You should always be on a dry area and have a plan to finish at an exit.

Stains are usually applied by wiping them on with a lint-free cotton cloth or paint pad. Thin stains can be brushed on. Be certain to thoroughly mix the stain to keep the ingredients of a uniform consistency. As you work, mix it every now and then.

Wipe the stain on with a folded rag using a sweeping motion. When using a brush, apply across the grain and then finish with the grain. Do not apply more than you can wipe off in the time allowed. A second person wiping will speed up the application. Generally the stain should be allowed to set only a few minutes. Now wipe off all the stain on the surface with a

358

lint-free cotton rag; replace it frequently. As you apply the stain, try to end the edge of the strip being stained along the edge of a board. Minimize the overlap between rows of stained strips. The overlap could become darker than the rest of the floor.

Remember that oil-soaked rags can be flammable, so remove them from the building. They can eventually ignite through spontaneous combustion. They should be placed in metal receptacles with lids and ventilated bottoms for laundering or disposal. It is also important to maintain proper fire extinguishing equipment and smoke detectors in all areas where flammable and combustible materials are being used or stored.

Some hardwoods, such as maple, do not accept stain well because they have a tight, closed grain. Sometimes penetration can be improved by thinning the stain. Since this will lighten the color, a second coat may be necessary.

Allow the stain to dry at least 24 hours before proceeding with the next finishing steps.

USING SANDING SEALERS

Sanding sealers can be used on bare or stained wood flooring. They create a smooth surface upon which the oil- or water-based polyurethane topcoating is applied. After application, allow to dry as specified by the manufacturer. Then lightly sand the surface, clean it with a tack rag, and apply the finish coating. Be certain the sealer used is recommended for use with your finish coat.

Selecting the Floor Topcoat Finish

There are a variety of finishes available and each has advantages and disadvantages. Consider the traffic it will receive, the possibility of spilled substances, and the overall appearance of the finished floor. Some finishes are easier to apply and perhaps give an inexperienced homeowner a better chance for success. Finishing is a difficult process that requires knowledge on how to handle the material and how to apply it. Some projects are best left to a professional floor finisher.

Consider the gloss of the finished coating. A high-gloss is very durable and possibly a bit tougher than the satin or other low-gloss finishes (16-44). But sometimes it is the look of a glass or plastic layer that some homeowners do not like. It also reflects light, which might be objectionable in some rooms. Satin and other low-gloss finishes are not as glossy looking and for most areas are quite durable.

Oil-modified urethanes are easier to repair than moist-cured urethanes and the acid-curing Swedish finish (refer to **Table 16-1**). If making a repair, it is important to try to ascertain what type of finish was used. You can check this by sanding and coating a spot in a closet where it will not show, to see if the materials are compatible.

Be aware that most floor-finishing materials contain toxic solvents and chemicals. Observe the manufacturer's safety recommendations; these are on the label of the container. It is important to observe the precautions listed as you apply the finish.

As finishes cure they produce odors and fumes that are harmful to anyone in the house. Some are more toxic and have a strong odor, making it difficult to breathe in the room. Moisture-cured, acid-curing Swedish, and oil-modified urethanes are especially troublesome. Water-based finishes do not have the heavy odor, but do bother some people. Stay out of the room 24 to 48 hours after the floor has been finished.

16-44 This southern pine floor has a durable glossy finish that will withstand considerable traffic.

Courtesy Southern Pine Council

359

Table 16-1 A Comparison of Floor Finish Surface Coatings.

Water-based urethane	• Little odor • Very durable • Recoatable • Nonflammable • Low VOC
Oil-modified urethanes (cures as solvent evaporates)	• Some odor • Good hardness and abrasion resistance • Dries slowly • Recoatable • Flammable • Room requires ventilation • Good chemical resistance
Moisture-cured urethanes (cures as it absorbs moisture)	• Strong odor • Outstanding abrasion resistance • Dries quickly if humidity is high • Recoatable • Flammable • Room requires ventilation • Good chemical resistance
Swedish finish (acid curing)	• Strong odor • Very durable • Dries rapidly • Recoatable • Flammable • Room requires ventilation
Seal and wax or an oil finish	• Some odor • Moderately durable • Drying time varies • Frequent recoating necessary • Flammable

The fumes from solvent-based finishes are flammable, so check to see that all pilot lights, electric motors, and other sources of sparks or flame are turned off. When applying finishes, you should wear rubber gloves, eye protection, and a NIOSH/MSHA-approved respirator. (NIOSH is the National Institute for Occupational Safety and Health. MSHA is the Mine Safety and Health Administration). Check the manufacturers recommendations for any additional protection that may be needed.

Floor finishes are composed of various ingredients, which, when combined, produce the characteristics of the product. **Pigments** are small particles providing color or opacity. **Polymers** or **resins** hold the pigment in the coating. **Solvents** are liquids in which the pigments, polymers, and resins are suspended. Some finishes will have additives influencing the properties, such as speeding up the drying time.

It should be noted that the solvents in solvent-based finishes contain various **volatile organic compounds** (VOCs). When the solvent evaporates, these are released into the air and combine with sunlight and oxygen, creating a low-level ozone. These harmful emissions have been regulated by the Environmental Protection Agency (EPA). While the allowable levels of VOC emissions will no doubt be adjusted over the years, and may be more restrictive in some states, the typical level is 250 grams per liter for floor-finishing materials. Be aware that there are many finishes that do not meet this standard.

Topcoat finishes fall into two major types—**surface coatings** and **penetrating coatings**.

SURFACE COATINGS & SURFACE SEALERS

Surface coatings, often referred to as **surface sealers,** do not penetrate the wood but form a wood-bonding surface layer. Those widely used include water-based urethane, Swedish finish (acid curing), oil-modified urethanes (solvent evaporates to cure), and moisture-cured urethane (absorbs moisture to cure). A comparison of the various surface coatings is shown in **Table 16-1.**

Water-based urethane finishes have the solid pigments suspended in water with several additives. They are a combination of urethane and an acrylic and have a catalyst mixed into the water base before it is applied to the floor. The more durable water-based coatings have a higher percentage of urethane. While they cost more they are worth the extra cost.

Water-based finishes usually take several more applications than the oil-based to get the same film thickness. They dry faster and are available in gloss and satin finish. They dry clear and do not change the color of the wood as much as solvent-based finishes.

The first application is a water-based sealer that is compatible with the water-based finish coats. Follow the manufacturer's recommendations. In addition to sealing the surface, the water-based sealer reduces the amount of absorption of the following coats so a layer finish is built up on the surface.

The water-based sealer will raise the grain of the bare wood, so it is necessary to sand it lightly or buff it with a fine screen after it is dry, as recommended by the manufacturer. Be careful not to cut through the sealer to the bare wood. After sanding or buffing, wipe the floor with a tack rag and apply the next coat.

Oil-modified urethane finishes have an oil base modified with urethane additives. Many manufacturers refer to this finish as a **polyurethane.** It is the most commonly used finish and is the easiest to apply. While it dries rapidly, it takes time to cure, so follow the manufacturer's directions. A light sanding is required between coats to provide a surface for the next coat to properly bond to. This urethane, like other urethane finishes, has a high VOC rating. Glossy and satin finishes are available.

Moisture-cured urethane finishes depend on the humidity in the air to cause a reaction that enables them to dry. The higher the humidity, the more rapidly they will dry. To be totally effective, it is best to try to maintain a stable humidity level. These finishes are flammable so precautions to avoid sparks from electric motors, gas stove pilot lights, and the like must be taken. They have a high VOC level and are difficult to apply. While these produce a very hard finish, the homeowner should most likely employ a professional finisher.

Acid-curing Swedish finishes contain formaldehyde and are cured by a hardener additive. They provide a clear, durable finish but are more expensive than the other types. Sometimes, when applied to raw wood, they raise the grain and have to be dressed with a buffer having a screenback or a lightly sanded by hand with a very fine-grit paper.

They have a high-VOC level, and are difficult to apply; the job should be left to professional finishers. Be certain to follow the safety recommendations on the label, if you do the job yourself.

PENETRATING WOOD-FLOOR FINISHES

Penetrating finishes are absorbed into the wood and seal the pores against dirt and moisture. Some penetrating sealers are finished with a topcoat. Others form the final coat.

Sealers penetrate the wood and reduce the absorption of the final topcoats. They are much like the primers used on bare wood that is to be painted. They provide a surface upon which the finish coats can build. Sealers must be compatible with the finish coat.

Linseed and **tung oils** penetrate the wood and serve as the final finish. They enhance the natural color of the wood. Some have a small amount of stain. Worn or damaged areas are touched up by rubbing oil over them and allowing it to soak into the wood. These are not durable, long-wearing finishes. They can be waxed with a compatible paste wax and lightly buffed as needed.

Stains are used to enhance the color of the bare wood. Regular stains do not provide a finish coat and must be covered with wax, oil, or urethane finish materials.

Stain-sealers enhance the color of the bare wood and provide a protective sealing film over which a urethane topcoat can be applied.

Wax finishes are easy to apply and require frequent maintenance. They involve applying the wax with a rag or a stiff brush. Work it well into the pores of the wood. If the manufacturer recommends a drying time, wait and then buff it with a power buffer that has a polishing pad.

The wax will often yellow or become brittle and must be removed and the floor recoated. Commercially available wax removers are available. They can be removed by power-sanding but this removes the beauty of the aging color of the floor. Be aware that many wax products have a high VOC level, so safety precautions would be in order. Check the instructions on the container.

OTHER FINISHING MATERIALS

Varnish and **shellac** have been used for years to finish floors. Traditionally varnish was made by mixing natural resins with linseed oil and turpentine, which were combined in different proportions for different uses.

Natural varnishes are seldom used today. They have been replaced with synthetic resins such as alkyd, phenoic, and polyurethane. Urethane finishes are now the dominant floor finish. Some refer to the urethane finishes as modern varnishes.

Shellac is produced using a resin made from the scale of the **lac bug** combined with denatured alcohol. It gives the wood a natural look much like a hand-rubbed finish. It is protected with a high-quality paste wax. It is not as durable as other finishes and spots when wet with water or other spills, such as food and various solvents.

Shellac is available as white or orange. White is used if you want a light finish. Orange is used to produce a brown hue.

Applying the Topcoat Finish

After the flooring has been sanded, filled, stained—and sealed, if desired—the topcoat finish is applied. Again, follow the recommendations of the finish manufacturers. The following suggestions are typical of what they recommend.

URETHANE TOPCOATS

After thoroughly cleaning the floor with a tack rag, begin by applying enough finish to complete a one-foot-wide strip along a wall running in the same direction as the flooring strips or planks. A lamb's wool pad is a good tool to use **(16-45)**. Apply an even coat, wiping off any excess finish. Start the next strip by allowing it to overlap the first by about 2 or 3 inches. Do this before the first strip dries. Finishers call this "working to a wet edge." Do not allow the finish to be thicker in the overlap. Stroke it until the finish is a uniform thickness so lap marks do not remain. An 8- to 10-inch natural bristle brush is used with urethane finishes **(16-46)**.

Plan the work so you have a way to leave the room as you approach the other side.

Allow the finish to dry as specified by the manufacturer; typically this is 24 hours. Maintain a normal room temperature. It will likely be necessary to ventilate the fumes from the room before applying the next coat. After the first coat has dried, power-buff it with a fiber buffing pad or No. 1 steel wool; clean the resulting dust from the floor. Apply a second layer in the same manner.

During the finishing process, stay off the floor as much as possible. If it is necessary to walk on it, lay out a fabric runner. Perspiration from hands, knee prints, footprints, and any water will cause discoloration, spotting, and uneven coating thickness.

Allow to dry; buff, clean, and apply a third coat, as necessary.

APPLYING WATER-BASED FINISH

Water-based finishes are applied with a short-nap applicator, a foam pad, or a wide synthetic bristle brush. A brush 8 to 12 inches wide is recommended.

The application process is the same as that described for urethane finish. Be careful not to oversaturate the bare floor. The excess moisture will raise the moisture level of the floor and may cause problems later on. Brush with the grain, and overlap each strip by about 4 inches. Work to a wet edge to get a smooth transition. Buff when dry and apply additional coats, as necessary.

APPLYING A SHELLAC FINISH

Shellac has an alcohol solvent, so it dries rapidly. Apply in strips along a wall, as described for urethane finishes. It should be diluted into a rather thin coating, allowing the alcohol to evaporate. If it is too thick, some alcohol may be trapped below the resin as it cures, giving a soft surface that may not harden for a long time. Apply the coats rapidly and overlap by an inch or so; but brush it out so the overlap does not become thicker than the rest of the finish.

Shellac will dry in about 2 or 3 hours. Then dress the surface with a fine steel wool, such as a 4/0 grade, or a 320-grit, open-coated abrasive paper. Wipe the dust off the surface with a tack rag and apply the next coat; three coats are often used. After it is thoroughly dry, dress it by rubbing with 4/0 steel wool and a rubbing oil or a paste wax. Carefully buff to get the final appearance desired. Again, follow any directions given on the container.

Wood-Flooring Materials

If you decide to remove the old flooring and replace it with new softwood or hardwood flooring, there are several choices available. As you consider softwoods or hardwoods, remember that these terms do

16-45 The finish is applied in the same direction as the grain of the flooring using a lamb's wool applicator. Overlap each run by about 3 inches.

Courtesy Southern Pine Council

not refer to the hardness of the wood. Some softwoods are harder than some hardwoods. **Softwoods** come from coniferous or cone-bearing trees. They retain their needles all year round. **Hardwoods** come from deciduous trees. They have broad leaves that fall off in the winter.

Major species of **softwoods** include Eastern white, slash, Southern yellow or longleaf, shortleaf, and loblolly pine. Douglas fir, Sitka spruce, and California redwood are other species of softwoods.

Major species of **hardwoods** commonly used for flooring include red oak, white oak, pecan, hickory, beech, yellow birch, and maple.

Grades for Hardwoods & Softwoods

As you visit the building supply dealer and make a choice of the type of wood flooring to be used, consider the grades available. Obviously the first grade and prime grades are best and will cost more. They may also be more difficult to find. A summary of softwood and hardwood grades is in **Table 16-2.**

16-46 **Finishes can be applied with a wide brush that has fibers suitable for the finish being applied. Carefully feather out the overlaps on the wet edges.** *Courtesy Southern Pine Council*

Wood-Strip Flooring

Solid-wood **strip flooring** is available as tongue-and-grooved strips. The tongues and grooves are machined on the edges and ends (**16-47**). Typical sizes are shown in **16-48**.

BUNDLES

Strip flooring is shipped in **bundles** that have the contents sorted by grade and length. Bundles may have

---

Table 16-2 Grades for Hardwood and Softwood Flooring.

Hardwood flooring grades of the National Oak Flooring Manufacturers Association

Unfinished oak flooring
Clear Plain or Clear Quartered—best appearance
Select and Better—mix of Clear and Select
Select Plain or Select Quartered—excellent appearance
No. 1 Common—variegated
No. 2 Common—rustic

Pecan
First Grade Red—face all heartwood
First Grade White—face all bright sapwood
First Grade—excellent
Second Grade—face all heartwood
Second Grade—variegated
Third Grade—rustic

Beech, birch & hard maple
First Grade White Hard Maple—face all bright sapwood
First Grade Red Beech—face all red heartwood
First Grade—best
Second and Better—excellent
Second—variegated
Third and Better—mix of First, Second, and Third

Prefinished oak flooring
Prime—excellent
Standard and Better—mix of Standard and Prime
Standard—variegated
Tavern and Better—mix of Prime, Standard, and Tavern
Tavern—rustic

Softwood flooring grades of the Southern Pine Inspection Bureau
B and Btr—best quality
C and Btr—mix of B and Btr and C
C—good quality, some defects
D—good economy flooring
No. 2—defects but serviceable

Maple flooring grades of the Maple Flooring Manufacturers Association
First—highest quality
Second and Better—mix of First and Second
Second—a good quality, some imperfections
Third and Better—mix of First, Second, and Third
Third—good economy flooring

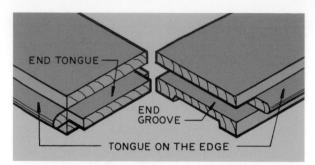

16-47 Tongue-and-groove strip and plank flooring are edge and end matched.

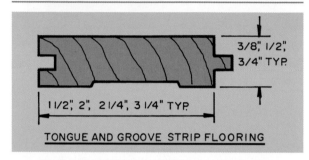

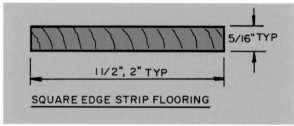

16-48 The most commonly used solid-wood strip flooring is tongue-and-grooved, but square-edge stock is available.

random or average length pieces, nested pieces or specified-length pieces (16-49).

Random- or average-length bundles contain strips from 9 inches up to around 8 feet. The short pieces are referred to as shorts. This is the most commonly used bundle.

Nested bundles have random-length pieces generally from 6 to 8 feet long.

Specified-length bundles are generally pieces two feet or less in length and are used when a special floor pattern of shorts and other lengths are required.

Wood Plank Flooring

Plank flooring is available as solid wood or laminated. It is available in widths from 3½ inches to 8 inches, but wider planks can be had by special order. Solid-wood planks are typically ¾ of an inch thick (16-50). Generally, planks 5 inches and wider are secured to the subfloor with screws driven through the face. These are counter-bored and covered with wood plugs.

Plank flooring is so wide that it is typically plain-sawn. Since they are wider and have fewer joints than strips, as they age they tend to develop wider gaps between planks than narrower strip flooring. Planks are often cut from beams salvaged from old buildings as they are torn down.

Plank flooring is usually shipped in boxes rather than bundled. This helps keep the planks flat and protects the shorts that are included.

16-49 Unfinished wood strip flooring is received on the job in bundles. Notice that it has been kept under plastic wrap to protect it on the job.

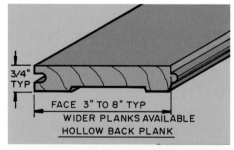

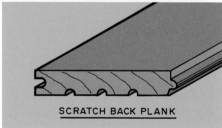

16-50 Solid-wood plank flooring is generally available up to 8 inches wide. Wider planks are sometimes available.

Prefinished Strip Flooring

Prefinished hardwood flooring arrives on the job carefully wrapped, ready to install (**16-51**). Once it is installed it is swept clean and the job is complete. No additional finishing is required. It is graded the same as unfinished wood flooring.

The factory finish is a tough, durable coating that is applied and cured in a dust-free environment so it provides the highest quality available (**16-52**).

Prefinished solid-wood flooring is available in the same sizes as unfinished flooring. It is installed the same way as described for unfinished flooring.

Installing Wood-Strip Flooring

If the old floor is beyond saving, it will have to be removed, and new wood-strip flooring installed. After the old flooring has been removed, go over the subfloor and replace pieces that are not sound. Glue and nail them to the joists. Renail the existing subfloor to the joist so there will be no squeaks. If the subfloor will need some areas replaced, use CDX plywood, oriented strandboard, or 1 x 6 solid-wood boards. They should be at least ⅝ inch thick but will have to come up flush with the old remaining subfloor. Run the long side of the sheets perpendicular to the floor joists.

Also check the old floor for sag. This can be done with a chalk line or laser. If there is a sag of more than ¼ inch, glue wood shims to the subfloor to bring the surface in that area up to level. Run them so they are perpendicular to the flooring. If a slight high spot is noted, you can sand it down with a drum sander or disc sander (**16-53**). Sag can also be corrected by the techniques discussed in Chapter 15.

If you notice bounce or deflection in the floor when under load, the floor structure will have to be strengthened. See Chapter 15 for information on stabilizing the floor structure. You can notice deflection, especially when you bring in the bundles of flooring and store them on the floor.

Addressing Sound Transmission

Rooms above living areas could have the transfer of sound the room reduced by adding structural, sound-deadening panels over the subfloor. Sound-deadening techniques are discussed in Chapter 15.

16-51 Prefinished hardwood flooring is shipped in bundles that are completely wrapped by the manufacturer. Damage to the finished surface must be avoided.

16-52 This prefinished oak flooring has been laid up to the area in the foyer to be finished with ceramic tile.

16-53 Small high spots can be removed with a disc sander. Larger areas will require a drum sander.

16-54 To check moisture, force the pins on the bottom of the moisture meter into the flooring and read the moisture content on the digital readout. Notice the flooring strips have been separated into layers by wood strips. *Courtesy Southern Pine Council*

16-55 A moisture meter is used to check the moisture content of the wood flooring and subfloor. This meter checks moisture content from 6 percent to 40 percent with a digital readout. *Courtesy Delmhorst Instrument Company*

15-56 The subfloor has been covered with red rosin paper. The seams are sealed with tape. This forms a vapor barrier, protecting the wood flooring from moisture below the floor.

Make a Moisture Check

Before bringing the unfinished wood into the house, the interior air should be at normal temperatures for a week or longer. The humidity in the room should be within limits expected for normal occupancy. This will vary depending upon the climate and time of year, but somewhere between 30 and 60 percent relative humidity is a comfortable range.

When the flooring is brought into the room, open up the bundles and check samples from each for its humidity content (**16-54**). Also check the moisture in the subfloor. They should be within four percentage points of each other. Since electrical characteristics of wood species vary, all species read differently on the moisture meter at the same moisture content. To get the actual moisture content, refer to a wood species chart supplied with the moisture meter (**16-55**). The flooring should have a moisture content of 6 to 10 percent.

Remember that the pins on the moisture meter measure the moisture only at the depth driven and in a line between the noninsulated portion of the pins. You can drive the pins into the wood at different depths in increments of as little as ⅛ inch, giving the moisture content at each depth.

Flooring Preparation

Once the subfloor is in condition to receive the new flooring, cover it with **red rosin paper (16-56)** or 15-pound **builder's felt (16-57)**. This forms a vapor barrier, protecting the wood flooring from moisture below the floor.

Now bring in the bundles of flooring. Open them up and spread around the room so their moisture content can equalize with the conditions in the room. Let the flooring condition for five to seven days before starting to lay it down. The flooring can have spacers placed between layers as shown earlier in **16-54**. This allows the air to circulate between the strips, allowing it to adjust to the moisture in the room.

Check the Room for Squareness

Now check the room for squareness. Generally the walls of a room do not meet exactly at 90 degrees. If all the corners are exactly 90 degrees the room is said to be square. One of the first things to do before making the layout plan is to check the room for squareness.

16-57 The subfloor can also be covered with 15-pound builder's felt instead of red rosin paper. Overlap the sheets at least 3 inches.

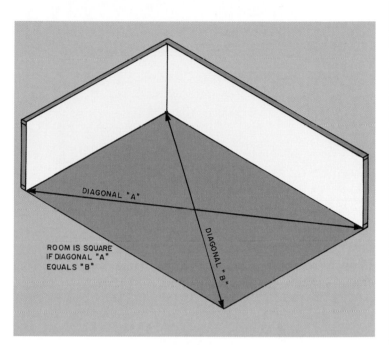

16-58 Check a room for squareness by measuring the diagonals. If they are the same length, the room is square. If it is not square, check each corner separately.

Begin by measuring the diagonals of the room **(16-58).** If they are the same length, the room is square. This simplifies the layout of the flooring. If the room is found to be not square, perhaps only one corner is out. An easy way to check this is to lay a large carpenter's square against the walls in a corner **(16-59).** Check each corner and note those not square. It will be along this wall that you will have a layout problem.

You can also check for squareness using the Pythagorean "3-4-5 technique." As shown in **16-59,** you measure 3 feet on one wall and 4 feet on the other. If the hypotenuse of the triangle formed measures 5 feet, the corner is square.

As shown in **16-60,** if two adjacent corners are not square, then one wall is out-of-square. If it is a small amount, you might consider setting the pattern parallel with the square walls, allowing the pattern to taper a little along the slanted wall. This is effective when the wall is in a part of the room that is not dominant. Another possibility is to lay the floor covering so that half the discrepancy is allowed on each of the opposite walls.

If a wall has a bow, you might consider trimming the edge of the flooring to fit around it and then using

a baseboard and shoe molding to try to cover up the irregular edge.

To check to find any bow in a wall, run a chalk line one inch out on each end and measure the distance from the chalk line to the wall along the length of the wall **(16-61).**

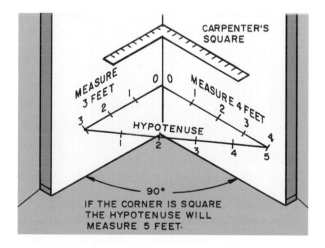

16-59 You can check a corner for squareness with a large carpenter's square or by using the Pythagorean "3-4-5 technique."

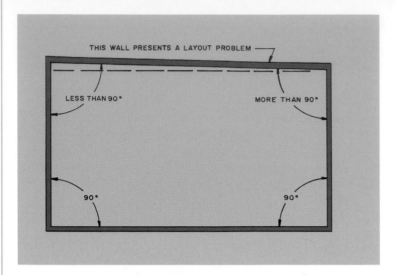

16-60 If one corner is not square it will throw the entire wall out of line with the opposite wall.

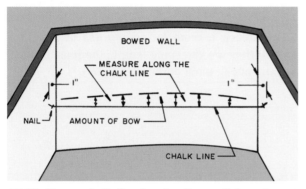

16-61 Use a chalk line to check a wall for a bow.

Planning the Layout

Generally the amount a room is out-of-square is small and not a major problem. When the flooring runs perpendicular to an out-of-square wall, the difference in length of the strips is not noticeable (16-62).

When the flooring runs parallel with an out-of-square wall, some adjustment is usually necessary(16-63); it depends on the amount of variance. Small amounts possibly can be ignored. If a correction is needed, it may require tapering a number of flooring strips to try to reduce the visual impact of the out-of-square wall. In both cases keep in mind that the flooring is run perpendicular to the floor joists.

Flooring installers have a number of ways to make a layout for installing strip and plank flooring. Lay out the starting line with a chalk line, and mark the location of the joists on the floor covering with chalk. Following are several layout techniques.

When the room is quite square, a starting line can be drawn ½ inch from the drywall. This leaves space for expansion of the floor and will be covered by the new baseboard and shoe (16-64). Snap the starting line on the builder's felt or red rosin paper with a chalk line. Also mark the location of each joist on them.

When the room has a wall out-of-square enough to present a problem getting the flooring to appear somewhat parallel with the wall, a layout is needed to allow for some of the flooring strips to be tapered.

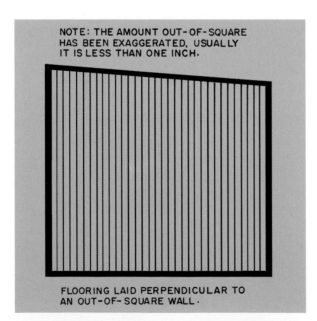

16-62 When the flooring runs perpendicular to the out-of-square wall, the difference is not noticeable.

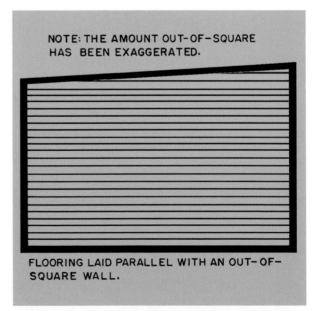

16-63 When the flooring runs parallel with the out-of-square wall, some adjustments must be made.

One way that is sometimes handled is shown in **16-65.** Here the room has a wall that is ¾ of an inch longer than the opposite wall. Locate the starting line from the wall ½ inch for the expansion space, plus the width of the flooring on one end. On the other end measure out ½ inch, plus the width of the flooring, minus half the out-of-square distance, which is ⅜ of an inch in this example. This locates the starting line. The piece next to the wall is tapered to the sizes shown. Repeat this on the opposite wall to handle the remaining ⅜ of an inch.

A technique that is easy to do—but which does leave the end flooring strip appearing tapered—is shown in **16-66.** With this technique find out how much the wall is out-of-square; in this example it is ¾ of an inch. Allow half of this on each end. Set one end of the starter line ½ inch from the wall for expansion. Set the other end along the out-of-square end, ½ inch

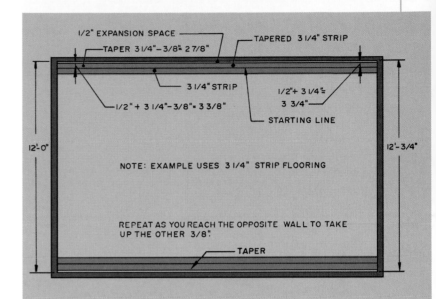

16-65 This room has one wall that is out-of-square. Some of the strips at opposite walls can be tapered; this reduces the amount of taper visible on each wall.

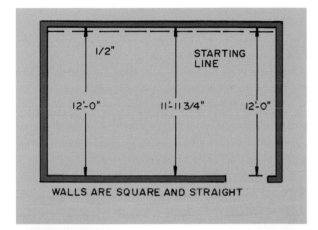

WALLS ARE SQUARE AND STRAIGHT

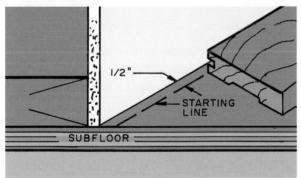

16-64 When the room is nearly square, the starting line can be located from a wall along which the flooring will run parallel.

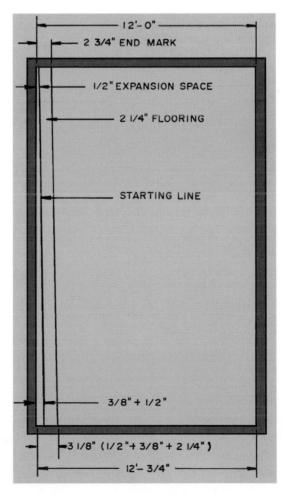

16-66 This is another way to handle an out-of-square wall. The strips by each wall will be visibly tapered, but the amount of taper is half the total amount.

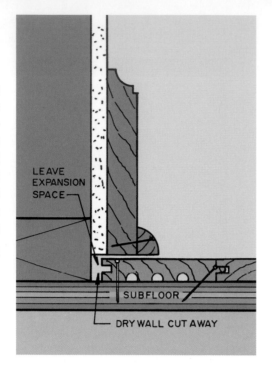

16-67 If the gap is too large to be adequately covered by the base and shoe, cut away the drywall and start the first strips to within ½ inch of the bottom plate.

plus ⅜ or ⅞ of an inch. Some installers also add the face width of the flooring strip to get the starter line out from the wall; it is easier to line up this way. If the baseboard and shoe will not cover the gap, ⅞ of an inch in this example, cut the drywall ¾ of an inch above the floor, and plan to set the first flooring strip under it. Keep the edge of the flooring ½ inch from the bottom plate (16-67).

As you lay the floor, the last strip on the opposite wall will have a similar gap. The gaps are covered by the baseboard and shoe. These end boards will show some taper, but it will be half of the total out-of-square amount. If the strip on the opposite wall is less than a full width strip, it can be cut on a taper.

Beginning the Installation

After the floor has been covered with builder's felt or red rosin paper, the starting line and joists have been located, and decisions made on how to handle an out-of-square room, installation can begin.

Begin the installation by laying the first piece of strip flooring or wood plank on the starting line (16-68). Set the end ½ inch from the side wall. Place the first strip with the **grooved** edge toward the wall. Face-nail and blind-nail this strip into each floor joist, and put one nail into the subfloor between the joists (16-69). Some installers might use 8d finishing nails. If there is a danger of splitting the flooring, drill small pilot holes for each nail. Typically, power-driven fasteners with barbed nails are used; they have greater holding power than finishing nails.

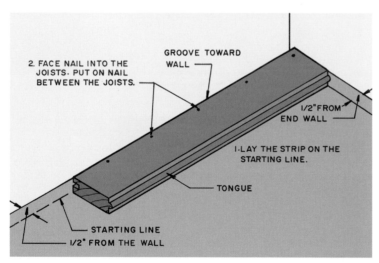

16-68 To begin the installation, lay the first piece of strip flooring along the starting line.

As you face-nail the starter strip, be careful it does not move off the starting line. If the wall is absolutely straight, you could use ½-inch-thick spacer blocks to help line up with the starter line (16-70). Begin face-nailing in the center of the strip, and work toward each end. Set the heads of the face nails below the surface of the flooring. If they are not covered by the baseboard and shoe, they can be filled. Recommended nailing schedules are shown in **Table 16-3**. Continue laying the first strip to the wall at the other end. Again, it must be carefully lined up with the starting line. The tongue-and-groove end joints must be tight. If a hairline gap cannot be corrected, it can be filled when the floor is finished. Sometimes a crack can be corrected by selecting another strip of flooring that fits better.

After the first strip has been installed, lay out several rows of flooring (16-71). This is called "racking the floor." Now is the time to watch for warped or defective strips. It is also the time to notice the color

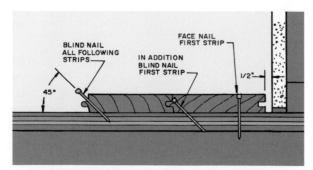

16-69 Face-nail the strip along the wall and blind-nail through the tongue on the other edge of each strip.

and grain, and to arrange the strips in the order that will produce the most uniform-looking floor (16-72). As the pieces are selected, also consider their length; the end joints should be as far apart as possible. End joints should be 4 to 6 inches apart; notice the spacing in 16-72.

The rest of the strips are blind-nailed through the tongue. They can be hand-nailed (16-73). Set the nails with a **nail set** (16-74) so the head will not block the closing of the joint. The use of a compressed-air power nailer is faster and less likely to damage the edge of the flooring (16-75).

As you lay each row, look ahead to see how it will come out on the opposite wall (16-76). Sometimes the last piece will have to be cut shorter. If it appears a short piece will be needed at the wall, it should be at least 8 inches long. Once the pieces are selected and laid out, nail the row through the tongue. Many flooring installers like to work as a team (16-77); teamwork helps in placing the pieces and holding them in place. It also greatly speeds up the work.

Table 16-3 Nailing Schedule for Wood Strip Flooring on ⅝-inch Subfloor.

¾" x 1½" or 2¼" or 3¼"	2" serrated-edge barbed fastener, 7d or 8d cement-coated spiral or cut nail, 15-gauge staple with ½ crown. Space 8" to 10" apart.
½" x 1½" or 2"	1½" serrated-edge barbed fastener, 5d cement-coated spiral or cut-steel or wire casing nail. Nail 10" apart.

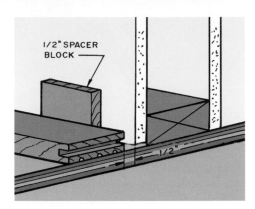

16-70 If the wall is straight, use spacer blocks to line up the first strip of flooring.

16-71 Keep several rows of flooring sorted and laid out ahead of the strip being installed. It is a good procedure for the installers to work in pairs.

16-72 This installation shows the overall appearance of strips selected for grain and color. Notice that the end joints have been staggered.

16-73 Strip flooring can be nailed by hand.

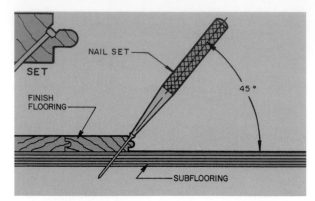

16-74 Set the nails with a nail set so the head of the nail does not hinder the closing of the joint.

16-75 A power-nailer has a cartridge holding the barbed nails at the correct nailing angle. When the shaft is struck with a mallet, the nails are driven the proper depth by compressed air. *Courtesy Southern Pine Council*

16-76 Select the pieces to make a row; consider their length, color, and grain.

There are many times the flooring must be fit around things protruding through the floor (16-78), or around openings (16-79). This requires careful fitting and cutting. A miter saw with a sharp, fine-toothed blade is used to cut flooring on angles and for crosscutting square ends (16-80).

As the pieces are laid, tap them into place so the edge joint is fully closed. Use a piece of scrap flooring as a tapping block (16-81). A faster and firmer way is to use a **flooring jack**; it holds the flooring together as you nail it, so both hands are free for using the nailing tools (16-82). If the strip has a slight bow, the

16-77 Working together helps speed the installation.

16-78 This plumbing installation requires the flooring to be cut and fit around it.

jack will straighten it and hold it as it is nailed. The flooring jack will also hold the last pieces (**16-83**).

It is important that the end matches between strips be tight. Check end joints for tightness before nailing. Very minute hairline cracks can be filled when the floor is finished.

As each row is laid and nailed, check for the possible development of a bow or waviness. While this often occurs, corrective action must be made as soon as it is noticed. It might be a defective strip of flooring that should be replaced. Possibly a very small amount might be planed off the next several strips. This takes careful work, and the planed pieces need to be carefully checked in place before they are nailed.

If the flooring at a **door opening** is met by wood flooring that is running **perpendicular** to it, the floor of the door opening has a header laid over it (**16-84**).

16-79 Openings, such as a heat register, require the flooring to be notched to fit around them.

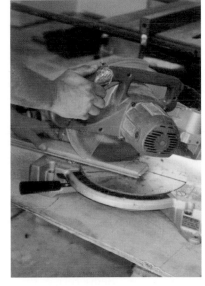

16-80 A miter saw is used to cut strip and other wood flooring.

16-82 A flooring jack holds the strips together as they are nailed. *Courtesy Cepco Tool Company*

16-81 A piece of scrap flooring can be used as a tapping block to close the tongue-and-groove joint.

16-83 A flooring jack will also close the joints of pieces next to the wall and hold them as they are nailed. *Courtesy Cepco Tool Company*

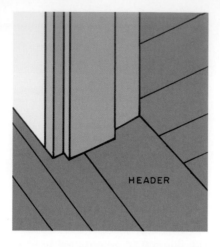

16-84 When flooring that meets in a door opening is perpendicular to the opening, a header is installed in the door opening.

HEADER

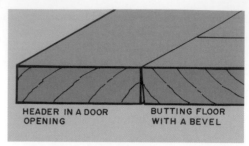

HEADER IN A DOOR OPENING

BUTTING FLOOR WITH A BEVEL

16-85 When flooring butts a header in a door, a slight bevel on the flooring will make a tighter joint.

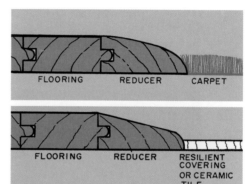

FLOORING   REDUCER   CARPET

FLOORING   REDUCER   RESILIENT COVERING OR CERAMIC TILE

16-86 These are two types of stock reducer available to blend a wood floor into one that is another material or is thinner.

16-87 A nicely installed strip flooring ready to be sanded and finished.

A tighter fit is produced if the butting edge of the flooring is cut on a very slight bevel (16-85).

When the wood flooring meets **flooring of a different type** or thickness, some type of reducer is used (16-86). There are a variety of stock wood reducers available.

You can look with pride at the finished installation that is now ready for the floor to be finished (16-87).

Installing Prefinished Wood Flooring

Prefinished flooring is installed in much the same way as unfinished flooring. Since the top face is the final finished surface, extra care must be taken to prevent damage. This includes the shoes worn, how tools are laid down, and being certain the miter saw is placed away, on the subfloor nearby (16-88).

Prefinished flooring is installed by nailing, as described for unfinished flooring. The starter strip is face-nailed along the starting line. The second strip is edge-nailed, as shown in 16-89. As succeeding rows are laid, the beautiful finished surface is exposed. Notice in 16-90 the installer is wearing soft-soled shoes to protect the surface. Also keep the surface clean; sweep off all wood chips and scraps, because stepping on them could damage the surface.

When flooring passes through a door and continues into another room or hall, the strips are carefully cut to fit through the opening and below the door stop and casing (16-91). If the finish carpenter did not allow a space for this, the strips would have to be trimmed off.

As mentioned, if the flooring at a door opening is met by wood flooring that is running perpendicular to it, the floor of the door opening has a header laid over it. When the wood flooring meets flooring of a different type or thickness, some type of reducer is used. There are a variety of stock wood reducers available.

16-88 Keep the miter and table saw off the prefinished flooring.

16-89 Prefinished hardwood strip flooring is started along the layout line and is edge-nailed.

16-90 Keep the flooring clean as you install it. Notice the installer is wearing shoes with soft soles.

16-91 The prefinished hardwood flooring is being laid through the door opening into the next room, which will use the same flooring. In this example the flooring runs in the same direction in both rooms.

16-92 This prefinished hardwood floor is ready to be cleaned and covered to protect it from other trades that may need to access the area.

In **16-92** is a finished installation of a prefinished floor; it is ready to be cleaned to remove all dust and footprints. If other trades, such as plumbing and electrical, are likely to have to walk on it, cover it with heavy paper designed for protective purposes. Do not cover with plastic sheeting because it will trap moisture and not allow the floor to breathe.

Installing Engineered-Wood & Laminate Flooring

Engineered-wood and laminate flooring are manufactured products, which, because of the multi-ply construction, are quite stable and experience less expansion and contraction than many other flooring materials. Engineered flooring is made completely from wood and provides the same texture and grain available with solid-wood flooring **(16-93)**. It is installed much the same way as solid-wood flooring. Laminate flooring is an assembly of several materials but contains no wood. The visual image is produced by a printed pattern on a plastic decorative sheet. The product is a tough, easy-to-clean flooring material that gives the grain and color of real wood **(16-94)**. It is used for the installation of a floating floor.

16-93 This beautiful, engineered oak plank flooring sets the tone for the entire room. *Courtesy Mannington Mills, Inc.*

Types of Engineered-Wood & Laminate Flooring

Engineered-wood flooring is a laminate of three layers of solid wood with a durable clear finish coating (16-95). The top veneer is available in a wide range of species and finishes. It is available as strip and plank flooring. Engineered flooring can be bonded to the subfloor with an adhesive, stapled, or placed over a cushion underlayment and installed as a floating floor. It is available in a range of widths, thicknesses, and lengths. Thicknesses of 5/16, 3/8, and 1/2 inch are available. Widths from 2¼ to 5 inches are common, as are lengths of 48 inches.

Laminate flooring is a rigid floor covering with a surface layer consisting of one or more thin sheets of a fibrous material (usually paper) impregnated with aminoplastic thermosetting resins (usually melamine). The two types are direct-pressure laminate flooring and high-pressure laminate flooring.

Direct-pressure laminate flooring fuses the wear layer onto the core material using pressure between 300 and 500 pounds per square inch (psi). The treated decorative sheet with the floor grain is considerably thinner than that used on the high-pressure laminate flooring (16-96). As well, the high-pressure laminate has additional layers of material.

High-pressure laminate flooring is fused at 1,400 pounds per square inch. The laminate and a high-pressure balancing backer are then bonded to a water-resistant, high-density fiberboard core using a urea-based adhesive (16-97). These laminates are installed over a cushion underlayment as a floating floor. Laminate flooring is available in a number of thicknesses and sizes. Thicknesses of 3/8 and 1/2 inch are available. Widths of 7½ inches and lengths up to 85 inches are typical.

Laminate flooring is also available with the high-pressure laminate surface representing tiles and various stone materials. Tiles 12 x 12 inches and larger are available.

16-94 The laminate flooring is durable and is used in an area where traffic and wear will occur. *Courtesy Armstrong World Industries, Inc.*

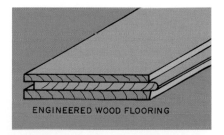

16-95 Engineered wood flooring is a laminate of three layers of wood and has a tongue-and-groove edge and end.

ENGINEERED WOOD FLOORING

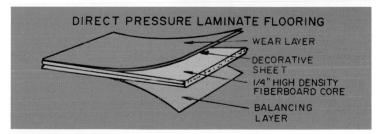

DIRECT PRESSURE LAMINATE FLOORING

— WEAR LAYER
— DECORATIVE SHEET
— 1/4" HIGH DENSITY FIBERBOARD CORE
— BALANCING LAYER

16-96 Direct-pressure laminating fuses the wear layer and decorative sheet onto the core material, using pressure between 300 and 500 pounds per square inch (psi).

16-97 High-pressure decorative laminate is manufactured at 1,400 pounds per square inch (psi). The laminate and a high-pressure balancing backer are bonded to a water-resistant, high-density fiberboard core.

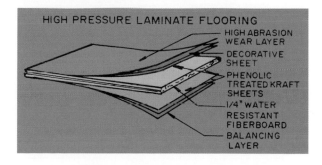

HIGH PRESSURE LAMINATE FLOORING

— HIGH ABRASION WEAR LAYER
— DECORATIVE SHEET
— PHENOLIC TREATED KRAFT SHEETS
— 1/4" WATER RESISTANT FIBERBOARD
— BALANCING LAYER

Installing Engineered Flooring by Stapling

The subfloor should be renovated as described earlier in this chapter. Engineered flooring can be secured to the subfloor by stapling or gluing.

The installation procedure is much the same as described for solid-wood flooring. If the starting wall is straight, measure out ¼ inch plus the width of one strip at each end wall, and mark this distance on the floor with a chalk line (16-98). If it is bowed or not perpendicular to the end wall, the first strip will have to be tapered or planed to fit these imperfections. Now check to see how the flooring strips will work out at the opposite wall. Measure the width of the room and divide by the width of the flooring. This will indicate how many rows of flooring are needed and the width of the last strip. If the last strip will be very narrow, cut some off the first strip so an equal-width strip will be at each wall.

STAPLE INSTALLATION

Cover the subfloor with red rosin paper as shown earlier in this chapter. It should be noted that while the flooring is generally stapled, it can be nailed.

Install the first row of flooring on the chalk line leaving a ¼-inch expansion space. Place the tongue edge on the chalk line and the groove side next to the wall (16-99). This installation is critical because all other rows depend on it for alignment; make certain it is straight and on the chalk line. If the wall is straight, ¼-inch blocks of plywood can be placed between the groove edge and the wall to set the space.

Now drill pilot holes through the face of the flooring next to the wall. The flooring can be secured with finishing nails or a **pneumatic brad tacker (16-100)**. Then toenail the strip through the tongue

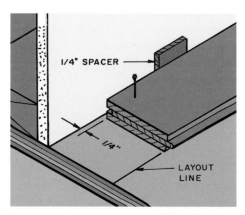

16-98 Begin the installation by marking a layout line on the sub-floor from the wall at ¼ inch plus the width of one strip of flooring.

16-99 Place the edge of the flooring with the tongue on the layout line. Face-nail the strip next to the wall.

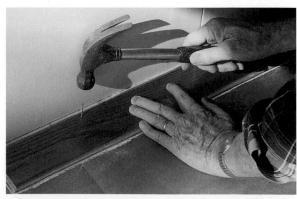

16-100 Position the first strip on the layout line and face-nail it next to the wall. The nail will be covered by the base and shoe.

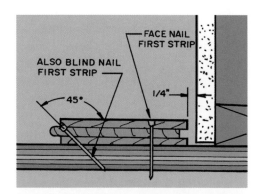

16-101 Toe-nail the first strip through the tongue; this helps stabilize it. This is necessary because the first strip sets the pattern for all the following strips.

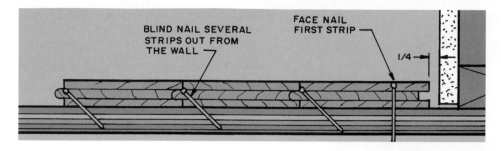

BLIND NAIL SEVERAL
STRIPS OUT FROM
THE WALL

FACE NAIL
FIRST STRIP

1/4

16-102 If you plan to use a power stapler, toe-nail the first three or four strips with finishing nails. This gets the edge away from the wall so the power stapler can be used. The entire floor can be hand-nailed with finishing nails if you prefer.

edge (**16-101**). Next install several rows by nailing through the tongue (**16-102**) with finishing nails set in drilled pilot holes (**16-103**). After several rows have been installed, a pneumatic stapler can be used—but not on the first couple of rows because there is not sufficient distance away from the wall to allow the stapler to be used. The air pressure on the stapler should be adjusted so the staple just nestles into the corner on top of the groove (**16-104**). If the staple is left too high, the butting piece will not close the joint. If it sits too low, it could split the tongue. Staple every 6 to 8 inches along the edge.

Use a tapping block to close the joints between strips (**16-105**). Stagger the end joints so that no

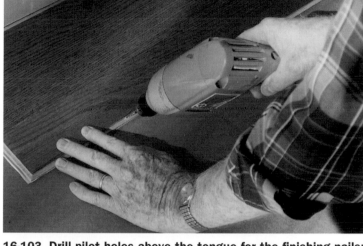

16-103 Drill pilot holes above the tongue for the finishing nails; this prevents splitting the tongue.

16-104 Adjust the power stapler so the staple is properly seated above the tongue.

16-105 Use a tapping block as needed to close the joints.

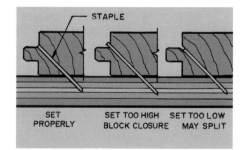

STAPLE

SET
PROPERLY

SET TOO HIGH
BLOCK CLOSURE

SET TOO LOW
MAY SPLIT

16-106 When the other wall is reached, put the last strip in on an angle and press to the floor. Face-nail next to the wall.

16-107 Mark the starting line a distance from the wall equal to two widths of the flooring strips plus ¾ inch for expansion.

16-108 Nail a solid-wood board along the layout line; it must be straight. the entire length of the wall.

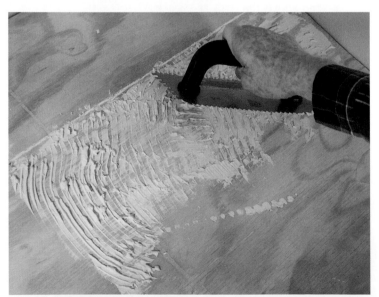

16-109 Spread a layer of adhesive from the layout line the width of two or three flooring strips.

adjacent strips are closer than 6 inches. The final strip will be fitted over the tongue and lowered into place (16-106). Then face-nail it along the edge by the wall. If the base and shoe will not cover the nails, set them and cover with a filler that matches the floor. The flooring will have to be cut to fit around pipes and posts, as discussed later in this chapter under laminate flooring.

Gluing Engineered Flooring

Engineered-wood strips can be installed by bonding to the subfloor with a manufacturer-recommended adhesive. Prepare the subfloor as described earlier. It is very important that it be clean and free of dust so the adhesive completely bonds to the subfloor.

Start by checking the width of the room, and calculate the width of the last strip. Again, if it is a narrow strip, consider cutting some off the width of the first strip so the strips at each wall are about the same. Lay out the starting line by measuring in the width of two pieces of flooring plus ¼ inch as shown in 16-107. If the wall is not straight, it will be necessary to trim the first piece of flooring.

Nail a straight piece of lumber, such as a 1 x 4, along the starting line. It is critical that this be straight. Check it with a chalk line. All other strips will be laid outward from it (16-108).

Now spread a layer of adhesive, from the layout line, a width of two or more flooring strips with a notched trowel (16-109). The manufacturer will specify the size of the notches to use. Usually ³⁄₁₆ x ⁵⁄₃₂-inch notches are adequate. They determine the thickness of the adhesive.

Let the mastic set for a while before starting to lay the flooring; typically 30 minutes is recommended. See the time on the label of the can.

Now place the first row of strips against the straightedge. Place the tongue against it. Place the second row against the first, and tap the joint closed with a tapping block. After the first two rows have been set, carefully press them into the adhesive (16-110). Then apply adhesive about two feet into the room along the full length. Do not put down more adhesive than you can cover before it sets; the can will give this information. Two to three hours is typical. Place planks in rows on top of the mastic. You can work several rows together. Stagger the end joints so no adjoining strips have end joints closer than six inches. After 4 or 5 rows have been laid, tie them together with strips of blue masking tape; place these about 12 to 14 inches apart **(16-111)**. Add more strips as the installation progresses. When you reach the other wall, fit the last piece in place. Trim it narrower, if necessary. Some installers face-nail this last piece on the edge next to the wall.

Return to the starting wall, remove the 1 x 4 starting strip, apply adhesive to the exposed subfloor. Allow it to set the required time and install the final two strips of flooring.

Scrape off any adhesive that may have worked up through the joint **(16-112)**. Then remove any surface residue with a foam pad and mineral spirits. Use a manufacturer-supplied cleaner to finish the cleanup. Then roll the floor with a 100-pound roller, available from most equipment rental agencies **(16-113)**. Remove the blue tape after the adhesive has set for 24 hours. The flooring will have to be cut to fit around pipes and posts, as shown later in this chapter under laminate flooring.

16-111 Tape the strips together after they have been firmly pressed into the adhesive and the joints tapped closed.

16-112 Carefully scrape off any adhesive that may have worked up through the joint with a plastic scraper.

16-110 After laying several strips press them firmly into the adhesive. Then apply more adhesive and install several more rows.

16-113 After the floor has been laid, roll it with a 100-pound floor roller.

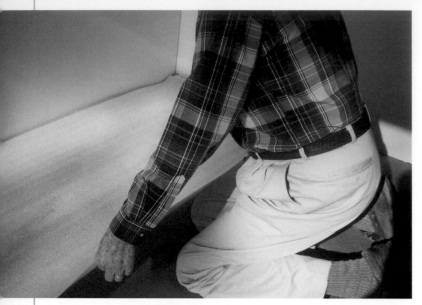

16-114 Lay the foam underlayment over the subfloor. Some recommend placing a plastic sheet or red rosin paper down first to serve as a moisture barrier.

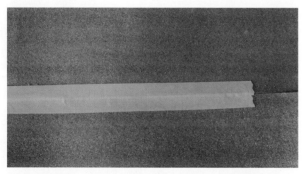

16-115 Butt the adjoining foam underlayment sheets and tape the joint with 2-inch-wide masking tape.

16-116 Lay the first three rows in a stair-step fashion. Continue spacing the end joints like this across the entire floor.

Installing a Laminate Floating Floor

Laminate floating floors are designed to be installed by gluing the tongue-and-grooved edges together, while other types use a special joint that does not require glue. The following discussion details the gluing technique.

Laminate flooring must always be installed over a **foam underlayment,** forming a floating floor. Some manufacturers recommend installing engineered flooring this same way.

Begin by installing a cushion-type underlayment; it also serves as a moisture barrier. Start laying the foam underlayment sheet along one wall, and spread it smoothly across the floor (**16-114**). Butt the edges together and seal with masking tape (**16-115**).

Now make a trial layout of the first three rows without glue. Lay with the groove edge facing the wall. Lay in a stair-step fashion, as shown in **16-116.** A typical layout plan is shown in **16-117.**

If the wall has a slight irregularity that will cause a problem, plane the edge of the first strip to match the curve of the wall. If it is straight, place the first strip next to it but insert a ¼-inch spacer block to provide clearance, as shown in **16-118** and **16-119.** Since the floor is on a pad, it could move up and down along the wall. If it touches the wall, it will cause a squeaking sound. If all is straight, disassemble the first three rows and reinstall with glue.

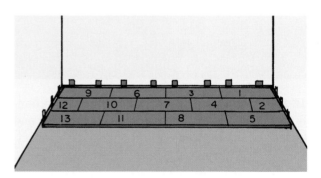

16-117 A typical plan for laying out the first three rows. Note that the end joints are staggered and spacer blocks are us along the walls.

As the trial layout is made, the end piece will have to be cut to fit the remaining space. Do not use pieces shorter than 8 inches. If this space is small, move the row over and shorten the piece against the right wall. With some adjustment the end pieces will fall above the minimum of 8 inches (16-120). Be certain to keep the end pieces ¼ inch from the end walls.

Now lay the first row tightly against the spacer blocks. Then start the second row, possibly using the piece cut off when fitting the first row.

To lay the second and third strips, begin by applying the manufacturer-recommended glue to the groove. Fill the groove the entire length of the strip and apply to the ends that butt another strip (16-121). Set the strip in place and press the tongue into the groove. Continue to install the second complete row. Use a tapping block to press the joint closed; tap with a hammer if necessary (16-122). Check the installation to be certain it is straight; use a chalk line or long straightedge. Then glue in the third row. Let the glue set before installing additional rows. All of the following rows depend on these first three rows being properly installed.

After the joint is pressed closed, some adhesive will be forced out of the joint. Remove any glue with a plastic scraper and wipe with a clean, damp cloth (refer to 16-112). Then wipe with a dry cloth.

16-118 Lay the first strip against the wall; place spacers along the edge and end of the strip.

16-119 Spacer blocks are placed on the edges by the wall and at the ends of each strip.

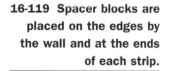

16-120 Adjust the length of the first piece in each row so that the piece on the other wall will be at least 8 inches long.

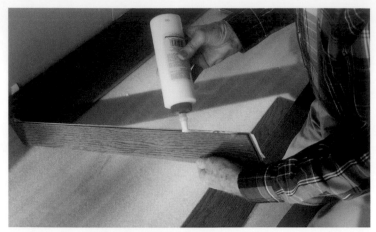

16-121 Apply the glue to the groove or as recommended by the flooring manufacturer. Apply to the entire length of the strip and to the grooves in the ends that butt another strip.

16-122 Continue installing the strips; press into place and tap with a tapping block when necessary. Notice that some of the glue is forced out of the joint.

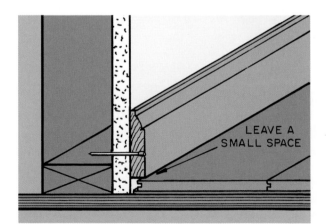

16-123 Install the baseboard to cover the space at the wall. Allow a space below the baseboard so the flooring can move/expand and contract without buckling.

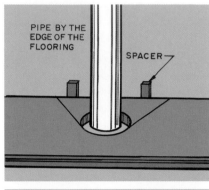

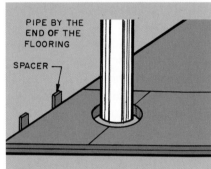

16-124 Two ways to fit engineered-wood and laminate flooring around a pipe or post.

Continue installing strips and work across the floor. Stagger the end joints. After 6 or 8 strips have been laid, tie them together with strips of blue masking tape (refer to 16-111). This tape is made to give a clean release from the surface of glass, wood, and metal if removed within seven days; it will leave no sticky residue on the surface. When you get to the opposite wall it may be necessary to cut a strip narrower to fill the remaining end space. Remember to leave a ¼-inch space at the wall.

Finally, remove the strips of masking tape, and clean off any glue residue using a cloth and mineral spirits, water, or a solution recommended by the manufacturer. This should be done within one hour after the floor has been laid.

Allow the floor to dry overnight before removing the spacers or walking on it. Then the baseboard can be installed (16-123).

FITTING AROUND PIPES

Often a pipe or post is in the field of the floor, requiring that the flooring be cut to fit around it. Two possible ways are shown in 16-124. After locating the center of the hole, bore it about ½ inch larger than the pipe. Cut across the flooring to the hole. Put glue on the edges and push the piece back in place. Hold it with spacers as the glue sets.

Installing Laminate Floating Floors without Glue

Some manufacturers supply laminate flooring that has a specially designed edge joint that snaps the strips together and requires no glue. The flooring is installed as described above except the gluing procedures are not needed.

To install this type of laminate flooring, place the joining strip at an angle to the groove on the first piece. Press down on it to snap the joint closed (16-125).

16-125 This laminate flooring has a special edge joint that does not require glue. When properly installed, it will snap the flooring strips together.

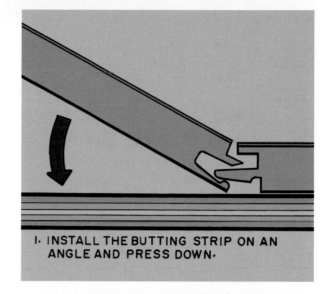

1. INSTALL THE BUTTING STRIP ON AN ANGLE AND PRESS DOWN.

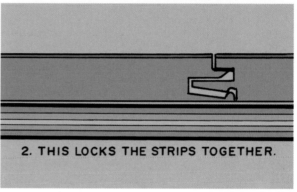

2. THIS LOCKS THE STRIPS TOGETHER.

CHAPTER 17

Finishing an Attic Room

Many homes have a large, unfinished attic area that could be converted into either a storage area or a habitable space (17-1). As you consider undertaking this project, there are many things to consider before you actually start to work. These factors will reveal many possibilities for use of the attic space and also will point out some limitations. Following are suggestions of things that should be investigated.

Evaluating the Structure of the Roof

First look in the attic. If the roof has been framed with **trusses**, you will most likely have to give up trying to convert the attic space into a livable area. You never cut or alter any part of a truss (17-2).

The **upper cord** serves as the rafter and carries the sheathing and shingles. The **lower cord** serves as the ceiling joist. It carries the drywall finished ceiling. It is designed to carry only the ceiling load and cannot be used to support the floor on the attic room. The **web members** carry tension and compression forces, providing the strength and rigidity of the assembly. Cutting any of these will lead to a truss failure. The various parts of a truss are joined with **gussets.** These can be wood or metal gusset plates. Metal plates are most commonly used.

If you must alter a truss, secure the services of a structural engineer to redesign the web structure so that when existing web members are removed, other framing is in place to provide the needed support.

17-1 This room over a two-car garage was made into a comfortable extra bedroom.

If you are adding a new room or garage to the house, the roof can be framed with trusses that are designed to provide the room cavity and carry the roof and floor loads (17-3).

Developing a Plan

Begin by making a scale drawing of the area on ¼-inch-square graph paper. Let each square represent a distance, such as one foot, or if the room is small, let each square represent 6 inches. Locate doors, stair, windows, electric features, as well as wall outlets, lights and switches, and any plumbing that may already be in the area or need to be added. If you plan to add a bathroom, try to locate it near an existing soil stack so a short connection can be made. Also realize a bathroom requires extra load-bearing capacity in the floor—so this must be part of the overall plan.

The following photos and drawings are for an existing conversion of a space over a large garage. The side view of the garage is shown in **17-4.** This shows the hip roof, which is very steep, and a skylight. The end of the garage is shown in 17-5. Notice how the

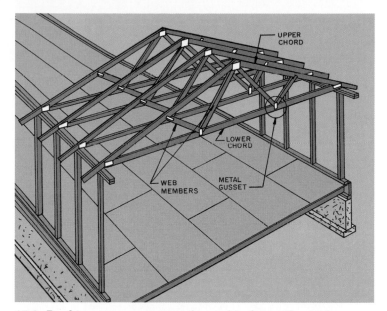

17-2 Roof trusses are commonly used to frame the roof; however, the web members cross the area where a room could be built, making it unlikely you can convert this space into a habitable area.

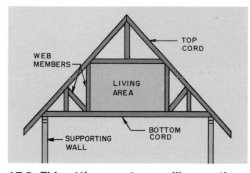

17-3 This attic room truss will carry the roof loads and provide space for a habitable living area.

17-4 This is a side view of the garage showing the skylight. Notice the hipped roof.

17-5 This is an end view of the garage showing the hip roof.

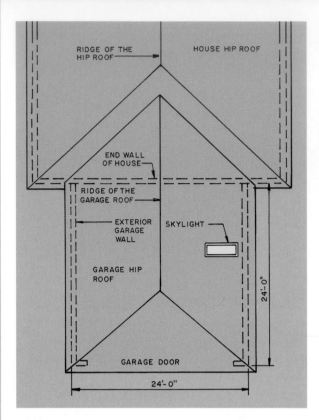

17-6 This is a drawing of the roof of the house showing the size of the garage, the type of roof, and the roof overhang.

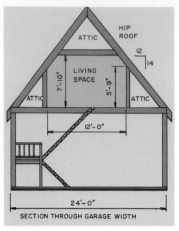

17-7 These drawings establish the width and length of the garage and the height of the ridge and proposed ceiling and knee wall. The drawings set the number of square feet of living area available with these dimensions.

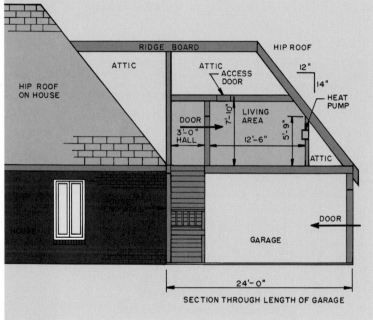

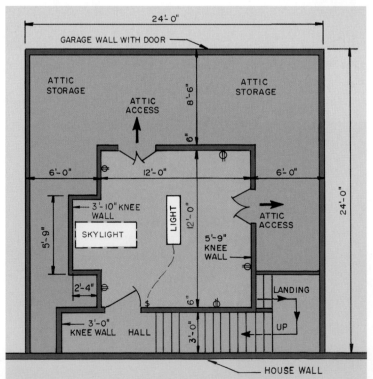

17-8 A floor plan of the proposed room over the garage. Notice how the floor was extended over to a low knee wall, allowing access to the skylight.

hip roof slopes back sharply from the exterior wall. This will reduce the amount of living space available in the room. A top view of the garage showing the extent of the hip and how the garage roof connects with hip of the house is shown in 17-6.

After careful measurements of the garage were taken, a section drawing was made through the width and length of the garage (17-7). These show the area available and the decisions that were made on this project. Notice the sloped roof which allows the ceiling of the room to be 7'-10", and makes it seem spacious. It was also decided to make the knee wall 5'-9" high. This allows easy use of the floor right up to the knee wall. How to develop a floor plan:The beginning plan in 17-8 locates the walls, access to the skylight, attic door, and the stairs that must be built to reach the room. Notice a small hall was laid out at the top of the stairs. The other features were added to the floor plan, including electrical outlets and lights.

Assessing Code Requirements

Check the local building code for the required minimum square feet of approved habitable area. Typical requirements indicate the minimum habitable area of a room is 70 square feet; however, any flooring area below a sloped ceiling less than 5 feet or a flat furred ceiling less than 7 feet is not considered habitable area. No habitable room can have more than 50 percent of the required floor area below a sloped ceiling that is less than 7 feet high.

Codes will require a glazed area equal to 8 percent of the floor area and an operable open window area of 4 percent of the floor area.

ACCESS TO ATTIC SPACE

Building codes require a permanently installed stairway to habitable living space on upper floors. They also specify the design requirements for it. For access to livable space in an attic, the stairs will have to be 36 inches wide and have a handrail. The maximum riser height allowed is 7¾ inches and the minimum tread width is 10 inches (17-9). There should be at least 6'-8" headroom over all parts of the stairs. A typical layout for a straight stairway is in 17-10.

If the house was built without a permanent stairway to the attic space, the stairs on the first floor will have to be found. The stairs in 17-11 were built to access livable space over a garage. The garage was deep enough to allow the stairs to be built on the end wall and still allow a small to midsize car to fit in the garage (17-12). After you design the stairs, you will

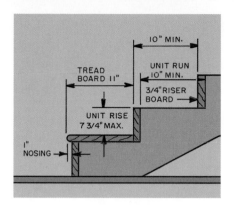

17-9 This shows typical code-specified minimum tread size and maximum riser sizes.

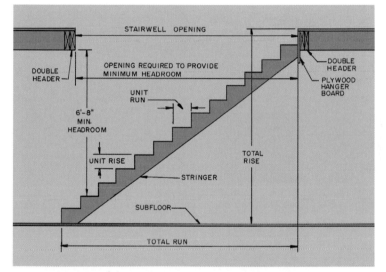

17-10 A typical layout drawing for a straight stair. Observe the tread and riser specifications in the local building code.

17-11 This U-shaped stair was built in the end of the garage to provide access to a room over the garage.

389

17-12 After the space required for the stairs is known, measure and find how much space will be left for a car. Will this be enough for your needs?

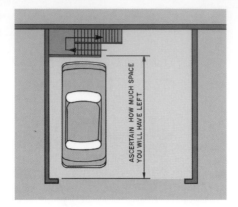

ASCERTAIN HOW MUCH SPACE YOU WILL HAVE LEFT

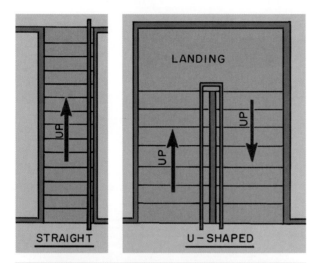

LANDING

UP

UP

UP

STRAIGHT

U–SHAPED

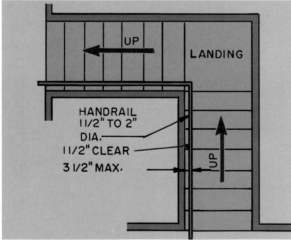

UP

LANDING

HANDRAIL
11/2" TO 2" DIA.

11/2" CLEAR

3 1/2" MAX.

UP

17-13 Frequently used stair layouts. The choice depends upon the space available on both floors. The stairs from the landings need not be the same length. The height of the landing will influence the number of treads on each section of stair.

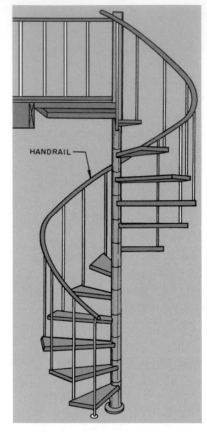

HANDRAIL

17-14 Circular stairs provide access to another level and require a minimum of floor space.

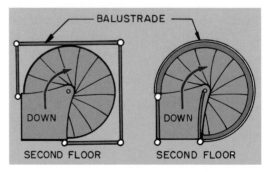

BALUSTRADE

DOWN

DOWN

SECOND FLOOR

SECOND FLOOR

17-15 Round stairwells can be protected with rectangular or round balustrades.

know how much space is taken up on the first floor. Since there are several stair designs available, this gives some help in finding a location. Typical examples are shown in **17-13**.

In some cases **spiral stairs** are permitted to be used to provide access to second floor living spaces, but they must meet detailed code requirements (17-14).

A spiral stair rises in the form of a helix. As it rises it turns around a large central newel post. It requires less floor space than other types of stair.

As the spiral stair enters the second floor, the opening is protected by a round or rectangular balustrade as shown in **17-15**, **17-16**, and **17-17**. Spiral stairs can be made to rotate in a clockwise or counterclockwise direction (17-18). Some size information is also shown in 17-18.

Disappearing stairs are suitable for access to attic space that is used for light storage. They are never approved for use as access to habitable space (17-19). Two types are available. One has a folding ladder (17-20), and the other a straight, fixed ladder that slides up into the attic (17-21). They are easy to install and provide safe access if used according to the manufacturer's load-bearing limits.

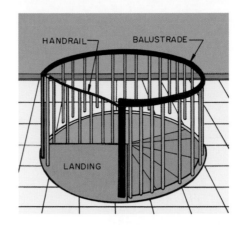

17-16 The stairwell opening must be fully protected to avoid accidental stepping into the opening. This round balustrade is strong and decorative.

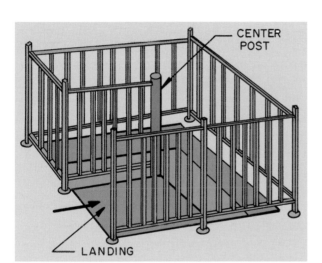

17-17 A rectangular balustrade can be installed around a square or rectangular floor opening at the second floor level.

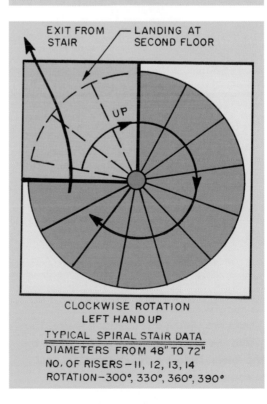

COUNTERCLOCKWISE ROTATION
RIGHT HAND UP

17-18 Spiral stairs are designed to rotate access clockwise (left-hand up) or counter clockwise (right-hand up).

17-19 Disappearing stairs provide an economical way to access storage space in the attic and do not permanently occupy floor space.

CLOCKWISE ROTATION
LEFT HAND UP
TYPICAL SPIRAL STAIR DATA
DIAMETERS FROM 48" TO 72"
NO. OF RISERS – 11, 12, 13, 14
ROTATION – 300°, 330°, 360°, 390°

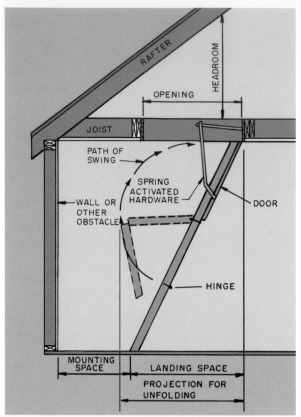

17-20 This is a location drawing showing the factors that must be considered as you decide where to install folding-type disappearing stairs.

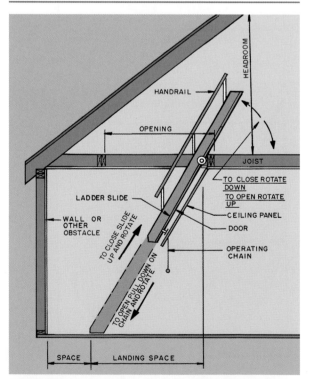

17-21 These are sliding disappearing stairs. They slide up into the attic and folds down against the top of the joists. Note the spacing factors that must be considered.

When planning to use disappearing stairs, check the attic in the proposed location to see if the roof there is high enough to give the necessary headroom. Also lay out the landing area, folding area, and mounting area required to lower and mount the ladder. (Refer to **17-20** and **17-21** for examples of these layouts.) If they can be placed near the center of the house, the maximum headroom in the attic will be available.

Disappearing stairs are available in several sizes and load-carrying capacities. A 300-pound maximum load is a common specification. Study the manufacturer's recommendations for installation. They describe how to prepare the opening and adjust the length of the ladder so it sits firmly on the floor.

ROOF SLOPE & PITCH CONSIDERATIONS

Consider the slope or pitch of the roof and the width of the house (**span**) as you decide if it is possible to actually make a room that is code-approved for a living area. The ceiling should be at least 7 feet high and the room should contain at least 70 square feet of area below this ceiling. Check your local building code.

Generally the angle of the roof is given in terms of slope. **Slope** is a ratio of the **rise** in inches to **12 inches of run**. The **run** is half the span. The roof in **17-22** has a slope of 10 inches of rise per 12 inches of run. In other words, it rises 10 inches for every 12 inches of run, so it has a 10/12 slope.

The **pitch** of a roof is the angle expressed as a ratio of the **rise** to the **span**. For example, the roof in **17-22** has a rise of 10'-0", therefore it has a pitch of 10/30 or ⅓. It has 10 inches of rise for every 2 feet of span.

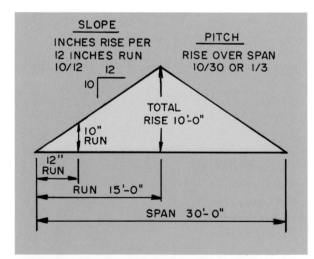

17-22 The angle of the roof can be expressed in terms of slope or pitch.

Attic Headroom

Before you get too far into planning an attic conversion, be certain there is enough height to permit the room to be built according to code.

To ascertain if there is enough headroom to finish the attic, you can measure 7 feet from the subfloor and run a chalk line, establishing the ceiling. Then drop lines from each end of the rafter to the floor to get the width of the proposed room. Multiply the width by the proposed length to get the usable square feet of livable floor area (17-23). The effect on available living area is shown by the illustrations in 17-24.

Assessing the Joist Size

Generally the ceiling joists of the first floor become the floor joists for the attic room. Usually the ceiling joists are smaller because they only carry the drywall and insulation, so if a floor load is placed on them they will have to be strengthened.

Building codes usually require the floor in sleeping areas to be able to support a minimum design **live load** of 30 pounds per square foot plus a design **dead load** of 10 pounds per square foot. Rooms other than sleeping rooms will typically have a minimum requirement of 40 pounds per square foot **live load.** The dead load is typically 10 pounds per square foot. If a bathroom is to be included, additional load considerations will be necessary.

If the attic is used for limited storage and is reached by a disappearing stairway, the joist size used for the ceiling joists is usually acceptable.

A **live load** is the weight of items to be brought into the room. The **dead load** is the weight of the materials making up the floor and ceiling assembly.

The size of the joist required to carry the specified load depends upon the unsupported span and the species of wood. Building codes contain tables for determining these sizes.

REINFORCING THE ATTIC FLOOR JOISTS

There are a number of ways the attic floor can be strengthened to meet the building code. Before you select one of the following suggestions, check with the local building inspector to see if the plan is acceptable.

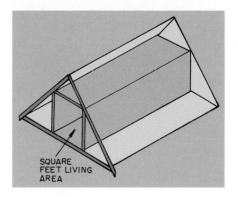

17-23 The livable floor area is noted in square feet, easily found by multiplying room width by length.

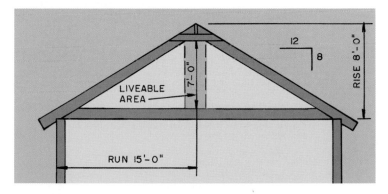

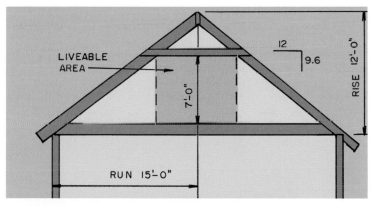

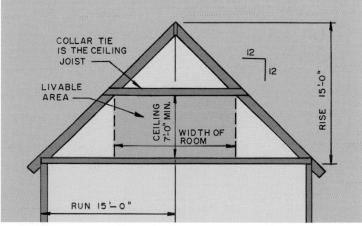

17-24 The slope of the roof influences the width of the room and the available livable area.

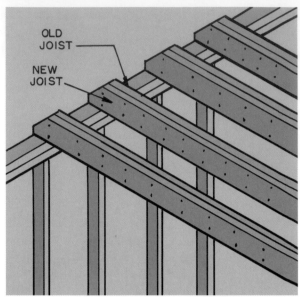

17-25 One way to strengthen the ceiling joists to carry the new floor is to double the joists.

Perhaps the most commonly used method is to double each joist by gluing and nailing one of the same size next to it (17-25). If necessary because of a long span, you could add a larger joist to each of the existing joists such as adding a 2 x 10 joist to an existing 2 x 6 or 2 x 8 joist (17-26). While this will give the strength needed, it will lower the ceiling height, so be certain you allow for this as you plan.

Depending on the situation in the rooms below, you could run a girder along the ceiling (17-27) or build a load-bearing partition (17-28). Both will materially affect the room below.

Taking Collar Ties into Consideration

As you check out the attic space, you will usually notice there are **collar ties** which run across the attic and are nailed to each pair of rafters (17-29). Measure

17-26 When joists wider than the existing ceiling joists are added, this reduces the ceiling height.

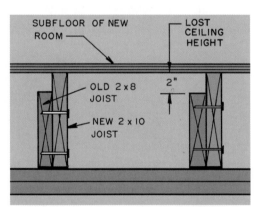

17-27 A girder can be added to support ceiling joists so they will carry the floor loads. This reduces the span of the joists.

17-28 A load-bearing partition can be installed below the ceiling joists to support the floor load. This has a great effect on the room divided by the partition.

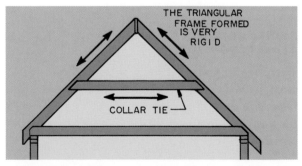

17-29 Collar ties are an important part of the roof framing structure. If you must move them, get professional advice.

from the proposed level of the new floor to the collar ties. If they will be 7 feet or more above the floor, they can serve as ceiling joists for the flat portion of the ceiling. If that distance is not available, you will have to check with an engineer to see what happens if you move them. They are an important part of the roof framing plan and cannot simply be knocked out and moved. It is possible you might be permitted by the building inspector to install shorter ties above the original ties. If you do this, be certain to install the new collar ties **before** you remove the original ties.

The addition of ¾-inch CDX plywood gussets on each side of the rafters at the ridge will add to the strengthening of the roof (17-30).

Allowing for Natural Light & Ventilation

If the attic room will be against the gable end, it is possible to cut an opening and install a window in it

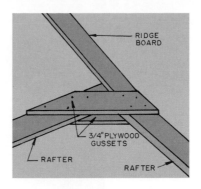

17-30 Plywood gussets on each side of the rafters at the ridge board add strength to the roof framing.

17-31 This window has been installed in the gable end and the sheathing has been covered with a moisture-protecting plastic sheet.

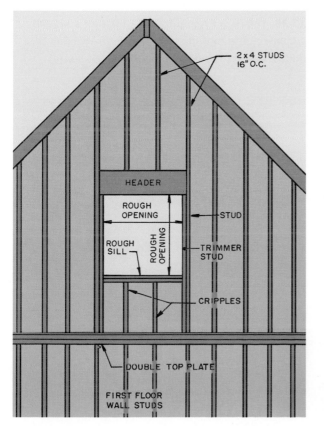

17-32 The siding is laid over the sheathing, completing the installation. This gable end was covered with wood shingles and painted.

17-33 Typical framing details for putting a window in a gable end.

17-34 These single-window dormers have been framed, and sheathed, and the windows installed. They are ready for the siding to be installed.

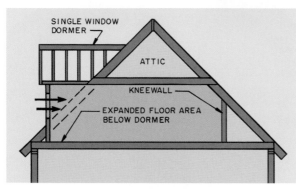

17-35 This section through an attic room shows how a single-window dormer adds a little additional floor area plus light and ventilation.

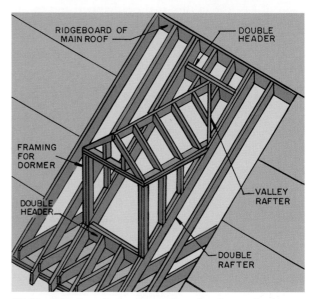

17-36 This is a typical framing plan for a single-window gable roof dormer. The ridge extends to a double header located below the ridge board.

17-37 A shed dormer is usually quite large and has several windows.

(**17-31** and **17-32**). This might require some extra bracing to support the ridge board of the roof and you will need to install a header over the window opening (**17-33**).

Another solution is to install one or more **dormers**. The most frequently used dormer is placed up on the roof and holds a single window (**17-34**). This provides light, ventilation, and a small floor area (**17-35**). Typical framing is shown in **17-36**.

A more elaborate dormer that provides considerable floor space, light, and ventilation is a **shed dormer (17-37)**. The section drawing in **17-38** shows how a shed dormer runs from the ridge to a wall that rests on top of the exterior wall of the house. A typical framing drawing is shown in **17-39**.

As you consider dormers, two factors must be evaluated: cost, and the exterior appearance of the house.

It is important to preserve the integrity of the exterior of the house. Some styles, such as Cape Cod and Early American, have dormers as part of their overall design (**17-40**). To place these on a design in

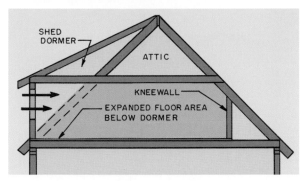

17-38 This section through a shed dormer shows how it adds considerable floor area plus light and ventilation. It usually has the outside wall set on top of the exterior wall of the house.

17-40 These dormers fit in with the overall design of this period house and complement the total appearance.

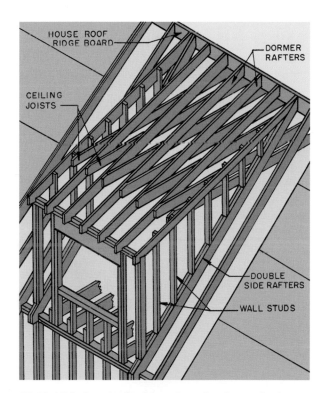

17-39 This is a typical framing plan for a shed dormer. The rafters have been run from the ridgeboard of the house roof.

17-41 The dormer is not in keeping with the architectural style of this house and is a distraction.

which they will clash is a mistake (**17-41**). Also it is important to get the dormers in proportion to the size of the house and the amount of roof. Oversize dormers make the house look top-heavy (**17-42**). An interesting design shown in **17-43** has a dormer specially designed to be compatible with a contemporary house. The large dormers in **17-44** provide considerable light and floor area in the attic, but they must be on a large house.

The **shed dormer** tends to be large, and if it is on the front of a house, affects the styling. The shed dormer in **17-45** is on a side roof. Notice the roof starts at the ridge, and the outside wall is set in above the knee wall. In **17-46** the roof does not go to the

17-42 This oversize dormer dominates the front elevation and makes the already busy house look top heavy.

17-43 The contemporary house has a large metal roof and an especially designed dormer that is compatible with the size of the roof and the architectural style.

17-44 This large dormer is used on a contemporary house that is large enough to visually accept the mass.

ridge, but its outside wall rests above the outside wall of the house.

Another way to get light and ventilation into an attic room is to use **skylights (17-47).** They are often placed on the rear roof because many people do not like them on the front but basically they are rather unobtrusive **(17-48).**

If the attic room is large and has a wide span and a steeply sloped roof, a series of skylights can provide a dramatic, well-lighted interior **(17-49).** Generally, in an attic the window can be at a level where it is easily reached and can be opened with a cranking

17-45 This shed dormer runs from the ridge of the house. The wall with windows is above the knee wall.

17-46 This shed dormer roof does not run to the ridge of the house, but the wall with windows is above the exterior wall of the house.

17-47 This fixed skylight lets in a lot of natural light but does not provide ventilation.

17-48 Skylights are inconspicuous and do not conflict with the architectural style of the house.

Courtesy Velux America, Inc.

17-49 A series of skylights can provide considerable natural light. Some types have a cranking mechanism that can be reached with a pole to open, providing ventilation.

Courtesy Velux America, Inc.

17-50 Skylights are opened with a cranking mechanism. A screen fits over the opening on the inside of the window.

mechanism (17-50). This provides a little more floor space and ventilation. The installation in 17-51 was made on a truss-framed roof. Notice the floor has been extended to a knee wall directly below the skylight. The skylight fits between two trusses. The opening to the skylight was widened to include three trusses.

Another approach is to install a **balcony roof window (17-52)**. The floor is extended to the exterior wall, widening the room. The window has two sashes. The top sash swings out, forming a roof over the balcony railing. The bottom sash swings out on a bottom hinge, forming the railing of the little balcony (17-53). The view provided is terrific.

Another attractive installation involves installing a manufactured **attic dormer sunroom** that provides

17-51 This skylight was installed between two roof trusses that were designed to form a room in the attic. The vertical members shown are truss cords. They were painted to match the interior trim.

17-52 This balcony roof window is manufactured as a complete unit and makes the sloping roof a window wall. It opens upon a balcony. *Courtesy Velux America, Inc.*

17-53 A section through the balcony roof window. The top sash opens up, providing ventilation. The lower sash hinges at the bottom, providing a balcony railing. The end railings are fixed. *Courtesy Velux America, Inc.*

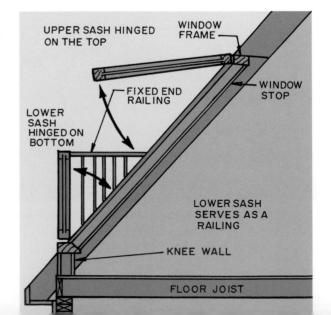

UPPER SASH HINGED ON THE TOP

WINDOW FRAME

FIXED END RAILING

WINDOW STOP

LOWER SASH HINGED ON BOTTOM

LOWER SASH SERVES AS A RAILING

KNEE WALL

FLOOR JOIST

considerable light and greatly increases the floor space in the attic room (**17-54**). It has a maintenance-free aluminum glazing track system on the exterior of laminated Northern white pine beams.

Roof Ventilation

Very likely the existing roof may already have provision for ventilating the enclosed attic space that surrounds the room framing. Basically, soffit vents flow air into the attic space and some provision is made to direct the flow up and out from below the roof. The existing roof may have ridge vents, gable end vents, or power fans. If this venting is not there, it should become part of this remodeling project; otherwise the new room and existing roof will be subject to high temperatures and moisture which can allow mold to take hold and cause wood to deteriorate.

Constructing the Attic Room

After you have worked out your plan, addressed your building codes, and obtained the necessary permits, you are ready to build out the space. Construction includes the floor, walls, and partitions, as well as the needed electrical, plumbing, and heating/air-conditioning features.

Installing the Subfloor

The subfloor can be ⅝-inch plywood or oriented-strandboard. Install the long side of the panels perpendicular to the floor joists (**17-55**). The panels should have tongue-and-groove edges, so blocking will not be required below the edges.

Secure the panels to the floor joists with 8d ring or screw-shank nails. They are spaced 6 inches apart on all edges and 12 inches apart on the joists within the panel. It is a good practice to lay a bead of adhesive on top of the joist and then lay the panel in place. Leave a ⅛-inch space between the edges and ends of the panels. Stagger the ends of the panels as shown in **17-55**.

Installing the Knee Wall

Knee walls are short partitions which run from the sloping rafter to the subfloor. They are typically 4 feet high; however, walls as low as 3 feet can be successfully used, though they limit access to the floor area along the wall (**17-56**).

17-54 This attic dormer sunroom provides considerable natural light and greatly increases the floor space of an attic room. *Courtesy Four Seasons Solar Products Corporation*

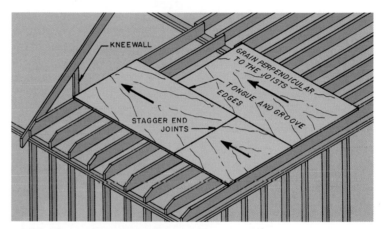

17-55 Plywood or oriented-strandboard subflooring is installed with the long direction perpendicular to the joists, and end joints staggered over a joist. Edges should be tongue-and-grooved.

17-56 This knee wall is about 6 feet high. Notice the access door to storage space behind the wall. This area is covered with subflooring.

17-57 Typical knee wall construction. The bevel on the top plate should be flush with the edge of the stud.

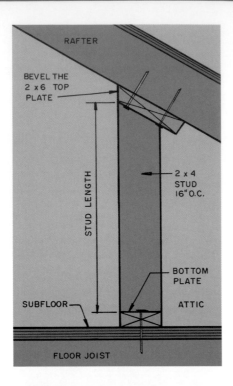

17-58 Locate the top edge of the knee wall with a chalk line.

17-59 Use a level and a piece of scrap 2 x 6 to mark the angle to be cut on the top plate.

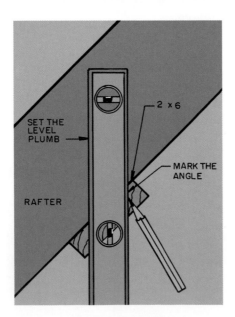

A knee wall consists of a top plate, studs, and a bottom plate (17-57). The first thing to do is run a chalk line along the rafters, locating the front edge of the wall (17-58). Be certain to measure the height perpendicular to the floor. This will also let you find any rafters that are not in the same plane. If a rafter is bowed and does not touch the chalk line, nail a thin wood shim in the area, bringing it flush with the chalk line. If a rafter is bowed out, you may have to jack it straight and nail another rafter to it to hold it straight. If the rafters are in bad shape, considerable structural renovation will be necessary before starting to frame the room.

The top plate is a 2 **x** 6 member with the edge facing the room cut on a bevel so it is in the plane of the wall. To get the angle, hold a piece of 2 **x** 6 along the rafters and place a level vertically until it is plumb. Mark the angle on the stock. Use this angle to cut a bevel on all the pieces of top plate (17-59).

Lay the bottom plate on the subfloor and measure the length of the wall stud on the long side as shown in 17-57. Mark the angle of the cut on the end of a scrap piece and use it to set the miter saw. Cut all the studs to length with the bevel on the top end.

There are several ways the knee wall can be installed. Possibly the easiest way is to assemble each wall flat on the subfloor (17-60). Space the studs 16 inches on-center. Then place it against the rafters and nail the bottom plate to the subfloor and the top plate to the rafters. Often the wall studs will not line up with the rafters so it is easy to nail to them (17-61). Be certain to check the wall for plumb before nailing.

As part of the preliminary planning for knee wall construction, consider making some of the attic space behind the wall storage. One way is to prepare a door

17-60 The knee wall can be assembled on the subfloor and raised into place.

(17-62) in the framing and lay the subfloor over the joists in the attic space (17-63). This provides good storage space but it is unprotected from the heat, cold, and dust that filters into the attic through the vents. The door opening should be weather-stripped to keep the cold and hot attic air from entering the room. Bond rigid plastic insulation to the plywood as shown in 17-63.

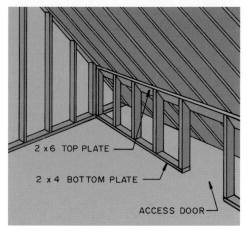

17-61 Place the knee wall against the rafters and sub-floor, check it for plumb, and nail to the rafters and sub-floor.

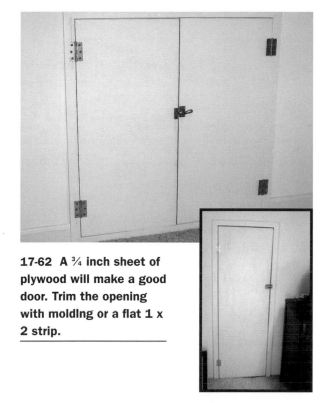

17-62 A ¾ inch sheet of plywood will make a good door. Trim the opening with molding or a flat 1 x 2 strip.

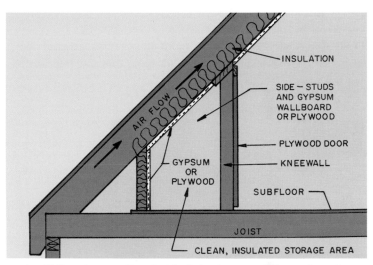

17-64 If the storage area behind the knee wall is enclosed and insulated, items stored will be fully protected.

17-63 The space behind the knee wall can be covered with the subfloor and exposed with a small door, providing usable storage space. Notice the rigid plastic insulation on the door.

17-65 The enclosed storage area can have shelving installed to handle many small items.

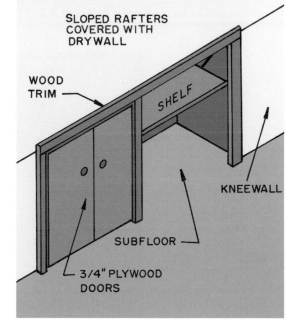

17-66 The end wall to an attic room has the door to enter the room. Notice that closets were built along this wall because it has the area with the highest ceiling.

17-67 The end wall of this attic room was moved over far enough to allow a bed to be positioned between it and the knee wall.

A much better plan is to frame in a storage enclosure and insulate it as shown in 17-64. An easy way to do this is to build the sides from ½-or ¾-inch plywood or 2 x 4 studs and cover with gypsum wallboard. Shelves could be installed if small items are to be stored here (17-65).

Installing the End Partition

The room will usually have a partition built across the end, running parallel with the rafters. And it will have a door providing access to the new room (17-66 and 17-67). Plan how the room will be used before locating the door.

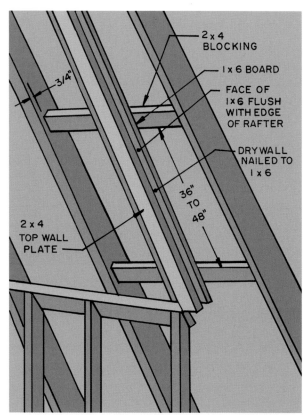

2 x 4 BLOCKING

3/4"

1 x 6 BOARD

FACE OF 1 x 6 FLUSH WITH EDGE OF RAFTER

DRYWALL NAILED TO 1 x 6

36" TO 48"

2 x 4 TOP WALL PLATE

17-68 Locate the end wall and then construct the top plate on the sloped rafters. If it falls between rafters, install with blocking.

17-69 After the top plate is installed, locate the bottom plate with a chalk line and nail it to the subfloor.

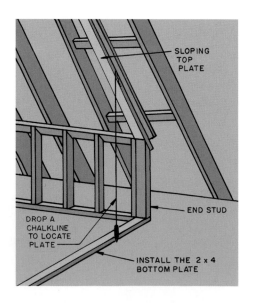

SLOPING TOP PLATE

DROP A CHALKLINE TO LOCATE PLATE

END STUD

INSTALL THE 2 x 4 BOTTOM PLATE

After locating the end wall, install the top plate parallel with the rafters. Usually this will fall between rafters, so nail 2 x 4 blocking between the rafters to which the top plate is nailed **(17-68)**. Then drop a chalk line to the floor and locate the bottom plate **(17-69)**. Nail it to the subfloor.

You can now proceed to frame the door opening. It will have the longest studs and the other studs can be located from each side of it. Then cut the short studs to length, cutting the top end on the angle of the top plate **(17-70)**. Again check each stud for plumb. After the wall has been completed, cut away the bottom plate that is across the door opening.

Installing Electrical & Plumbing Features

Once the framing is complete, have the electrical features installed and any plumbing piping put in place. Since these are carefully regulated by the building code, it might be best to have licensed contractors do the work. If you plan to try it, first visit with the building inspector to see what requirements must be met. It may be that licensed contractors have to be hired to check your work before the building inspector signs off on that part of the project.

Heating/Air-Conditioning the Room

Providing heat and cooling to the room will often not be possible using the existing system. Getting ducts to the room and providing return air ducts is usually not possible due to the location of the furnace. A visit to a heating/air-conditioning dealer will provide you with alternate choices.

HEAT PUMPS

In geographic areas where electric heat pumps will function satisfactorily, a small unit such as that shown in **17-71** and **17-72** is a good choice. It is efficient and both heats the room in the winter and cools it in the summer. The outdoor unit is connected to the indoor unit with small-diameter copper tubing and electricity is run to both units. The heat or cooling from the coils in the indoor unit is moved into the room by a fan.

The heat pump dealer should visit the room and take measurements so the proper-size unit will be supplied.

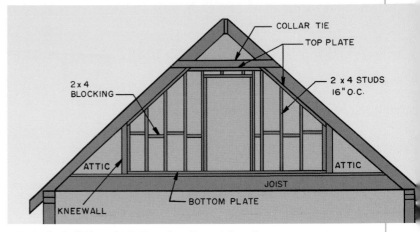

17-70 Install the studs framing the end wall.

17-71 The indoor unit of a small-room heat pump mounts in the wall. It is operated by a remote control which directs it to heat or cool and sets the desired room temperature.

17-72 The outdoor unit sits outside the house. In the winter the outdoor unit takes heat from the air to heat coils in the indoor unit. In the summer the room heat is taken up by the coils in the indoor unit and discharged outside by the outdoor unit.

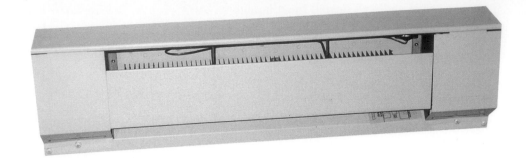

17-73 An electric baseboard provides an excellent way to heat a room. It requires only

ELECTRIC BASEBOARD HEATERS

Various types of **electric room heater** are available. One type is in the form of a baseboard heater **(17-73)**. These types of heater require no duct work or vents and are relatively easy to install. As the heating elements heat the air, it rises, and the cooler air near the floor flows into the grill at the floor, is heated, and exits out the grill along the top of the unit **(17-74)**. The heat flow is produced by convection. **Convection** is the motion of air due to a heat differential. In this case, cold air is dense and therefore heavier, and settles to the floor. As it is heated it becomes less dense, lighter, and then rises.

The baseboards should be installed on exterior walls because they are the coldest, and the heat from the baseboard will flow up the wall and warm this cool surface, making the room more comfortable.

Since it will require several baseboard units to heat the attic room, a separate electric circuit should be run in that is large enough to handle the required power load.

First determine the number of watts needed to heat the room. A **watt** is a unit of electrical power. Meet with your dealer, presenting information about the room, such as the cubic-foot measurements of the room, the size and number of windows, and the amount of insulation. With this information the dealer can determine about how much heat the room will need. This is measured in British thermal units (BTUs). A **British thermal unit** is the amount of heat required to raise the temperature of one pound of water by one degree Fahrenheit. Then the baseboard units can be selected. They are available in lengths from 2 feet to 10 feet and require from 200 to 2,500 watts of power. Once these decisions have been made, the electrician can figure the power needed to operate the baseboard and the size wiring required. For example, a 3-foot baseboard could require 300 to 750 watts, depending upon the heating elements. To get the BTU rating of an electrical heater when you know the wattage, multiply it by 3.4. For example, a 300-watt heater will produce 1,020 BTU.

Depending on the situation, the electrician will decide if the baseboard heaters will operate on 120V or 208V power. This must be decided before you purchase the baseboard units. Also, select units that provide thermostatic control.

EXTERIOR WALL

HEATED AIR

METAL FINS

ELECTRIC ELEMENT

COOL AIR IN

FLOOR JOIST

17-74 Electric baseboards heat the cool air on the floor and the heated air rises into the room.

ELECTRIC WALL HEATERS

Another type of electric heat is a fan-force **wall heater** **(17-75)**. It has electric heating elements surrounded by steel fins. A blower forces air over the elements and heated fins direct it into the room. The entire mechanism is mounted first in a metal case, and then mounted on the wall.

The electrical considerations are much the same as discussed for electric baseboard heaters.

GAS SPACE HEATERS

Natural gas or propane heaters are another choice for heating the new attic room **(17-76** and **17-77)**. Should you live in an area where a little heat may be needed occasionally, the gas-fired log fireplace **(17-78)** will provide some heat to remove the chill and be very decorative. As you choose a unit, give the dealer information about the space to be heated so the number and size of wall heaters can be determined. You will have a choice between vented and unvented heaters. Typically building codes require gas heaters

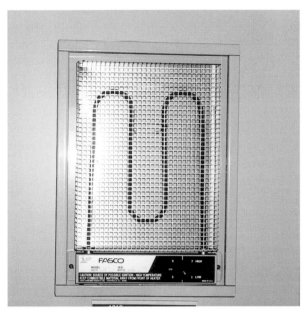

17-75 This is one type of wall-mounted electric heater. There are a variety of sizes and designs available.

17-76 Wall-mounted gas heaters provide considerable heat. They can operate on natural gas or propane.

17-77 This gas-fired heater can be mounted on the wall or rest on the floor as shown.

17-78 This gas-fired fireplace is made as a total unit and set in place on the floor. It will provide quite a bit of heat and is very decorative.

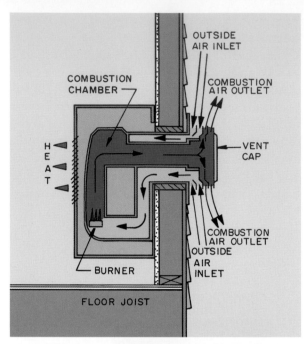

17-79 This is one system used to vent gas-fired wall heaters. The inflow of outside air is an important part of the venting system. Notice the double pipe used to exhaust combustion gases and bring in fresh outside air to support combustion.

17-80 Window air conditioners can be installed in windows or through an exterior wall.

to be vented unless a particular unit has been approved for nonvented operation.

If you expect to rely on several wall-mounted gas heaters to provide continuous heat over a period of several cold months, the dealer will most likely recommend using a vented unit. It is safer and free from possible air contamination. In **17-79** is an illustration showing a section through a typical vented wall-mounted gas heater. The vent goes through an exterior wall. Notice in this example the vent is a pipe within a larger pipe. The combustion gases are vented from the combustion chamber through the center pipe directly to the outdoors. Fresh outside air is run into the combustion chamber through the larger outside pipe. This unit will not rely on the indoor air and thus will not reduce the indoor oxygen content. In all cases observe the manufacturer's installation instructions. In some areas the code will require the heaters to be installed by approved technicians employed by the dealer supplying the unit.

Unvented noncatalytic gas heaters are illegal in most places. They pose great dangers.

There are, however, **vent-free gas heaters** that meet the standards of the American Natural Standards Institute (ANSI). These are **catalytic gas heaters** which in the catalytic combuster provide complete combustion of all flue products released from the primary fire at the burner. The room requires fresh air to support combustion, and ventilation to make the room a safe place to live. Catalytic heaters have specific requirements for admitting fresh air into the room where the heater is installed. Observe the manufacturer's recommendations and contact the local building inspector as you install this type of heater.

AIR CONDITIONERS

Since electric and gas heaters do not provide cooling air in the summer, a small air conditioner will be required in warmer climates.

A small room can be cooled by installing a room-size window air conditioner (**17-80**), which comes in various cooling capacities. So take information about the size of the room to the dealer as you choose a unit. The air in the room is circulated into the unit where it is cooled, dehumidified, and filtered. Air conditioners operate on electricity; small units use 120V current, while larger units use 220V current. Also check the amperage required. It may require a separate circuit if the amperage is near 20 or other appliances are connected to the circuit you plan to use.

I. SCREW THE EXPANDABLE PANEL TO THE WINDOW SASH AND SILL AS DIRECTED BY THE MANUFACTURER.

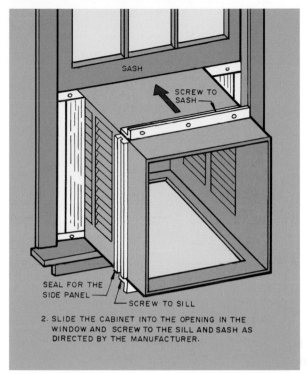

2. SLIDE THE CABINET INTO THE OPENING IN THE WINDOW AND SCREW TO THE SILL AND SASH AS DIRECTED BY THE MANUFACTURER.

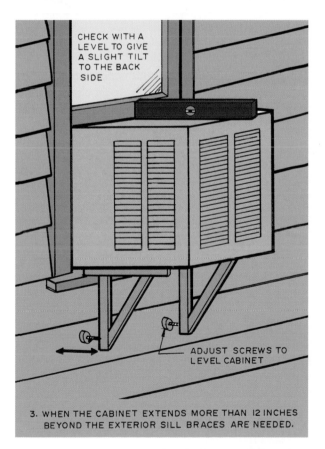

3. WHEN THE CABINET EXTENDS MORE THAN 12 INCHES BEYOND THE EXTERIOR SILL BRACES ARE NEEDED.

4. INSTALL THE CHASSIS IN THE CABINET. SECURE IT TO THE CABINET AS DIRECTED BY THE MANUFACTURER.

17-81 Typical steps for installing a room air conditioner in a window. Follow the manufacturer's instructions.

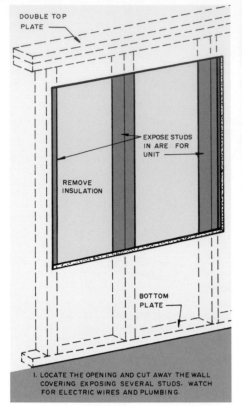

DOUBLE TOP PLATE

EXPOSE STUDS IN ARE FOR UNIT

REMOVE INSULATION

BOTTOM PLATE

1. LOCATE THE OPENING AND CUT AWAY THE WALL COVERING EXPOSING SEVERAL STUDS. WATCH FOR ELECTRIC WIRES AND PLUMBING.

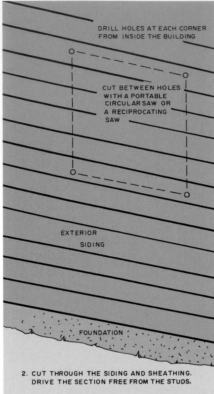

DRILL HOLES AT EACH CORNER FROM INSIDE THE BUILDING

CUT BETWEEN HOLES WITH A PORTABLE CIRCULAR SAW OR A RECIPROCATING SAW

EXTERIOR SIDING

FOUNDATION

2. CUT THROUGH THE SIDING AND SHEATHING. DRIVE THE SECTION FREE FROM THE STUDS.

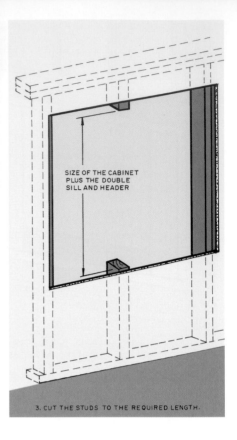

SIZE OF THE CABINET PLUS THE DOUBLE SILL AND HEADER

3. CUT THE STUDS TO THE REQUIRED LENGTH.

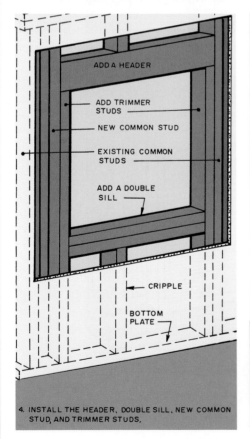

ADD A HEADER

ADD TRIMMER STUDS

NEW COMMON STUD

EXISTING COMMON STUDS

ADD A DOUBLE SILL

CRIPPLE

BOTTOM PLATE

4. INSTALL THE HEADER, DOUBLE SILL, NEW COMMON STUD, AND TRIMMER STUDS.

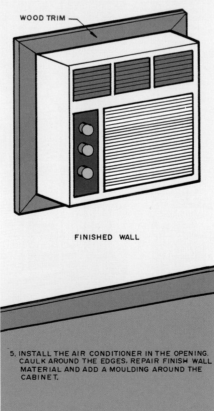

WOOD TRIM

FINISHED WALL

5. INSTALL THE AIR CONDITIONER IN THE OPENING. CAULK AROUND THE EDGES. REPAIR FINISH WALL MATERIAL AND ADD A MOULDING AROUND THE CABINET.

17-82 Room air conditioners can be installed by cutting an opening in the wall. Be careful when cutting to first find and move any electrical wires or plumbing that may be in the area to be cut open.

INSTALLING AN AIR CONDITIONER IN A WINDOW

To install an air conditioner in a window be certain the window is wide enough to hold the unit. It comes with side panels that expand to fill the space between the unit and the sides of the window. Typical installation details are in **17-81.** Always review and follow the manufacturer's instructions.

If the unit is small, it can be slid through the side panel and the flanges can be screwed to the window frame and sill. If it is a larger, heavier unit, it is easier if you remove the chassis holding the coils and compressor and mount the empty cabinet **(17-81).** If the cabinet extends more than 12 inches outside the window sill, it should be braced with metal brackets. Adjust the leveling screws on the bracket so the cabinet slopes about ¼ inch toward the outside, so moisture condensing in the cabinet will drain to an outside drain tube. Then slide the chassis into the cabinet from the room side and screw it to the cabinet. Be certain that all functions connecting the cabinet, sill, window frame, and side panels are sealed. Rubber-gasket-type material is available for this purpose.

INSTALLING AN AIR CONDITIONER THROUGH A WALL

While a small air conditioner can be installed in a window, this blocks natural light and ventilation. An alternative is to install it through an exterior wall. This could be the gable end or the outside wall of a shed or gable dormer. If the unit is placed high on the wall it will give better distribution of the cooled air.

Study the installation directions from the manufacturer and use the brackets or sleeves provided.

To install a window air conditioner through an exterior wall **(17-82),** follow these steps:

1. Mark the opening on the inside wall. Many prefer to place it higher on the wall than when the air conditioner is in a window because cold air is heavy and sinks to the floor; this gives the air a longer reach into the room. Cut the opening with a utility knife, drywall cutter, or saber saw. Cut it back until the studs to hold the unit have been exposed.

   Be very careful while cutting that the electric wiring or plumbing is not damaged. It may be helpful to cut the area into small pieces, to remove them, so that the wall cavity can be checked. Pull out the insulation that may hide wires or plumbing. If these are in the opening area, the opening can be moved or the wires or plumbing rerouted.

2. Drill holes in each corner of the desired opening through the sheathing and exterior siding. Then, from the exterior, cut the hole with a portable circular saw. A few nails may be hit in the process.

3. Cut the studs within the opening, leaving space for the unit plus the double sill and header. The opening should be about ¼ to ½ inch larger than the cabinet. Remember, this is a load-bearing wall, so if the ceiling needs temporary support, block it up with several 2 x 4 members joined to a horizontal member next to the ceiling.

4. Install the framing to form the finished size of the opening and to support the ceiling or second-floor load.

5. Install the cabinet and chassis as shown and described in **17-81.** Caulk around the sides of the opening and the cabinet. Repair damage to the finish wall and nail a molding around the unit to give a finished appearance.

Installing the Insulation

While there are a number of different types of insulation that could be used, **fiberglass blankets** with a treated paper vapor barrier on one side are most commonly used. They are efficient and easy to install **(17-83).** The vapor barrier keeps moisture that may develop in the room from penetrating the fiberglass insulation. It also provides ¾-inch flanges, used to staple the insulation to the wood framing **(17 84).**

The thickness of the insulation required will vary a great deal depending upon the climate. First realize the ceiling always requires more insulation than the

17-83 Fiberglass insulation blankets provide excellent results and are easy to install. *Courtesy Owens Corning*

Table 17-1 Typical R-Value Ratings for Fiberglass Blankets and Batts.

R-Value	Thickness (inches)
R 11	4 to 5 1/4
R 19	7 to 8 3/4
R 30	11 1/4
R 38	14 to 17 3/4
R 49	18 to 23

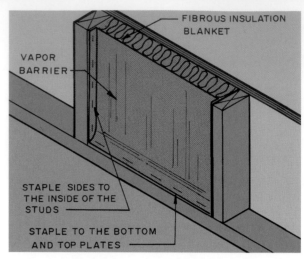

17-84 Staple fiberglass insulation blankets to the inside face of the studs, the rafters, ceiling joists, and the top and bottom plates.

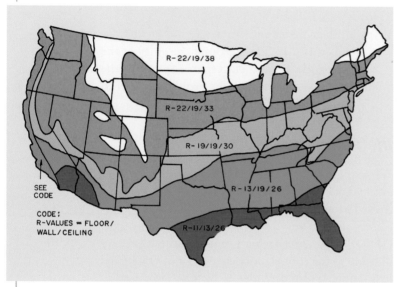

17-85 Suggested R-values for various parts of the United States.

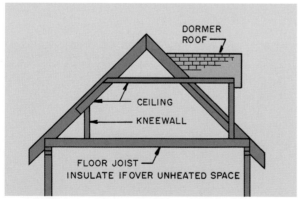

17-86 Insulate the knee wall, ceiling, other exterior walls and the floor if the new room is over an unheated space, such as a garage.

floor and walls. The floor and wall requirements are generally close to the same thickness.

Insulation is specified by its R-value. **R-values** are a measure of a material's capacity to resist the flow of heat through it. The higher the R-value, the greater the insulation value. In **Table 17-1** are R-values for various thicknesses of fiberglass insulation. In **17-85** are examples of recommendations for broad geographic areas. The data is shown as follows: R=22/19/33. This gives the recommendations for the floor (22), wall (19), and ceiling (33). Check with the local insulation dealer for specific recommendations for your space's needs.

Once the framing is complete, install insulation blankets in the knee wall, the sloped side ceilings, and the flat ceiling (**17-86**). Remember to place the side with the vapor barrier **toward** the room. The insulation in the sloped ceiling should leave a clear space of at least 2 inches between the roof sheathing and the insulation (**17-87**). This is necessary so the air flow from the soffit vents can flow to the vents high in the roof, thus ventilating the attic space around the room. Staple to the inside of the rafters and studs, and to the top and bottom plates (**17-88** and **17-89**).

Be certain to stuff loose pieces of fiberglass in the openings around the windows and doors and the wood framing. Do not pack it solid but leave it loose and open. It works better if air spaces between the fibers exist. Then staple a vapor barrier over this insulation. This could be a piece of polyurethane sheet material (**17-90**).

After the insulation is in place, install the drywall or other paneling on the ceiling and walls. Then install the finish flooring and interior trim.

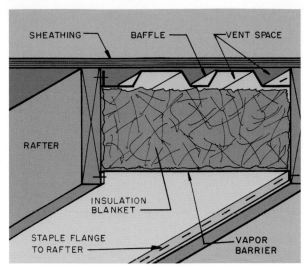

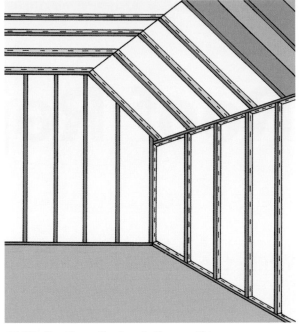

17-87 When insulating the sloped ceiling, leave 2 inches of clear space between the roof sheathing and the insulation. A baffle can be used to ensure the space will be maintained.

17-89 Continue the insulation up the slanted ceiling and over the flat ceiling across the room. Then do an exposed end wall.

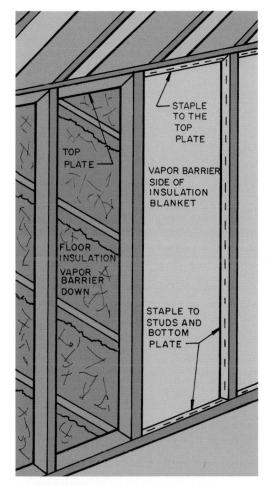

17-88 Begin by stapling the insulation blankets to the knee wall studs and the top and bottom plates.

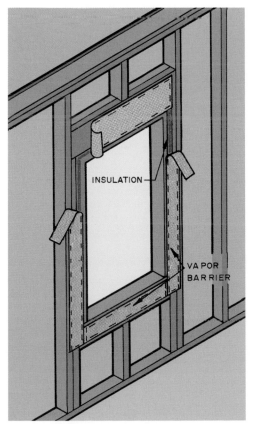

17-90 Openings around door and window frames should be filled with insulation and covered with a vapor barrier.

413

CHAPTER 18

Roof Repairs

While the exterior of the house gets considerable wear from the elements, the roof seems to get the very worst. It is raised to high temperatures, exposed to rain, ice, snow, and subjected to freezing temperatures, and high winds. It should be checked regularly for signs of deterioration that indicate some renovation is needed (18-1).

Check the shingles for loose nails or damaged areas, or if the roof covering is 15 to 20 years old, consider if it might need replacing. If it has asphalt shingles, are there spots where the granules have worn away, reducing its ability to resist sun, wind, and rain? Check in the attic for signs of leaking. This will often show on the ceiling of the room below. Check all flashing to see if it is still firmly secured and if it is rusting and needs a protective coating or replacement. Obviously, if you have gutters they should be cleaned frequently, especially in the fall when the trees lose their leaves. Clogged gutters can eventually cause the fascia to rot, so check it as well. The water will also get the edge of the roof sheathing wet if a metal drip edge has not been installed. Rotten sheathing along the edge of the roof is a major repair problem. The soffit has vents used to flow air through the attic to vents high on the roof. They should be free of clogs and have the protective screen intact to keep out insects and squirrels.

18-1   This fiberglass shingle roof has a 25-year warranty and has been on for 15 years, shedding surface granules. Take a real close look at it to see if it appears to be suitable for continued use.

Finally, look for sags in the roof. This indicates some type of structural problem requiring some reinforcement of the rafters, or possibly installation of additional rafters.

Fixing Sagging Rafters

Sagging rafters can be caused by deteriorated wood rafters, undersize rafters, or rafters spanning a distance beyond their structural capabilities. Check for the amount of sag by running a chalk line from one end of each rafter to the other end. The amount of sag can be measured from the string to the edge of the rafter at the highest point on the bow (18-2).

If the rafter is sound and the amount of sag is small, it can sometimes be straightened by installing a 2 x 4 or 2 x 6 strut between the rafter and a 2 x 6 plate nailed to the top of the ceiling joist (18-3). A purlin is notched about ¾ inch into the strut and provides bracing for the entire length of the roof. The length of the strut should be long enough to force the rafter straight. If the strut is a 2 x 4 it should not be over 8 feet long. If the length exceeds this, go to a 4 x 4 or 2 x 6 strut. The strut should be on an angle of at least 45 degrees. If the bow is larger you may have to use a jack to straighten the rafter and run a 2 x 4 or 2 x 6 strut to the top plate of an interior load-bearing wall (18-4). If a load-bearing wall is not close, consider

18-2 Use a chalk line to check the rafters for bow. The amount of bow in this illustration is greatly exaggerated.

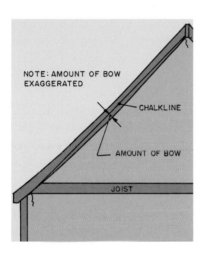

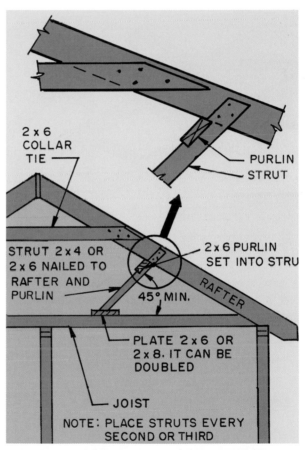

18-3 If the rafters need just a little straightening and stiffening, braces can be run to a plate on top of the joists. Anything with greater stress should go to a load-bearing interior wall.

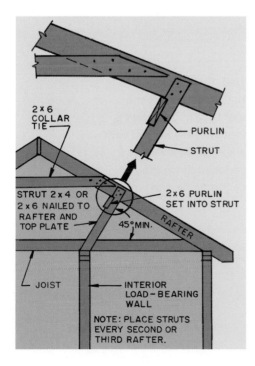

18-4 Greater stress can be placed on the strut if it is run to a load-bearing wall.

running the strut to a 4 x 6 girder placed on top of the ceiling joist (refer to **18-3**). This is a matter of judgment. You do not want to crack the drywall ceiling below. Try to place the struts above each ceiling joist if possible.

The struts installed as just shown not only straighten the rafter but reduce the distance it must span, so if the rafter is undersize, they will also help correct this problem because they reduce the span of the rafter.

Another way to stiffen and support rafters that are a little undersize is to build a knee wall the length of the roof. Nail it to the rafters and ceiling joist (**18-5**).

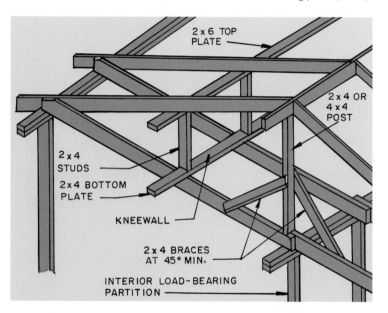

18-5 Sagging rafters can be straightened and held straight with knee walls. A sagging ridge board can be raised and propped up with posts.

It may require a little jacking or temporary braces to hold the rafters straight as the knee wall is installed. See Chapter 17 for information on building knee walls. If it would help to stiffen the ridge board, posts can be run from a load-bearing interior wall or from 4 x 6 plates on top of the rafters as shown in **18-3**. Brace the posts with side struts. The size of the post will vary depending upon the length. The short rises, such as 3 to 5 feet, can usually use a 2 x 4 post. Longer rises require a 4 x 4 post. Place every 4 feet or closer if needed.

Once the rafters are straight and braced, the roof can be stiffened by adding collar ties and gussets if they are missing. Collar ties are placed in the upper third of the roof height (**18-6**). Some add a 2 x 4 brace on top of the collar ties if they are more than 8 feet long.

Gussets are also used to stiffen rafters. They are made from ¾-inch CDX plywood and are glued with waterproof glue and secured with drywall screws to each side of the rafters (**18-7**).

If the rafters are seriously undersize or spaced wider than 16 inches on center, they can be strengthened after they have been straightened. This requires nailing another rafter to each or a 2 x 4 as shown in **18-8**. it is referred to as "sistering" the rafter, and can also be done if the rafter has started to deteriorate. The addition of gussets plus the sister rafter greatly strengthens the roof. If the rafters are spaced too far apart consider adding new rafters between them (**18-9**). Run from the ridge to the top plate of the exterior wall.

When rafters run perpendicular to the ceiling joists they can be braced to 2 x 6 or wider stock nailed

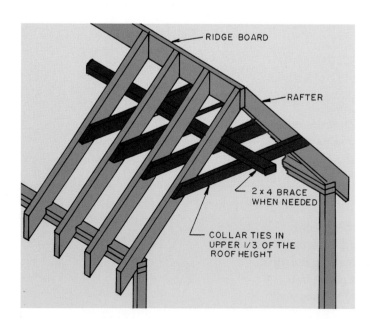

18-6 Install collar ties in the upper third of the roof height. Ties over 8 feet long are braced with a horizontal 2 x 4 installed down the center of the ties.

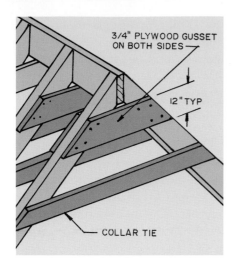

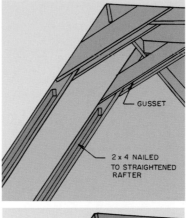

18-7 Plywood gussets installed on both sides of each set of rafters helps stiffen the roof frame.

18-8 Weakened rafters can be strengthened by installing a new rafter next to them. This is called "sistering."

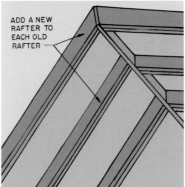

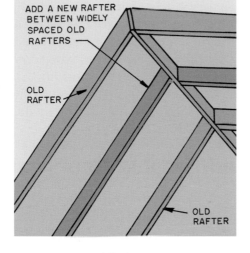

18-9 If rafters are sound but spaced wider than 16 inches on center, a new rafter can be installed between them.

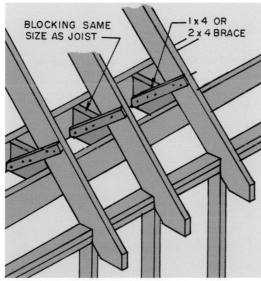

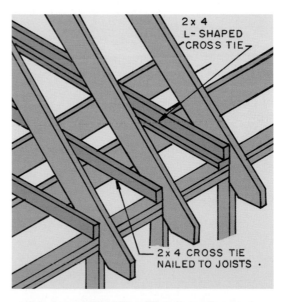

18-10 When the rafters run perpendicular to the ceiling joists, the rafters can be braced by first running blocking between ceiling joists and then nailing a brace to it from each rafter.

18-11 A crosstie will stiffen the rafter and ceiling joist framing and serve as support to tie the rafters across the house together.

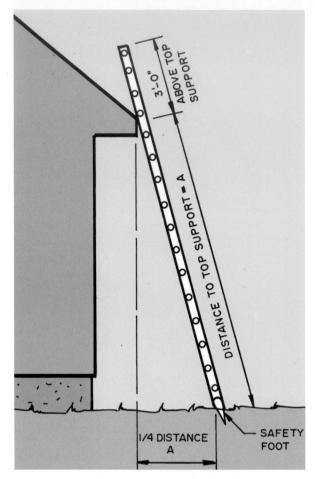

18-12 It is important to correctly position the ladder and have the feet firmly placed on level ground.

between the joists (**18-10**). A crosstie to the other wall nailed to the ceiling joists adds stability to the ceiling joists (**18-11**) and ties the rafters at the top plate.

Working on the Roof

Working on the roof is an especially dangerous situation. If the roof is very steep it would be wise to call a roofing company to make repairs.

If you plan to do the repairs, first be certain the ladder is in good shape. Do not use a ladder that has a broken part or is bent. Do not use a ladder that has had a temporary repair made on a bent or broken part. Position the ladder so the feet are about one-fourth its length from the wall and the top extends three feet above the edge of the roof (**18-12**). The ladder should have feet that permit it to be placed firmly on the ground, and they should be level. If the ground slopes, build up the lower side with large pieces of plywood so the foot is on a firm base (**18-13**). As an extra precaution you can drive stakes into the ground to assure the feet will not slip away from the house. This is especially important if the ground is wet (**18-14**).

Addressing Roof Leaks

Roof leaks can cause serious damage to the sheathing, rafters, trusses, ceiling joists, insulation, and the interior ceiling and walls. The first sign of a leak may

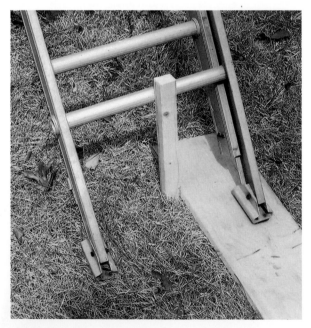

18-13 If the ground is not level, build up the low side with large pieces of plywood and stake the ladder so it will not slip.

18-14 A couple of stakes help to keep the feet of the ladder from sliding away from the house.

be minor damp spots on the ceiling. You may notice dampness on a wall or a moldy smell. Mold can develop in the attic and inside a wall or ceiling. When the leak gets to this point it has probably been wetting the sheathing and rafters for a long time and they have possibly been damaged and need repair or replacement. Repairing leaks at a chimney is discussed later in its own section.

Where to Look for Leaks

Go on the roof and check the following places for damage. Look carefully because the opening or crack may be very small, yet will allow water to penetrate below the shingles.

1. Damaged or missing shingles, slates, or tiles.
2. Flashing around skylights, chimneys, and vent pipes.
3. Valley flashing and flashing where the roof butts a wall.
4. Clogged or damaged roof drains and gutters.
5. Ice dams.
6. Damaged hip and ridge shingles.
7. Chimney masonry that absorbs water or has damaged mortar joints.
8. Deteriorated sheet roofing on a flat roof or one with a little slope.

FINDING THE LEAK

Begin by going into the attic and looking for water stains on the sheathing and rafters and for wet insulation. The wet spot on the ceiling will help you know in a general way which part of the attic to explore. The moist area will appear darker than the natural color of the sheathing. Generally, the moist area will be some feet away from the wet spot on the ceiling because the moisture tends to run down the sheathing before it drops off to the ceiling (18-15). Remember, a wet spot on the ceiling could be caused by a leaking water or waste-disposal pipe. If it is a two-story house and the wet ceiling is on the first floor, it is not likely to be from a leaking roof but from pipes in the ceiling.

Once the leak is located in the attic, measure the distance from the ridge board and an outside wall. Go on the roof and lay out these distances to find the area that's leaking. Now plan to make the needed repair. Be careful when you walk on the roof because you may cause considerable damage with your shoes. Wear soft-sole shoes. These help protect the shingles and also give some protection from slipping. If the roof is wet it is especially slippery, and if it is steep, consider staying off until it dries. Remember, if you

walk on cold asphalt shingles they may crack and if they are hot the surface can be damaged when you step on it.

Preventing & Dealing with Ice Dams

If you live in areas that have considerable snow and long periods of cold, it is important that the roof be prepared to resist leakage due to ice dams. An **ice dam** is caused when heat from the attic melts the snow on the roof and the water trickles down the roof to the

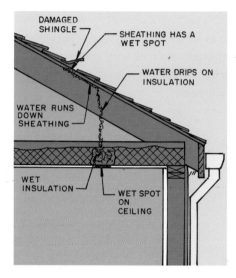

18-15 A roof leak often runs down the sheathing before it drops on the insulation, so the wet spot is often not directly below the leak.

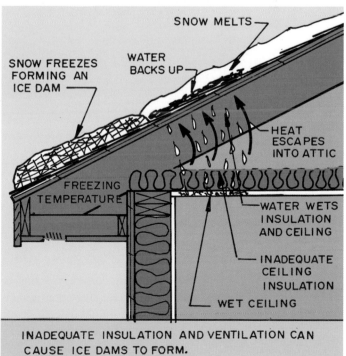

18-16 Inadequate ceiling insulation and attic ventilation can lead to the formation of ice dams at the eave.

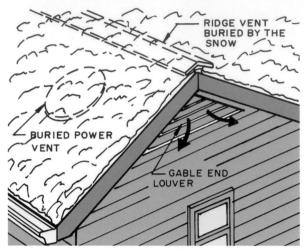

18-17 In areas of heavy snowfall, ridge vents and power vents often are covered with snow, eliminating this means for ventilating the attic. Gable end vents will remain open.

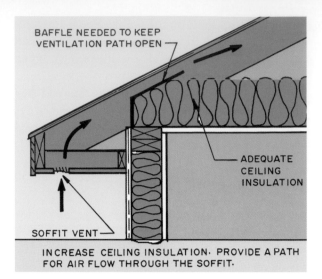

18-18 Increase ceiling insulation to reduce heat loss into the attic. Provide a baffle to ensure that the air flows freely from the soffit vents into the attic.

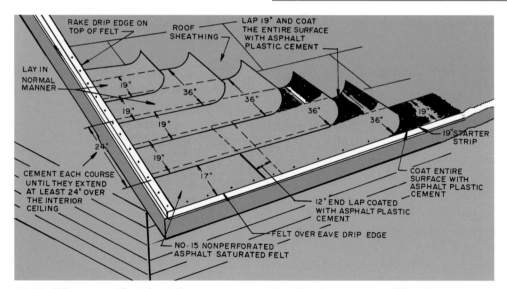

18-19 When re-roofing, install proper eave protection to prevent ceiling damage, should an ice dam occur.

eave, only to freeze on the overhang. As this layer of ice builds under the snow, water will eventually find its way under the shingles and onto the interior ceiling (18-16). If you notice icicles hanging from the eaves, this signals you have an ice dam problem.

If the attic has adequate ventilation, heat lost to the attic will not build up and cause the snow on the shingles to melt. The air in the attic should have the same temperature as the outside air. If you notice the snow sealing off the ridge vents and electric fan vents, consider adding louvers in the gable ends. Fans can be mounted on the louver to increase the volume of air flow (18-17).

Also increase the amount of insulation in the ceiling to reduce heat loss to the attic. The amount recommended will vary with the climate. Adequate soffit vents are also needed to provide a good air flow (18-18).

Some place electric heat tape along the eaves. It will tend to melt the snow and allow the water to drain off the roof. Sometimes this will not solve the problem but only move the ice dam up the roof. Consider good ventilation and insulation.

If the shingles are to be replaced, that is the time to install eave protection to seal the roof so if an ice dam occurs, the water will not find its way into the house. Along the eave and **at least two feet** over the interior of the house, install a synthetic rubber shield or cemented double layers of No. 15 asphalt roofing felt, as shown in **18-19.**

Quick Temporary Repairs

There will be times when a storm causes some damage or you just noticed a leak and need to make a quick repair as you plan for a permanent fix. If some shingles have become curled or even have pieces broken off, lift the shingle and slide a waterproof sheet material under it. A piece of aluminum or sheet metal makes a good repair (**18-20**). You can bond the shingle to the repair piece with roofing cement if needed to close up the edge.

A large damaged area can be covered with a durable plastic sheet, builder's felt, or some other water resistant material. Slide it under the shingles on top and seal the edges with roofing cement (**18-21**). If necessary to hold it in place, wood strips can be nailed across the repair (**18-22**).

Two other temporary repairs are shown in **18-23** and **18-24.** These are satisfactory enough to last quite a while. They do produce a large black spot on the roof which, because it is a different color than the shingles, is quite noticeable. If this is a problem, plan to replace the shingle as soon as possible.

Small cracks or splits in fiberglass and asphalt shingles can be repaired with roofing cement as follows:

1. Lift the shingle and spread a layer of roofing cement underneath it on both sides of the split (**18-23**).
2. Press the shingle into the cement.
3. Fill the split with more roofing cement. Be sure to spread the cement about 1 inch (25 millimeters) on each side of the split on top of the shingle.

This procedure produces a good temporary patch. It will last longer if a piece of fiberglass cloth is pressed into the roofing cement.

Make a temporary repair on a badly torn shingle as follows:

1. Put a layer of roofing cement under the torn shingle as described above.
2. Press the shingle into the cement and nail the loose edges to the sheathing with 1½-inch galvanized roofing nails.
3. Apply a layer of fiberglass cloth and more roofing cement on top of the tear (**18-24**). If the wind raises the ends of the shingles but has not broken them, they can be recemented. Put 1-inch dabs of roofing

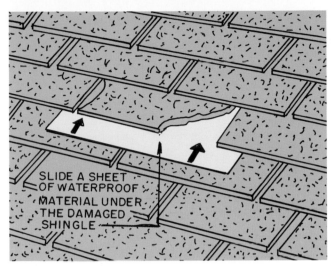

18-20 An emergency repair can be made by sliding a waterproof sheet under the damaged shingles.

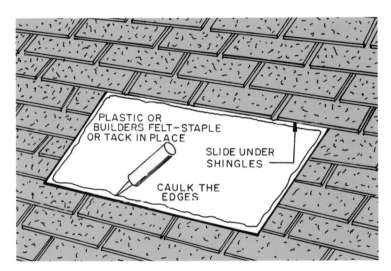

18-21 Cover the damaged area with plastic sheet or builder's felt. Tack or staple in place then caulk around the edges.

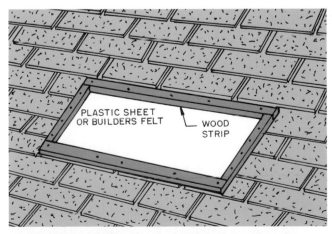

18-22 If it is windy, nail wood strips along the edges.

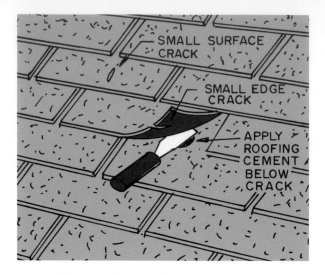

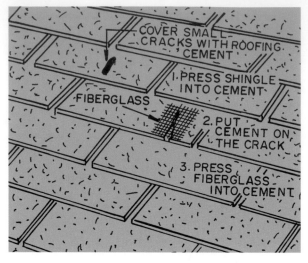

18-23 Minor surface cracks can be sealed with roofing cement. Edge cracks should be bonded to the shingle below and sealed on top with roofing cement and a piece of fiberglass mesh.

cement under the edge and press the shingles in place **(18-25)**.

Shingles can be worked more easily when they are warm. Shingles may crack if they are worked on when cold. If the roof is very hot, the shingles can be damaged when they are walked on. In the summer, work early in the morning before the shingles get hot.

Repairing or Replacing Roofing

Once any immediate problems have been addressed, you may want to plan for the careful replacing or repairing of the roofing material. How the repair is made depends on the type of roofing material.

Asbestos-Cement Shingles

If the house is an older one that has **asbestos-cement shingles**, they cannot be removed or repaired by the homeowner alone. Asbestos fibers are a health hazard and the shingles must be removed by an approved asbestos abatement contractor. Contact the local building inspector for assistance.

Replacing a Damaged Asphalt or Fiberglass Shingle

Fiberglass and asphalt shingles are available in a number of designs. The one shown in **18-26** has small pieces bonded to the face of full tiles, giving a look something like wood shingles, and providing an interesting shadow-line.

18-24 A badly torn shingle can be repaired temporarily using roofing cement, a few roofing nails, and a piece of fiberglass mesh.

If a shingle has come loose and slides down but is in good condition, it can be replaced. Remember to put dabs of roofing cement along the bottom edge to prevent the wind from raising it and possibly breaking it off. To replace a damaged shingle, use the following procedure:

1. Lift the shingles above it and remove the nails with a pry bar **(18-27)**. Slide out all of the pieces of the damaged shingle.
2. Slide the new shingle in place. Make certain the edge lines up with the other shingles in its course **(18-28)**.
3. Nail the new shingle by placing nails in the recommended locations **(18-29)**. Begin by raising the

shingle on top and pressing the nail into the shingle with your thumb. If the shingles are warm and flexible, raise the top one carefully and drive the nail (18-30). If they are cold or old and brittle, they may crack if lifted, so the nail will have to be driven by placing a pry bar or large screwdriver on the nail and striking the pry bar or screwdriver as shown in 18-31. Drive the nail so that the head rests firmly on the top of the new shingle. Do not drive the nail hard enough to break the surface.

18-26 The fiberglass shingle has narrow pieces bonded to a full shingle, giving the appearance of a wood shake shingle.

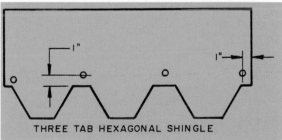

THREE TAB SQUARE BUTT SHINGLE

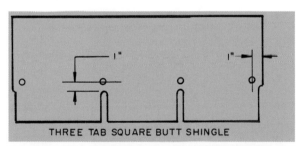

THREE TAB HEXAGONAL SHINGLE

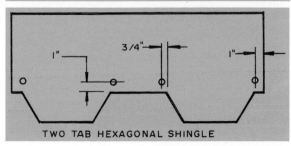

TWO TAB HEXAGONAL SHINGLE

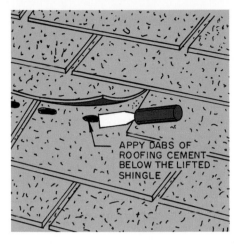

APPY DABS OF ROOFING CEMENT BELOW THE LIFTED SHINGLE

18-25 If the wind lifts the edges of the shingles, place dabs of roofing cement below the edge and press the shingle into it.

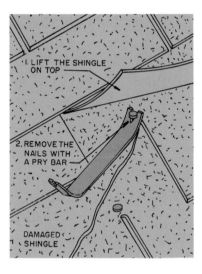

1. LIFT THE SHINGLE ON TOP

2. REMOVE THE NAILS WITH A PRY BAR

DAMAGED SHINGLE

18-27 Remove a damaged asphalt or fiberglass shingle by lifting the shingle above it and pulling the nails with a pry bar. Be careful not to crack the top shingles while lifting the damaged one out.

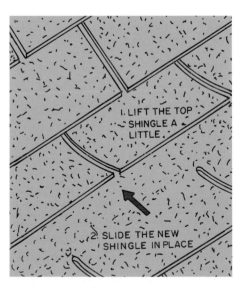

1. LIFT THE TOP SHINGLE A LITTLE.

2. SLIDE THE NEW SHINGLE IN PLACE

18-28 After removing the damaged shingle, lift the top shingle and slide the new one under it. Be certain the edge lines up with the other shingles in the course.

18-29 Recommended nailing patterns for several styles of fiberglass and asphalt shingles.

4. Stick the replacement shingle to the top shingle and to the one below it with dabs of roofing cement. **Ridge shingles** cover the shingles where they meet at the ridge and **hip shingles** cover the meeting of shingles on the hip rafter (**18-32**). If the ridge or hip shingle has developed a small surface crack, it can be repaired by applying roofing cement into the crack (**18-33**). If the damage is major, the damaged area will have to be coated with roofing cement. Cut a piece of shingle to overlap the damaged shingle and nail it

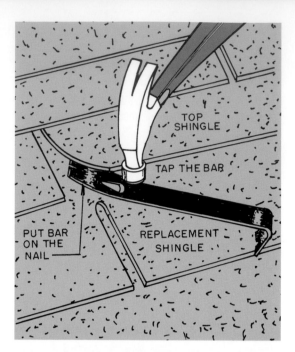

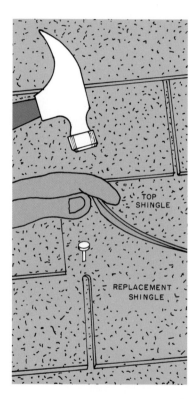

18-30 If the shingles are warm and flexible, it may be possible to lift the top shingle enough to allow a new nail to be driven without cracking the lifted shingle.

18-31 If the shingles are cold or brittle, carefully lift the top shingle enough to set the nail in place with your thumb. Then drive it in place by placing a pry bar on the nail and striking the bar with a hammer.

18-32 The shingles on a hip roof meet at the hip rafter. The joint is covered with hip shingles.

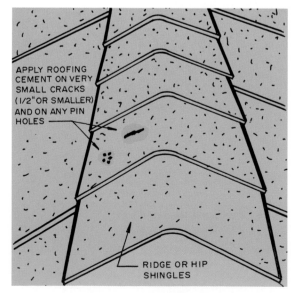

18-33 Very tiny surface cracks and pinholes on hip and ridge shingles can be sealed with a coat of roofing cement.

on all four corners with 2½-inch roofing nails. Before nailing, apply a spot of roofing cement where the nail will be installed and cover the head of the nail after it has been driven into place (**18-34**).

Repairing Roll Roofing

Roll roofing (asphalt roofing products manufactured in roof form) is often used on flat roofs or shed roofs that have only a slight slope. This type of roofing is available in 36-inch-wide rolls that have a part covered with a granular mineral and a part with an exposed asphalt surface (**18-35**).

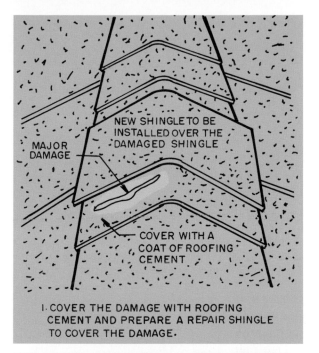

I. COVER THE DAMAGE WITH ROOFING CEMENT AND PREPARE A REPAIR SHINGLE TO COVER THE DAMAGE.

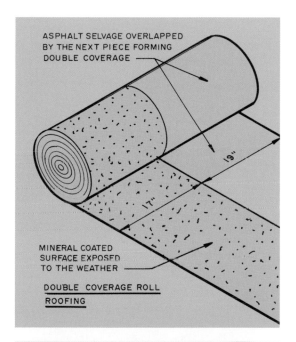

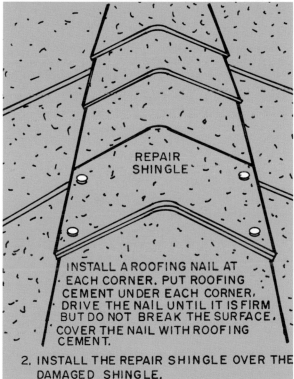

2. INSTALL THE REPAIR SHINGLE OVER THE DAMAGED SHINGLE.

18-34 Hip and ridge shingles with large cracks should be repaired by bonding and nailing a new shingle over the damaged shingle.

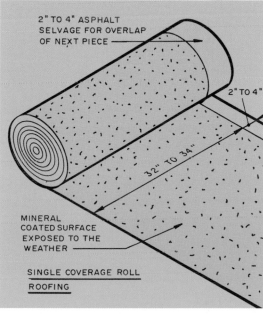

18-35 Two commonly used types of roll roofing—one providing double coverage, the other single coverage.

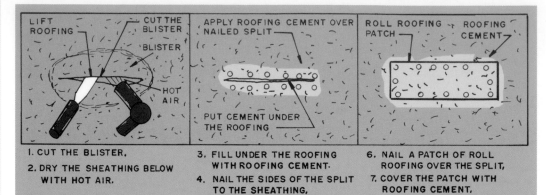

1. CUT THE BLISTER.
2. DRY THE SHEATHING BELOW WITH HOT AIR.

3. FILL UNDER THE ROOFING WITH ROOFING CEMENT.
4. NAIL THE SIDES OF THE SPLIT TO THE SHEATHING,
5. COVER WITH ROOFING CEMENT.

6. NAIL A PATCH OF ROLL ROOFING OVER THE SPLIT,
7. COVER THE PATCH WITH ROOFING CEMENT.

18-36 Small cracks and pinholes in roll roofing can be repaired by coating them with roofing cement and nailing a patch cut from roll roofing over the break. Cover the patch with roofing cement.

18-37 Large breaks in roll roofing require that the damaged area be cut away. First the sheathing must be dried, then two roll roofing patches can be applied.

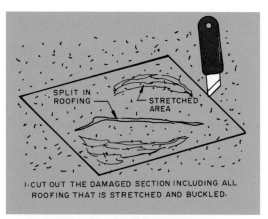

1. CUT OUT THE DAMAGED SECTION INCLUDING ALL ROOFING THAT IS STRETCHED AND BUCKLED.

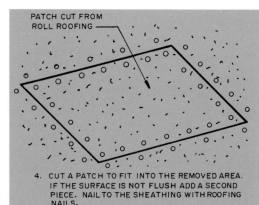

4. CUT A PATCH TO FIT INTO THE REMOVED AREA. IF THE SURFACE IS NOT FLUSH ADD A SECOND PIECE. NAIL TO THE SHEATHING WITH ROOFING NAILS.

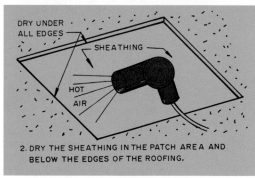

2. DRY THE SHEATHING IN THE PATCH AREA AND BELOW THE EDGES OF THE ROOFING.

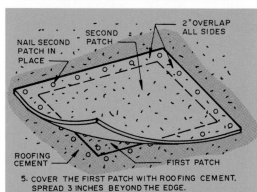
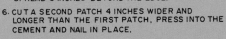

5. COVER THE FIRST PATCH WITH ROOFING CEMENT. SPREAD 3 INCHES BEYOND THE EDGE.
6. CUT A SECOND PATCH 4 INCHES WIDER AND LONGER THAN THE FIRST PATCH, PRESS INTO THE CEMENT AND NAIL IN PLACE.

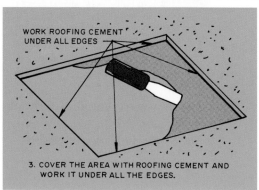

3. COVER THE AREA WITH ROOFING CEMENT AND WORK IT UNDER ALL THE EDGES.

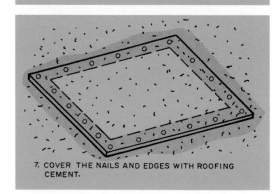

7. COVER THE NAILS AND EDGES WITH ROOFING CEMENT.

Heat can cause a blister in this kind of built-up roof if water gets below it. The roll-roofing material will often crack at the blister, causing a major leak. If it is a **small** blister, slit the blister, lift the roofing, and dry the sheathing with a hair dryer or other blower. Force roofing cement under the roofing around the slit, press the roofing into the cement, and nail on both sides. Cover the slit with a coat of roofing cement about 2 inches around the slit. Cut a patch, nail it over this area, and cover the entire patch with cement (18-36).

If it is a large blister and the roofing around it is stretched and damaged, proceed as follows (18-37).

1. Cut away the blistered roll roofing and dry the exposed sheathing. The drying can be sped up by using a heated blower such as a hair dryer.
2. Be certain the roofing left around the cut-away area can be pressed flat. If it is irregular and buckled, cut this area away until what is left will lie flat on the sheathing. Be certain the sheathing below is dry.
3. Spread roofing cement below the edges of the roll roofing, press it flat against the sheathing, and nail the edges. Spread roofing cement over the entire area of the exposed sheathing.
4. Cut a patch of roll roofing the same size as the exposed sheathing. The piece removed can be used as a pattern. Nail this patch in the exposed area with galvanized or aluminum roofing nails.
5. Spread roofing cement over this patch and extend it at least 2 inches beyond the edge.
6. Cut a second patch of roll roofing 4 inches longer and wider than the first patch. Place it over the first patch, press it into the cement, and nail around the edges.
7. Apply a coat of roofing cement over the nails and carefully seal the edge of the patch.

Replacing Wood Shingles & Shakes

Wood shingles are sawn on both sides, so they are smooth (18-38). **Wood shakes** are hand-split from the log and are rough on both sides (18-39). Some types are split and then sawn; this forms two shakes, so one side is smooth.

Generally, No. 1 grade (blue label) shingles and shakes are used for roofing. They are made from cedar, spruce, and pine. It is recommended they be given a water-repellent coating every few years to help reduce the absorption of moisture.

18-38 This roof has been finished with wood shingles. It has a rather smooth look.

18-39 Wood shakes are split rather than sawed. They produce a rough appearance and vary in thickness.

Wood-shingle roofs can have mold and discoloration removed by power washing. Simply washing with bleach also cleans them up.

Over the years, wood shingles and shakes curl or cup due to being wet and then drying. A temporary repair can be made by slipping a piece of flashing under a defective shingle. Since the metal repair will be obvious, eventually the shingle or shake will have to be replaced. To replace the shingle or shake, do the following (18-40):

1. Remove the old shingle or shake. Drive a thin wood wedge under the shingle above, raising it just a little off the damaged shingle.
2. Split the damaged shingle with a chisel and remove the splintered parts.

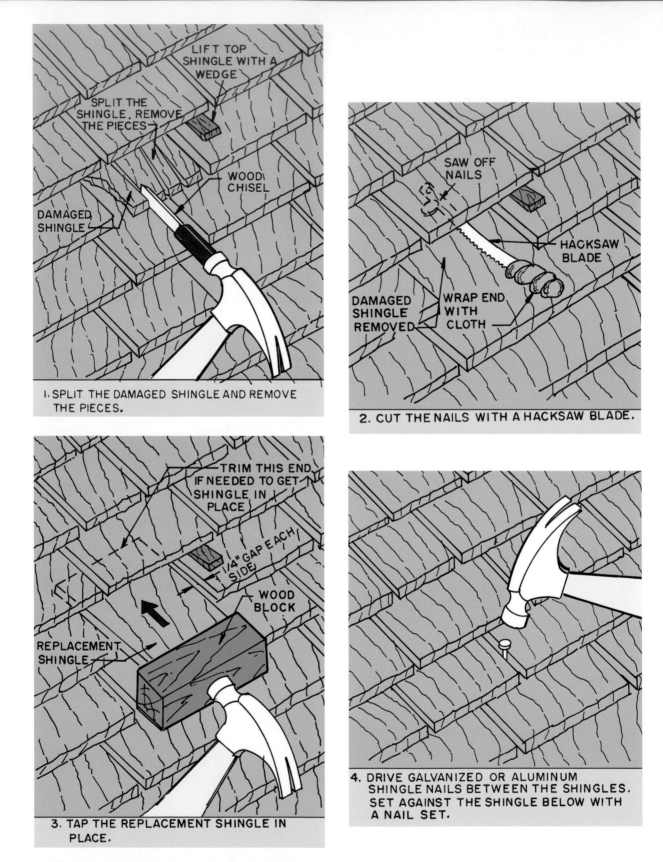

LIFT TOP SHINGLE WITH A WEDGE

SPLIT THE SHINGLE, REMOVE THE PIECES

WOOD CHISEL

DAMAGED SHINGLE

1. SPLIT THE DAMAGED SHINGLE AND REMOVE THE PIECES.

SAW OFF NAILS

HACKSAW BLADE

DAMAGED SHINGLE REMOVED

WRAP END WITH CLOTH

2. CUT THE NAILS WITH A HACKSAW BLADE.

TRIM THIS END IF NEEDED TO GET SHINGLE IN PLACE

1/4" GAP EACH SIDE

WOOD BLOCK

REPLACEMENT SHINGLE

3. TAP THE REPLACEMENT SHINGLE IN PLACE.

4. DRIVE GALVANIZED OR ALUMINUM SHINGLE NAILS BETWEEN THE SHINGLES. SET AGAINST THE SHINGLE BELOW WITH A NAIL SET.

18-40 Work carefully when replacing damaged wood shingles or shakes. Since they do not bend, old shingles have to be split and removed and the nails sawed off before the new shingles or shakes are installed.

3. Cut off the nails left by running a hacksaw blade or metal-cutting keyhole saw under the shingle.

4. If several courses must be replaced, start with the lowest course. Cut the replacement shingles to fit the space and match the pattern of the roof. The shingles should have a ¼-inch gap on the edges, so each replacement shingle will be ½ inch narrower than the space.

5. Slide the shingle in place. If it will not go all the way under, cut a little off the thin end. Tap it in place by putting a wood block on the edge and hammering on the block.

6. Nail the shingles to the sheathing with 4d galvanized or aluminum shingle nails. If several courses are to be installed, nail the lowest courses so the shingle above covers the nail about one inch. The final shingle will have to be nailed between the shingles. Set it against the shingle with a nail set. Cover the exposed nails with roofing cement. Drive the nails so their heads rest firmly on top of the shingle but do not break the surface (**18-41**).

Repairing a Metal Roof

The most commonly used metal roofing materials are copper, galvanized steel, and aluminum. **Terne** roofing is another type, but is not commonly found. It is steel-coated with an alloy composed of lead and tin and is generally painted. Copper is usually left unfinished and over the years develops a green patina

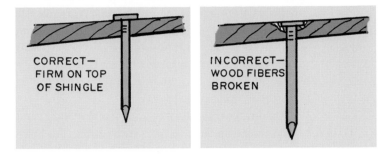

18-41 Set the shingle nails firmly on the surface of the wood shingle or shake, but do not break the wood fibers on the surface.

18-42 This multilevel, coated aluminum roofing handles water coming from several directions. It provides a long-lasting watertight roof.

18-43 Metal roofing is available in a wide range of colors. This roof blends smoothly with the light-color wood shingle siding.

(coating). Galvanized steel and aluminum roofing do not need painting, but often are coated to provide the color desired **(18-42)**. This also protects steel from rusting. After years of exposure, the galvanizing tends to wear away. All metal roofs have a long life, however, and require little maintenance except for an occasional repainting **(18-43)**.

In addition to sheets of roofing metal, shingles and shakes with embossed surfaces are available. The manufacturers of these products have designed methods for connecting the material and securing it to the sheathing.

It is important to remember that metal roofs can corrode if different metals are touching. For example,

copper roofing installed with steel nails will corrode at each nail. Always use nails or screws of the same material as the roofing.

STEEL & COPPER ROOFS

Steel and **copper roofs** can have minor damage repaired as follows **(18-44)**:

1. Make certain the metal is flat against the sheathing. Clean around the damaged area; this includes the removal of any paint. Steel wool is good to use to clean and polish the surface. Also clean the bottom edge of the patch.
2. Cut a patch of the same material. Make it around 2 inches wider than the damaged area on each side.

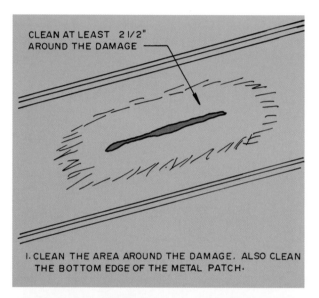

1. CLEAN THE AREA AROUND THE DAMAGE. ALSO CLEAN THE BOTTOM EDGE OF THE METAL PATCH.

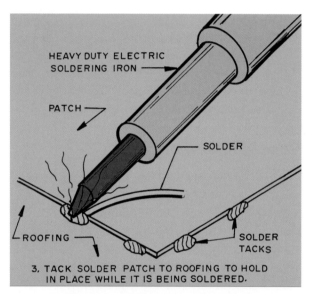

3. TACK SOLDER PATCH TO ROOFING TO HOLD IN PLACE WHILE IT IS BEING SOLDERED.

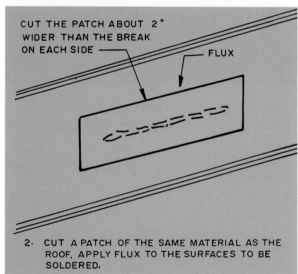

2. CUT A PATCH OF THE SAME MATERIAL AS THE ROOF. APPLY FLUX TO THE SURFACES TO BE SOLDERED.

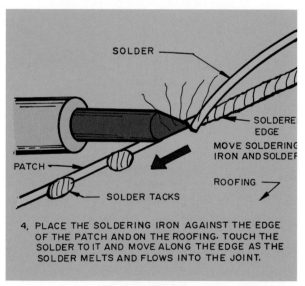

4. PLACE THE SOLDERING IRON AGAINST THE EDGE OF THE PATCH AND ON THE ROOFING. TOUCH THE SOLDER TO IT AND MOVE ALONG THE EDGE AS THE SOLDER MELTS AND FLOWS INTO THE JOINT.

18-44 Copper and steel roofs can be patched by soldering a patch made from the same material as the damaged area. Be certain the surfaces to be soldered are clean, free of paint, and coated with the proper flux.

3. With a flux, coat the edges of the patch and roof surface to be soldered. Use a flux recommended for the material being soldered. This removes any oxide that may be left on the surface.

4. Place the patch over the area. Tack solder in several places to hold it in position. (**Tack-soldering** consists of soldering the edges in several places with short strips of solder.)

5. Now, using a large soldering iron, place the face flat against the edge of the patch and down on the roofing. Touch the solder to the soldering iron. As the patch and roofing heat up, the solder will melt and flow between them. Move the soldering iron and solder along the edge, melting and flowing solder between the patch and the roof.

A small electric soldering iron may not produce enough heat to get the patch and roofing hot enough to melt the solder. A large soldering iron or a propane torch is needed. If using a propane torch, be aware that the metal could be overheated and a fire started on the roof.

ALUMINUM ROOFS

Aluminum roofs cannot be soldered with equipment that is normally available to the nonprofessional. However, small breaks can be repaired by placing an aluminum patch over the area **(18-45).** The following steps show one way to make this repair:

1. If the edges of the damaged area protrude above the surface of the aluminum roof, drive them flat with a hammer.

2. Spread a layer of roofing cement about 2½ inches beyond the edges of the damage. If the roof is over sheathing, coat the sheathing also.

3. Cut the aluminum patch about 2 inches wider than the damaged area on each side.

4. Press the patch into the cement. Secure the patch with aluminum **sheet-metal screws.** Drill the proper-diameter hole for the size screw being used. Tighten each screw until it is firm. Do not over-tighten or it will be stripped out of the metal roofing.

5. Apply a coat of roofing cement around the edge of the patch. Extend it about 2 inches beyond the sides of the damaged area. After it dries a day or so, apply a second coat.

Repairing Slate Roofing

Slate is a quarried stone that is split into thin sections that are cut to size. The colors vary depending upon

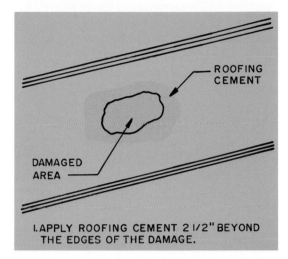

1. APPLY ROOFING CEMENT 2 1/2" BEYOND THE EDGES OF THE DAMAGE.

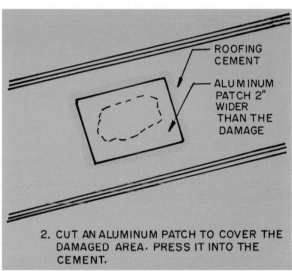

2. CUT AN ALUMINUM PATCH TO COVER THE DAMAGED AREA. PRESS IT INTO THE CEMENT.

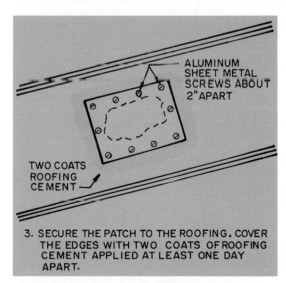

3. SECURE THE PATCH TO THE ROOFING. COVER THE EDGES WITH TWO COATS OF ROOFING CEMENT APPLIED AT LEAST ONE DAY APART.

18-45 One way to patch damage on aluminum roofing is to combine layers of roofing cement below and over an aluminum patch, and secure with aluminum sheet-metal screws.

the location of the quarry. It is an excellent, long-lasting roofing material, but very heavy. The roof structure and supporting walls must be designed to carry the load (18-46).

While slate is very durable and lasts for many years, occasionally it will be cracked due to falling debris, or come loose when the slate nails begin to fail. Generally it is best to replace damaged slate rather than try to patch it. Since slate roofs are slippery and slates easily crack when walked upon, it is best to have repairs made by a roofing company. Tools used tend to include a **slate ripper** and a **slate hammer** (18-47).

REMOVING & REPLACING DAMAGED SLATE

Following is one way damaged slate can be removed and replaced. Begin by sliding the ripper under the broken slate tile and placing the end against a nail. Hit the ripper several times to loosen it (18-48), and do the same for each nail. Then hook the ripper on each nail and drive it down to pull each nail (18-49). Cut a new slate tile to fit into the opening, but size it to leave a ¼-inch space on each side. Then nail a slate hook into the sheathing in the space between the underlying slates (18-50). Use thin wood wedges to raise the top slates a little, making it easier to slide in the new slate. Be careful not to crack the raised slates. Slide the new slate down on the hook (18-51). It is held in place by the hook.

Holes are punched in the slate with the tip of the slate hammer. Tap the hole from the back of the slate (18-52). This will produce a beveled opening into which the head of the roofing nail will fit. It will be

18-46 Slate is heavy but produces a rough, textured roof that is very appealing.

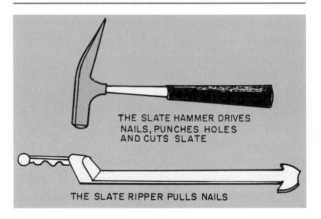

18-47 The slate ripper and slate hammer are used to replace and install slates.

18-48 Slide the ripper under the top slate and drive it against each nail to loosen it. Use thin wood wedges to slightly raise the top slate, if necessary, to get the ripper in place.

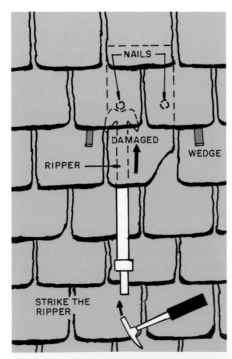

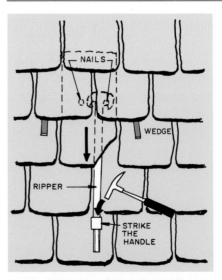

18-49 Hook the ripper on each nail and drive it down, pulling the nail out of the sheathing.

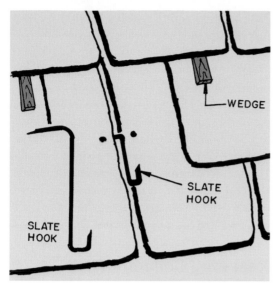

18-50 Drive a slate hook into the sheathing.

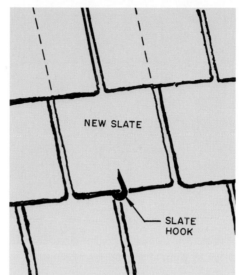

18-51 Slide the new slate in place and down onto the slate hook.

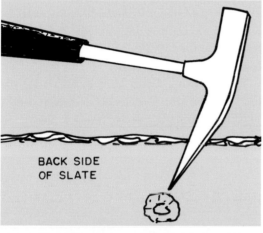

18-52 Holes in slate are cut from the back side with the slate hammer. This forms a recessed area on the front to hold the head of the nail.

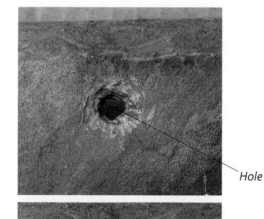

Hole

Nail Is in Hole

18-53 The hole on the top of the slate will have a cone-shaped recess into which the head of the roofing nail will fit.

18-54 Here is a pair of slates with the holes prepared, ready for installation.

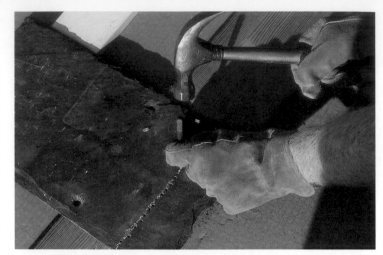

18-55 Slate can be hand-cut by punching a series of small holes with a nail set.

18-56 To break the slate on the line of cut, place it on a board, lay a board on top, press firmly together, and strike the scrap piece with a hammer.

flush with the top of the slate (**18-53**). In **18-54** are a pair of slate tiles with holes that are ready for installation.

Slate can be **cut** by punching a series of holes along the line of the cut (**18-55**) and then placing the slate between two pieces of wood and lightly tapping the part to be removed (**18-56**). The rough edge produced can be smoothed by lightly tapping off the irregular spots (**18-57**). Wear gloves to protect your hands.

Since slate roofs are slippery and usually steep, it is recommended that the homeowner not work on them to try to make repairs. Slate tiles also crack easily when walked on, and additional damage could occur.

Repairing Concrete Tiles

Concrete tiles are a mix of cementitious materials such as portland cement, hydraulic cements, sand, fly ash, pozzolons, and fine aggregates. Color is produced by adding an iron-oxide pigment to the mix, which produces the color throughout the entire thickness of the tile (**18-58**). The tiles are finished with a coat of clear acrylic sealer.

The following repair procedures are for flat-profile concrete tile. Concrete tile is also available in

18-57 After snapping off the scrap piece, the edge can be smoothed and trued up by tapping the irregular areas with a hammer.

18-58 Concrete tiles are made to look like slate or clay tile; wood shakes are lighter and less expensive than slate and clay products. *Courtesy Vande Hey Raleigh Architectural Roofing Tile. www.vhrt.com*

curved profiles, and is repaired as explained for clay tile in the next section.

REPAIRING FLAT-PROFILE CONCRETE TILE

To repair flat-profile concrete tile:

1. Break the damaged tile into small, removable pieces. While you might hit it directly with a hammer, using a heavy chisel will often give close control and possibly be less likely to damage the adjacent tiles (18-59). Always wear eye protection and gloves.

2. Notice if the tiles are nailed directly to the sheathing (18-60) or attached to 1 x 2 wood battens (18-61).

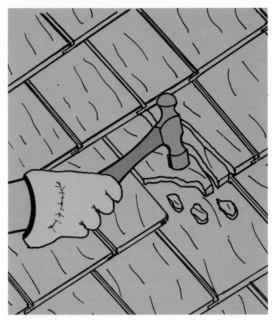

18-59 Break the damaged tile with a hammer or large chisel. Be careful not to damage the adjacent tiles.

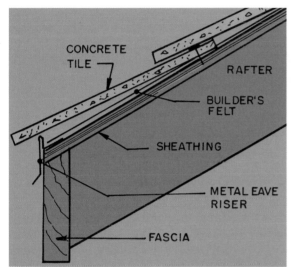

18-60 Some concrete tiles are manufactured so they are nailed directly to the sheathing.

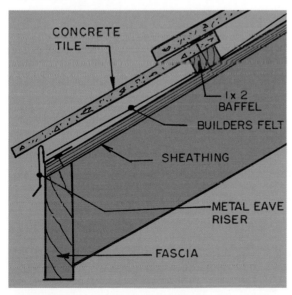

18-61 Many manufacturers make concrete tiles with a hook that lays over a 1 x 2 wood batten that is nailed to the sheathing.

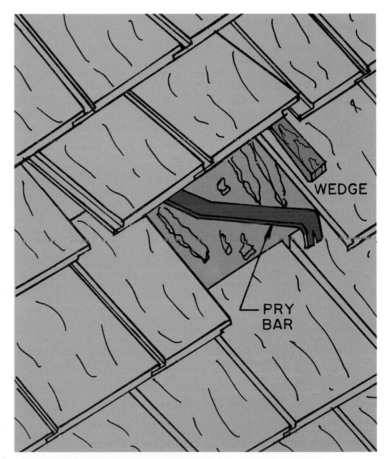

18-62 Raise the tiles up a little above the damaged tile area and remove all damaged pieces. Clean the area and remove all nails. Patch the felt if it has been damaged. Be very careful not to crack the tiles being lifted.

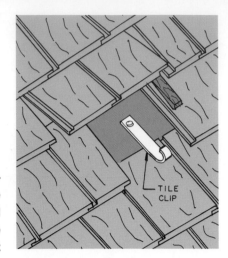

18-63 Nail a tile-retaining clip to the sheathing. The end should hold the new tile in line with the adjacent tiles.

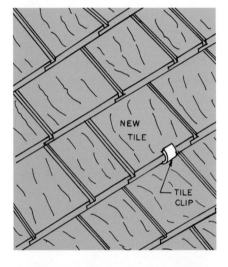

18-64 Raise the tiles over the cleared area and slide the new tile in place. Lower it down into the clip.

18-65 This wet-cutting masonry saw has a diamond blade which will cut concrete and clay tile. Be certain to wear eye protection or, better still, a full face shield.

3. Use a pry bar or wood wedges to raise the tiles on top of the damaged piece and remove all remaining pieces (**18-62**).

4. Now pull any leftover nails, clean the exposed area, and repair the roofing felt if it has been damaged. A patch can be secured with roofing cement and small holes or tears can be covered with it.

5. Install the new tile. First nail a retaining clip to the sheathing. Position it so it will hold the end of the tile in line with the tiles on each side (**18-63**).

6. Now raise the tiles that are over the repair area and slide the new tile in place, letting it slide down into the retaining clip (**18-64**). Be careful you do not crack the tiles that have been raised.

To cut concrete tile you will need to rent a wet power saw with a masonry blade. Always wear safety glasses, or better still, a full face shield (**18-65**).

Remember, concrete tiles are slippery and it is recommended a roofing company be employed to make repairs. The concrete tiles also crack easily and you may break some as you walk on them.

Repairing Clay Tile Roofing

Clay tile has been used for hundreds of years in countries around the world. They are manufactured in a variety of shapes and colors (**18-66**). They are made from various clays, shale, and other natural earthly materials. Colors can be varied by mixing different types of clay. Some have a colored glaze fired on the surface, which produces a glass-like coating. You may have to search around to find replacement tiles that are the color and profile found on older roofs.

Commonly used undulating tile shapes are pan and cover shapes (also referred to as barrel/missions) and S Shape (referred to as Spanish S) (**18-67**).

The following repair procedure is for clay tile with the mission profile. The exact profile and installation can vary some depending upon the manufacturer. Concrete tiles are also made in these tile profiles. Clay tiles are available in a variety of flat tiles and are repaired much like described for flat concrete tile.

REPLACING CLAY CAP TILE

Following is a typical repair procedure for replacing cap tile on mission tile roofs.

1. Carefully raise the cap tile above the broken tile and hold it up with a board. Use a pry bar to pull the nail (**18-68**).

2. Remove all scraps of the broken tile so the felt is clean. If the felt has damage, coat small holes or

18-66 Clay tile roofs are beautiful and last for many years.

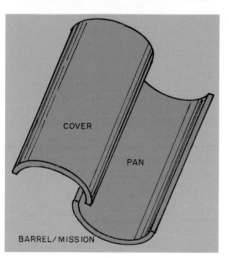

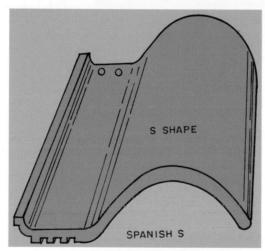

18-67 Two of the frequently used clay tiles. Concrete tiles are also made in these profiles.

18-68 If the cover tile is broken on a mission tile roof, wedge up the tile above it and remove the nail. Clean away all pieces of tile.

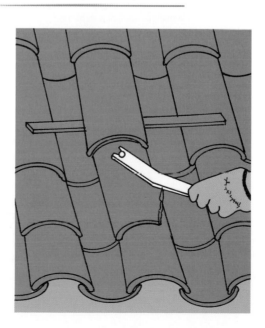

437

18-69 Slide a new tile in place and nail it to the sheathing.

18-70 Lower the top tile by removing the wood wedge and make adjustments so it fits over the new tile.

18-71 Base flashing is bonded to the down side of the chimney with roofing cement. It lays over the roofing that has been installed up to the chimney.

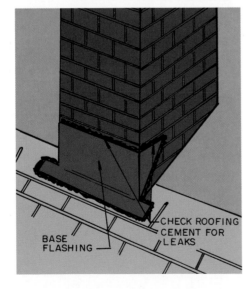

tears with roofing cement. Large tears should be covered with a roofing-felt patch bonded with roofing cement.

3. Lay the new tile in place. Before nailing it, lower the old top tile over it to see if it will fit properly. Since there could be a difference in the tiles, it may be necessary to trim the new tiles with a power wet saw before you actually nail them in place. All cutting of clay tile is done with a power wet saw with a masonry cutting blade. Always wear eye protection when cutting.

4. When the tiles fit properly, nail the new ones in place (**18-69**).

5. Then lower the old top tile over the new tile, completing the installation (**18-70**).

Repairing Leaks at a Chimney

If moisture appears in the attic or on the ceiling by a chimney, it may mean the **flashing** has come loose, the mortar between some of the bricks has deteriorated, or the bricks are absorbing moisture and allowing it to flow inside the house.

Examine the chimney flashing to see if it has come loose somewhere. Most often it has pulled out of the mortar joint because the mortar has deteriorated. The front of the chimney has a base flashing bonded to the chimney and roofing with roofing cement (**18-71**). It is laid over the shingles that have been laid up to the chimney. Counterflashing is set in a mortar joint above the base flashing (**18-72**). If the mortar has

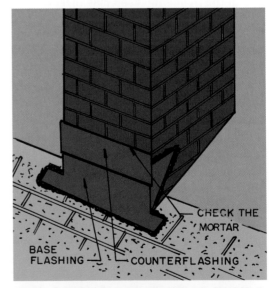

18-72 A strip of counterflashing is installed in the mortar joint above the base flashing and is laid down over it.

deteriorated and is crumbling, clean it out and refill with fresh mortar. While it is loose, check the roofing cement along the base flashing.

Step flashing is installed in the mortar joints up the side of the chimney as shown in **18-73**. It is then covered with counterflashing (**18-74**). The counterflashing is set 1½ inches into the mortar joint (**18-75**). If the counterflashing has come loose in the mortar joint, clean out the mortar and reset the flashing in the joint.

The back of the chimney may have a flat piece of flashing (**18-76**) or a cricket (**18-77**). Examine the parts set in the mortar joint. If they are loose, clean out the old mortar and reset. The edges resting on the shingles can be sealed with roofing cement.

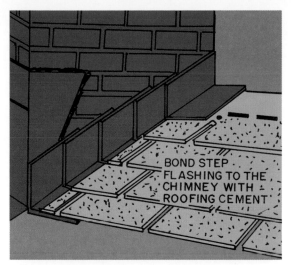

18-73 Step flashing is laid along the side of the chimney. It is laid on top of each course of shingles. It is bonded to the bricks with roofing cement.

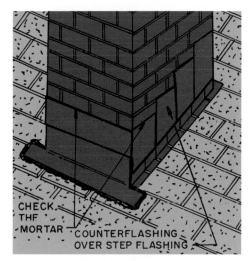

18-74 The step flashing is covered with counterflashing that is set in a mortar joint.

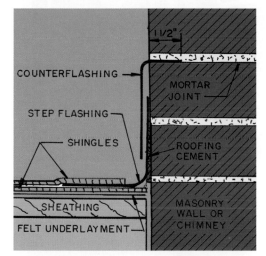

18-75 The counterflashing is set 1½ inches into the mortar joint.

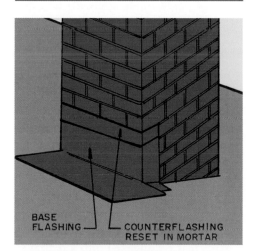

18-76 The back of the chimney is often flashed with a flat flashing bonded to the chimney. The shingles are laid over it. Counterflashing waterproofs the junction with the masonry.

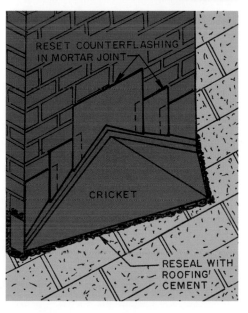

18-77 A cricket is an excellent way to divert water at the back of a chimney. Check the edges on the roof, and seal with roofing cement if it appears to have deteriorated.

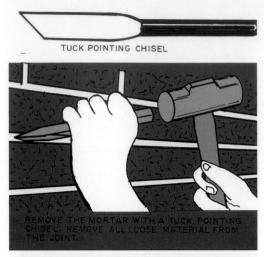

TUCK POINTING CHISEL

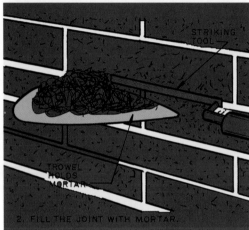

1. REMOVE THE MORTAR WITH A TUCK POINTING CHISEL. REMOVE ALL LOOSE MATERIAL FROM THE JOINT.

STRIKING TOOL

TROWEL HOLDS MORTAR

2. FILL THE JOINT WITH MORTAR.

CONCAVE JOINTING TOOL

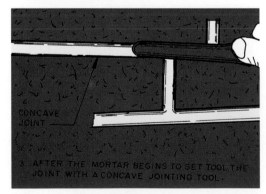

CONCAVE JOINT

3. AFTER THE MORTAR BEGINS TO SET TOOL THE JOINT WITH A CONCAVE JOINTING TOOL.

18-78 If the mortar joints have deteriorated, moisture can penetrate the masonry. Remove the old mortar and repoint with the appropriate mortar.

If the flashing has developed holes it will have to be replaced. If it is galvanized steel and has rusted but is sound, sand it and coat with a paint designed for use on rusted metal.

Now check the mortar joints in the chimney. If they are sound and the flashing is all right, the masonry may be absorbing water. When the chimney has had time to thoroughly dry, coat it with a water-repellent material. This is a clear liquid available at the local building supply dealer. It will not discolor the bricks.

If a few minor cracks between the mortar and the masonry are found, they can be filled with caulking designed especially to seal cracks in masonry and concrete. It is highly weather-resistant and flexible.

Repointing the Chimney

If the mortar joints are badly deteriorated, the chimney will have to be tuck-pointed. **Tuck-pointing** consists of removing the defective mortar joints and filling the area in with fresh mortar.

If it is a small job and the roof is not too steep, you may want to try it yourself. Be certain to cover the shingles on the roof so they are not damaged.

Following are the basic tuck-pointing steps **(18-78)**.

1. Chisel out the old mortar about ¾ to 1 inch deep. Use a tuck-pointing chisel. Be certain to wear eye protection.
2. After chiseling out the old mortar, clean all loose material out of the joint. Compressed air is best for this, but a good brush can be used.
3. When preparing to apply the new mortar, wet the joints with water.
4. The new mortar should be as similar to the old mortar as possible. If the chimney is very old, the exact mortar will most likely not be available or you will not be able to determine its composition. New mortars expand at different rates than older mortars and may produce enough pressure to cause the faces of the bricks to spall (break up) and fall off. When in doubt, some homeowners use the currently available type-O mortar. Mix it as follows: one part type-O portland cement, two parts lime, and seven parts fine mortar sand. Add as little water as possible, so that the mix is rather stiff. If it is squeezed into a ball, it should retain that shape.
5. Apply the mortar into the joints with a striking tool. After it has begun to set, go over the joints with a jointing tool, forming a slightly concave surface.

Repairing Flashing

Hopefully the flashing on your house is copper or aluminum because they do not rust. Galvanized steel flashing will last many years, but eventually will break down and rust. You can prevent this by washing it with a liquid detergent to remove all dirt and then painting it with a paint designed for use on galvanized steel. This will last quite a while but watch it and repaint as necessary to maintain a protective coating.

Vent Pipe Flashing

Possibly vent stack flashing will be the first to leak (18-79). It consists of a metal base plate and a flexible neoprene collar that fits around the pipe. If it starts to leak you can place roofing cement around the joint with the pipe, but this is only temporary. The flashing unit should be replaced as shown in 18-80.

18-79 Vent pipes are flashed with specially designed flashing units.

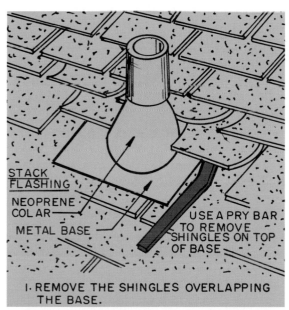

I. REMOVE THE SHINGLES OVERLAPPING THE BASE.

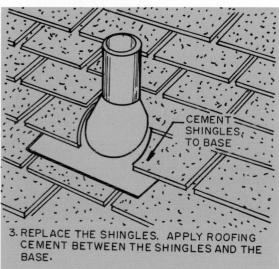

3. REPLACE THE SHINGLES. APPLY ROOFING CEMENT BETWEEN THE SHINGLES AND THE BASE.

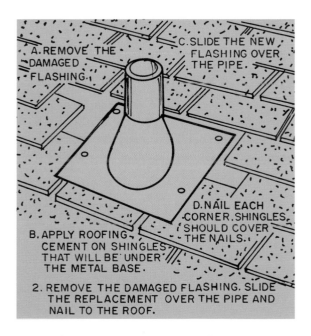

2. REMOVE THE DAMAGED FLASHING. SLIDE THE REPLACEMENT OVER THE PIPE AND NAIL TO THE ROOF.

18-80 When vent stack flashing deteriorates it is best to replace the flashing. Temporary repairs can be made with roofing cement.

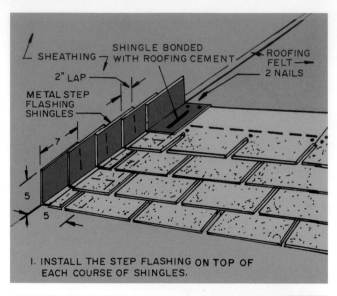

I. INSTALL THE STEP FLASHING ON TOP OF EACH COURSE OF SHINGLES.

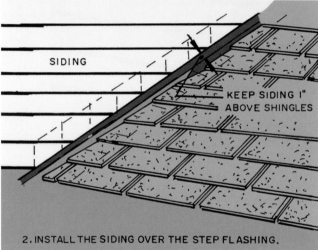

2. INSTALL THE SIDING OVER THE STEP FLASHING.

18-81 When a roof butts a wood framed wall to be finished with wood, vinyl, aluminum, or some other type of siding, step flashing is installed over each course of shingles. After the shingles have been laid, the siding is installed over the flashing. It serves as the counterflashing.

Flashing on a Masonry Wall

A roof butting a masonry wall is flashed as described earlier for flashing a chimney. Two layers of flashing are used. The first layer is **step flashing.** It is nailed to the roof sheathing and bonded to the masonry wall with roofing cement (refer to **18-73**). The second layer, called **counterflashing,** is bent in an L-shape. One end is placed in a mortar joint, and the other overlaps the first layer. The mortar will have to be chiseled out at a depth of about 1½ inches, the edge inserted into this opening, and the joint refilled with mortar (refer to **18-75**).

Flashing on a Frame Wall

When a roof butts a frame wall, the flashing is placed behind the wood siding. This is usually aluminum or copper and requires no maintenance. It is referred to as step flashing. Each piece of flashing is placed on top of a course of shingles and nailed to the wall sheathing. Then another course of shingles is laid on the wall and another piece of flashing placed on top of it. It is nailed to the sheathing with two nails in the top edge. The next flashing overlaps the first 2 inches of the original flashing and covers the nails **(18-81)**.

After the flashing and shingles are in place, the siding is installed. It covers the flashing, but should be kept about 1 inch above the shingles. The siding serves as counterflashing. If for some reason it is suspected that water may be leaking between the flashing pieces, the seams can be caulked with high-quality exterior caulking. If the flashing has rusted and has pinholes, it should be replaced. A temporary repair can be made by coating it with roofing cement.

Valley Flashing

Most roofing shingles use **open-valley flashing** to seal the union between butting roofs. Wood shingles and shakes, slate, clay and concrete tile, and metal roofs each have specific installation techniques that must be followed **(18-82)**. Asphalt shingles generally flash a valley by weaving the shingles as shown in **18-83**. However, open-valley construction can be used with asphalt shingles.

If the valley was built using the open-valley technique, it may have used mineral-faced asphalt roll roofing **(18-84)** or copper, aluminum, or galvanized steel flashing **(18-85)**. If the mineral-faced valley begins to leak it can be coated with roofing cement as a temporary repair. However, eventually it will have to be replaced. The metal flashing will last many years and only the galvanized steel is likely to begin to rust. Keep it painted with a rust-inhibiting paint designed for use on steel. If for some reason it has been allowed to rust through, it will have to be replaced. This involves laying back the shingles along each side, removing the old flashing, and installing new aluminum or copper.

Currently the use of woven valley flashing when asphalt and fiberglass shingles are installed is common practice. A key to success is the installation of a satisfactory underlayment up the valley **(18-86)**. If you think it is leaking try sticking down the edge of the shingles with roofing cement. If the shingles are old

18-82 Various types of roofing material typically use an open-valley flashing.

18-83 Asphalt and fiberglass shingles typically use a woven-valley flashing.

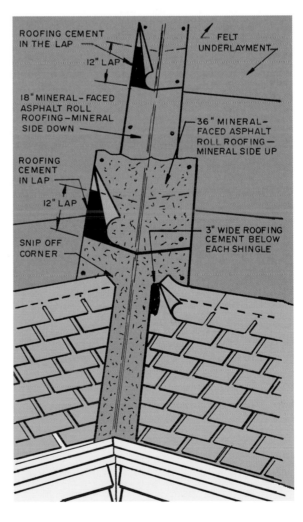

ROOFING CEMENT IN THE LAP

FELT UNDERLAYMENT

12" LAP

18" MINERAL-FACED ASPHALT ROLL ROOFING—MINERAL SIDE DOWN

36" MINERAL-FACED ASPHALT ROLL ROOFING—MINERAL SIDE UP

ROOFING CEMENT IN LAP

12" LAP

SNIP OFF CORNER

3" WIDE ROOFING CEMENT BELOW EACH SHINGLE

18-84 This open-valley flashing is composed of two layers of mineral-faced asphalt roll roofing.

and cracked, the best solution is to reshingle the roof and install a new underlayment in the valley.

Rake & Eave Protection

If the rake and eave are not protected by a metal dry edge, the water running off the roof will run back under the shingles—causing the sheathing, fascia, and rake board to rot (18-87). The drip edge should be installed before the shingles are in place. The drip edge is nailed to the sheathing along the edge of the roof at the eave. The felt underlayment is placed over the drip edge and is laid up the roof. The drip edge on the rake

is laid on top of the felt and nailed to the sheathing (18-88).

It is difficult to install a drip edge after the shingles are in place. Sometimes it is possible to force the edge under the shingles and face-nail it to the fascia. Be certain to use nails made of the same metal as the drip edge, otherwise an electrochemical reaction will occur between the dissimilar metals and corrosion between the nail and drip edge will occur, eventually destroying one or both of them. You can get some relief by laying a bead of caulking along the joint where the shingles meet the fascia (18-89). Often the fascia has already started to rot and split and should

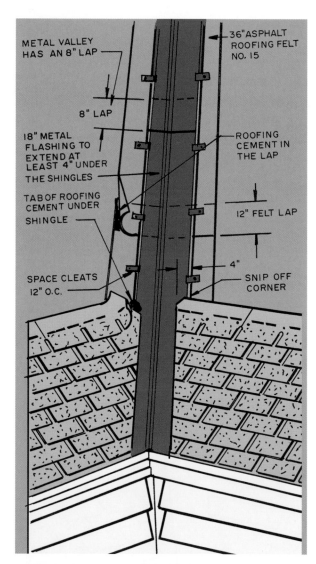

18-85 Metal valley flashing is held to the sheathing with cleats that fit into the curved edge of the flashing.

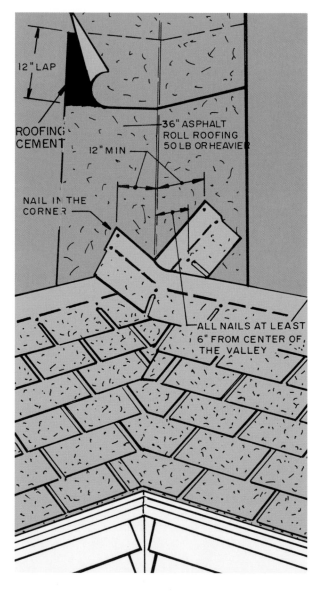

18-86 Woven-valley flashing uses 50-pound roll roofing over the roofing-felt underlayment. The three-tab asphalt or fiberglass shingles are laid over each other in the valley.

be removed and replaced. Remember to paint both sides of the new fascia before installing it **(18-90)**.

Gutter & Downspout Repair

Blocked drains cause gutters to overflow, which soaks the fascia. As the paint ages, it may also go into the wood and cause rot. Placing a screen of some sort keeps leaves out and reduces the number of times the drains must be cleaned **(18-91)**. While aluminum gutters are now commonly used, older homes may have galvanized steel gutters. Over the years the galvanized coating wears away and they start to rust. If

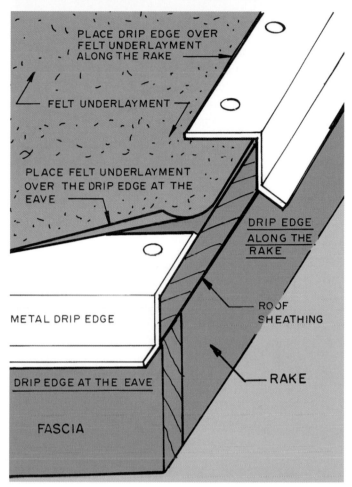

18-87 The sheathing at the fascia and rake boards should be covered with a drip edge.

18-88 Metal drip edges are placed on the sheathing along the fascia and rake boards. They prevent water from getting into the sheathing and causing it to rot.

18-89 A heavy bead of caulking can be laid along the top of the fascia where the shingles overlap. This will help keep moisture from running up under the shingles at the eave.

18-90 Rotted fascia should be removed and replaced. Temporary repairs can be made by caulking the damage and painting the surface.

the rust is extensive, replace them with aluminum gutters.

If the damage has not gone through the metal, sand off the rust and use paint designed for galvanized steel. If there are small holes they can be sanded and wiped clean with paint thinner. When dry, apply a coating of epoxy over the area and press a piece of fiberglass cloth in it. This is the same cloth as used in auto body repair. When this sets up, put a second coat of epoxy over it **(18-92).** Some use roofing cement instead of epoxy.

A large hole can be repaired with a metal patch. Sand the rust off the area, coat the area around the hole with epoxy, and press the metal patch into it. The patch should be of the same metal as the gutter. Now secure the patch by blind-riveting it to the gutter. Finally, seal the edges of the patch with epoxy.

There are other products on the market that are in the form of a caulking. After cleaning the area apply this material around and over the hole.

Algae on Shingles

In many parts of the country algae growth appears on roofs as dark streaks **(18-93).** These are common in areas where it is warm, with high humidity, such

18-91 Screen covers made especially to keep debris out of gutters help reduce the need for frequent cleaning.

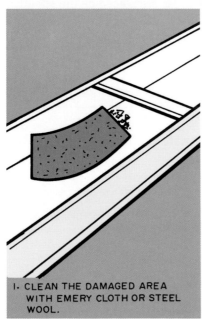

1. CLEAN THE DAMAGED AREA WITH EMERY CLOTH OR STEEL WOOL.

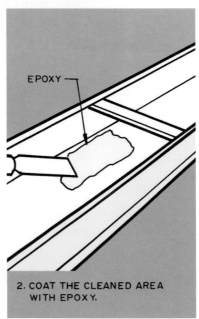

EPOXY

2. COAT THE CLEANED AREA WITH EPOXY.

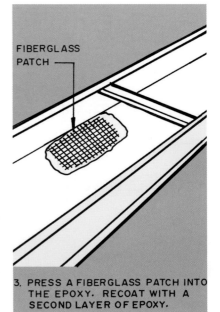

FIBERGLASS PATCH

3. PRESS A FIBERGLASS PATCH INTO THE EPOXY. RECOAT WITH A SECOND LAYER OF EPOXY.

18-92 Holes in gutters can be patched using epoxy and a fiberglass patch.

as the southeastern United States; however, algae is appearing in northern states. The algae appears mainly on the north and east sides, which receive the least sun. Roofs that are shaded by trees also tend to be stained by algae.

Specific research has not shown that algae actually damages asphalt shingles; however, it is a common perception that it does. Whether algae does or does not harm the roofing, there is no doubt that the greatest damage occurs as it is removed. Walking on the roof, spraying chemicals, cleaning with a brush, and using high-pressure sprays most likely shorten the life of asphalt shingles by removing the surface granules.

It is helpful if the roof is kept free of leaves and debris which, if left on it, hold moisture that promotes the growth of algae. Keep the spaces between the shingles, shakes, and tiles clear so the water runs smoothly to the eave. Also remove any tree branches that overhang the roof and consider thinning the trees at least on the north and east sides so the sun can keep the roof dry.

Over the years the fillers and surface granules used to make asphalt and fiberglass shingles have changed. Some manufacturers add copper-bearing granules to the shingle surface, which helps retard the development of algae. For the most effective algae-resistant shingles, at least 10 percent of the roof coating granules will be copper. When you select algae-resistant shingles, check the amount of copper granules indicated on the label. The higher the percentage of copper granules, the more effective the shingle will be to resist the growth of algae.

An effective way to reduce the growth of algae is to install pure zinc strips along the ridge of the roof (18-94) and below anything that breaks the flow of water from the ridge such as a chimney or dormer (18-95). The rainwater must flow continuously from the ridge cap to the eave. Zinc strips can be used on asphalt, wood, metal, and flat clay and concrete tile. They are not as effective on the rolled Spanish tile roofs because the water does not flow continuously from the ridge cap to the eave.

Installing Zinc Strips

Zinc strips are 2½ inches wide and available in rolls up to 50 feet long. To install them, first have the roofing thoroughly cleaned. Then lay the zinc strip against the ridge cap or the edge of the ridge vent and nail every two feet with manufacturer-supplied neoprene-washered roofing nails. If you live in a part of the country where algae is especially problematic, or

18-93 In warm, humid climates algae forms on the roof, producing an unsightly appearance. This algae may also appear in northern areas.

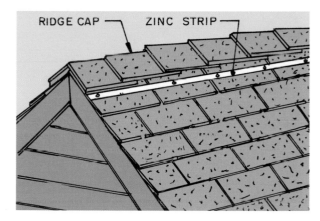

18-94 Pure zinc strips are installed along the edge of the ridge cap or ridge vent.

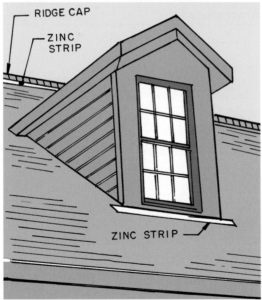

have an unusually large roof, install a second row of zinc strips halfway down the roof.

Removing Algae

Since roofs are slippery when wet, it is wise to hire a roof-cleaning contractor to remove the algae on steep-sloped roofs. Wood shingles and shakes, and clay and concrete tile, are more dangerous to clean than asphalt and fiberglass shingles. If you have a low-slope roof, you can clean it by scrubbing the shingles with a soft- bristle brush using a solution of 1 gallon household bleach, 2½ gallons water, and 1 cup ammonia-free detergent. Start cleaning at the ridge and work down toward the eaves. Pour some of the solution on the area to be cleaned, and scrub it. Be careful not to damage the shingles. After you clean a section, rinse it with a hose.

Be aware that since you work from the top down, you will be standing on wet shingles the entire time. Wear shoes with rubber soles so you have some grip.

18-95 Any items on the roof that interrupt the flow of water from the ridge should have zinc strips installed on the down side.

CHAPTER 19

Repairing & Replacing Windows

Windows not only let in natural light and provide ventilation but are a major architectural feature of the exterior of a house (19-1). As you consider remodeling, several major issues arise. First, if the existing windows are very old, do not open and close properly, let air leak into and out of the house through worn or warped sections, or have damaged or rotted sash, they usually should be replaced (19-2). New windows available are made much more durable than the older windows found on many houses and are very energy efficient. The second issue is using new windows to improve the appearance of the house.

19-1 Windows are a major part of the architectural design of a house.

19-2 When a wood window shows signs of rot spreading across the sill and sash, it is best to replace the entire unit.

19-3 The Early American-style house typically has windows with small panes of glass set in vertical and horizontal muntins.

19-4 A contemporary-style house typically features large glass areas.

If the house is a particular architectural style, such as Early American, the windows should be double-hung and the glazing in the sash divided into small panes with muntins (**19-3**). This style occurred because at that time it was difficult to produce large, strong sheets of glass. A contemporary-style house is more natural, with large, fixed-glass areas and large casement windows (**19-4**). The old farmhouse was also built with double-hung windows with small panes (**19-5**). A study of books showing architectural styles will give some clear impressions about window choice. Should it be necessary to replace windows, choose a style to reflect the architecture of the house.

A third possibility to consider when deciding to replace deteriorated windows is to dramatically change the size and even the location of windows. A dark room would benefit if larger replacement windows were used, or if a single window were replaced with several windows which could be grouped hor-

19-5 The old American farmhouse had windows with many small panes divided by muntins.

izontally (**19-6**) or vertically (**19-7**). Or you can install an entire window wall (**19-8**), add a bow or bay window (**19-9**), or add a large fixed glazed area (**19-10**), sometimes called a picture window.

A fourth possibility can occur when the house is not a particular style, and an extensive remodeling on the exterior is planned. The windows are a major

19-6 A dark room can have natural light increased greatly by replacing a single window with two or more windows and even adding a transom.

19-7 This large window assembly expands the glazing horizontally and vertically, allowing natural light to fall far into the room.

19-8 A wall with large fixed units provides an excellent way to view the scene outside.

19-10 Large, single-pane glazed wall areas provide a terrific view and natural light but usually have no provision for ventilation. Some will have side lights or sash below that will open. This unit has casement windows on each end.

19-9 A bow or bay window not only provides natural light and ventilation but is a major architectural feature. This is a bay window.

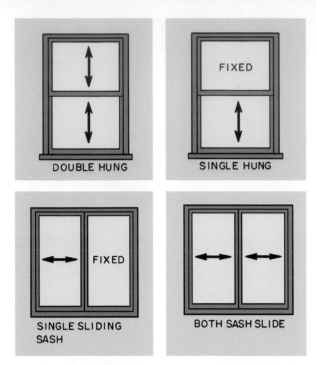

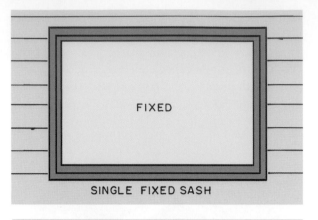

SINGLE FIXED SASH

19-11 Double-hung and single-hung windows slide vertically while sliding windows move horizontally.

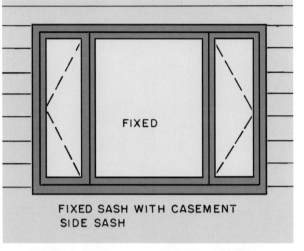

FIXED SASH WITH CASEMENT SIDE SASH

19-12 Swinging windows are available that are hinged on the side, top or bottom of the sash. These include casement, awning, and hopper windows.

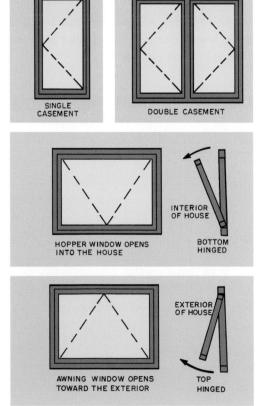

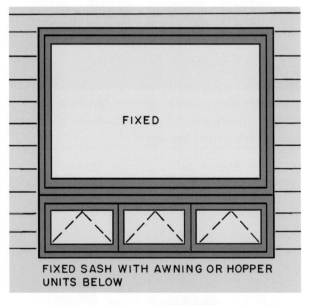

FIXED SASH WITH AWNING OR HOPPER UNITS BELOW

19-13 Fixed windows are primarily used to open up a room to a great view. They are often combined with other windows that open, providing ventilation.

architectural feature and the type and size chosen can dramatically alter the appearance of the finished structure. Often 25 to 35 percent of the exterior wall area is covered by windows. Since the siding tends to blend into the scene, the windows become the dominant feature. Consider the type of window to be added. Choices between the many standard types—double hung, casement, sliding, and fixed—give another opportunity to change the exterior appearance of the house. If you plan to sell this house after remodeling, this is an especially important consideration.

Remember, while it is often possible to find replacement windows that will fit into the opening left by the old window if the ones you choose do not fit, some carpentry may be needed to change the size of the opening anyway, which presents the opportunity to change the size and even the location of the new windows. Some manufacturers will make windows to fit your rough opening.

Selecting New Windows

As new windows are selected, a choice will be made between **sliding (19-11)**, **swinging (19-12)**, **fixed (19-13)**, and **glass and acrylic block** units **(19-14)**. There are many other special types available such as **bay** and **bow windows (19-15)**. Each of these provides different features which must be considered. For example, sliding windows provide only half the wall opening for ventilation, while casement windows open the full length to ventilation. Some have

19-14 Glass- and acrylic-block windows provide natural light and privacy. *Courtesy Pittsburgh Corning Corporation*

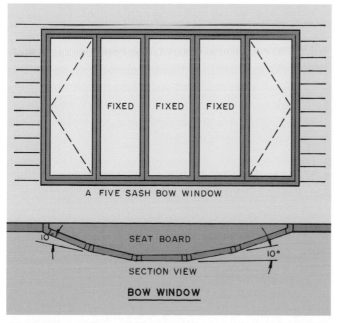

FIXED FIXED FIXED

A FIVE SASH BOW WINDOW

10°   10°

SEAT BOARD

SECTION VIEW

BOW WINDOW

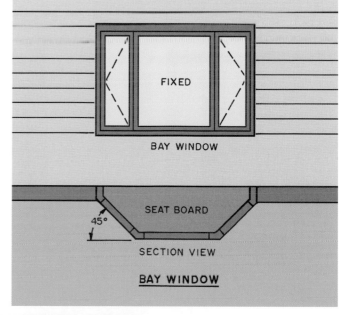

FIXED

BAY WINDOW

45°   SEAT BOARD

SECTION VIEW

BAY WINDOW

19-15 Bay and bow windows open up a room and provide natural light and ventilation.

19-16 This wood-framed window is clad with a vinyl covering.

muntins which seriously cut the view and natural light, while a fixed glazed area provides an unobstructed view and full natural light but no ventilation. Some awning windows swing from the top; they can provide ventilation even if it is raining but do not provide a great view. Examine each and make your selection based on light, view, or ventilation—or all three.

Consider the **ease of operation.** Sliding windows will have to be unlocked and pushed horizontally or vertically. Casement windows will be unlocked and open and close with a cranking mechanism. Awning windows also operate with a crank.

Consider the **material and construction** of the window. **Wood-framed units** have been standard for many years. They are treated with water-repellent preservatives and are factory primed, which provides an excellent base for the exterior paint. These units are also good insulators and are energy efficient. They are available clad with vinyl or aluminum that is maintenance-free **(19-16).**

Windows with **vinyl frames** are widely used in new construction and as replacement windows. They operate well, will not swell when exposed to moisture, and are maintenance-free. Low-density cellular vinyl frames have the same thermal resistance as solid wood **(19-17).**

Fiberglass frames will not shrink, warp, swell, or rot. They are strong and can hold large panes of glass. They have excellent thermal resistance.

Aluminum frames require little maintenance and operate easily. Exposure to moisture does not affect their operation. They have poor insulating properties, though some types are made with thermal breaks separating the inside and outside parts of the frame.

Wood-composite frames are made from shredded, dried, seasoned hardwoods mixed with resins and preservatives and fused under heat and pressure. They look a lot like wood and have good thermal resistance. They are prefinished during manufacture and do not require painting.

19-17 Vinyl-framed windows resist weathering and have good thermal resistance.

Glass & Acrylic Block Windows

Acrylic-block window units are made from light-weight, square acrylic units that are set in resilient polymer clips, forming a complete window unit (**19-18**). They have dead air space between the exposed surfaces that increases their energy efficiency.

Glass-block glazing units are made by fusing two formed glass halves together, leaving a hollow core which has a partial vacuum (**19-19**). They have good thermal properties and are very strong. These units are available in a variety of shapes.

Repairing Existing Windows

Sometimes existing windows may be basically sound yet not operate easily and show signs of deterioration. Often the defects can be repaired and the windows continue to function satisfactorily. However, if the windows are very old, they will most likely not be very energy efficient.

Repairs to Wood Windows

Older wood windows were not pressure treated to resist rot as they are today. If you see signs of minor rot or checks, clean out the area so fresh wood is exposed. Then fill the opening with a wood crack filler (**19-20**). Check the information on the can to be certain it is recommended for repairs exposed to the weather. When it has hardened, sand it smooth, prime, and then apply several coats of exterior paint as used on the rest of the window.

19-18 Acrylic block windows are easy to install, provide light and privacy, and are light enough to be used to glaze casement sash. *Courtesy Hi-Lite Products, Inc.*

19-19 Glass blocks are durable, have insulation value, and are available in a variety of shapes. *Courtesy Glashaus, Inc.*

19-20 If the wood frame has minor rot or cracking, clean it back to solid wood and fill with a wood crack filler.

19-21 Most sliding sash can be removed from the frame by lifting them up and pulling the bottom into the room.

Horizontal Sliding Window Problems

Horizontal sliding windows run on a track. Because the track is exposed to the elements, it will collect some dust and dirt which will hinder its free movement. When you wash the window, clean the tracks at the same time. Also be certain that when the sash was painted, paint did not get down into the track. This is difficult to scrape out but it can be done. It may be necessary to remove the sash, clean off the paint, and seal the wood sash with a clear sealer coat. Most sliding sash can be removed by lifting them up and pulling the bottom out into the room (19-21).

19-22 Regularly lubricate the casement window pivots and exposed cranking arms.

Thieves can also remove them unless you provide a means to keep them from being lifted out.

After the track is clean it helps to lubricate it with a wax, paraffin, or a silicone spray.

One last possible problem is if the metal track has gotten bent. Use a pliers to straighten small dents. A wood block could be forced into the track and it could be hammered straight by tapping the bend against the block.

Casement Window Problems

Wood casement windows swing on pivots usually located at the top and bottom corners on one side of the sash. This hardware is secured to the frame with wood screws. If these pivots seem loose it means the screws are not secure in the frame. Try to tighten them. If they will not tighten, remove it and glue small sticks in each hole or fill the screw holes with a wood filler that will hold screws. Check the can to be certain it will hold screws. When the glue or filler has dried, reinstall the screws.

Lubricate the moving parts of the pivot mechanism and the crank arm. A lightweight machine oil is good, as are some of the lubricating materials in spray cans (19-22).

If the wood frame has become swollen and the sash will not close easily, let it dry thoroughly. Then sand and scrape the edge until the window closes. Seal the edge with a clear sealer, such as a lacquer. Check to see if paint was smeared over the edge of the sash and this is what is causing the problem. Scrape off the paint and seal the edge. If the windows are clad with vinyl or aluminum, swelling will not occur unless there is a break in the cladding.

Repairing Wood Double-Hung Windows

If the sash sticks when you try to open it, most likely it has been painted so many times that the paint has glued it tight. The easiest way to clean this is to cut through the paint between the sash and the window stops around sides of the window with a utility knife (19-23). Then force a putty knife or broad-blade wall scraper between the sash and the stop. Move it all around the window, breaking the paint seal (19-24). Use it to scrape the paint off the sash. If the opening is wide enough you might be able to insert a piece of sandpaper folded over a piece of metal in the space and smooth the surfaces some more (19-25).

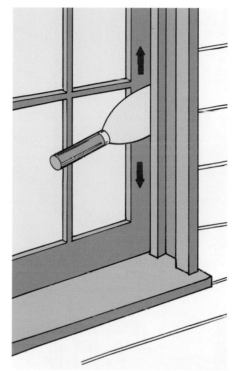

19-23 If the sash is painted closed, first cut the paint film with a utility knife.

19-24 After cutting the paint film, insert a putty knife or broad-blade wall knife into the space between the sash and stop. Work it around all sides of the window.

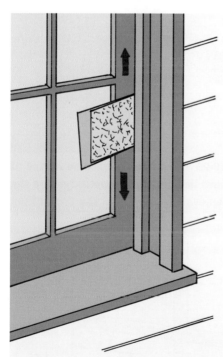

19-25 After separating the sash and stop, slide a piece of sandpaper folded over a piece of metal into the space and move it around the window, smoothing both surfaces.

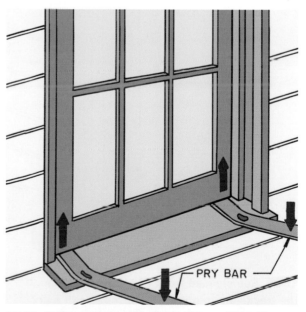

19-26 If the sash is still stuck after cutting the paint, put pry bars under each side and carefully press down. Do not overdo it or you may damage the sash or break the glass.

457

If the sash is still not moving you might try placing a pry bar below each side and pressing down to try to move it. Put a piece of scrap stock on the sill (19-26). Be aware that excess pressure could crack the glass.

When the sash can be moved use a wood chisel to remove all the paint on the exposed stops and on the sides and bottom of the track in which it slides (19-27). Sand the track and sides of the stop, then coat them with paraffin wax. If the track is metal or plastic, spray it with a dry silicone spray (19-28).

There will be times when the sash is stuck so tight the only way to free it is to remove the stop. Begin by cutting it free from the paint (refer to 19-23 and 19-24) and then pry it loose (19-29). If it is held with screws, remove them before prying it loose. Then

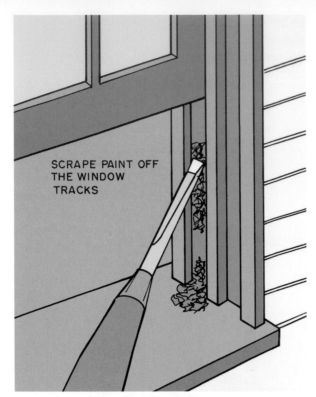

SCRAPE PAINT OFF THE WINDOW TRACKS

19-27 Use a wood chisel to remove all traces of paint on the stops and tracks.

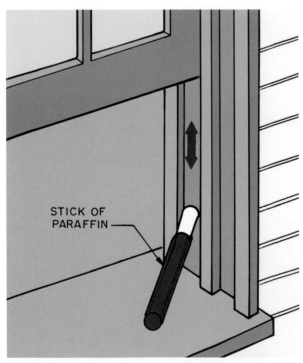

STICK OF PARAFFIN

19-28 Rub cleaned wood tracks with paraffin. If the tracks are metal or plastic, coat with a dry silicone spray.

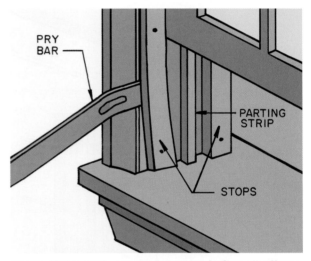

PRY BAR

PARTING STRIP

STOPS

19-29 If the sash cannot be moved after all efforts have been made to free it, remove the stop and remove the paint from the contact surfaces.

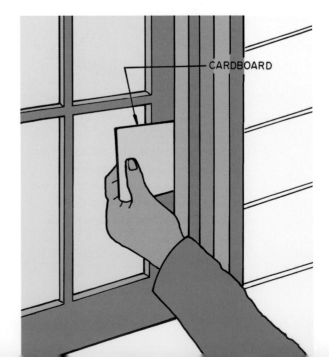

CARDBOARD

19-30 When you reinstall the stops, place a piece of cardboard between the sash and the stop to give the needed clearance.

scrape off all the paint in the track and clean it off the stops and the sides of the sash. If you need to touch up the paint on the sash and stop, do it before you replace them. Let it dry thoroughly. Coat with wax or silicone spray and reinstall the stops. You can leave a small space between the sash and the stop by placing a piece of cardboard between them. Then nail or screw the stop in place (19-30).

REPAIRING DOUBLE-HUNG WINDOW SASH CORDS

Many older homes have double-hung windows that are held open by weights on sash cords run over pulleys in the window frame. The weights run in a pocket alongside the window frame (19-31). When the cord breaks, the weight falls to the bottom of the pocket and the sash will not stay at the height desired. To replace the sash cord, do the following:

1. Remove the **interior trim** around the window and the **sash stops** on the inside of the window. (Try to avoid damaging the stops. However, they can be replaced.) This exposes the weights. Some older windows may have a board below the interior trim serving as a cover over the pocket.
2. Now pull the lower sash out of the frame. The cord is connected to the sash and the weight, as shown in **19-32.**

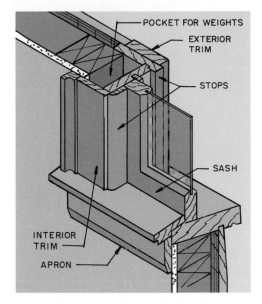

19-31 **Windows with sash weights have a pocket prepared in the wall behind the window frame for the weights to move up and down.**

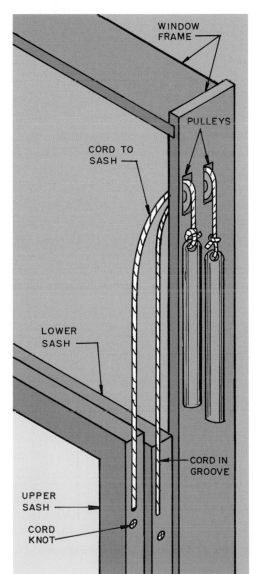

19-33 **Run the new cord over the pulleys and connect it to the sash and weights.**

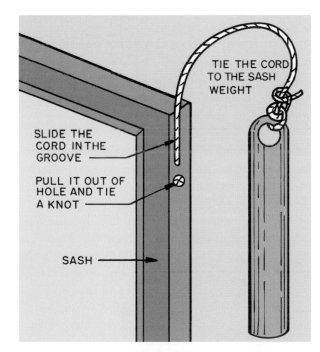

19-32 **The sash cord is tied to a weight. The cord slides in a groove cut through a hole in the sash where a knot is tied in the end of the cord.**

19-34 When replacing the sash cord, consider using sash chain instead.

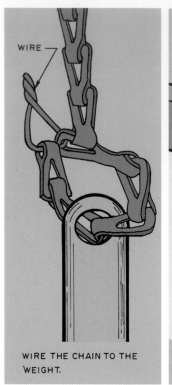

WIRE

WIRE THE CHAIN TO THE WEIGHT.

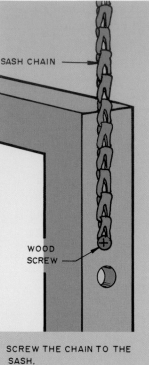

SASH CHAIN

WOOD SCREW

SCREW THE CHAIN TO THE SASH.

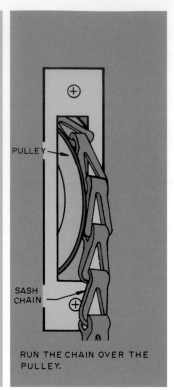

PULLEY

SASH CHAIN

RUN THE CHAIN OVER THE PULLEY.

3. Remove the broken or damaged cord from the weight and the other end from the sash. It is recommended that all the cords on both sash be replaced while repairs are being made.

4. Run the new cord over the pulley and connect it to the sash. The correct length of cord can be determined by measuring the old cord. Tie it to the weight and slide it into the pocket (19-33).

5. If the upper sash has a broken cord, the lower sash and its stops will have to be removed before the upper sash is removed. Replace the upper sash as described for the lower sash.

6. Reinstall the stops and the interior trim.

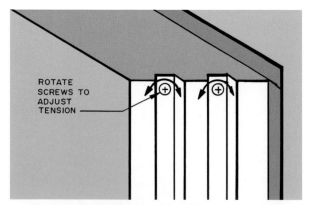

ROTATE SCREWS TO ADJUST TENSION

19-35 Some spring-balanced sash can have the tension adjusted by turning adjustment screws at the top of the channel.

It is recommended that broken sash cords be replaced with **sash chain.** The installation is the same as for replacing cords, except that the chain has to be wired to the weight and secured to the sash with a screw as shown in **19-34.**

REPAIRING SPRING-BALANCED SASHES

Newer windows typically have spring-balanced sashes. These work satisfactorily for many years. However, if the window is not operating properly its spring tension may need adjusting.

There are several different types of spring mechanism, and the one on the window will have to be carefully examined. One type has an adjustment screw in the window track. Turn the screw right or left until there is enough tension to permit the window to operate easily and stay open (19-35). Another type requires the loosening of the screw that holds the spring tube to the side jamb (19-36). Be certain the screw stays through the tube, because it holds the spring in place. Rotate the screw and tube counterclockwise to reduce tension, making it easier to raise and lower the sash. Rotate the screw clockwise if more tension is needed to hold the sash in place. Then replace the tube and screw it to the window frame.

If the spring is broken, remove the inner stops and the screw at the top, pry out the tube, and remove the spiral rod from the sash. Install the new spring and adjust the tension.

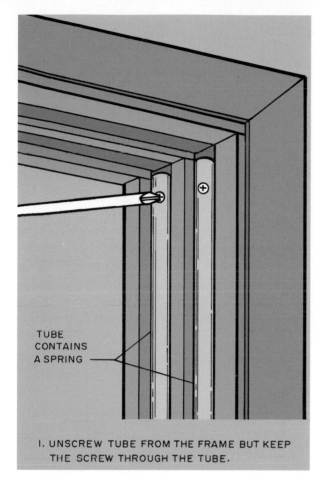

I. UNSCREW TUBE FROM THE FRAME BUT KEEP THE SCREW THROUGH THE TUBE.

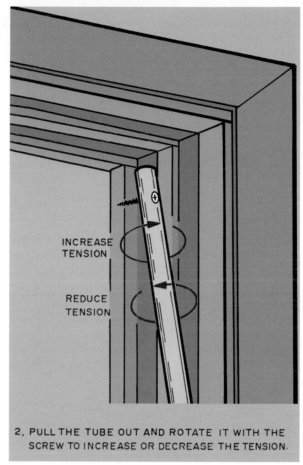

2. PULL THE TUBE OUT AND ROTATE IT WITH THE SCREW TO INCREASE OR DECREASE THE TENSION.

19-36 Some spring-balanced sash are controlled by turning the spring tube. If the sash needs more tension, turn the tube clockwise. If it is too tight, turn it counterclockwise to loosen it.

Deciding to Replace Windows

The time may come when the old windows are in bad shape and renovation is not the best thing. Consider installing replacement windows. Windows now available are made from materials that resist weathering and are highly energy efficient. Old wood windows permit air to filter into and out of the house, increasing heating and cooling costs. Replacement windows can often save 15 to 20 percent. They also add value to the house and could pay off when the house is sold.

Replacement Window Considerations

As you visit building supply dealers and look at the various replacement windows available, try to select a style that is suitable for the architectural style of the house. You will find a good choice available and can usually match the style of the old window. If you plan a total makeover of the house exterior, it may be possible to change the type of window.

Consider the material and how this will appear on the house. For example, a narrow-sash white vinyl window might not be great on a house that traditionally has wide brown wood sash. Replacement windows will have double-glazing, low-E glass, and this should be seriously considered as you choose a product.

Consider the size of the opening between the existing window side jambs and the top jamb and sill. Windows that will fit into the existing openings are considerably easier to install than if you have to cut back to the wall studs forming the rough opening.

Replacement windows are made in a wide range of sizes and some manufacturers will make them to fit the existing opening. Should no window be found to fit the opening it will be necessary to do a little carpentry work to adjust the rough opening.

461

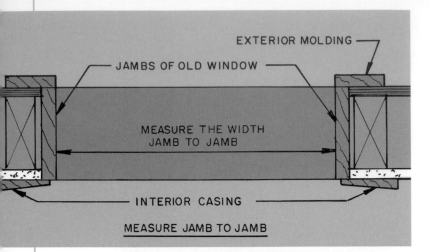

EXTERIOR MOLDING

JAMBS OF OLD WINDOW

MEASURE THE WIDTH
JAMB TO JAMB

INTERIOR CASING

MEASURE JAMB TO JAMB

Replacement windows are available in **replacement sash kits** and **full window replacements units**. A replacement sash kit is less expensive and uses the existing window frame and puts new channels and sash in it (**19-37**). A full window replacement unit has a fully assembled frame into which new sash are installed. The completely assembled unit is installed in the opening left by the jamb framing of the old window (**19-38**).

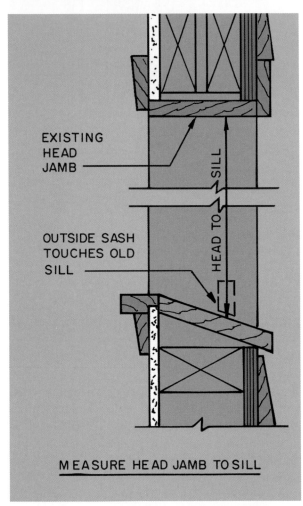

EXISTING HEAD JAMB

OUTSIDE SASH TOUCHES OLD SILL

HEAD TO SILL

MEASURE HEAD JAMB TO SILL

19-37 When ordering a sash replacement kit, measure between the existing jambs in three places, and the space between the head jamb and the sill, where the outside sash touches the sill when closed.

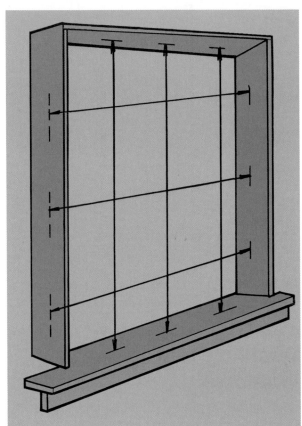

19-38 When ordering full replacement windows, measure the length and width of the frame opening in three places and order using the smallest dimension.

INSTALLING WINDOW REPLACEMENT KITS

Older houses with double-hung windows balanced by weights hanging on a cord that runs over a pulley that is set in the window frame and connected to the sash can have broken cords or deteriorated sash be easily replaced with a window replacement kit. Begin by removing the stops and sash. Then cut away any hanging cord and unscrew the pulleys, leaving the frame clean of all obstructions (**19-39**). If windows using a spring-balance system are to be replaced, then remove the stops, sash, and tracks.

The replacement window kit has vinyl jamb liners with weather-stripping and new sash. The upper and lower sash slide in the jamb liners. The sides of the vinyl channel are spring loaded to apply pressure on the sash, which hold it in the position desired (**19-40**).

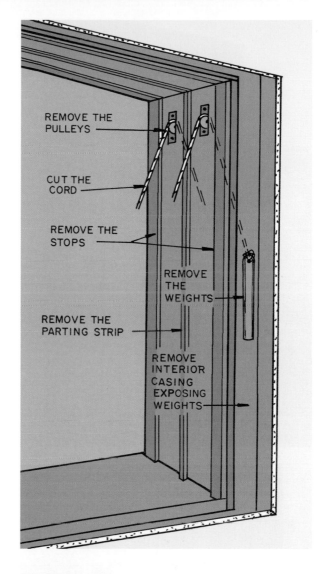

REMOVE THE PULLEYS

CUT THE CORD

REMOVE THE STOPS

REMOVE THE WEIGHTS

REMOVE THE PARTING STRIP

REMOVE INTERIOR CASING EXPOSING WEIGHTS

19-39 When replacing old double-hung windows, remove the old sash, stops, pulleys, cord, and weights.

19-40 This is a sash replacement kit. The jamb liners are installed on the inside of the old window frame and the new sash are inserted into the jamb liners. *Courtesy Kolbe and Kolbe Millwork Co., Inc.*

To order a replacement window, carefully measure the width and height (**19-41** and refer to **19-37**). Then from the sizes available select the one that best fits the opening. If necessary, install thin wood strips on the jambs to adjust the size of the opening, for a better fit.

Begin by removing the stops and the old sash as shown in **19-42** through **19-45.** Once the old window sash, stops, cord, weights, and pulleys are removed and only the frame and the clear opening are left, you can begin to install the replacement window.

Begin by installing the clips holding the jamb liner in place **(19-46).** Install the sash stops and end pad as shown in **19-47** and **19-48.** Now snap the jamb liner

19-42 Carefully remove the sash stops with a putty knife or pry bar. Save them because they will be reapplied after the installation of the replacement sash. *Courtesy Kolbe and Kolbe Millwork Co., Inc.*

19-41 Measure the width between the faces of the side jambs and from the head jamb to the sill at a point where the outside face of the bottom sash touches the sill when the window is closed.

19-43 Lift out the bottom sash. If it has rope-and-pulley operation, cut the rope and remove the pulleys and counterweights. *Courtesy Kolbe and Kolbe Millwork Co., Inc.*

19-44 Remove the side parting stops and discard them. Now remove the upper sash, cut the rope, and remove the pulley and counterweights. *Courtesy Kolbe and Kolbe Millwork Co., Inc.*

19-45 Remove the head jamb parting strip and discard it. *Courtesy Kolbe and Kolbe Millwork Co., Inc.*

19-46 Nail the jamb liner clips to the frame. Space as specified in the directions by the manufacturer. *Courtesy Kolbe and Kolbe Millwork Co., Inc.*

465

19-47 Slide the vinyl sash stops into the top inside track of each jamb liner. *Courtesy Kolbe and Kolbe Millwork Co., Inc.*

19-48 Apply the self-sticking foam pads to each jamb liner head.

Courtesy Kolbe and Kolbe Millwork Co., Inc.

in place over the jamb liner clips **(19-49)** and install the head parting strip **(19-50)**. Insert the top sash into the jamb liner and insert a clutch pivot above the clutch in the track. Lay it flat and press the jamb liner while pushing the sash into the track. Raise the sash to the top position **(19-51)**. Repeat the steps to install the lower sash **(19-52)**. Finally, replace the original sash stops **(19-53)**.

19-49 Snap the jamb liner over the clips in the old window frame. *Courtesy Kolbe and Kolbe Millwork Co., Inc.*

19-50 Reinstall the wood parting strip at the head and cover it with a vinyl parting strip.

Courtesy Kolbe and Kolbe Millwork Co., Inc.

the manufacturer **(19-56).** Then insulate
old frame and the frame of the new wind
opening with a bead of caulk to block
Then replace the existing inside stops (

REPLACING A WINDOW WITH A
ONE

Typical framing for a window openir
frame wall is shown in **19-58.** If this wi
replaced by a smaller window, ren
window including the frame. This leave
the size of the original rough opening
ence in size is only a few inches, 1-ir
wood strips can be nailed to the trin
narrow the opening, and to the rough
a shorter opening. Do not place strips
This will lower the top of the windov
longer line up with the tops of the do
windows in the room **(19-59).** If th
greater, install new wall framing as sh
After the window has been installed
the exterior siding and interior wallbc
quired. Remember to place insulatio
new window frame and the studs and

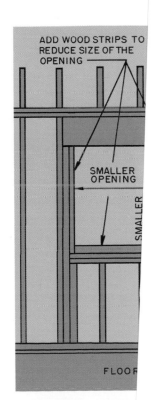

**19-59 If the new wind
little smaller than nee
sides of the opening
framing to produce th
opening.**

**19-51 Install the upper sash by slanting it so the
guides at the top fit into the jamb liner, then lay
the sash flat. Be certain the guides are in the jamb
liner. Press in on the jamb liner and press the sash
into the track. Then slide it to the top of the frame.**

Courtesy Kolbe and Kolbe Millwork Co., Inc.

**19-52 Repeat the installation steps for the lower
sash.** *Courtesy Kolbe and Kolbe Millwork Co., Inc.*

**19-53 Reinstall the original sash stops. If they were
damaged, replace them with new stops.** *Courtesy Kolbe and
Kolbe Millwork Co., Inc.*

19-54 Measure the op[...] frames and the top fra[...] window manufacturer [...] replacement window [...]

Courtesy Weather Shield Windo[...]

19-56 Slide the repl[...] opening by putting [...] slowly tilting the wi[...] Secure as directed [...]

Courtesy Weather Shield Wi[...]

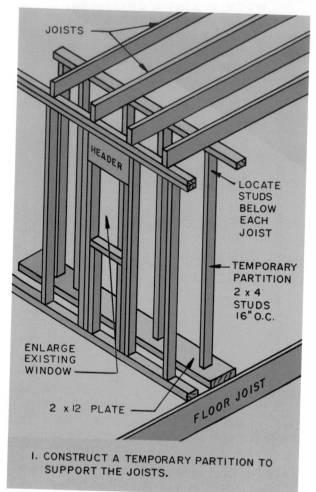

1. CONSTRUCT A TEMPORARY PARTITION TO SUPPORT THE JOISTS.

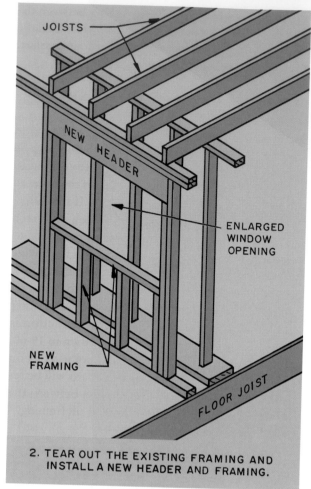

2. TEAR OUT THE EXISTING FRAMING AND INSTALL A NEW HEADER AND FRAMING.

19-61 A temporary partition can be built to support the overhead joists as the old headers and some studs are removed and a new header is installed.

19-62 If the floor below the temporary partition needs support, install a beam supported on concrete blocks.

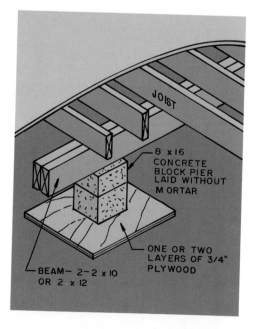

REPLACING A WINDOW WITH A LARGER ONE

If the rough opening framed in **19-58** has to be enlarged to receive a larger window, considerable carpentry work is required. First remove the old window completely. Since the old header and some of the studs will be removed, it is necessary to support the ceiling joists in the window area. Under normal loading conditions this can be done by constructing a short supporting partition as shown in **19-61**. If there is concern about the load on the floor, a temporary short beam could be installed below the floor directly under the partition **(19-62)**.

Cut out any studs that are in the new rough opening and remove the old header. As this is accomplished, consider what has to be done to the exterior siding and drywall. Try to disturb these as little as possible. The reconstruction is shown in **19-63.** Establish the location of the new window and install full studs on each side. This distance is equal to the rough opening plus 3 inches for the trimmer studs.

Install the new header on the trimmer studs. Finally the double sill and cripples are installed. Keep a stud or cripple every 16 inches so sheathing and siding can be nailed with minimum cutting.

The header is typically two 2 x 12 members assembled with a ½-inch piece of plywood between them (19-64). For normal ceiling construction this header will span 8 to 10 feet.

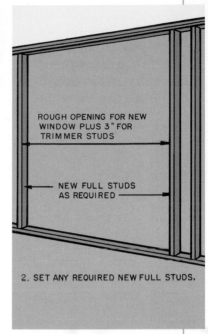

1. REMOVE THE EXISTING HEADER, SILL, CRIPPLES AND ANY STUDS IN THE NEW OPENING.

ROUGH OPENING FOR NEW WINDOW PLUS 3" FOR TRIMMER STUDS

NEW FULL STUDS AS REQUIRED

2. SET ANY REQUIRED NEW FULL STUDS.

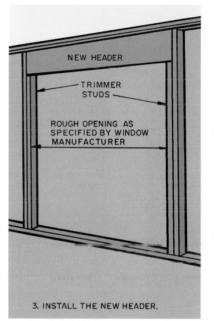

NEW HEADER

TRIMMER STUDS

ROUGH OPENING AS SPECIFIED BY WINDOW MANUFACTURER

3. INSTALL THE NEW HEADER.

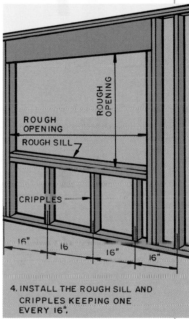

ROUGH OPENING

ROUGH SILL

CRIPPLES

16" 16 16" 16"

4. INSTALL THE ROUGH SILL AND CRIPPLES KEEPING ONE EVERY 16".

19-63 Typical framing for an enlarged rough opening.

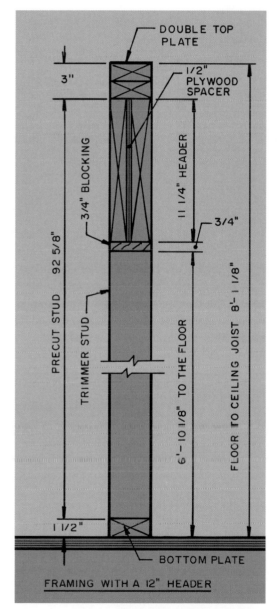

DOUBLE TOP PLATE

3"

1/2" PLYWOOD SPACER

3/4" BLOCKING

11 1/4" HEADER

3/4"

PRECUT STUD 92 5/8"

TRIMMER STUD

6'-10 1/8" TO THE FLOOR

FLOOR TO CEILING JOIST 8'-1 1/8"

1 1/2"

BOTTOM PLATE

FRAMING WITH A 12" HEADER

19-64 A typical widely used header design.

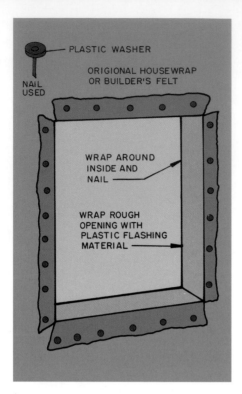

19-65 Wrap the new rough opening with plastic flashing material. Wrap it inside and nail to the inside of the studs.

PLASTIC WASHER

NAIL USED

ORIGIONAL HOUSEWRAP OR BUILDER'S FELT

WRAP AROUND INSIDE AND NAIL

WRAP ROUGH OPENING WITH PLASTIC FLASHING MATERIAL

19-66 The black plastic flashing material is wrapped inside the rough opening and nailed to the studs. This shows the window installed.

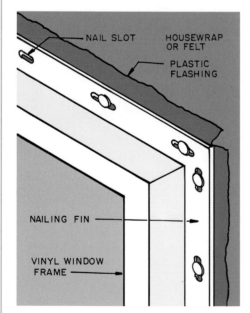

NAIL SLOT

HOUSEWRAP OR FELT

PLASTIC FLASHING

NAILING FIN

VINYL WINDOW FRAME

19-67 This vinyl replacement window is secured to the rough opening framing with a nailing fin.

19-68 This window has been secured to the rough opening framing by nailing the nailing fin to it.

Flashing the Rough Opening

When replacing a window with one the same size, the rough opening will most likely have been flashed before the original window was installed. If it is in good condition install the replacement window over it. If it is damaged repair the damage so the sides of the rough opening are sealed as shown in the next section.

If you have changed the size of the rough opening it is necessary to flash it. The existing situation can vary depending upon the original construction of the exterior wall. In any case sides of the rough opening will need to be flashed as shown in **19-65.** The material used is typically a strong black plastic flashing material. Builder's felt is also used. The material is wrapped around the rough opening and nailed to the studs and rough sill inside the house as shown in **19-66.** Then install the replacement window following the instructions provided by the manufacturer.

If the replacement window is held in place with a nailing fin nail **(19-67),** nail it over the flashing **(19-68).** In mild climates the nailing fin is then sealed to the housewrap or felt with an adhesive-backed window wrap tape **(19-69).** In colder climates and areas with harsh weather, flash as shown in **19-70.** Builder's felt flashing is secured over the nailing fin on all sides.

19-69 After the nailing fin has been secured to the framing, it is sealed with a layer of adhesive-backed window tape.

19-70 It is better to flash the nailing fin with wide strips of building paper, starting by inserting the first layer under the sill nailing fin. The top strip is inserted in a slit cut in the housewrap or builder's felt.

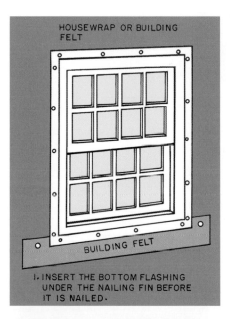

1. INSERT THE BOTTOM FLASHING UNDER THE NAILING FIN BEFORE IT IS NAILED.

2. INSTALL THE FLASHING ON THE SIDE NAILING FINS, LAP IT OVER THE BOTTOM FLASHING.

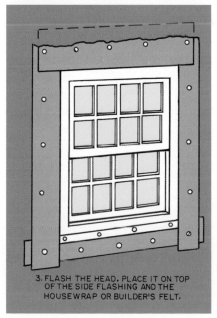

3. FLASH THE HEAD. PLACE IT ON TOP OF THE SIDE FLASHING AND THE HOUSEWRAP OR BUILDER'S FELT.

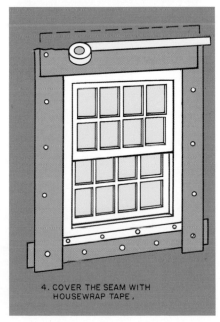

4. COVER THE SEAM WITH HOUSEWRAP TAPE.

Cutting Glass

Generally, glass can be cut to size at the local building supply dealer. If not, buy a piece larger than needed and cut it yourself.

First measure the size of the opening into which it will be placed very accurately. Then cut the piece ⅛ inch narrower, length and width. A mark can be made on the glass with a china marker or a fine-point felt-tip marker.

Place the piece to be cut on a thick cardboard or newspaper pad. Lubricate the wheel on the glass cutter with a light machine oil or kerosene. Line up a straightedge along the line, positioning it so the cutter wheel will run along the marked line (19-71).

Place the cutter against the straightedge, press down on it with even pressure, and draw it across the glass. The cutter wheel will score the surface of the glass (19-72). Run the cutter wheel across the glass only once. A double or triple score could cause the glass to break with an uneven edge.

If a large piece is to be removed, place the glass on the edge of a thin piece of wood or a steel rule with the part to be removed sticking over the edge. Press down on this part and it will snap on the scratch line (19-73). Wear gloves to protect hands.

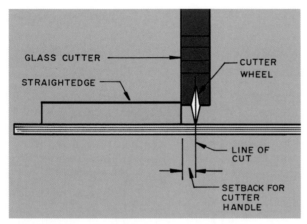

19-71 When starting the cut, remember to line up the cutter wheel with the line to be cut. Since the cutter handle has thickness, the straightedge is not placed on the line.

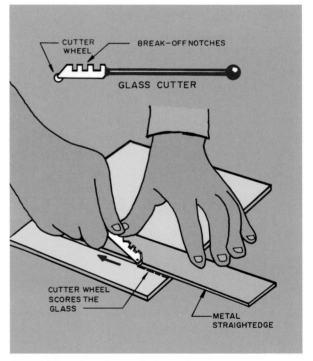

19-72 To cut a piece of glass, line up a straightedge along the line of cut and pull the cutter wheel along it, pressing hard enough to score the surface of the glass. Run it across only once.

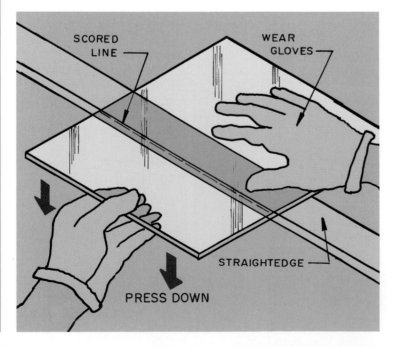

19-73 After scoring the glass, place the scored line over the edge of the straightedge and press down. It will break along the scored line. Wear gloves to protect your hands.

If the piece to be removed is a narrow strip, position the glass so the part to be removed sticks over the edge of a table (19-74). Tap the underside of the glass along the crack with the ball end of the glass cutter. Tap along the line as the crack moves to the bottom side of the glass.

Another way to remove a narrow strip is to score it, place one of the break-off notches on the cutter over the edge of the glass, and pull the cutter down (19-75). This technique breaks away small pieces. Work along the scored line. Pliers can also be used to pull off the narrow strip.

Wear gloves when handling and cutting glass. The edges are always sharp and even small shards can pierce the skin.

Glazing Compound

At the hardware store examine the glazing compound cans available and choose the one that best fits your situation. The example shown in 19-76 is suitable for face-glazing wood and metal window frames. It forms a watertight seal, resists cracking, and allows for expansion and contraction. It can be painted after it has set for 7 to 14 days, depending upon the temperature. Clean the putty knife immediately after finishing with mineral spirits.

Replacing Broken Window Panes

You can replace single-pane standard window glass if it is in older wood sash, or some metal sash, if it is held in place with glazier's points and sealed with glazing compound. If the broken pane is made up of two or more pieces of glass with air spaces between, the air space is sealed and often filled with gas to pro-

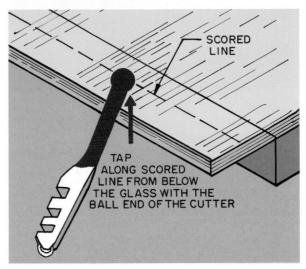

19-74 Narrow strips can be removed by scoring on the cut line, extending the glass over the edge of a tabletop, and tapping the score line from below with the ball on the end of the glass cutter. Wear gloves.

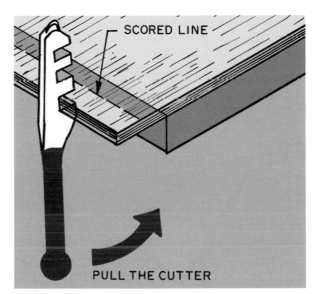

19-75 Narrow strips can be removed by scoring the glass and breaking the strip off using the break-off notches on the glass cutter. Place the notch that fits the glass thickness over the edge and rotate the ball end, providing the leverage to break the glass on the scored line. Wear gloves.

19-76 Examine the information on the can of glazing compound and select the one that best suits your needs.

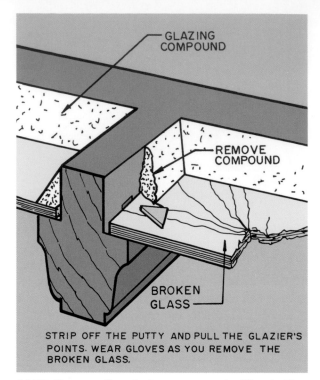

19-77 Remove the glazing compound with a knife or chisel but be careful not to damage the wood sash. Wear gloves so the broken glass will not cut your hand.

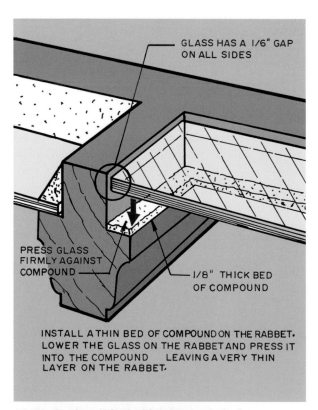

19-78 After the broken glass has been removed, scrape off all paint and glazing compound. Sand, if necessary. Then seal the wood rabbet.

vide increased insulation. It will have to be replaced by the window dealer. Other glass that will require professional service includes **low-E glazing,** which creates a heat barrier; **insulating glass,** which is a double- or triple-pane unit sealed around the edges and gas filled; **laminated glass,** which has layers of glass bonded with interlayers of plasticized polyvinyl butyral so it can resist breakage; and **acoustical glass,** which is laminated with sound-absorbing plastic between layers of glass.

Single- and **double-strength sheet glass** can be easily cut and installed by the homeowner. **Plastic sheet glazing** is cut with fine-toothed woodworking tools and is installed the same way as described for single-glazed glass panes. These broken panes can be replaced without removing the sash from the window.

The replacement glass should be cut ⅛ inch smaller in width and length than the opening.

Begin by removing the putty and glazier's points, and lift off all the shards of glass. Wear gloves when removing the broken glass and cleaning the sash and muntins (**19-77**). Now scrape away all paint and putty

19-79 Apply a ⅛-inch-thick layer of glazing compound to the surface of the rabbet. Lay the glass in place and press into the compound. A thin layer of compound should remain between the glass and the surface of the rabbet.

in the rabbet into which the new glass will be laid
(19-78). Use a brush to remove all dust and loose particles. The new glazing compound will not stick to
the rabbet if it is not clean. Then coat the rabbet with
a light coat of linseed oil, which helps preserve the
wood to allow the compound to stick. Some prefer
to prime the rabbet with a primer used to paint the
window instead.

Once the rabbet is clean and sealed, lay a thin layer
of glazing compound (19-79). Some roll the compound into a long ribbon and lay it into the rabbet.
It can then be pressed thin with a putty knife. The
goal is to get a thin layer of compound between the
glass and the surface of the rabbet. Now lay the glass
into the rabbet and press it against the compound. A
thin layer of compound should remain under the glass
and along the end. Scrape off any that gets on the inside and outside of the glass.

Next install the glazier's points. Two types are commonly available (19-80). They are laid on the glass
and driven into the side of the rabbet, usually with a
screwdriver. Place about 6 inches apart (19-81).

Finally, make another roll of compound and place
it on the edge of the glass along the rabbet (19-82).
Smooth it out with a putty knife, forcing it firmly
against the glass and the rabbet (19-83). Remove any

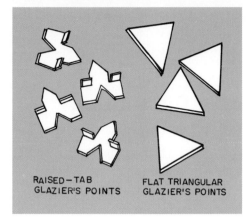

**19-80 Raised-tab
and flat triangular
glazier's points are
commonly used to
hold the glass to
the wood sash.**

RAISED-TAB
GLAZIER'S POINTS

FLAT TRIANGULAR
GLAZIER'S POINTS

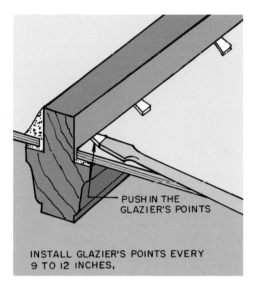

**19-81 Tap the
glazier's points
halfway into the
wood sash.**

PUSH IN THE
GLAZIER'S POINTS

INSTALL GLAZIER'S POINTS EVERY
9 TO 12 INCHES.

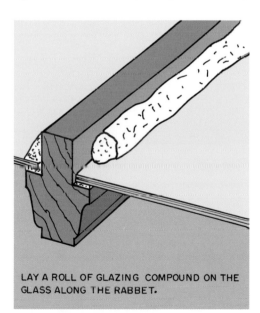

LAY A ROLL OF GLAZING COMPOUND ON THE
GLASS ALONG THE RABBET.

**19-82 Lay a roll of glazing compound in
the rabbet.**

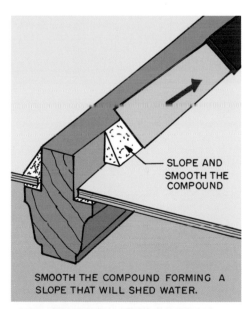

SLOPE AND
SMOOTH THE
COMPOUND

SMOOTH THE COMPOUND FORMING A
SLOPE THAT WILL SHED WATER.

**19-83 Smooth the glazing compound
with a putty knife, forming a tight seal
to the glass and frame. Scrape off
excess compound from the sash and
glass.**

excess compound. If it sticks to the knife, wet the knife with water. Let the compound cure as directed on the can before painting it. This is typically 7 to 14 days.

Replacing Old Glazing Compound

After many years if the glazing compound has not been kept painted it will begin to crack and pieces will drop off the frame. Years of exposure to heat and cold also cause deterioration. When this happens you can seal the cracks with caulking; however, this is a short-term solution. The pieces still in place will most likely drop off over time. Also the wood frame may have been damaged by moisture leaking through the deteriorated glazing compound. The best solution is to remove the compound, treat the wood frame, and install new compound.

Use a wood chisel, knife, and putty knife to cut away the remaining old glazing compound. Be careful not to cut into the wood frame. If the frame has some damage, scrape away the rotted material and fill the hole with a crack filler. After it has hardened, work it so it is flat and smooth. Paint a light coat of linseed oil on the exposed wood and let it soak in for a while. Then replace the glazing compound as explained in the section on replacing broken glass.

Additional Information

For more information, see these Sterling Publishing Co., Inc. books by William P. Spence, *Windows & Skylights,* and *Carpentry & Building Construction.*

CHAPTER 20

Repairing & Replacing Exterior Doors

Exterior doors get a lot of wear from the weather as well as frequent use. Some

doors are protected by a porch or are in a recess set into the house, and therefore weather slowly.

Some doors are used more times a day than others, which contributes to wear on the hinges, threshold,

and frame. Some problems, such as loose hinges, are minor and easily repaired. Others, such as a

rotten threshold, are major and present a difficult repair process.

While doors are constructed in several ways, the example in 20-1 will establish the commonly used terminology.

Fixing Doors That Stick

Some problems can be eliminated if the doors are given proper care. Possibly the most frequent problem is doors that stick. This is often caused by moisture entering the top, bottom, and sides of the door. Before a door is installed these edges should be primed and painted or protected with several coats of exterior spar varnish. Check the existing doors in your house and see if they need sealing. After a few years this seal coat may wear thin and need to be reapplied. Apply the seal coat when the door is thoroughly dry and the humidity is low so you do not trap moisture

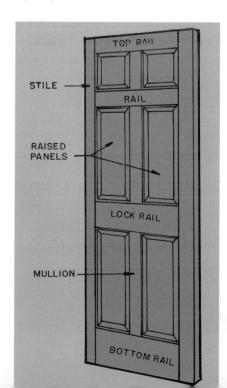

20-1 Terms used to identify parts of an outside panel door.

20-2 This carefully finished fiberglass door provides the focal point for the front elevation of the house. Notice the hardwood threshold below the door.

in the wood. This same problem occurs with interior doors, so check them also.

Wood doors get nicks and scratches on the inside and outside surfaces. This is another repair needed to keep moisture out of the wood. It also restores the attractive appearance (20-2). If you plan to refinish the entire door after filling dents and cracks, it is best to remove the hardware.

While you are sealing and refinishing the exterior doors, examine the jambs and sill. If they are sound but the finish is worn, refinish them and recaulk along the edges that meet exterior siding. Again, it is important to keep moisture out of the wood jambs.

Other doors, such as metal and fiberglass, can present similar problems. A metal exterior door may have a wood or steel frame as shown in **20-3**. When the exposed wood framing is not sealed it may begin to rot and the metal facing will start to rust from the inside (**20-4**). How do you repair this damage? Try to cut away the bottom wood framing and replace with a pressure-treated wood piece. This can be secured in place with screws driven through the metal

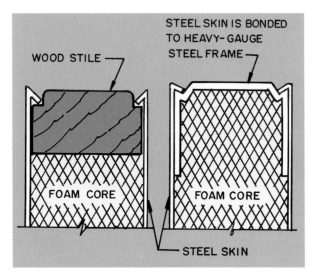

20-3 Typical steel door construction.

20-4 The bottom rail of this steel exterior door has rotted and must be replaced. This will also cause the metal facing to rust.

face. This is not a beautiful repair but might save the door. If the door has a metal edge it must be kept painted so it will not rust.

Fiberglass doors also have a wood frame and are filled with a polystyrene or polyurethane core which provides insulation (20-5). The edges need to be sealed, as just described. The fiberglass skin is bonded to the frame (20-6).

Sticking doors are also caused by some shifting in the frame of the house. The floor could have developed a sag or a partition a bow or twist. Remedies for these problems are covered in other chapters in the book.

Sticking doors may be caused when old wood doors start to separate at the joints. The stiles and

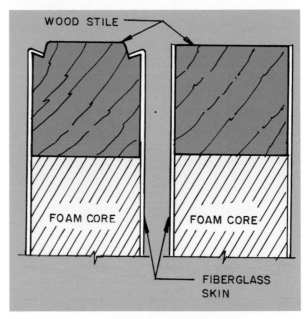

20-5 **Typical construction of a fiberglass door.**

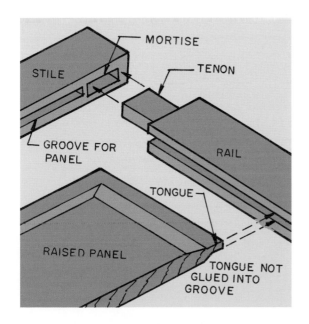

20-6 **This view shows the top rail of a fiberglass door before it has been sealed. The stile on the edge has been stained and sealed.**

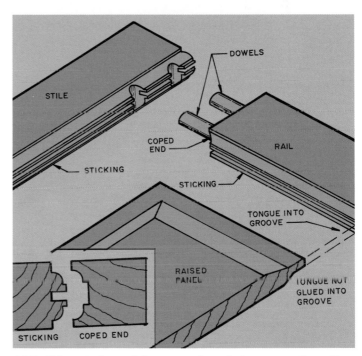

20-7 **This wood-panel door uses dowels to secure the rails to the stiles.**

20-8 **Older doors often used a mortise-and-tenon joint to secure rails to the stiles.**

481

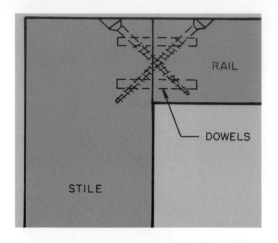

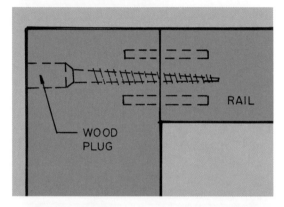

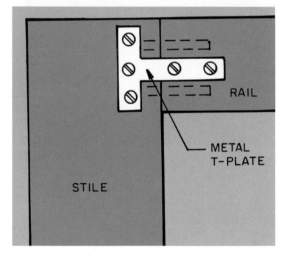

20-9 If a rail-to-stile connection is held with dowels and becomes loose, it can sometimes be secured with screws or T-plates.

rails may be held together with dowels (20-7) or tenons (20-8).

If the door joints are really loose you might be able to disassemble it, scrape the old glue off the dowels or tenons, and reglue. Use a waterproof glue. Clamp the door back together with long bar clamps. Do not glue the panels in the grooves. They need to be free to allow for expansion and contraction.

If a door with dowels cannot be disassembled or possibly only one or two joints are loose you might be able to get enough strength to continue using it by setting wood screws through the rail and stile as shown in 20-9. If the door is heavily used this may be inadequate. If the door is in a location where it is not generally seen, it can be secured by screwing metal T-plates on each side of the door. While this is not a beautiful repair, when it is painted to match the door it is not very noticeable (20-9).

If the door has tenons, pull the joint closed and drive several wood screws through the stile into the tenon (20-10).

If a door has **warped** it can never be straightened to its original condition. If it is an exterior door it should be replaced.

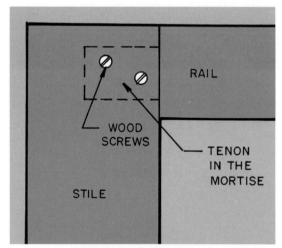

20-10 If a rail-to-stile connection is held with tenons, it can be secured by driving several screws through the stile into the tenon.

Loose hinges can also cause a door to rub on the jamb and stick. Exterior doors are heavy and require screws that penetrate the frame; one should be long enough to pass through the airspace behind the frame into a stud (20-11). Usually a 3½-inch wood screw will be long enough. If after the screw is driven into the stud the door still rubs a little, plane off a slight amount where it rubs. This can be located by marks on the edge of the door and the jamb (20-12). When you dress the edge of the door, plane it on a bevel of about ¹⁄₁₆ to ³⁄₃₂ inch so it will move by the jamb without hitting if the fit is close (20-13). Be certain to seal the edge immediately after planing it.

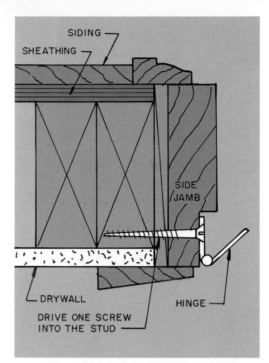

20-11 Exterior door hinges should have one long screw run through the jamb into the stud.

20-12 The very top corner of this door is rubbing on a small area of the jamb. If the door and jamb are solid, plane a little off the door. Then reseal the door and jamb surfaces.

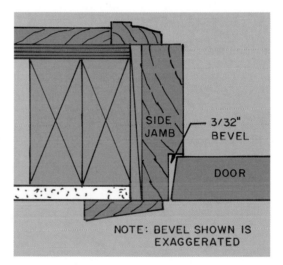

20-13 When fitting a door, plane the edge on the swinging side on a slight bevel facing toward the inside of the door.

20-14 When hinge screws cannot be tightened, remove them and glue wood plugs in each hole. When dry, trim flush with the jamb and reinstall the screws.

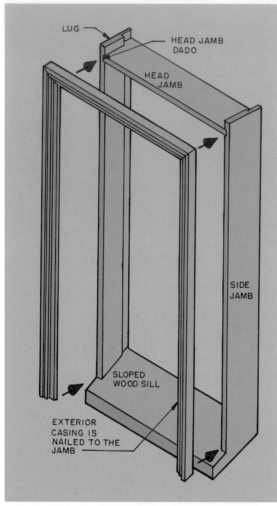

20-15 A typical exterior door frame and casing found in older home construction.

If the hinge is loose and the screws will not tighten, remove them and glue a round stick in each hole or fill the hole with a crack filler that will hold screws. This information will be found on the can (**20-14**).

Damaged Thresholds & Jambs

A typical exterior door frame for an older home construction is shown in **20-15**. If this was not primed or carefully sealed before installation it may over years begin to rot. Small amounts of rot on the exposed surfaces can be repaired by scraping away the rotted material and filling the cavity with exterior-quality wood filler. After this has dried it should be sanded and possibly have additional coats applied. To help it remain in place, drill several holes in the cavity and be certain to force wood filler in them. This provides an anchor for the patch (**20-16**).

If the damage is beyond patching, cut away and replace the door frame and sill. While a damaged sill can be replaced often, it is necessary and best to replace the entire door frame and sill. Exterior doors are heavy and require a sound frame to function properly.

Sills and jambs are also damaged by termites and wood-boring beetles. Usually the damaged wood has to be removed. Then have an exterminator treat the area to prevent reinfestation. Next replace the damaged and rotten parts or entire door frame if necessary.

20-16 Small rotted and damaged areas can be repaired by cleaning the area to raw wood, drilling a couple of holes to tie in the filler, and filling the opening with a wood filler suitable for exterior use.

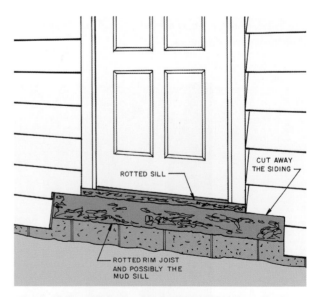

20-17 If the sill has rotted, the rim joist may also need to be removed and replaced. See Chapter 15 for structural floor repairs.

Remember to check the rim joist, mud sill, and floor joists in the area by the door sill. If all three have rotted, the header and ends of the floor joists also may be rotted (20-17). They must be repaired before a new door frame can be installed. Floor repairs are covered in Chapter 15.

Sills in Older Houses

When you begin to replace an old damaged sill, be aware that over the years door frame construction has changed and this will change your approach a little. One such situation found in older houses is shown in 20-18 where the wood sill is set on a slope which makes it extend below the top of the joists. The rim joist is notched to get the sill flush with the finished floor, and the subfloor is cut back to make room for the sill (20-19). When necessary, blocking can be installed between the joists to support the back edge of the threshold and the unsupported edge of the subfloor as shown in 20-18. A threshold is installed along the back edge of the sill and the finished floor (20-20). The threshold can be wood, metal, or a composite material.

More recent installations have a threshold that rests flat on the subfloor. There are quite a number of different designs available. When choosing a new door and frame, get the advice of the door manufacturer on what you plan to do and how to install it.

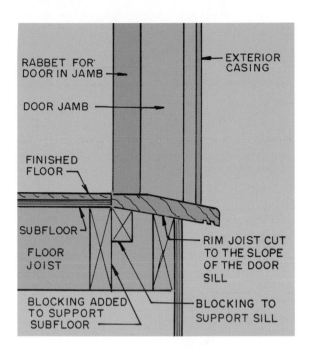

20-18 Older homes will often have the rim joist and floor joists cut to receive a wood door sill.

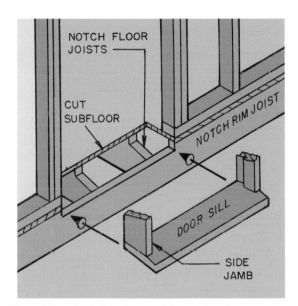

20-19 The rim joist, floor joists, and subfloor have been cut back to receive the door sill.

20-20 The threshold is installed over the space between the sill and the finished floor.

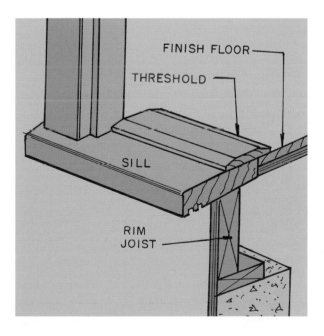

If the wood sill needs to be replaced but the old jambs and door are sound, follow the next set of directions. If the jamb is also bad and the door not worth saving, consider tearing them out and replacing with a new completely assembled jamb, threshold, and door unit.

Replacing a Damaged Wood Sill

Begin by removing the door. On some hinges on exterior doors, the pin is secured and cannot be removed; in this case, unscrew the hinges from the jamb. Older hinges can be separated by removing the pin.

If this is an older house that used wood sills as shown in **20-18** and **20-19**, accurately measure the width of the sill. Check to see if the jambs are actu-

ally at right angles to the edge of the sill. Cut the new sill to these measurements, including (if necessary) cutting a slight angle on an end where the jamb is not at right angles to the edge of the sill.

If the doorstop rests firmly on the sill, it will have to be loosened enough to permit the sill to be removed. First pry up or unscrew the threshold (**20-21**). It helps if it is cut into two or more short pieces. Then cut the sill into two or more pieces and pry it up (**20-22**). If there is a possibility that sawing through the entire width of the sill might damage the finish floor, try boring some large-diameter holes on the line of the cut, and cutting the web between them with a chisel (**20-23**).

Repair any damage to the floor joists and rim joist and install a layer of flashing below the sill area

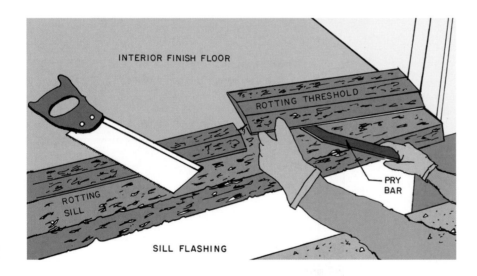

INTERIOR FINISH FLOOR

ROTTING THRESHOLD

ROTTING SILL

PRY BAR

SILL FLASHING

20-21 Cut the rotted threshold into two or more pieces and pry loose from the wood sill. Be careful not to damage the finish flooring.

20-22 After the threshold has been removed, cut the rotting sill into two or more pieces and pry it loose from the joists. Again, be careful not to damage the finish floor.

FINISH FLOOR

PRY BAR

ROTTED SILL

FLASHING

20-23 One way to separate the sill into several pieces is to bore a series of holes and cut the web between them with a chisel.

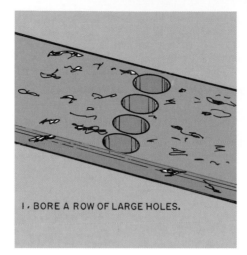

1. BORE A ROW OF LARGE HOLES.

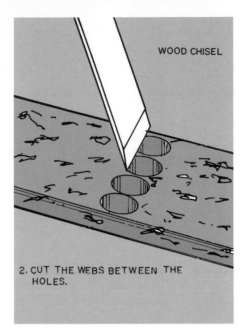

WOOD CHISEL

2. CUT THE WEBS BETWEEN THE HOLES.

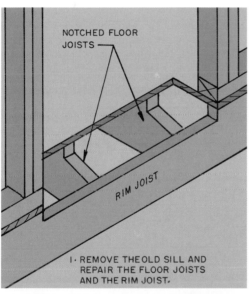

NOTCHED FLOOR JOISTS

RIM JOIST

1. REMOVE THE OLD SILL AND REPAIR THE FLOOR JOISTS AND THE RIM JOIST.

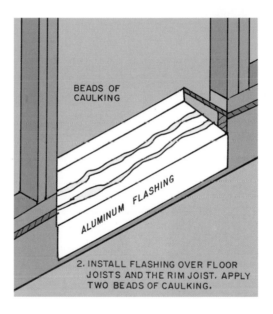

BEADS OF CAULKING

ALUMINUM FLASHING

2. INSTALL FLASHING OVER FLOOR JOISTS AND THE RIM JOIST. APPLY TWO BEADS OF CAULKING.

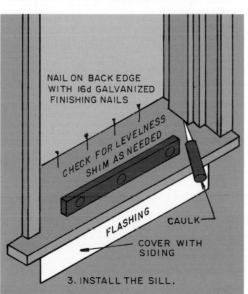

NAIL ON BACK EDGE WITH 16d GALVANIZED FINISHING NAILS

CHECK FOR LEVELNESS SHIM AS NEEDED

FLASHING

CAULK

COVER WITH SIDING

3. INSTALL THE SILL.

20-24 After the old sill has been removed, repair any damaged floor joists on the rim joist. Then install the flashing and the new sill.

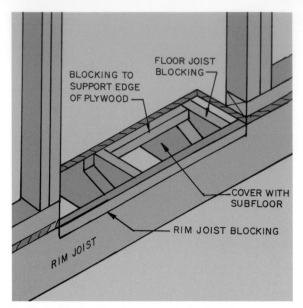

20-25 When converting an old sill that has notched joists, nail blocking as needed to support the added subfloor.

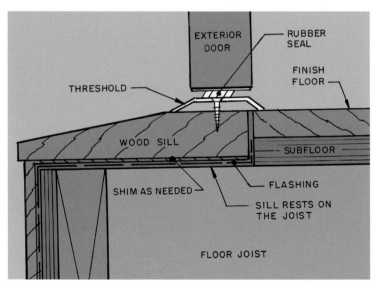

20-26 The installation places a wood sill flat on top of the floor joists. Notice the flashing below it.

(20-24). Lay a double bead of caulking on the flashing and place the new sill over it. Nail to the blocking along the back edge. It will be inside the house and not exposed to the weather. Check the sill to be certain it is level. Shim it when necessary. Caulk the joint between the sill and the side jamb.

If you prefer to install a totally new door and frame, it will have a flat threshold that rests on the subfloor. This will require blocking the joists and rim joist and installing a subfloor over them (20-25).

Installing a New Exterior Door

When you buy a new exterior door and door frame, the threshold will be designed to be installed flat on the subfloor as shown in 20-26 and 20-27. The door frame will be assembled and the threshold secured on the bottom. The door will also be installed in the frame.

One such threshold is a composition wood-like material that butts the jamb, while the door stop fits on top (20-28). It is secured to the subfloor by screws through the raised back edge. The screws are recessed and covered with plugs matching the material in the threshold. A metal threshold is shown in 20-29. It is

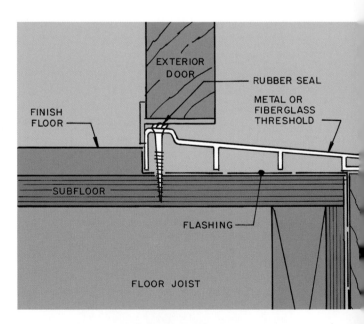

20-27 This one-piece metal or fiberglass threshold is placed directly on the subfloor and secured with screws through the back raised edge.

20-28 This sill is made from a hard composition material and is secured to the subfloor by screws through the back edge. The screws are covered with special plugs.

20-29 This attractive metal threshold is secured to the subfloor by a fiberglass molding on the back edge.

20-30 This copper threshold is secured to the subfloor with the copper molding along the back edge.

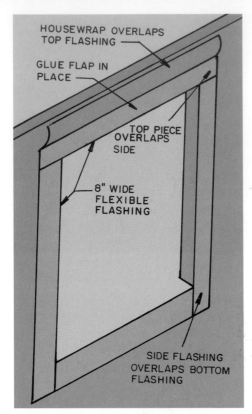

20-31 Install flexible flashing at the head. It should fit under the housewrap and overlap the side flashing.

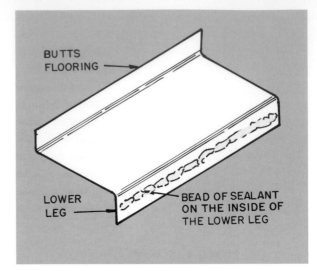

20-32 An aluminum sill pan is installed on the subfloor between the sides of the door opening.

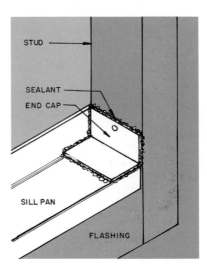

20-33 Set the sill pan into the caulking between the studs. Secure to the subfloor with rust-resistant screws.

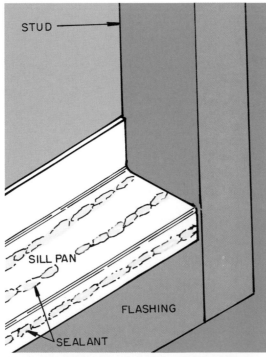

20-34 Seal the sill pan to the studs with an end cap.

secured with screws on the back edge. They are covered with wood molding. A copper threshold is shown in 20-30. It was installed on the subfloor and the jambs set on it. It is held with screws on the back edge which are covered with a plastic strip. There are other designs available.

Before installing the new door frame, be certain the exposed edges of the rough opening are properly flashed. They can be wrapped with housewrap or plastic weather-resistant sheeting (20-31). Then install an aluminum sill pan (20-32) between the studs of the rough opening (20-33), and install the end caps next to the outside studs (20-34). Two beads of caulking are laid on the pan and the door frame is lowered onto it.

Some prefer to lay aluminum flashing over the subfloor and let it extend down the exterior wall 12 to 15 inches as shown in 20-35. This replaces the flex-

20-35 This floor flashing for an exterior door opening is made from aluminum and extends down over the sheathing, providing continuous flashing. It is secured with rust-resistant screws.

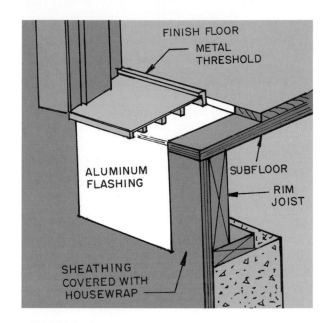

20-36 The aluminum sill flashing is laid back on the subfloor and down on the wall 10 to 12 inches.

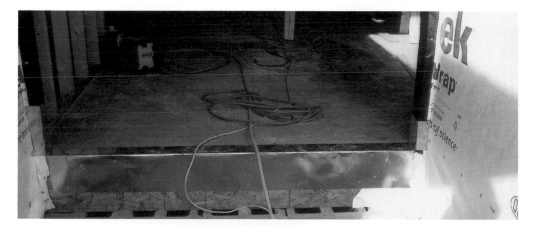

20-37 The sill flashing is lapped up and around the trimmer stud.

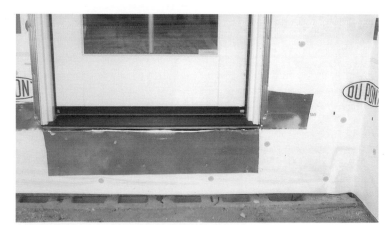

20-38 This finished installation shows the end of the fiberglass threshold extending beyond the door and the aluminum flashing running below the opening over the sheathing that has been covered with housewrap. Notice that adhesive-backed aluminum foil has been used to flash the sides of the rough opening.

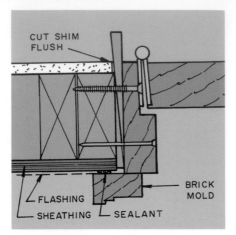

20-39 Rough openings for exterior doors in older homes were typically made $^3/_4$ to 1 inch wider than the door frame so wedges could be used to true it up and secure in position.

ible flashing that is used with a sill pan. In **20-36** the aluminum sill flashing has been laid across the door opening. The corners at the studs are bent up and around the studs (**20-37**). A finished installation is shown in **20-38**. Notice that the flashing extends well below the door threshold.

Securing the Door Frame

Exterior door frames are secured to the trimmer stud by nailing through the brick molding or using a nailing flange. Since you are replacing an older door frame, the size of the rough opening will most likely be larger than if it were newer construction. Older door frames were set between studs spaced at least ¾ inch wider than the door frame, to allow shims to be installed to level and to plumb the frame (**20-39**). The head jamb was also shimmed and nailed to the header

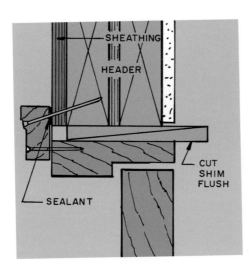

20-40 Shim and nail the head jamb to the header. This prevents it from bowing, over time.

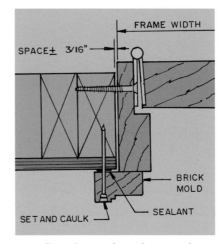

20-41 Rough openings for exterior doors are sized so there is just enough room to slide in the door frame. It is important that the trimmer studs be straight and perpendicular.

20-42 If a new door frame is to be installed in the rough opening of a door in an old house, add blocking on one or both side jambs to reduce the size needed for the new frame.

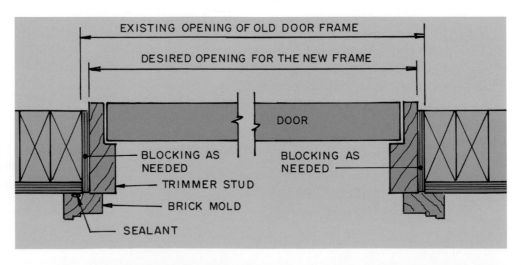

(20-40). The current door frames are installed in rough openings that leave just enough space to slide in the frame (20-41).

If the new frame is to be installed in a rough opening for an older door, it may be necessary to narrow the opening to suit the new door frame. This can be done by nailing a thin strip of plywood on one of the trimmer studs (20-42). The new door frame is then secured to the wall by nailing through the brick molding. It is also essential that a 3-inch screw be driven through each hinge into the stud. This greatly strengthens the installation and provides security. Without these screws it is easy to tear off the door and enter the house.

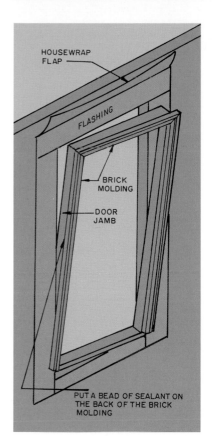

20-43 Place the bottom of the door frame on the sill pan and raise into position. Press it firmly in place to seal with caulking.

HOUSEWRAP FLAP

FLASHING

BRICK MOLDING

DOOR JAMB

PUT A BEAD OF SEALANT ON THE BACK OF THE BRICK MOLDING

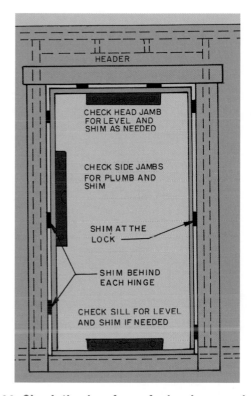

HEADER

CHECK HEAD JAMB FOR LEVEL AND SHIM AS NEEDED

CHECK SIDE JAMBS FOR PLUMB AND SHIM

SHIM AT THE LOCK

SHIM BEHIND EACH HINGE

CHECK SILL FOR LEVEL AND SHIM IF NEEDED

20-44 Check the door frame for levelness and plumb. Shim as necessary, being careful not to overtighten the shims, because this can cause a bow.

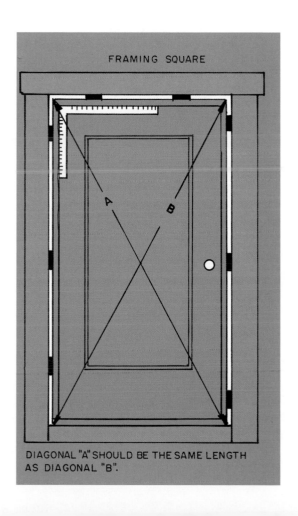

FRAMING SQUARE

A   B

DIAGONAL "A" SHOULD BE THE SAME LENGTH AS DIAGONAL "B".

20-45 Check the door frame for squareness with a carpenter's framing square and measure the diagonals. If they measure the same, the frame is square. Install the door and check to see if it closes without rubbing the frame.

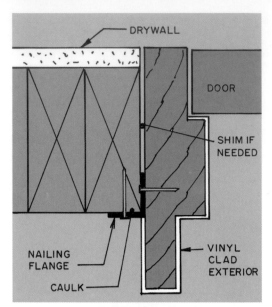

20-46 This is a typical vinyl-clad exterior door jamb that is held to the studs and header with a nailing flange.

Place a bead of caulking on the back of the brick molding, put the threshold on the floor, and raise the door in place (20-43). Tack it in two places but do not drive the nails home. Check the frame for plumb and levelness (20-44). See if the door closes properly. It should not rub on the frame. Measure the diagonals to verify if it is square (20-45). If it needs adjusting, drive wedges where needed to hold it in place. Then nail the brick molding to the studs and install the long hinge screws that are supplied with the door. Use steel nails plated with cadmium, zinc, nickel, or chrome to resist rust. Countersink them and cover with caulking.

20-47 The nailing flange runs on the side and head jambs and is nailed to the trimmer studs and the header. Be certain the frame is level and plumb before driving any nails flush.

20-48 Cover the nailing flange with a special adhesive-backed aluminum foil made for this purpose.

Another type of exterior door has a vinyl-clad jamb and is secured with nailing fins just like most windows (20-46) have. The frame is installed in the same manner as described for door frames with brick molding; however, the nailing flange is nailed to the trimmer stud (20-47) and then covered with an adhesive-backed aluminum foil tape to waterproof the connection (28-48). Lay a bead of caulking behind the flange before the door is set into the opening.

Flashing the Head

After the door frame has been installed, flash the brick molding at the head with aluminum or copper. After the metal head flashing is installed, lay the housewrap down over it and seal it with caulking (20-49).

The head of a door frame installed with nailing flanges is shown in 20-50. Notice the nailing flange has been sealed to the sheathing with caulking. A 12-inch piece of flexible flashing is laid over the nailing flange and sealed with caulking. The housewrap is laid over this and is also sealed with caulking. The flange actually serves as a metal flashing. This is one example of jambs that are available. Other designs are manufactured.

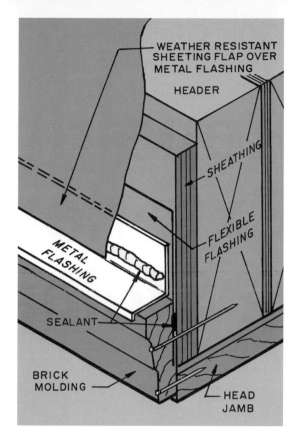

20-49 The door frame was installed using brick molding at the head. It has aluminum flashing laid over the molding and is covered with the housewrap.

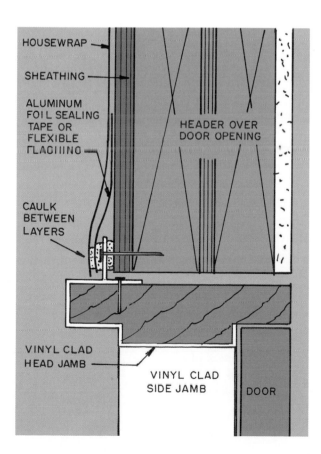

20-50 The head of a door frame installed with nailing flanges can be flashed with flexible flashing or adhesive-backed aluminum sealing tape and the housewrap is laid over it. Notice how each layer is caulked along the flange.

CHAPTER 21

Repairing Exterior Trim & Siding

Exterior siding protects the interior of the house 24 hours a day and

is subjected to heat, cold, wind, and rain **(21-1)**. Over the years it may begin to fail due to deterio-

rated caulking which allows water to penetrate the siding. Bumps and scratches damage all kinds

of siding. If the damage affects the watertight barrier it will have to be repaired or replaced. Often

a simple repair can restore the siding to its original condition and the repair painted to match the

existing siding. Other times the damaged section must be removed. This is a decision that must be made as renovation is considered.

There are times when the siding on the entire house is in such bad condition that it should be removed and completely replaced or covered with new siding. Exterior walls that are brick or stucco must also be inspected and maintained.

21-1 The siding is the dominant element of a house and the watertight envelope that encloses the living area.

Checking Siding for Needed Repairs

The homeowner would be wise to make regular inspections of the siding, roof, and exterior trim so developing damage can be easily corrected and the cause eliminated.

Eliminating Mold & Mildew

You should do your best to keep the exterior free of mold and mildew. Rent a pressure washer and use the chemicals they recommend (21-2). If the paint is bad and peeling, the pressure washer may strip off the paint. Another technique is to make a solution consisting of three parts water and one part chlorine bleach. This can be brushed on the house (21-3) or sprayed with a garden sprayer (24-4). Fill the sprayer jar with chlorine bleach and set the dial on the top to produce a 3-to-1 mix. Even with this technique a soft brush on a long pole will help. After it soaks a bit, rinse with fresh water.

Inspecting the Caulking

As the house siding is checked be certain to look at the caulking. Quality caulking will last many years; however, over time it will pull loose and need to be replaced. Check inside and outside corners and any

21-2 A power pressure washer with chemicals will remove mold and dirt from the siding and roof.

21-3 Mold and mildew can be removed from siding using a bleach solution and a soft bristle brush.

21-4 Chlorine bleach can be sprayed on the siding to kill mold and mildew. If not too bad, most will wash off when the siding is flushed with fresh water from a garden hose. It may be necessary to hit some spots with a brush.

21-5 There are many joints in the exterior siding that must be kept watertight with caulking.

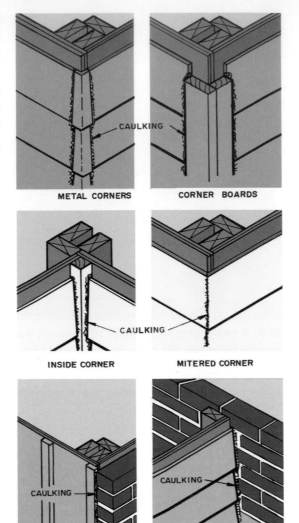

METAL CORNERS

CORNER BOARDS

CAULKING

INSIDE CORNER

MITERED CORNER

CAULKING

CAULKING

areas where metal corners have been used. Caulk as shown in **21-5.** Also check around the windows and doors. Expansion and contraction of materials as well as slight movements in the wall can open the caulking around windows and doors.

REPLACING CAULKING

It is best to remove the old caulking, clean the surfaces down to the original material, and lay in a bead of caulking.

Silicone and **polysulfide** caulking have a long life and are elastomeric (will stretch) but are not paintable. **Polyethylene** and **polyurethane** caulking have a long life and can be painted. In areas that will be painted, be certain to choose a caulk that will hold paint. Clear caulking is almost unnoticeable and is used where caulking cannot be painted.

The surface to be caulked should be dry and the air temperature at least 50°F (10°C). It should set at least 3 hours before getting wet, so do not caulk on a rainy day. Check on the container to see what times are recommended for painting.

The caulking comes in a tube that is placed in a **caulking gun (21-6).** The end of the tube is cut on an angle. The size of the extrusion is controlled by where you cut the plastic tip **(21-7).**

To apply the caulking squeeze the trigger. The tip can be pushed along the joint to be sealed or pulled along it. Pushing seems to work better. After the bead has been laid, smooth it out with your finger so the joint is full and the caulking is firmly sealed to the material. A plastic picnic spoon also works well to seal the joint.

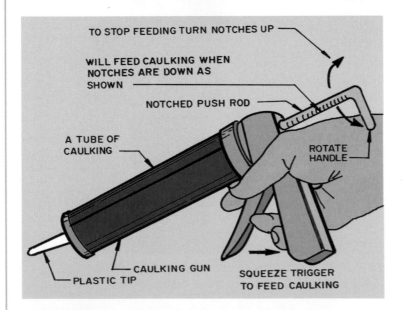

TO STOP FEEDING TURN NOTCHES UP

WILL FEED CAULKING WHEN NOTCHES ARE DOWN AS SHOWN

NOTCHED PUSH ROD

A TUBE OF CAULKING

ROTATE HANDLE

PLASTIC TIP

CAULKING GUN

SQUEEZE TRIGGER TO FEED CAULKING

21-6 Caulking is sold in tubes placed in a caulking gun. To operate the gun, squeeze the trigger which moves the push rod forward against the end of the tub. Squeezing some more forces the caulking out the tip of the tub.

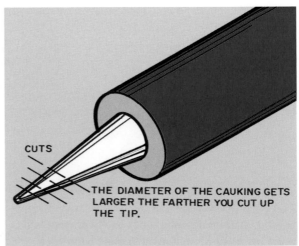

CUTS

THE DIAMETER OF THE CAUKING GETS LARGER THE FARTHER YOU CUT UP THE TIP.

21-7 The diameter of the extruded caulking depends upon where the end of the tip is cut.

Inspecting the Flashing

If the paint is blistering and peeling and the siding is moist, there is a big leak somewhere. Besides looking for damaged siding, check the flashing. This situation occurs frequently around doors and windows. The siding will have to be removed and the flashing replaced. Use new aluminum flashing. Older homes used galvanized steel but exposure to moisture will eventually cause it to rust and fail. See Chapters 19 and 20 for information on window and door repairs.

Checking Soffits for Renovation

The soffit is a member that is not checked very often because it is not very visible. However, if a roof has very poor ventilation, moisture will collect and condense in the attic and some will work its way to the soffit, causing it to rot. Some older houses have little or no ventilation access through the soffit into the attic. This access is needed to move the air in the attic out through large vents on the gable end or through roof louvers (**21-8**). Newer houses use ridge vents (**21-9**). A section through a ridge vent on a house with asphalt shingles is shown in **21-10**.

If the soffit is rotting or warped, and the paint is blistered, it should be replaced. Provision should be

21-8 Gable end vents work well because they are up next to the ridge of the roof. Several will be needed to provide adequate ventilation.

21-9 Most new houses use a ridge vent because it provides a continuous exit for attic air at the ridge.

21-10 Ridge vents are a good way to vent an attic. Since hot air rises this is the best place to have vents.

CourtesyCor-A-Vent,Inc.

499

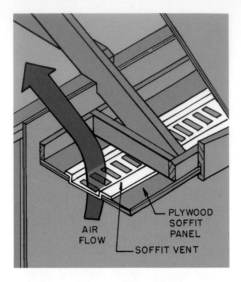

21-11 Typical soffit construction where aluminum continuous vent strips are used.

AIR FLOW

PLYWOOD SOFFIT PANEL

SOFFIT VENT

made to add additional vents high on the roof and gable end to carry the moisture out of the attic. While vent construction can differ, the example in **21-11** is typical. Remove one or both rotted boards. Exterior plywood is also widely used for soffit material. Install the back piece of soffit board, set the metal vent strip in place, and nail the second board to the lookouts.

Another way to increase the amount of airflow is to install round aluminum or plastic vents (**21-12**). These can be installed by boring 2½-inch-diameter holes through the soffit board. Then glue these in place. The hole can be bored with an expansive bit in a brace (**21-13**) or with a hole saw mounted in an electric drill (**21-14**). Be careful when using the hole saw. If it binds, the drill will kick and can hurt your wrist.

Apply a bead of glue or caulk around the edge of the vent and press in place (**21-15**). Clean up any excess glue or caulk that squeezes out around the vent. These small round vents do not provide adequate ventilation alone, but can be a good supplement to existing but inadequate venting.

Another approach is to remove the damaged wood soffit and replace it with a vinyl or aluminum soffit (**21-16**). These materials have small holes spaced across the entire surface. They do not rot or warp.

Large rectangular soffit vents are available and are installed in the existing soffit if it is sound. Mark the outline for the opening on the soffit (**21-17**), drill a

21-12 The 2¹/₂-inch round soffit vents can provide additional venting when the existing vents are proving inadequate. On the left is the exposed side. On the right is the back with insect screen.

21-13 Holes for round soffit vents can be bored with an expansive bit.

21-14 A hole saw is a good tool to use to bore holes for small, round soffit vents.

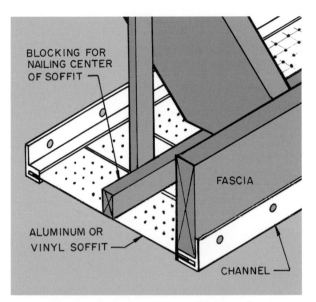

21-15 Secure the vent in place with a bead of glue or caulk around the edges.

21-16 The deteriorated soffit can be removed and replaced with a vinyl or aluminum perforated soffit panel.

BLOCKING FOR NAILING CENTER OF SOFFIT

FASCIA

ALUMINUM OR VINYL SOFFIT

CHANNEL

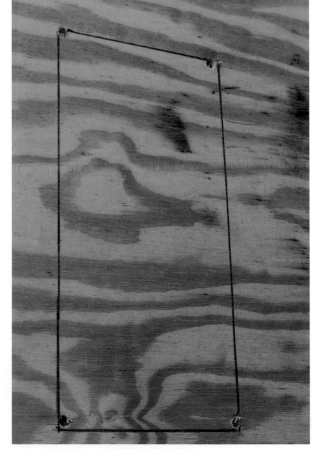

21-17 Make a template for the hole and mark the outline for the vent on the wood soffit.

21-18 Drill a hole in each corner and cut the opening with a saber saw.

21-19 Place caulking on the lip of the vent, place it in the opening, and secure with rust resistant screws.

starter hole in each corner, and cut the opening with a saber saw (**21-18**). Lay a bead of paintable caulk along the back of the flange, place the vent in the hole, and secure with stainless steel screws (**21-19**). Again, since a wide continuous vent will provide more uniform airflow and probably more air, and is preferred.

Repairs to Siding

Various siding materials and the details of their construction call for different approaches to their repair. Once you determine that the siding needs repair, you may find as you undertake the repair that the damaged siding must be replaced.

Repairing Wood Lap Siding

Lap siding may simply overlay the strip below or have a rabbet in the bottom edge which fits over the top of the piece below (**21-20**). It is important that it be kept painted, and any cracks or damage be rapidly repaired, because it does absorb moisture and will eventually rot. Wood lap siding is also subject to damage from insects, such as termites. It is available in several wood species; the most commonly available are pine, cedar, redwood, and spruce.

Check the entire wall for nails that have worked themselves out. If the siding is sound, reset the nails

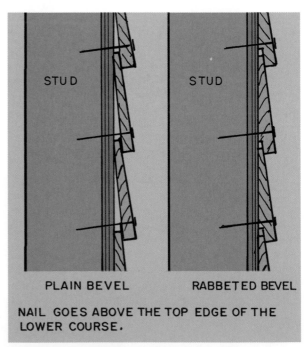

21-20 Two commonly available types of lap wood siding. Notice the nail is placed above the top edge of the lower course.

21-21 Reset popped nails. Pull the old nail and use a new one that is longer so it goes in solid wood.

below the surface with a nail set and fill the holes with an exterior crack filler. Some like to pull the nail and use a new one that is a little bit longer. Be careful you do not split the siding as you set the nail (**21-21**).

Minor damage or small splits can be repaired by filling with an exterior crack filler. Larger splits should be glued and the siding nailed as shown in **21-22**.

Bowed pieces of siding can usually be flattened enough to look satisfactory by driving wood screws

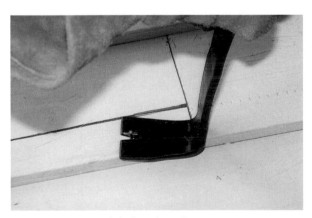

(A) Remove any nails below the split.

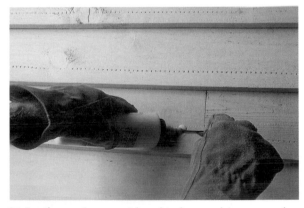

(B) Pry the crack open with a chisel, screwdriver, or pry bar and fill with polyurethane glue.

(C) Hold the crack closed as the glue dries by tacking a wood strip below it. Do not drive the nails all the way in. Scrape off any glue that has squeezed out of the joint before it can dry.

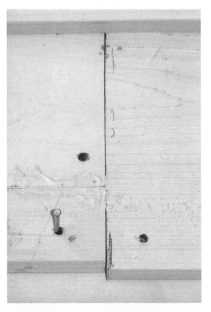

(D) After the glue dries remove the block and fill the nail holes with an exterior wood filler. Then drive 8d finish nails above and below the crack. Set them and cover with exterior wood filler. Finally, caulk the end joint.

21-22 Repairing large splits in wood lap siding.

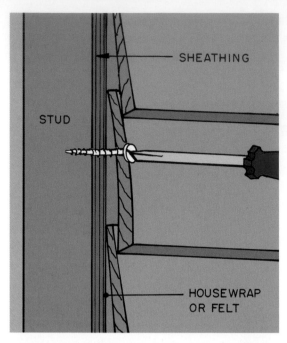

21-23 Small bows in lap siding can sometimes be pulled up enough to be nearly flat. Drill a pilot hole in the siding for the screw. Do not pull up enough to split it.

21-24 Mark the cuts on each end of the damage perpendicular to the edge. Longer replacement pieces blend in better than very short lengths. The ends should fall over a stud. Look for existing nails or use an electric stud finder. If replacing several pieces stagger the end cuts.

through them into the studs **(21-23)**. Drill pilot holes through the siding and small anchor holes into the sheathing and stud. If the siding is thick enough, use a counterbore so the head of the screw can be recessed and covered with exterior crack filler. If this is not possible, it is very important to use screws that will not rust. You can locate needed studs by finding the locations of the nails originally used to install the siding. Large bows will require the siding to be replaced.

If the wall shows these signs of deterioration it means there is some place where moisture is getting behind the siding. This should be found and fixed before making repairs.

REPLACING A DAMAGED PIECE OF SIDING

If it appears the damage cannot be repaired, then plan to remove the damaged piece. The first thing to do is to try to find a piece of siding that matches the one on the wall. It may be that an exact match is not available. Buy the closest-width piece you can find and cut it to the needed width. While it will not match perfectly, this will not matter except perhaps on the high visible front of the house. If you are a pretty good carpenter you might be able to remove a good piece from the back of the house and use it for the repair. Then use the new replacement piece on the back.

Many prefer to replace the entire piece of siding that has damage; however, only the damaged section need be removed. Begin by marking the ends of the section to be removed. The removed section should be 32 to 48 inches long because this will allow the new piece to be nailed to several studs **(21-24)**. If you are replacing several pieces, stagger the end joints.

Next remove any nails in the bottom edge of the piece to be removed **(21-25)**. Drive some shims on each side of a cut, thus raising it a little above the piece below **(21-26)**. Then cut on the line. A backsaw gives

21-25 Remove nails in the bottom edge by lifting the edge with a pry bar. Then tap the siding back against the sheathing and pull the raised nail. Protect the siding below the pry bar.

a smooth cut which helps produce a tight joint. You can place a wood scrap on the siding below so it will not be damaged by the saw. Cut up close to the siding above the damaged piece but be careful not to hit it (21-27).

Now remove the shims below the damaged piece and tap them up under the top piece. Then use a keyhole or compass saw to finish the cut (21-28). Remove or cut off with a hacksaw blade any nails that will be in the way when you slide the new piece up under the top siding. If the damaged piece does not come out easily, split it into several narrow pieces with a chisel and pry them out.

Next check the sheathing behind the opening. If it is sound, make certain it is covered with a piece of housewrap. If it was covered with builder's felt, repair any damage. If the sheathing is rotten, more siding will have to be removed so it can be replaced.

Finally, cut the replacement piece, allowing a ⅛-inch gap on each end so it can expand as needed. Paint both sides and all edges with a primer recommended for the paint used on the exterior of the house. Slide it in place and nail to each stud, like the piece it's replacing (21-29). If several pieces were replaced install the lowest piece first and work up the wall. Set the nails and fill the depression with exterior crack filler. Caulk the end joints.

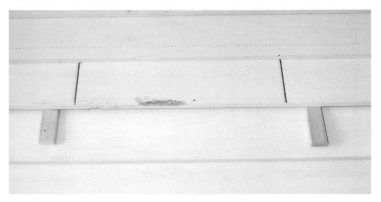

21-26 After pulling any nails on the lower end of the damaged piece, tap several wood shims beneath it to raise it a little above the siding below it.

21-27 Place a piece of thin plywood over the piece below the cut. Cut as much as possible with a backsaw. Be careful not to damage the piece above by hitting it as you saw.

21-28 Pull out the shims and tap them under the siding above the piece to be removed. Finish the cut with a keyhole or compass saw. If the piece will not slide out, split it into sections with a wood chisel and pry out the pieces.

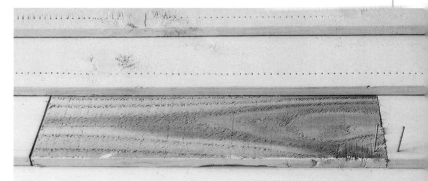

21-29 Cover all sides, edges, and ends of the repair piece with a wood preservative or prime it to receive the exterior paint. Then slip it under the siding above. It should be cut ⅛ inch shorter than the space so the ends can be caulked and to allow for expansion. Face-nail on the bottom edge at each stud. Set and fill the nail holes with exterior wood filler.

21-30 This house has beautiful wood shingles finishing the exterior walls. Wood shingles can be stained, painted, or left to weather naturally.

Repairing Damaged Wood Shingles & Shakes

Wood shingles and shakes are commonly used as siding and provide a unique textured and finished wall (21-30). **Shingles** are sawn and are smooth on both sides. **Shakes** are typically split, forming a very rough irregular surface. Some are sawn on the back side only. They are available in four grades. Use grade No. 1 on the walls. An undercoursing grade is used as backup when shingles are laid in double course, as shown in **21-31** and **21-32.** Wood shingles and shakes are available in lengths of 16, 18, and 24 inches. Red cedar and Southern pine species are available.

As you plan to replace one or more damaged wood shingles, be aware that they may be laid in single or double course (**21-31** and **21-32**). This will make some difference as you replace them. After you remove a damaged wood shingle, take a piece with you to the building supply dealer so you get replacement shingles of the same species and length. The width can be changed by cutting them narrower. Leave a ¼-inch space between the sides of the shingles.

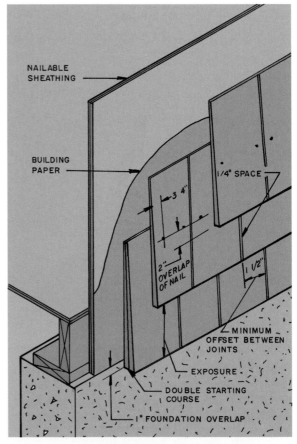

21-31 Recommendations for applying single-course wood shingles and shakes.

Begin the replacement by splitting the shingles into small strips with a wood chisel (21-33) and prying them out from below the shingle above. You can drive thin wedges under the shingles above to help remove the pieces. If a piece will not move, it probably has a nail in it and must be split again. Be careful not to damage the shingles above or below the damaged one.

Once the pieces are removed cut off the nails under the shingle above (21-34). Tap in the wedges enough to enable you to slide a hacksaw blade under the shingle above and cut the nails flush with the sheathing or undercourse shingle.

Now prepare the replacement shingle to fit the opening. If the shingle is the correct length, cut it to width to fit in the opening. Cut it so it leaves a ¼-inch expansion space on each side.

**21-32
Recommendations
for applying
double-course wood
shingles and shakes.**

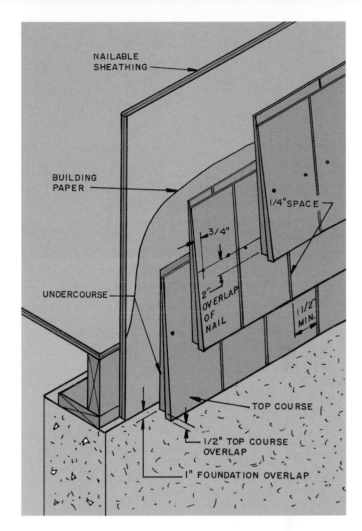

21-33 Raise the shingles above the damaged one with wedges and split the damaged shingle into small strips with a wood chisel. Pry out the strips.

21-34 Raise the shingles above the damaged area with the wedges and cut the nails from the old shingle with a hacksaw blade. Cut flush with the shingle below.

21-35 Slide the replacement shingle under the top shingle and get the butt flush with the shingles on each side.

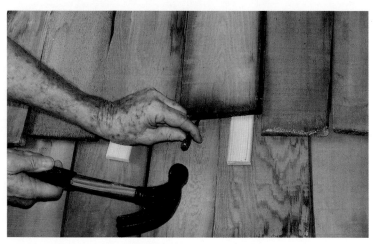

21-36 Drive two 5d shingle nails on an angle below the butt of the new shingle. Set and caulk.

Slide the shingle in place with the thin end under the shingle above. Tap it in place until the butt is flush with the other shingles (**21-35**). Then drive two 5d galvanized shingle nails on an angle up under the replacement shingle. Set the head and caulk as shown in **21-36**.

Repairing Damaged Vinyl Siding

Vinyl siding is tough and resists damage but will crack if hit hard enough, especially when the air temperature is low. While it will not shatter, it may crack. If this happens small cracks can be patched but larger damaged areas should be removed and replaced. Remember, vinyl does fade with age and the replacement will most likely not be exactly the same; however, over time it will fade and come closer to blending in with the old siding.

If the damage is small and you do not want to replace the piece, a repair can be made by removing the strip with the damage as explained for replacing a piece. The repair consists of a patch glued on the back of the siding.

If the crack is small and the face of the siding is not discolored, glue the patch on the back. If the face around the damage is fragmented or stained, cut out the damaged area and glue the patch on the back over the hole.

Cut a patch from the same material as the siding. Cut it about ½ inch wider and longer than the damaged area. Then coat the area around the damage with a two-part epoxy and coat also the face of the patch, but do not coat the patch in the area that will be visible after the repair has been made (**21-37**). Follow

21-37 Place the vinyl siding back side up on a firm surface. Clean the area where the patch will be bonded. Sandpaper if necessary to get a suitable surface. Apply the epoxy glue over the bonding area and to the face of the patch.

the directions on the epoxy container. Generally the patch can be set in place as soon as it has been coated. Press it firmly against the siding and let the epoxy harden (**21-38**). Now reinstall the panel.

Some vinyl siding manufacturers do not recommend this repair. Vinyl siding expands and contracts a great deal due to temperature variations. The patch will expand and contract at a different rate than the total panel, and this may cause the panel to bulge at certain temperatures.

If several walls covered with vinyl siding have wide damage such as from a hail storm or high winds the replaced panels will most likely be a slightly different color, so a varied appearance occurs. Consider replacing all the siding on one wall and using the faded but good pieces from that wall to make repairs on the damaged wall.

Another possibility is after making repairs paint the entire wall with an acrylic latex paint or a product recommended by a siding dealer or a paint dealer. This solution is good for a few years but generally the siding cannot be repainted and will have to be replaced. Part of the problem is that vinyl siding and the paint film expand and contract at different rates, leading to failure in adhesion.

REPLACING A SECTION OF VINYL SIDING

To remove a section of vinyl siding, a siding removal tool, often referred to as a **zip-tool**, is used (**21-39**). They are available at many building supply dealers. An emergency tool can be made with heavy wire

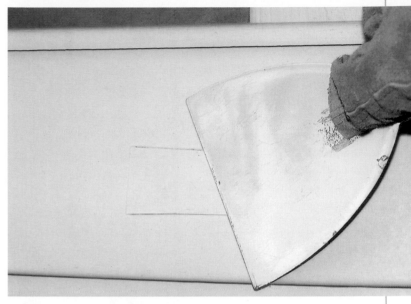

21-38 Lay the patch over the damaged area and press it firmly in place. Once it has dried the panel can be put back on the wall.

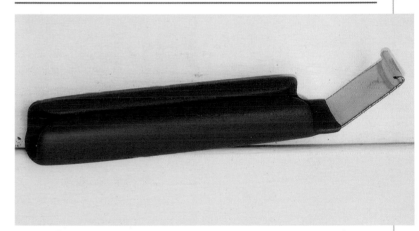

21-39 The vinyl siding removal tool is used to open the joint between panels.

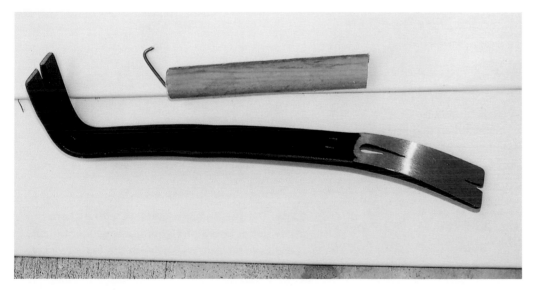

21-40 These can be used instead of the siding removal tool to open the joint between panels. If using the pry bar protect the surface of the panel below.

21-41 To open the joint between the panels, wedge the blade of the siding removal tool under the panel and hook it onto the back lip. Pull the lip downward and slide the tool along the length of the panel. This unlocks the connection and exposes the nail row. Use the tool to relock the overlapping panel.

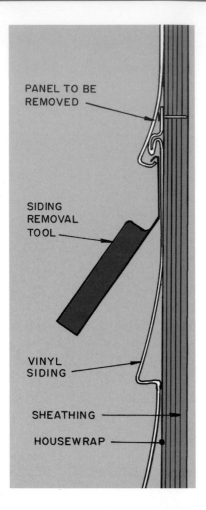

inserted into a wood handle, or you can even use a pry bar (**21-40**). A pry bar is harder to use, so the siding removal tool is recommended.

The siding removal tool is used to unlock the bottom edge where two panels are joined with a lock joint. A typical lock joint is in **21-41**. Insert the tool under the bottom edge and hook it under the lip and slide it along the edge, pulling the lip out of the piece above. Place blocks below the loose piece to keep it from slipping back in place (**21-42**).

Slide a pry bar under the nailing flange and pull its nails. Place a piece of wood on the siding below to prevent damage and help the prying process (**21-43**).

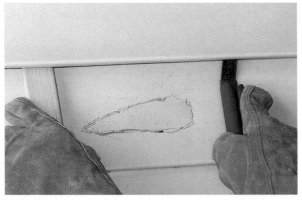

21-42 The damaged vinyl panel is unlocked from the joining panel with a siding removal tool. Hook the tool into the joint, pull it down and slide it along the edge. Place a wood block below the loosened edge to keep it from slipping back into the joint.

21-43 Slide a pry bar under the raised siding and pry out the nails in the nailing flange of the damaged section. Use a wood block to give additional prying force.

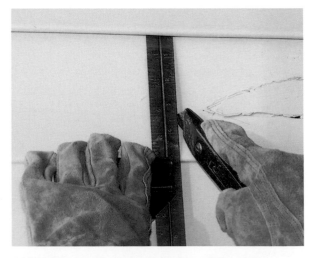

21-44 Cut the damaged section with a sharp utility knife. Use a straightedge to guide the cut. It will take several passes with the knife to make the cut.

When a damaged section is loose, mark the location of the side cuts. Then cut along the square with a utility knife (21-44).

Now cut the replacement piece 2 inches longer than the opening. This gives a 1 inch overlap on each end. Cut the nailing edge back 1 inch on each end (21-45).

Slide the piece in place and nail the top nailing edge to the sheathing with 3d galvanized siding nails. Place the nails in the center of the nailing slot and do not drive too tight. The piece needs to be able to expand and contract and slide on the nails. The nails can be pressed into the sheathing enough to stick. Then place a pry bar on the head and strike the bar with a hammer (21-46). Finally, relock the lower lip to the course above with the zip tool (21-47).

21-45 Cut the repair piece 2 inches longer than the open space. Notch the nailing flange a little over 1 inch on each corner. The cut can be made with a power miter saw that has a fine-tooth blade or a sharp utility knife.

21-46 Slide the repair piece under the raised course above. Snap the bottom lip of the repair piece over the nailing lip of the siding strip below the repair area. Nail it to the sheathing through the nailing flange. Place the pry bar on the nail and hit it, driving the nail into place. Do not set tight. The siding needs to expand and contract and must slide on the nail.

21-47 Lock the lip of the piece above to the top flange of the replacement piece with a siding removal tool.

21-48 Dents in aluminum siding can be leveled by sanding the damaged area to roughen the surface and coating with auto body filler. Feather it out over the area.

Repairing Damaged Aluminum Siding

Aluminum siding is more difficult to remove and replace than vinyl because it easily kinks and is dented by the tools used to remove and replace it. It can be removed with a siding removal tool (zip tool) as discussed earlier for removing vinyl siding. Be careful because aluminum siding will dent, so protect it with a wood scrap. If the panels are buckling they may have been improperly installed. Consult the contractor responsible for the installation.

Small dents can sometimes be repaired by applying layers of auto body filler over the depressed area. Sand the area to cut the finish and roughen it up so the auto body filler will bond. Then apply the filler over the dent and feather it out over the surrounding surface (**21-48**). Apply additional coats if necessary to fill a deep depression. After it hardens, sand smooth (**21-49**), prime, and finish with several coats of an aluminum siding paint (**21-50**). If the house is old it may be difficult to match the color exactly.

21-49 When the filler has hardened, sand it smooth. If necessary apply additional coats.

21-50 After auto body filler has been sanded smooth and feathered out on the surface, prime and paint the repair with a recommended aluminum siding paint.

Larger dents or a tear in the siding can be repaired by marking off a rectangular area around the damage, leaving 2 inches clear surface on each side of the damaged area and 2 inches below the bottom of the course that is just above the damaged area (21-51).

Now cut out the damaged area. Light-gauge aluminum siding can be cut with a heavy-duty utility knife (21-52). Heavier gauge will have to be cut starting with a hacksaw (21-53) and finishing with a straight blade or aviator metal snips (21-54). If the felt or housewrap has been damaged, repair it (21-55).

Make the patch from a scrap piece of aluminum siding. Hopefully you saved some pieces when the house was covered; otherwise the installer may have

21-51 Lay out the cut lines around the damaged area. Allow 2 inches between the butt of the top strip of siding and the top edge of the cut and on each side of the damage.

21-52 Light-gauge siding can be cut with a utility knife. Metal snips will be needed for heavier-gauge material.

21-53 The cut through the folded bottom edge can be made with a hacksaw.

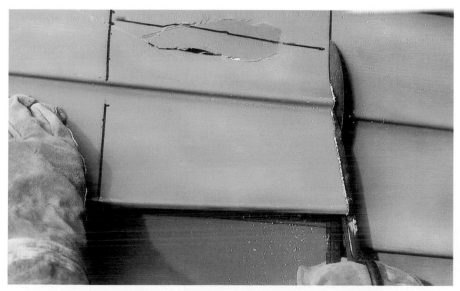

21-54 Cut the damaged area out with aviator snips or tin snips.

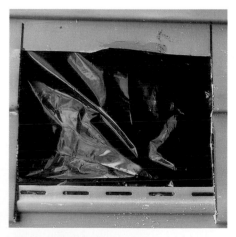

21-55 If the builder's felt or housewrap has been damaged, repair it.

scraps you can use. Cut the patch three inches wider than the cut-out area and cut off the nailing flange (21-56). The top edge of the patch should be long enough to fit up against the bottom edge of the course that is just above the patched area.

Clean and roughen the finish on the back edges of the patch. A coarse sandpaper is often used. Clean the area around the opening to which the patch is to be bonded. Lay a wide bead of clear silicone caulk around the edges of the patch but not on the channel on the bottom (21-57).

Lay the patch over the open area and slide it up against the bottom of the course that is just above the patched area. Slip the J-channel on the bottom edge of the patch over the top edge of the course that is just below the patched area. Press firmly in place and hold until it remains bonded. Carefully smooth the caulking that may come out along the edge so the edge is covered with a uniform layer (21-58). Also work the caulk at the top of the patch so it smoothly seals the patch with the bottom of the course above.

21-56 Cut the patch 3 inches wider than the opening and cut off the nailing flange but be certain the cut edge will fit up against the bottom edge of the course above, when it is installed.

21-57 After sanding the back edges of the patch, make certain the patch, and the area to which the patch will be bonded, are clean. Apply several rows of clear silicone caulk on the back of the patch.

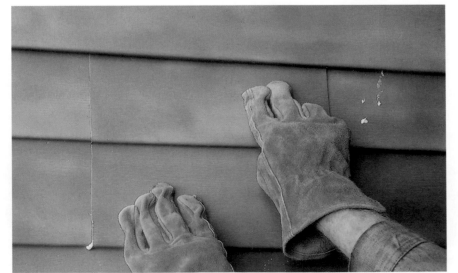

21-58 Place the patch over the open area. Slide it up until the bottom flange engages on the course below and the top edge is against the course above. Press firmly so all edges are bonded. Wipe away any caulk that may squeeze out around the edges with a cloth dampened with mineral spirits.

Repairing Portland Cement Stucco

Stucco wall finishing has been used for many years. It consists of a number of cementitious coats applied over a metal mesh or lath which is nailed or stapled to the wood sheathing or masonry wall (**21-59**). Over the years small cracks may appear or pieces may have been broken off by a blow. Narrow cracks can be filled with a polyurethane concrete crack sealant or siliconized acrylic caulk. They are sold in tubes and applied with a caulking gun. Some building supply dealers have a caulk designed especially for repairing portland cement stucco. If possible get a color close to the color of the stucco.

To repair the small crack, first clean it out by running a small wire inside it to loosen any particles and then clean it out with a brush or compressed air (**21-60**). A vacuum cleaner will also remove any loose particles. Then lay the caulk into the crack, forcing it so the crack is fully covered (**21-61**). Press the caulk

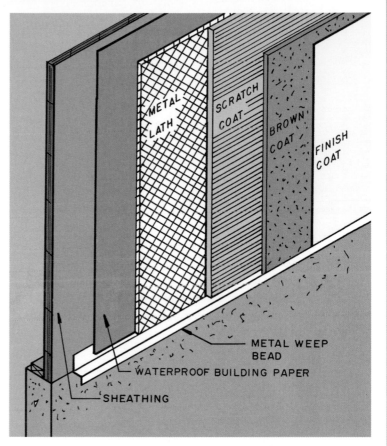

21-59 Typical construction of a portland cement stucco exterior wall.

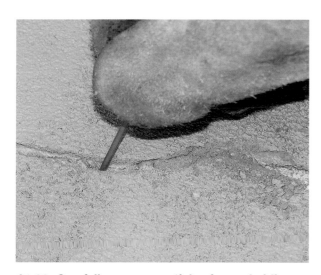

21-60 Carefully remove particles from a hairline crack with a wire, and clean with a soft brush or compressed air.

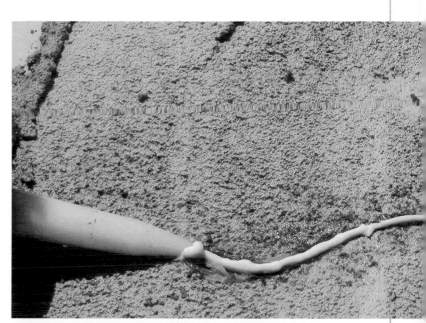

21-61 After cleaning the hairline crack, fill it with a high-quality exterior caulking. Keep the caulking off the surface of the stucco.

21-62 Press the caulk into the crack with a putty knife. Lay it so it is flush with the surrounding surface. Immediately wipe off any caulk that has gotten on the surface of the stucco on each side of the crack. Sometimes a finger or the tip of a plastic spoon does a better job of smoothing the caulk.

into the crack with a putty knife (**21-62**). If the caulk sticks to the putty knife, wipe it with mineral spirits. Work the caulk flush with the surface, noting any texture around it that may be included. Be certain to remove all excess caulk. After the caulk has completely dried, prime and paint the repaired area.

To repair a wide crack, cut away all loose stucco with a mason's chisel. Undercut the opening, forming a wedge which helps hold the patch in place. Clean all loose particles from the crack and lay in a layer of stucco as shown in **21-63**.

A larger damaged area will require careful work because the patch installed must be watertight (**21-64**). Since the repair involves chipping out old stucco, be certain to wear gloves and eye protection. A full face shield would be especially helpful.

Begin by chipping away all loose cementitous material (**21-65**) around the edges and clean down to the metal lath (**21-66**). In some cases it is not necessary to get to the lath if the scratch and brown coats have not been damaged.

If the metal lath has rusted or is badly deteriorated, cut away the bad part, pry it loose from the sheathing with a pry bar, and remove all nails (**21-67**). Nail a new piece of metal lath in the opening. If the sheathing has deteriorated it will have to be replaced. This will mean a larger area of stucco will have to be torn off.

If the sheathing and metal lath are sound and the opening is clean, begin filling it with stucco patch

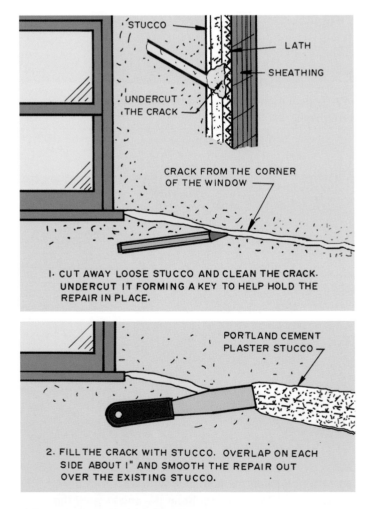

21-63 Large cracks are cut wider at the bottom, forming a wedge-shaped repair. Thoroughly clean the opening before filling it with stucco.

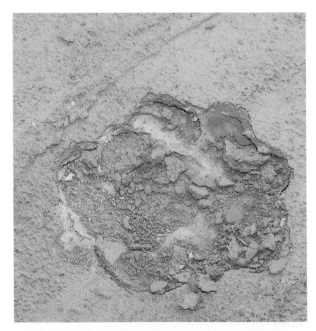

21-64 A hard blow can cause a larger area to be broken, requiring removal of the stucco and brown coat down to the metal lath.

21-65 Cut away all loose stucco and brown coat over and around the damaged area. Cut the edges sloping inward, providing a wedge-shaped opening. This helps hold the patch in place. Be careful not to damage the lath or the builder's felt below. If the exposed scratch or brown coats are sound, it is not necessary to cut down to the lath. Plaster over these sound coats.

21-66 Remove all loose particles so the patch bonds to firmly fixed stucco. A brush or vacuum is generally used.

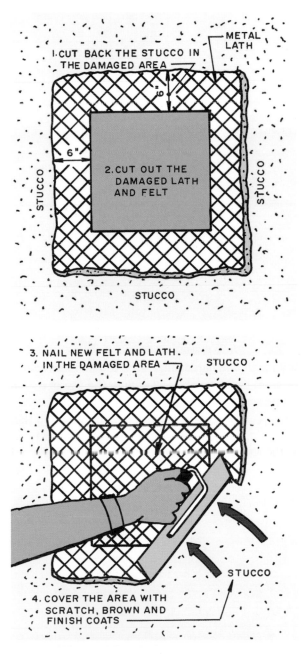

21-67 If the lath has rusted or was damaged by the blow, the entire damaged area needs to be cut away. The builder's felt and metal lath will be cut away and replaced. Then cover the area with the scratch coat, brown coat, and finish coats.

21-68 Fill the
opening with a
1/4-inch-thick layer,
letting each harden
before the next
layer is added.

material. It is available at the local building supply dealer in premixed and powder form. Using premixed compound simplifies things (21-68).

Fill the hole with several ¼-inch-thick layers. Most repairs will require the application of two or three thin layers of compound. Let each dry before laying on the next coat. When the last layer is hard, trowel a finish coat over the entire area. Feather it out on the surrounding surface. Large repairs will have this coat smoothed with a float or masonry-finishing trowel (21-69). Small patches can be finished with a pointing trowel (21-70). You can use a bristle brush to get a textured surface or use a trowel to get swirls. When the repair is thoroughly dry, paint it with product recommended for coating stucco.

When completely dry, prime and paint.

Repairing Brick Walls

Brick and **stone** exterior walls last for many years; however, the mortar usually does not. It will be the thing to check occasionally. If you see places where the mortar has fallen out, you may have waited too long. Water could be seeping inside the wall, causing the sheathing and possibly the studs in that area to rot. This also produces mold inside the wall which can work through the insulation and the drywall. If this has occurred, probably the only solution is to remove all the brick in this area; replace the sheathing, insulation, and drywall; and treat, to kill all evidence of mold. Then clean the old mortar off the bricks and re-lay the wall.

Check the mortar by probing it with a knife, ice pick, or even a sharp nail. If it cuts easily and sheds particles, it needs to be replaced. This replacement process is called **tuck-pointing**.

21-69 After the repair has hardened, trowel a thin finish coat over the entire area. Large repairs will require the use of a float or masonry finishing trowel. Feather the coating out over the repair and try to create a surface texture to match that on the original wall. When the finish coat is thoroughly dry, paint to match the color of the existing stucco.

21-70 Small
repairs can
have the finish
coat applied
with a pointing
trowel.

Tuck-pointing Brick Walls

Remove the deteriorated mortar with a **plugging chisel** or a **cold chisel (21-71).** Work carefully so all particles of the old mortar are chipped loose yet the bricks are not damaged. Clean it out about ¾ to 1 inch deep. The cleaned area should be square on the back **(21-72).** Remember to wear eye protection or a full face shield and gloves. Clean the joint with a brush and vacuum away any fine particles **(21-73).** Before applying the new mortar, wet the joint **(21-74).**

21-71 Deteriorated mortar can be removed with a cold chisel. If a large area needs renovating the mason will clear the joints with a power grinder.

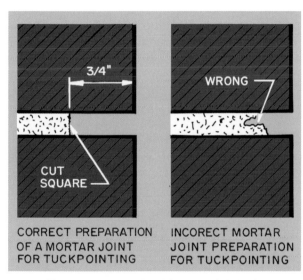

CORRECT PREPARATION OF A MORTAR JOINT FOR TUCKPOINTING

INCORECT MORTAR JOINT PREPARATION FOR TUCKPOINTING

21-72 The old mortar should be cut back about ³/₄ inch and the back should be cut clean and square so a full replacement mortar layer can be inserted.

21-73 Remove all loose particles from inside and around the edges of the joint with a brush. A powerful shop vacuum is then used to clean away any fine particles remaining.

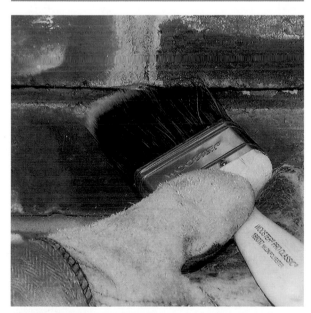

21-74 Before applying the new mortar, wet the joint thoroughly.

21-75 Small repairs require only a minimum amount of mortar to be mixed. Read the helpful directions on the bag.

Mix the mortar as directed on the bag. Add enough water to create a thick, plastic mix. It should be able to stand in a pad and not flow or creep. Since the new mortar will most likely be a different shade than the old mortar, pigments can be added to the mix. To check the color, mix a small trial batch and let it cure to see what color develops (**21-75**).

To produce the strongest mortar, the correct consistency must be produced during mixing. The amount of water added should be just enough to produce a plastic-like mixture that will hold its shape and stay on the trowel and not run off or fall into a crumbled pile. It can be checked by building a small pile and cutting it with a trowel as shown in **21-76**.

Place a pad of mortar on a piece of plywood or metal and hold it up to the open joint. Move it into the joint with a grooving tool (**21-77**) and pack it firmly into the joint in several thin layers (**21-78**).

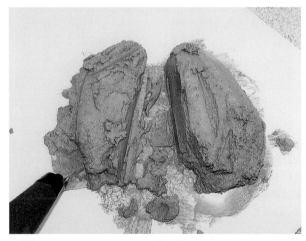

21-76 A properly produced mortar will stand when cut and stick to the trowel and not run off or crumble.

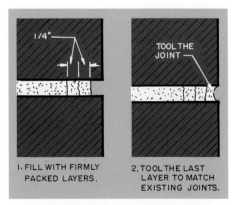

21-77 Place a small pad of mortar on a piece of metal or plywood and press it into the joint with a pointing trowel or the grooving tool if necessary. Pack it in 1/4-inch layers until the joint is completely full. Wipe mortar off the face of the brick with a wet cloth.

1. FILL WITH FIRMLY PACKED LAYERS.

2. TOOL THE LAST LAYER TO MATCH EXISTING JOINTS.

21-78 Pack the mortar in the joints in several thin layers, working it all the way to the back.

Constantly wipe off any mortar that gets on the face of the bricks. A wet rag will do for this step (**21-79**).

Smooth and form the contour of the mortar to match that on the wall. Generally it is a concave surface made with the grooving tool (**21-80**). If the joints are flush, use a trowel to cut the mortar flush with the face of the brick. Other commonly found mortar joints are shown in **21-81**.

After the mortar has set up a little, any that is still on the surface of the bricks can be brushed off with a fiber bristle brush. You can wipe any remaining on the surface with a wet abrasive pad like those used in the kitchen (**21-82**).

21-79 Wipe the fresh mortar off the face of the bricks with a wet cloth.

21-80 This filled joint is being smoothed with a grooving tool which forms a concave profile.

21-82 After the mortar has set a short while and begins to harden, brush off any remaining particles along the edges of the joint and wipe the surface free of mortar.

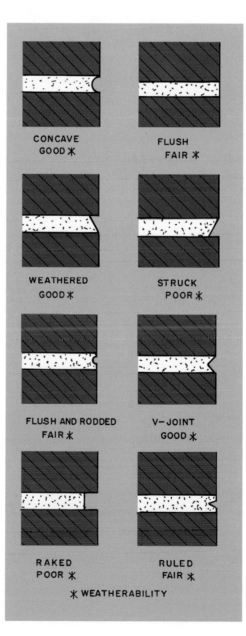

CONCAVE
GOOD ✳

FLUSH
FAIR ✳

WEATHERED
GOOD ✳

STRUCK
POOR ✳

FLUSH AND RODDED
FAIR ✳

V-JOINT
GOOD ✳

RAKED
POOR ✳

RULED
FAIR ✳

✳ WEATHERABILITY

21-81 Commonly used finished mortar joints. Notice some are rated higher for weatherability. Over time they are less likely to let water penetrate the joint.

REPLACING A DAMAGED BRICK

If a brick has started to fail but the cracked area is small or narrow, it probably has been placed under stress or excess moisture. If it is not in a highly visible area, the repair can be made by sealing all openings and filling any cavities flush with concrete and masonry sealer (21-83). Small cavities are best filled by using several thin layers of sealer over a period of hours.

If the brick appears to be totally defective, it should be replaced. Begin by cutting away the mortar with a cold or mason's chisel until the brick pieces can be removed. Clean out all loose mortar particles (21-84). If you have to replace more than two or three bricks in a course do not do them all at the same time, because they may be under great weight and bricks above may fall. Replace two or three and let the mortar cure several days. Then do a couple more.

Once the opening is clean wash it with a hose. This helps not only to clean it, but puts some moisture in the surrounding bricks which will help the new mortar to cure properly.

Mix the mortar as directed on the sack. Add water to produce a putty-like mixture. Since you need only a small amount a plastic tub or bucket will be large enough (refer to 21-75). To test the mix lay a large lump on a flat surface and cut it with a trowel. It should stand firmly on the trowel (refer to 21-76).

21-83 Small mortar cracks that may let water enter the wall can be sealed by caulking with concrete crack and masonry sealant.

21-84 Completely remove a damaged brick and chisel and scrape away all the old mortar.

21-86 Put a layer of mortar on the top and sides of the brick.

21-85 Lay a bed of mortar on the back, bottom, and sides of the opening.

Put a ⅜-inch-thick layer of mortar on the back, bottom, and sides of the opening (21-85). Put a layer of mortar on the top and sides of the brick (21-86) and slide it into the opening. Carefully work the brick into the opening. Tap it with a mallet or the handle of the trowel (21-87). Excess mortar will build on the sides. Press it in with a grooving tool until it is firmly packed. Scrape away all excess mortar with the grooving tool (21-88). Brush off any mortar still on the face of the bricks with a soft bristle brush (21-89). Several times a day for two days lightly spray the area with water. This will produce a harder mortar.

21-87 Carefully slide the brick into the opening. Keep it centered so the mortar on the bottom of the opening is not scraped away. This is not an easy thing to try.

21-88 Pack the mortar into the joints with a grooving tool. A firm, wide bed of mortar is needed on all sides. Then finish troweling the joint so it matches the existing joints in the wall.

21-89 After the mortar sets up a short while, brush off all loose particles along the edges of the joints.

21-90 The Exterior Insulation and Finish System provides an attractive, tough, textured exterior finished wall.

Repairing Exterior Insulation & Finish Systems (EIFS)

Exterior insulation and finishes (21-90) are also commonly referred to as **modified** or **synthetic stucco.** The EIFS system is a multilayer installation consisting of a water-resistant sheathing material that is covered with a waterproof weather barrier. Over this a sheet of expanded polystyrene insulation board with moisture-drainage panels is applied and held with mechanical fasteners. A coat is troweled over the insulation board, and a fiberglass reinforcing mesh is embedded into it. After this hardens the acrylic finish coat is troweled over the base coat (21-91). The acrylic finish coat is a special mixture available from the company supplying the various components of the system. It is available in a range of colors.

Repairing Physical Damage

If the wall is damaged from a hard blow, the expanded polystyrene panel can be cut away and replaced. The base coat, reinforcing mesh, and finish coat can then be installed over it. It is recommended that the homeowner not attempt to make this type of repair. The materials needed are not available on the general building materials market and have to be purchased from the manufacturer. The homeowner should contact a qualified contractor authorized to make such repairs. It is important to have the damage repaired properly so a watertight seal is produced.

Applying a Cleaning Solution

There are things that should be done to maintain this exterior finish material. If the surface is badly soiled or a new color is desired, it may be recoated with a manufacturer-supplied coating material. (The following suggestions were made by Dryuit Energy Systems, Inc., manufacturer of EIFS wall systems.)

The following cleaning products are recommended for use on Dryuit EIFS wall systems; Enviro Clean, EIFS Clean n Prep, MP2107, Wash Down, and Building Wash 3.

Test the cleaning compound on a section of the wall to see if it is effective. If stains do not clean easily, try other products or contact the cleaning product manufacturer. Never use acidic cleaners on EIFS finishers.

Always wear eye protection, rubber gloves, and a NIOSH-approved dust-mist respirator.

The cleaning solution may be applied by a **low-pressure** sprayer, 30 to 50 psi, or through a pressurized-water-cleaning unit. The pressure should be enough to **just wet the surface. Never use high pressure to apply the cleaning solution.** The cleaner should provide the cleaning action. A light scrubbing with a long handled soft bristle brush can help. Follow the cleaner manufacturer's instructions. Some recommend application from the bottom upward in vertical sections. Allow the chemicals to remain on the wall the time recommended by the manufacturer and then rinse thoroughly.

Pressure-Rinsing the Wall

The rinse on the area where the cleaning solution has been applied should be done with large amounts of clean, **pressurized water** working from the top to the bottom before the cleaning solution can dry. Then rinse the foundation and other areas below the wall. The higher pressures are needed to remove surface contaminants lifted by the chemical action of the cleaner. Be careful the higher cleaning pressure is such that it does not damage the finish coating.

Removing Mildew & Algae

The following mildew/algae cleaners are recommended for use on Dryuit EIFS Wall Systems: Jomax, Mildew Creek, and Olympic. You can also use a mixture of 1 part household bleach and 3 parts water to clear mildew. **Never add ammonia to a bleach solution.**

Problems with some of the Early EIFS Installations

The Exterior Insulation and Finish Systems must be installed by authorized contractors. Even with this requirement some of the early installations after a few years ran into trouble. The problem was that moisture was able to leak in around windows, doors, soffits, louvers, and other items in the wall. The flashing and caulking were not sufficient to keep water from leaking behind the polystyrene insulation panel, which eventually got around the waterproof barrier on the sheathing. The sheathing began to deteriorate and the wall covering failed. This made it necessary to strip off the entire system and install a new one. The system shown in **21-91** has drainage channels cut in the back of each polystyrene insulation panel so any leakage now drains to the bottom of the wall and weeps out of the drain track onto the ground. This is much the

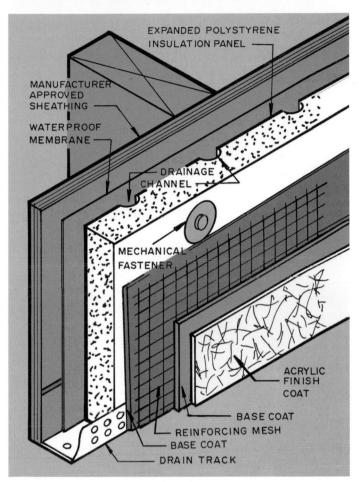

21-91 This Exterior Insulation and Finish System (EIFS) is designed for use on residential buildings. It provides insulation plus a hard, tough exterior finish coating that is troweled over the reinforcing mesh.

same as is done with brick veneer walls which weep at the bottom through open mortar joints.

Damaged Asbestos Shingles

Many older homes were finished with **asbestos shingles.** They provided a hard, durable wall finish and were easy to apply. However, asbestos fibers have been found to be a major health hazard and asbestos is now banned from use in most construction materials. If an older house has asbestos siding or roof shingles, do not try to repair or replace. Consult the local building inspection department for advice. They can refer you to contractors qualified to remove and dispose of the damaged materials. It requires considerable safety equipment such as respirators and protective clothing. Additional information can be had by contacting the Consumer Product Safety Commission.

Metric Equivalents

[to the nearest mm, 0.1cm, or 0.01m]

inches	mm	cm	inches	mm	cm	inches	mm	cm
⅛	3	0.3	13	330	33.0	38	965	96.5
¼	6	0.6	14	356	35.6	39	991	99.1
⅜	10	1.0	15	381	38.1	40	1016	101.6
½	13	1.3	16	406	40.6	41	1041	104.1
⅝	16	1.6	17	432	43.2	42	1067	106.7
¾	19	1.9	18	457	45.7	43	1092	109.2
⅞	22	2.2	19	483	48.3	44	1118	111.8
1	25	2.5	20	508	50.8	45	1143	114.3
1¼	32	3.2	21	533	53.3	46	1168	116.8
1½	38	3.8	22	559	55.9	47	1194	119.4
1¾	44	4.4	23	584	58.4	48	1219	121.9
2	51	5.1	24	610	61.0	49	1245	124.5
2½	64	6.4	25	635	63.5	50	1270	127.0
3	76	7.6	26	660	66.0			
3½	89	8.9	27	686	68.6	inches	feet	m
4	102	10.2	28	711	71.1			
4½	114	11.4	29	737	73.7	12	1	0.31
5	127	12.7	30	762	76.2	24	2	0.61
6	152	15.2	31	787	78.7	36	3	0.91
7	178	17.8	32	813	81.3	48	4	1.22
8	203	20.3	33	838	83.8	60	5	1.52
9	229	22.9	34	864	86.4	72	6	1.83
10	254	25.4	35	889	88.9	84	7	2.13
11	279	27.9	36	914	91.4	96	8	2.44
12	305	30.5	37	940	94.0	108	9	2.74

Conversion Factors

1 mm	=	0.039 inch	1 inch	=	25.4 mm	mm	=	millimeter
1 m	=	3.28 feet	1 foot	=	304.8 mm	cm	=	centimeter
1 m²	=	10.8 square feet	1 square foot	=	0.09 m²	m	=	meter
						m²	=	square meter

Index

About the Author

WILLIAM P. SPENCE is a do-it-yourself expert who has authored more than three dozen books for Sterling Publishing Co., Inc. and for other publishers. He also has written articles extensively for magazines and journals. Spence has been a professor of technical arts and applied sciences, including Chairman, Department of Industrial Education and Art and Dean, College of Technology, both at Pittsburg State University, Pittsburg, Kansas. He has also worked in industry as General Manager and Design Draftsman at manufacturers in Virginia, has worked as a cabinetmaker, and became involved in real estate sales and land development with Sandhill Properties, Inc. of Whispering Pines, North Carolina. He earned a doctorate in education and a masters in education at the University of Missouri as well as his bachelor of science and bachelor of science in education at Southeast Missouri State University. He also served in the U.S. Navy. He makes his home in Pinehurst, North Carolina where he is currently writing technical books full time.

Other Sterling Publishing Co., Inc. Books by William P. Spence

Building Your Dream House
Carpentry & Building Construction: A Do-It-Yourself Guide
Constructing Bathrooms
Constructing Kitchens
Constructing Staircases, Balustrades & Landings
Doors & Entryways
Encyclopedia of Construction Methods & Materials
Encyclopedia of Home Maintenance & Repair
Finish Carpentry: A Complete Interior & Exterior Guide
Installing & Finishing Drywall
Installing & Finishing Flooring
Insulating, Sealing & Ventilating Your House
Interior Trim: Making, Installing & Finishing
Residential Framing: A Homebuilder's Construction Guide
Roofing Materials & Installation
Windows & Skylights
Woodworking Basics: The Essential Benchtop Reference (with L. Duane Griffiths)